Mars: A Volcanic World

Giovanni Leone
Editor

Mars: A Volcanic World

UNIVERSIDAD
DE ATACAMA
VICERRECTORÍA DE INVESTIGACIÓN
Y POSTGRADO

INCT

Springer

Editor
Giovanni Leone
Instituto de Investigación en Astronomía
y Ciencias Planetarias
Universidad de Atacama
Copiapó, Chile

ISBN 978-3-030-84102-7 ISBN 978-3-030-84103-4 (eBook)
https://doi.org/10.1007/978-3-030-84103-4

This Springer imprint is published by the registered company Springer Nature Switzerland AG
The registered company address is: Gewerbestrasse 11, 6330 Cham, Switzerland

Preface

This book was born from the need to tell a realistic story after more than two decades of one-way narrative depicting Mars as a fluvial world in which liquid water was present and active on its surface.

Although there are people who today still believe in this story, it is the duty of science to make a rigorous clarity about the real situation of the Red Planet because the word "belief" does not belong to science but rather to religion. Many people, although acknowledging that Mars is today an arid and cold world, still assess that was a fluvial world in the past claiming different environmental conditions. Available data returned by many missions, both orbiters and landers, have been interpreted in a way that supports such a narrative all over the past 20 years.

However, those same data have also been interpreted without any preconfectioned narrative to show that Mars in reality is a volcanic world and that liquid water was never able to run on the surface of the planet. This was clear since the first Mariner mission put its "eyes" on the planet but that reality was too hard to accept for people who have hoped to find liquid water out there for so long time.

The large volcanoes appearing in all their majesty in the first images did not leave any space to imagination, Mars immediately appeared as a volcanic world. The lack of water and the tenuous atmosphere, largely composed of CO_2, also showed an inhospitable world that would have been quite hard to colonize.

Today a new paradigm is possible and must be pursued. Yes, Mars is not the world that many people expected it was but still we can make a good deal with it. We can explore it to test new technologies that will be useful on Earth and on other worlds. We can see if there are valuable minerals that one day will be inexorably depleted on Earth. We can learn how to develop our activities in a situation of scarcity or absence of water. And much more.

But, first, let's see how Mars was since the beginning....

Copiapó, Chile

Giovanni Leone

Contents

History of Scientific Studies and Current Views of Mars

Giovanni Leone

Abstract

The history of the studies of Mars shows how controversy always existed between two main parties: one in favour and one against the existence of liquid water on the surface of the planet. Such a controversy could be understood in past time when the absence of better data than today could not support one party or the other. Today, after the wealth of existing data showing how dry is the planet sounds rather incredible that such a controversy still exists. This chapter shows why existing data do not support the reasons of the pro-water party and why volcanism takes the main role in shaping the Red Planet.

1 From the First Telescope Maps to the First Spacecraft Photos

The first interpretation of the dark regions of Mars, observed in the telescopes of the time as possible "seas", was made by Johann Hieronymus Schröter during the nineteenth century but Giovanni Virginio Schiaparelli was the first in 1877 to see the dark areas as thin lines (Fig. 1), which he called "canali" because interpreted as filled with water [1]. This first interpretation was later dismissed by Lowell for a new one involving plant life or minerals but the view of water canali survived until the Space Age only supported by a minority of astronomers [1]. Although the Mariner 4 mission returned images of the surface of Mars since 1965 [2], only in 1975 a first comparison among the maps of Schiaparelli and Antoniadi with the first photos of the Mariner 9 mission to Mars was made. Only the Aeolis region appeared quite similar to the one shown in the map of Antoniadi and at least consistent with that of Schiaparelli [3]. This comparison is a first strong example about how the different resolution of an image gives a different interpretation. After Aeolis, only Valles Marineris, and probably Cerberus and Nepenthes, basically had a correspondence to the original drawings. Today, all the canali revealed to be just an illusion resulting from observations at limited resolution [3].

2 History of Robotic Missions to Mars

The first mission to Mars was the Soviet 1 M launched the 10th of October 1960, just 5 years later than the official beginning of the Space Race between the U.S.S.R and the U.S.A. in August 1955. This mission was followed by other four Soviet (1 M-2, 2MV-4, Mars 1 or

G. Leone (✉)
Instituto de Investigación en Astronomía y Ciencias Planetarias, Universidad de Atacama, Copiapó, Chile
e-mail: giovanni.leone@uda.cl

G. Leone (ed.), *Mars: A Volcanic World*,
https://doi.org/10.1007/978-3-030-84103-4_1

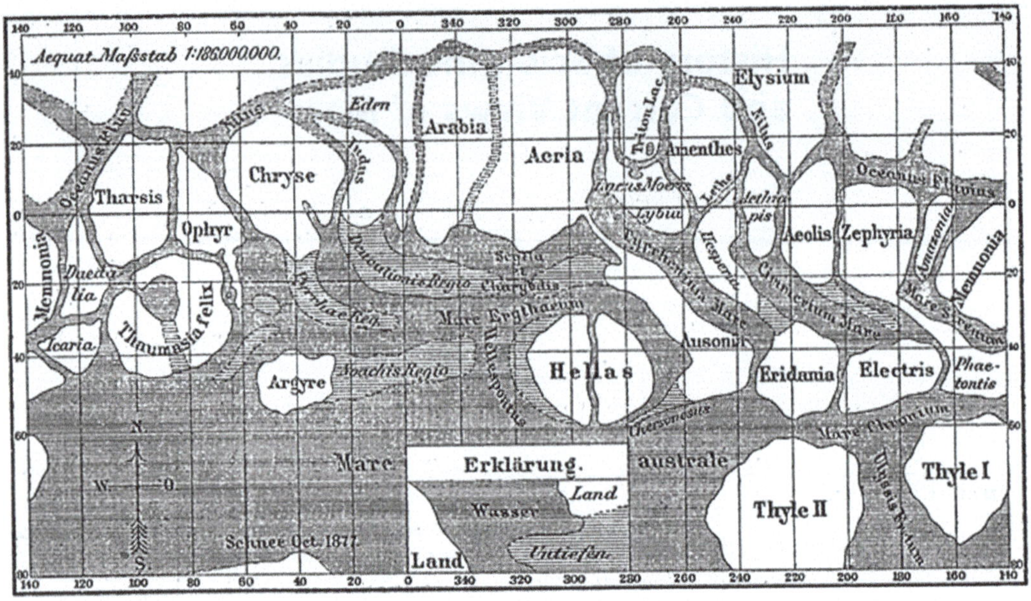

Fig. 1 Schiaparelli map of Mars. Public domain

2MV-4 nr. 2, 2MV-3) and one US (Mariner 3) launch and spacecraft failures until that US Mariner 4 reached Mars for a first successful flyby on July 15th, 1965. The first results of the Mariner 4 were 22 photographs successfully sent to Earth with which was possible to draw a first Schiaparelli-Antoniadi style map and, most remarkably, the inference that the thin atmosphere of Mars and the scarcity of water characterized the planet since its formation [2]. The results of the radiometer experiments of Mariner 6 and Mariner 7 showed how the polar caps are essentially formed by frozen CO_2 with a minimum measured temperature of 150°K at the south pole of Mars [4]. After the loss of Mariner 8, all the scientific objectives planned for this mission had to be integrated in the new Mariner 9 mission [5]. The Mariner 9 mission to Mars essentially confirmed what was already seen with all the previous Mariner missions.

The unfriendly environment for liquid water [6] as well as the igneous composition of the surface were then confirmed by the Viking 1 mission [7, 8]. A situation that was confirmed again years after both Viking missions were over [9–11]. As a result, the missions to Mars were stopped for a couple of decades from the end of the 70's to the end of the 90's. Nearly twenty years after the Vikings, the Pathfinder mission was launched to Mars and essentially found the same results at the mouth of Ares Vallis, a valley first thought to have discharged catastrophic floods of water to Chryse Planitia. The fluvial misinterpretation occurred upon the observation of rounded pebbles thought to be of sedimentary origin [12], which later showed to be of basaltic or andesitic composition, but likely deposited by flood lavas [13, 14].

Other missions were sent to craters thought as filled with water for long time: Spirit to Gusev crater [15]; Opportunity to Victoria and Endeavour craters [16]; Curiosity to Gale crater [17]. However, the data returned by all these missions showed a volcanic mineralogy dominated by basaltic composition and followed by andesitic and rhyolitic compositions [18–21].

3 Lava Versus Water: Controversial Arguments and Current Debate

Today the surface of Mars is recognized as extremely dry. However, the observation of anastomosing fluvial networks and huge outflow channels has historically fueled speculations that once liquid water may have flowed abundantly on the surface of Mars [22]. The low pressure of the atmosphere [10, 11, 23, 24], between 2 and 6 mbar (<1% of that of Earth at sea level) confirmed by Curiosity rover measurements at the border between the highlands and the lowlands [25], and its average surface temperature of −30 °C do not allow liquid water to be stable on the surface of Mars [22, 24, 26]. Water is in the order of 1 ppm during winter [10], with an average of 20–30 precipitable microns [24], and with a maximum of ~200 ppm (consistent with the maximum amounts of 60–70 precipitable microns) found in the atmosphere during the northern summer [27]. If all the atmospheric water condenses on the surface, it would form a layer only 10 μm deep [28]. Although such scarcity, coupled to instability, is generally acknowledged today, there are authors who still believe that liquid water may currently exist to form gullies [29] and recurring slope lineae (RSL) [30]. Other authors believe that water existed in the past for long periods of time from the Noachian to the Hesperian [31–35] or even in the Amazonian [36] claiming different and favourable climatic conditions. Other authors believe that conditions for liquid water were transient or time-limited [26, 37–39]. Others that water never existed in its liquid state [40–42]. There is a number of papers that focused only on liquid water [43–71], some of them claiming evidence, but missed to justify such "evidence" against its instability on the surface of Mars. Water may potentially exist as ice in the northern and southern polar subsurface, it was claimed to be detected from orbit by the SHARAD instrument onboard the Mars Reconnaissance Orbiter [72] and by the GRS instrument on the Mars Odyssey orbiter, although it was never directly seen and may just be related to hydrated minerals [73], or simply misinterpreted with tephra deposits that have similar dielectric constant to that of water [74]. The polar ice caps, which grow and shrink with the seasons, are mixtures of predominant CO_2 with a few traces of water ice [75]. The Phoenix lander detected CO_2 frost-cemented soil encaging small amounts of pure water ice within ~5 cm of the surface at a latitude of 68.219°N (Mellon et al. 2009), not certainly at the equator where it was never found [76], but sublimated within four Martian sols when exposed to the surface [77]. The putative presence of snowpacks and glaciers on top of Tharsis [78–81] and Elysium volcanoes [82] is at odds with the height of volcanoes located within the stratosphere of Mars where sublimation is favoured to glaciation [83]. Subsurface ice was also claimed to have been exposed by recent meteorite impacts poleward of ~40° latitude [84], but it was not explained how it could survive to the heat produced by the impact. Other important aspects that were not always taken in the due consideration in all the pro-water literature: the dehydrating effect of the primordial impacts [85]; the loss of water since the early past [86–90]; the results of the rovers showing igneous instead of sedimentary mineralogy (further discussed in the next sections); and, last but not least, the worldwide lack of alteration of the olivine into serpentine [91] that would be caused by the global presence of water otherwise [92, 93].

Absolutely less problems with lava. The erosional effect of lava was already known since long time [94–100], as well as its favoured role over liquid water in the low-pressure environment of Mars [41, 101–106]. The findings of the rovers [19, 107] and the constraints on the upper limit for the alteration of olivine into serpentine at relatively low temperature [93] now raised the first doubts about the total duration of the stability of liquid water on the surface of Mars. As a result, the estimates were brought down to 10,000 years at Gale crater [108]. Worth to notice that these estimates coincide with the upper limit of olivine alteration in extremely cold conditions [93]. The jarosite found at the Opportunity landing site indicates total dry conditions on Mars because it is a mineral that rapidly decomposes in ferric oxyhydroxides in humid climates [109] favouring volcanic processes for the formation of the hematite found in Sinus Meridiani. The scarce serpentine found on the surface of Mars might be explained with a sporadic alteration of olivine by circulation of hydrothermal fluids underground before that magma reached a pressure low enough to exsolve the typical amount of water naturally contained in it [42], an hypothesis consistent with the geochemistry of the scarce carbonates found on the surface of Mars [110]. Hydrothermal systems may have existed in Mars' subsurface during its early volcanically active past but once water reached the vicinity of the surface would have degassed already at depth [111] with little chance to remain liquid on the surface to carve the fluvial networks, even less for the outflow channels.

4 Arguments in Favour of Water

The past existence of water was claimed on the basis of misinterpretation of topographic features: shorelines [112], fluvial networks [31], gullies [29], recurring slope lineae (RSL) [30], tear-eyed shaped structures (TESS) [113–115], polygonally patterned ground [116], lobate debris aprons [117, 118], rootless cone fields, and deltas [119], or the mineralogical interpretation of phyllosilicates [120], hematite, carbonates, and secondary veins of calcium sulphate, or radar prospections of the surface of Mars [72, 121, 122].

5 Arguments Based on Geomorphology

The hypothesis of an ocean on Mars was interpreted from the observation of the Arabia and Deuteronilus putative shorelines [112, 123]. Tthe topographic profiles traced along the putative shorelines provided little support to the potential surface of a sea level [124, 125] showing how a volcanic origin would have been more plausible [35, 123, 126]. Such a mismatch between topography and putative shoreline was later explained with a true polar wander (TPW) [127], emplacement of Tharsis [35], or emplacement of the Vastitas Borealis Formation (VBF) [128], the latter being essentially a volcanic process. The paucity or absence of coastal landforms [129], or the possibility that such an ocean would have been completely frozen [34, 130], suggested the alternative explanation that the putative shorelines might also be ascribed to the original emplacement of the Martian dichotomy [123].

The fluvial networks were uniquely interpreted as cut by flows of water originated either by ground sapping or surface runoff (i.e. rain) [131–133]. Although the formation of these fluvial networks requires significant amounts of water, no explanation was given on how such amounts of water were indeed available and how they were released from the ground [133]. Volcanism from below (i.e. dykes) [134] or environmental warm up due to favourable obliquity [135, 136] were later suggested as possible heating sources, probably during the Noachian at the end of the late heavy bombardment (LHB) [32, 136–138]. Impact melt hydrothermal systems with ice covered rivers during the Noachian were also invoked to explain the origin of the fluvial networks, an hypothesis that does not require any significant climate change from the current situation [137]. Several other studies confirmed such a view of a drier and cold environment during the Noachian [139–143]. Whatever the climate might have been during the Noachian, the fluvial networks and the main outflow channels of Mars spread from the high volcanoes of Tharsis and Elysium [41]. Logically, these have a higher probability to be carved by lava as heated ice would sublime rather than melt at the stratospheric pressure encountered on top of the volcanoes [83, 136]. Most Tharsis and Elysium volcanoes have heights ≥ 14 km corresponding to the stratosphere of Mars, an altitude with such a low atmospheric pressure where only sublimation or evaporation can occur [83]. Even at the Phoenix landing site ice sublimated in about 4 sols [77]. In any case, it was acknowledged how it is quite difficult to replenish groundwater at hydraulic heads located at ≥ 10 km of altitude [41, 144].

The origin of the gullies on Mars was explained through an ice barrier protecting liquid water from the low pressure of the atmosphere [29]. However, other studies involving different possible processes like insulation [145], geothermal heating [146], cryovolcanism [147], brine seeps [148, 149], eruptions of liquid CO_2 [150], and granular flows [151], concurred unequivocally that liquid water was not the erosional agent forming the gullies [151].

Tear-eye shaped structures (TESS) have been invoked as evidence of water flooding on Mars [113, 115] although successive observations of rootless cones associated with TESS have shown involvement of lava flooding on a possibly "wet" substrate for their formation [152–154]. TESS have also been observed within the volcanic channels Mangala Valles, where no rootless cones have been found, and within Valles Marineris, where just a few rootless cones were found, thus confirming a volcanic formation of TESS by flowing lava [42, 101].

Rootless cones, sometimes called pseudocraters, are produced when active lava flows interact explosively with surface or near-surface volatiles [155]. The vigor of the resulting explosive interactions depends on many factors, including the shape of the flow (narrow vs. wide), the nature of the volatiles (shallow groundwater vs. a river vs. a lake or ocean), the volumetric flow rate of the lava, and others [156–158]. Some rootless cones are morphologically similar to primary vents and even in the field determining whether external water was involved can be difficult [159]. Unequivocal identification of a rootless origin via remote imagery is only possible if it is clear that the explosive interaction took place in locations that are distal from the source vent [160]. Fields of 10s to 100s of small randomly spaced conical hills on Mars were first interpreted to be rootless cones [161]. Since then, they have received a good deal of interest because, if that interpretation is correct, it implies considerable amounts of near-surface ground water or CO_2 ice during geologically recent times [160, 162, 163]. Water interactions at primary vents on Mars (i.e., hydromagmatic eruptions) have received less interest, although a family of volcanic features called pitted cones were recently interpreted to be tuff rings [154]. However, they do not necessarily suggest compelling evidence of groundwater or ice on the surface but probably some combination of underground hydrothermal activity [164]. In any case, their sporadic presence all over Mars associated with the scarcity of pyroclastic activity on Tharsis [165, 166], which would be the main source of the northern putative ocean [34], suggests a general lack of volatiles at global level.

Polygonally patterned grounds have been interpreted as desiccation patterns in impact crater supposedly filled with water [167] or generically suggested as "consistent with the thermal contraction of an ice-rich mantling unit" [116]. These polygonal patterns were already observed at the time of the Viking missions and were interpreted alternatively as typical cooling fractures in lava or fractures produced by thermal contraction in permafrost [168, 169]. Taking Gusev crater as a "famous example" of water filled crater [167], but being at subequatorial latitude (14°S) and shown as filled with lava containing unaltered olivine [107, 170], it suggests more lava cooling rather than permafrost or lake desiccation. The process of polygonal cracking in lava was already known and studied under the basic assumption of a zero temperature of the contact surface, but it occurred anyway without the involvement of water [171]. In fact, six out of seven classes of polygons were interpreted as formed through the effects of thermal cracking and differential sublimation of ice [116]. However, the predominance of CO_2-frost on Mars [4, 75] and the equatorial position of a consistent number of crater floor polygons [167] suggests that little or no water could have been involved in the process of cracking.

A group of features called lobate debris aprons were initially interpreted from orbital images to at least partially consist of water ice based on their geomorphology and restriction to latitudes between 52–28° N and 57–32° S [117, 118]. Further studies suggested an higher percentage of water, pure ice with only "a few 10s percent" of lossy contaminants as revealed by the attenuation of the dielectric constant of water [172], which left topographic depressions upon sublimation due to climate change [173]. These "high concentrations" of ice were detected at current- day environmental conditions of low atmospheric pressure which allows any ice to sublimate very quickly, 4 sols as seen at the Phoenix landing site [77], on volcanic areas mantled with highly porous deposits of basaltic sand moved by the wind [102]. How ambiguous are these radar detections will be further discussed ahead.

At last, the observation of deltas along the dichotomy boundary was taken as support of the existence of an ocean of water in the northern hemisphere of Mars, the idea was that the deltas were formed by sediments transported by rivers of water debouching into the lowlands [119]. The argument used in support of the origin of the water for the rivers is basically the usual, a warmer climate favouring liquid water [48, 131, 132, 174].

6 Arguments Based on Mineralogy

Phyllosilicates and clays were ascribed to past aqueous processes occurred on the surface of Mars [120, 175–177]. The spectra of submarine terrestrial clays were compared to available spectra of Martian clays, smectitic samples with FeO/MgO ratios ≈ 10–30 formed in submarine or low-submarine environment were found on the Noachian highlands of Mars [178] where an ocean was never reported. The presence of smectite is definitely not the proof of formation in ponding water [179, 180] as it might also be ascribed to the thermal alteration of lava [181]. Furthermore, the presence of unaltered olivine in Noachian units within the outflow channels and the fluvial networks, other than at global level [91], suggests that liquid water was not present or very limited in time at best [41, 42, 105].

Crystalline hematite was detected in basaltic sediments of Sinus Meridiani, Aram Chaos, and along Valles Marineris and interpreted as the result of chemical precipitation from aqueous fluids, "under either ambient or hydrothermal conditions" [182]. Thermal oxidation of magnetite-rich lavas is an alternative and viable hypothesis but large-scale water interactions was the favourite interpretation [183]. Again, these findings cannot justify a large-scale or global layer of water so far away from the lowlands where it was postulated. In fact, a subsequent study confirmed that the crystalline hematite in Terra (Sinus) Meridiani may have been formed by thermal oxidation precipitated from circulation of hydrothermal fluids in a 600 m-thick stack of pyroclastic deposit [184]. The Opportunity rover discovered that the hematite signature was associated from 1 to 5 mm concretions in aeolian deposits dominated by pyroxene-rich basaltic material [185].

The search for carbonates was first carried out with ground-based observations of the Syrtis Major-Arabia-Hellas regions of Mars [186]. The analysis of the dust on the surface of Mars with the Kuiper Airborne Observatory (KAO) [187], and the analysis of Mariner 6/7 spectral data [188], found little or no evidence. The observation of a 5.4 micron spectroscopic feature, interpreted as consistent with hydrous magnesium carbonate [188], and the putative evidence of carbonates on the surface of Mars revealed quite uncertain and elusive [189–191]. Many years later, the rover Spirit analysed some light-toned outcrops located at Paso Robles within Gusev crater finding that they are sulphates likely formed as volcanic hydrothermal fumarolic condensates [192]. Furthermore, the amount of 16 to 34 wt% of carbonate found in the 5 m Comanche outcrop of the Columbia Hills within Gusev crater, with grain sizes ranging between 0.5 and 1 mm, [193], is too small to support an extensive aqueous process (i.e. a paleolake) within the crater. Also, the amount of carbonate found at the Phoenix landing site is even smaller than that found at Gusev crater with a content of $CaCO_3$ of 3–5 wt% in the Wicked Witch sample [194]. The minerals found at the Phoenix landing site showed smectite and montmorillonite, phyllosilicates forming at high temperatures [77], justifiable only with the presence of lava flows on the cold surface of Mars. The spectral features found at Nili Fossae are consistent with Mg-carbonate scattered over an area of 50 km, however the low spatial resolution of the data excludes the presence of large quantities and such a finding shows no correlation at global scale on Mars [195] that would have been largely expected with the presence of an ocean. Such a scarcity of carbonate at global scale on Mars was explained through possible volcanic emissions of SO_2 producing acidic waters [196]. Although past volcanic activity is largely evident, the global lack of carbonates may indicate the presence of localized hydrothermal environments underground [110].

Experiments with sounding radar instruments like MARSIS [121] and SHARAD [72, 122] claimed evidence of underground ice. However, these claims have yet to pass a direct test on the surface, the dielectric constant for water used in the experiments (k = 3.15) is comparable to that of porous lava (k = 3.80). This method of investigation remains highly uncertain because depends too much on the porosity of the target layers, the higher is the porosity and the lower is the value of the dielectric constant for the same material [74]. The radar sounding is a useful tool but remains ambiguous at best. In any case, the targets chosen for these investigations were not located in water equivalent hydrogen (WEH) optimal locations. The regions selected for the experiments of radar sounding were the Mamers Valles (23° E–39° N and 28° E–40° N) and Reull Vallis (103° E–41° S and 105° E–43° S) [122], where the WEH is no more than 2–4% [197]. These regions are located at mid-latitudes and characterized by volcanic terrains covered by an eolian blanket of volcanic ashes transported by the wind. The expected porosity would be that of tephra with a few cm to hundreds of meters of thickness depending on the topography [102]. Similar characteristics could also be found in the Medusae Fossae Formation [121], likely formed as accumulation of tephra coming from Apollinaris Patera [198].

7 Arguments in Favour of Lava

The arguments in favour of lava are supported from the geomorphological observations of the main outflow channels spreading directly from the largest volcanoes of Mars [41, 42, 97, 102, 103, 105, 136, 199]. The advantage of lava over water is that lava is stable on the surface of Mars, a well-known fact since the Mariner missions [2, 23]. Also, there are minerals that are good environmental indicators, unaltered olivine and jarosite are strongly indicative of a dry environment since the Noachian [109, 200, 201]. The mineralogy found by the various rovers in craters interpreted as past paleolakes revealed very small or no sedimentary outcrops and prevailing basaltic composition instead [18–20, 170, 202–204], nothing that would suggest the infilling of large volumes of water. In addition to these minerals, the discovery of tridymite at Gale crater, a mineral that forms at temperatures above 850 °C [205], suggests once more infilling of lava rather than water. Fluid lava can form dendritic networks [206], can carve its own channels [96, 98, 100], and could thus form the outflow channels [41, 42, 101, 207]. After the initial post-Mariner [106] and post-Viking hypotheses [208], there is now a growing literature that shows evidence of lava as the main fluid carving the outflow channels and the fluvial networks [41, 42, 102–105, 199, 207, 209]. Mixed solutions as water first and lava later [207, 210], or catastrophic floods of water [211], failed to explain the lack of volumetric match between the required water supposed to be at the hydraulic head of the outflow channels and their final carved volumes observed today [42]. The hypotheses of sudden catastrophic floods of water [134] failed also to explain the unrealistic porosity that would be necessary to host the huge volumes of water that should have formed outflow channels much smaller than giants like Valles Marineris [211].

8 Sources and Sinks of Water: An Unfavourable Balance

The balance between sources and sinks all over the history of the planet is important to estimate its total water inventory at global level. In order to estimate such a balance, both hypotheses of local source (planetesimals along the orbit) and distal source (asteroids and comets coming from the distance) were taken in consideration. It was suggested that Mars may have had less initial water than the Earth, a total global equivalent layer (GEL) ranging from 600 to 2700 m [212]. By the way, a study of the volatiles present in the coma of the comet 67P/Churyumov-Gerasimenko suggested that the overall contribution to Earth's and Mars' water from comets was minor than 1% [213], including a possible different impact flux between them [214, 215]. A range of HDO/H_2O (D/H) ratio (δD) relative to standard mean ocean water (SMOW) around 1 was observed on Earth [216–218], which retained its water. The observation of a D/H ratio ~150 times bigger than that of the Earth either supports a strong escape of water from Venus or that it was dry since the beginning [219–221]. A D/H ratio of 5.5 SMOW was estimated for Mars [222] and was almost similar to the value of ~6 SMOW obtained by in situ geochemical measurements of the lava filled crater Gale [223]. A range between 6.2 and 7.1 SMOW, corresponding to a loss of water to space of 1200 m GEL, was estimated occurring during the strong volcanic peak that Mars experienced in the first 500 Ma [89, 224]. A higher D/H ratio of 8 SMOW was found in polar reservoirs substantially confirming the loss of water occurred in the first 500 Ma [90]. The loss of water during the pre-Noachian ranged between 41 and 99 GEL at ~6 SMOW, a loss much higher than the 10–53 GEL occurring during the remainder of the Martian history, with a global remaining inventory estimated in 20–30 GEL on the surface and 100–1000 GEL underground [88]. The latter upper limit thought as enough water to carve the outflow channels and the fluvial networks [126, 212]. All these isotopic values of SMOW suggest that even mantle water was quite scarce due to the strong volcanic degassing thus supporting other studies that reached the same conclusion [225–227]. Today the remaining inventory of water, estimated in 35 m GEL, is thought to be distributed among the polar caps, the atmosphere, and the permafrost [197]. However, such an amount of water was never proved to be really existing and, in any case, it would be absolutely insufficient to carve the outflow channels and the fluvial networks [199]. Another problem for liquid water is the one related to dehydrating impacts, regardless where and when occurring. Despite the misconception that impacts brought water to the planets, indeed we have already seen how the cometary water budget is not so significant. In reality, impacts eroded the atmospheres and dehydrated the terrestrial planets because of their high temperatures [228]. Giant impacts, in particular, seriously dehydrate planets [229]. Dehydration of serpentine already occurs at impacts of ~3 km s^{-1} [85]. The current estimates about the global water inventory of Mars might even be too optimistic because a low atmospheric pressure might already be existing since the accretion of the planet due to the erosion of significant impacts [230, 231]. Most of all, the giant impact that formed the Martian dichotomy, either from the north [232–234] or from the south [224, 235, 236]. Indeed, Mars lost much of its water already in its first 500 Ma [86, 88–90] thus suggesting that its combined low atmospheric pressure and low gravity were unable to retain any water. The southern polar giant impact (SPGI) was probably produced by either an iron giant impactor of 1600 km or a stony impactor of 2000 km of radius hitting Mars at ~5 km s^{-1} in a range of time between 4 and 15 Ma after CAI [224]. Its devastating effect may have reduced further any previously existing water, if any survived the high heat of the short lived radiogenic elements ^{26}Al and ^{60}Fe in the first 3 millions of years since accretion. In this desolating scenario the idea that a single impactor of 250 km of diameter may have freed 50 m GEL of water in the atmosphere to fall back down as rain for decades to millennia [237] appears unlikely because the planet proved already unable to retain any primordial water. Furthermore, impactors of 250 km of diameter were recorded for most of the pre-Noachian and Noachian [32] so that rain should have been available for long time but this would be at odds with the lack of alteration of the minerals, particularly jarosite and olivine, found and observed on the most ancient surface of Mars [91, 109]. The lack of bauxite, which requires abundant water for its formation, is another evidence of scarce or no rain [238]. At last, the erosion of the atmosphere by solar wind must be added to the erosion due to the impacts. Solar wind erosion accounts for the removal of 0.2 to 4 mbar of CO_2 and a few cm of water [239]. In total, the combined effect of impact and sputtering processes may have removed from 95 to 99% of the primordial atmosphere [240]. As consequence of the SPGI, volcanism may have provided some

CO_2 to the atmosphere but, at the same time, degassed any remaining water from the dehydrated mantle as a final process [102].

9 Conclusions

The fluvial networks and all the outflow channels are located along the slopes of high volcanoes [144] where lava and not water should be the most logical fluid with any erosional capability that could have any chance to survive in the cold and low pressure environment of Mars, from the pre-Noachian to current times. Several studies assuming a wide range of atmospheric pressures, a faint young Sun, and a denser CO_2 atmosphere, concluded that the climate of early Mars was cold [141–143]. The warmer climate expected during ancient Mars revealed fallacious [141–143, 241]. Even the remaining budget of water estimated for Mars is nothing more than a highly speculative idea until compelling evidence will be found, all the claims for liquid water on Mars were very far from being "compelling evidence". Some astronomic (obliquity) and geologic processes (volcanism) occurring during the history of Mars were believed as favourable to the presence of liquid water on the surface of the planet, the favourable obliquity periods should have supposedly warmed the planet [135] as well as the peak volcanism of the pre-Noachian [224] should have replenished the atmosphere with water and other volatiles. However, such processes did not increase the low pressure of the atmosphere and the scarcity of water already existing after the primordial impacts. This is the evidence inherited and observable today. In any case, water could not be the main erosive agent for the formation of all the fluvial networks and the outflow channels because sublimation and evaporation are the most likely conditions at the stratospheric heights of the Tharsis Montes [83], where the outflow channels originated. Other exotic solutions like water eventually produced from the polar caps basal melting and transported to hydraulic heads located at heights of more than 10 km already revealed impossible [41]. A mineralogical map of the equatorial regions of Mars, where most of the outflow channels and fluvial networks are also located, shows scarce serpentine and widespread unaltered olivine [91]. This occurs already in Noachian terrains as shown by the geo-chrono-stratigraphic map of Mars [242].

For all these reasons, lava remains the most likely fluid that carved the outflow channels and the so-called "fluvial" networks of Mars. For all these reasons, these should be renamed the volcanic channels of Mars.

References

1. Jones BW (2008) Mars before the Space Age. Int J Astrobiol 7:143–155
2. Leighton RB, Murray BC, Sharp RP, Allen JD, Sloan RK (1965) Mariner IV photography of Mars: initial results. Science 149:627–630
3. Sagan C, Fox P (1975) The canals of Mars: an assessment after Mariner 9. Icarus 25:602–612
4. Neugebauer G, Münch G, Chase SC, Hatzenbeler H, Miner E, Schofield D (1969) Mariner 1969: preliminary results of the infrared radiometer experiment. Sci 166(80):98–99
5. Masursky H (1973) An overview of geological results from Mariner 9. J Geophys Res 78:4009–4030
6. Hess SL, Henry RM, Leovy CB et al (1976) Mars climatology from viking 1 after 20 sols. Science 194:78–81
7. Baird AK, Clark BC (1981) On the original igneous source of Martian fines. Icarus 45:113–123
8. Toulmin P, Clark BC, Baird AK, Keil K, Rose HJ (1976) Preliminary results from the Viking X-ray fluorescence experiment: the first sample from Chryse Planitia, Mars. Science 194:81–84
9. Arvidson RE, Gooding JL, Moore HJ (1989) The Martian surface as imaged, sampled, and analyzed by the Viking landers. Rev Geophys 27:39
10. Lewis JS (1996) Mining the sky: untold riches from the asteroids, comets, and planets. Addison-Wesley Pub. Co
11. Bridges NT, Lackner CN (2006) Northern hemisphere Martian gullies and mantled terrain: implications for near-surface water migration in Mars' recent past. J Geophys Res 111:9014
12. Golombek MP, Cook RA, Economou T et al (1997) Overview of the Mars pathfinder mission and assessment of landing site predictions. Science 278(80):1743–1748
13. Rogers AD, Christensen PR, Bandfield JL (2005) Compositional heterogeneity of the ancient Martian crust: analysis of Ares Vallis bedrock with THEMIS and TES data. J Geophys Res 110:E05010

14. Wilson JH, Mustard JF (2013) Exposures of olivine-rich rocks in the vicinity of Ares Vallis: implications for Noachian and Hesperian volcanism. J Geophys Res Planets 118:916–929
15. Cabrol NA, Grin EA, Carr MH et al (2003) Exploring Gusev Crater with spirit: review of science objectives and testable hypotheses. J Geophys Res Planets 108 (E12):8076.
16. Wray JJ, Noe Dobrea EZ, Arvidson RE, Wiseman SM, Squyres SW, McEwen AS, Mustard JF, Murchie SL (2009) Phyllosilicates and sulfates at Endeavour Crater, Meridiani Planum, Mars. Geophys Res Lett 36:L21201
17. Le Deit L, Hauber E, Fueten F, Pondrelli M, Rossi AP, Jaumann R (2013) Sequence of infilling events in gale crater, Mars: results from morphology, stratigraphy, and mineralogy. J Geophys Res E Planets 118:2439–2473
18. Sautter V, Toplis MJ, Wiens RC et al (2015) In situ evidence for continental crust on early Mars. Nat Geosci 8:605–609
19. Sautter V, Toplis MJ, Beck P et al (2016) Magmatic complexity on early Mars as seen through a combination of orbital, in-situ and meteorite data. Lithos 254–255:36–52
20. Cousin A, Sautter V, Payré V, et al (2017) Classification of igneous rocks analyzed by ChemCam at Gale crater, Mars. Icarus 288:265-283.
21. Payré V, Fabre C, Cousin A et al (2017) Alkali trace elements in Gale crater, Mars, with ChemCam: calibration update and geological implications. J Geophys Res Planets 122:650–679
22. Carr MH (1987) Water on Mars. Nature 326:30–35
23. Ingersoll AP (1970) Mars: occurrence of liquid water. Science 168:972–973
24. Conrath B, Curran R, Hanel R, Kunde V, Maguire W, Pearl J, Pirraglia J, Welker J, Burke T (1973) Atmospheric and surface properties of Mars obtained by infrared spectroscopy on Mariner 9. J Geophys Res 78:4267–4278
25. Guzewich SD, Newman CE, de la Torre JM, Wilson RJ, Lemmon M, Smith MD, Kahanpää H, Harri A-M (2016) Atmospheric tides in Gale Crater, Mars. Icarus 268:37–49
26. Richardson MI, Mischna MA (2005) Long-term evolution of transient liquid water on Mars. J Geophys Res 110:3003
27. Trokhimovskiy A, Fedorova A, Korablev O, Montmessin F, Bertaux J-L, Rodin A, Smith MD (2015) Mars' water vapor mapping by the SPICAM IR spectrometer: five martian years of observations. Icarus 251:50–64
28. Maltagliati L, Montmessin F, Fedorova A, Korablev O, Forget F, Bertaux J-L (2011) Evidence of water vapor in excess of saturation in the atmosphere of Mars. Science 333:1868–1871
29. Malin MC, Edgett KS (2000) Evidence for recent groundwater seepage and surface runoff on Mars. Science 288:2330–2335
30. Ojha L, Wilhelm MB, Murchie SL, McEwen AS, Wray JJ, Hanley J, Massé M, Chojnacki M (2015) Spectral evidence for hydrated salts in recurring slope lineae on Mars. Nat Geosci 8:829–832
31. Baker VR, Hamilton CW, Burr DM, Gulick VC, Komatsu G, Luo W, Rice JW, Rodriguez JAP (2015) Fluvial geomorphology on Earth-like planetary surfaces: a review. Geomorphology 245:149–182
32. Carr MH, Head JW (2010) Geologic history of Mars. Earth Planet Sci Lett 294:185–203
33. Clifford SM, Parker TJ (2001) The evolution of the Martian hydrosphere: implications for the fate of a primordial ocean and the current state of the northern plains. Icarus 154:40–79
34. Carr M, Head J (2019) Mars: formation and fate of a frozen Hesperian ocean. Icarus 319:433–443
35. Citron RI, Manga M, Hemingway DJ (2018) Timing of oceans on Mars from shoreline deformation. Nature 555:643–646
36. Cabrol NA, Grin EA, Landheim R, Kuzmin RO, Greeley R (1998) Duration of the Ma'adim Vallis/Gusev crater hydrogeologic system, Mars.
37. Tosca NJ, Knoll AH (2009) Juvenile chemical sediments and the long term persistence of water at the surface of Mars. Earth Planet Sci Lett 286:379–386
38. Borg L, Drake MJ (2005) A review of meteorite evidence for the timing of magmatism and of surface or near-surface liquid water on Mars. J Geophys Res 110:12–15
39. Wade J, Dyck B, Palin RM, Moore JDP, Smye AJ (2017) The divergent fates of primitive hydrospheric water on Earth and Mars. Nature 552:391–394
40. Liu L-G (1988) Water in the terrestrial planets and the moon. Icarus 74:98–107
41. Leverington DW (2011) A volcanic origin for the outflow channels of Mars: key evidence and major implications. Geomorphology 132:51–75
42. Leone G (2017) Mangala Valles, Mars: a reassessment of formation processes based on a new geomorphological and stratigraphic analysis of the geological units. J Volcanol Geotherm Res 337:62–80
43. Adeli S, Hauber E, Kleinhans M, Le Deit L, Platz T, Fawdon P, Jaumann R (2016) Amazonian-aged fluvial system and associated ice-related features in Terra Cimmeria, Mars. Icarus 277:286–299
44. Banks ME, Lang NP, Kargel JS, McEwen AS, Baker VR, Grant JA, Pelletier JD, Strom RG (2009) An analysis of sinuous ridges in the southern Argyre Planitia, Mars using HiRISE and CTX images and MOLA data. J Geophys Res 114: E09003

45. Fassett CI, Head JW (2005) Fluvial sedimentary deposits on Mars: ancient deltas in a crater lake in the Nili Fossae region. Geophys Res Lett 32, L14201, doi:10.1029/2005GL023456
46. Gallagher C, Balme M (2015) Eskers in a complete, wet-based glacial system in the Phlegra Montes region, Mars. Earth Planet Sci Lett 431:96–109
47. Goudge TA, Mohrig D, Cardenas BT, Hughes CM, Fassett CI (2018) Stratigraphy and paleohydrology of delta channel deposits, Jezero crater, Mars. Icarus 301:58–75
48. Hargitai HI, Gulick VC, Glines NH (2017) Discontinuous drainage systems formed by highland precipitation and ground-water outflow in the Navua Valles and southwest Hadriacus Mons regions, Mars. Icarus 294:172–200
49. Hiesinger H, Head JW III (2002) Topography and morphology of the Argyre Basin, Mars: implications for its geologic and hydrologic history. Planet Space Sci 50:939–981
50. Hobbs SW, Clarke JDA, Paull DJ (2016) Analysis of crater valleys, Noachis Terra, Mars: evidence of fluvial and glacial processes. Geomorphology 261:244–272
51. Irwin RP, Howard AD, Craddock RA, Moore JM (2005) An intense terminal epoch of widespread fluvial activity on early Mars: 2. Increased runoff and paleolake development. J Geophys Res E Planets 110:S15
52. Irwin RP, Lewis KW, Grant JA (2015) Paleohydrology of Eberswalde crater, Mars. Geomorphology 240:83–101
53. Jaumann R, Reiss D, Frei S et al (2005) Interior channels in Martian valleys: constraints on fluvial erosion by measurements of the Mars express high resolution stereo camera. Geophys Res Lett 32: L16203
54. Loizeau D, Carter J, Bouley S, Mangold N, Poulet F, Bibring J-P, Costard F, Langevin Y, Gondet B, Murchie SL (2012) Characterization of hydrated silicate-bearing outcrops in Tyrrhena Terra, Mars: implications to the alteration history of Mars. Icarus 219:476–497
55. Bernhardt H, Hiesinger H, Reiss D, Ivanov M, Erkeling G (2013) Putative eskers and new insights into glacio-fluvial depositional settings in southern Argyre Planitia, Mars. Planet Space Sci 85:261–278
56. Mangold N, Kite ES, Kleinhans MG, Newsom H, Ansan V, Hauber E, Kraal E, Quantin C, Tanaka K (2012) The origin and timing of fluvial activity at Eberswalde crater, Mars. Icarus 220:530–551
57. Marra WA, Braat L, Baar AW, Kleinhans MG (2014) Valley formation by groundwater seepage, pressurized groundwater outbursts and crater-lake overflow in flume experiments with implications for Mars. Icarus 232:97–117
58. Martha TR, Jain N, Vamshi GT, Vinod Kumar K (2017) Evidences of early aqueous Mars: implications on the origin of branched valleys in the Ius Chasma, Mars. Planet Space Sci 148:56–65
59. Molina A, López I, Prieto-Ballesteros O, Fernández-Remolar D, de Pablo MÁ, Gómez F (2017) Coogoon Valles, western Arabia Terra: hydrological evolution of a complex Martian channel system. Icarus 293:27–44
60. Nicholson BG, Hancock GR, Cohen S, Willgoose GR, Rey-Lescure O (2013) An assessment of the fluvial geomorphology of subcatchments in Parana Valles, Mars. Geomorphology 183:96–109
61. Pajola M, Rossato S, Baratti E, Mangili C, Mancarella F, McBride K, Coradini M (2016) The Simud-Tiu Valles hydrologic system: a multidisciplinary study of a possible site for future Mars on-site exploration. Icarus 268:355–381
62. Quantin C, Allemand P, Mangold N, Dromart G, Delacourt C (2005) Fluvial and lacustrine activity on layered deposits in Melas Chasma, Valles Marineris, Mars. J Geophys Res 110:E12S19
63. Rosenberg EN, Head JW III (2015) Late Noachian fluvial erosion on Mars: cumulative water volumes required to carve the valley networks and grain size of bed-sediment. Planet Space Sci 117:429–435
64. Weitz CM, Bishop JL, Grant JA (2013) Gypsum, opal, and fluvial channels within a trough of Noctis Labyrinthus, Mars: implications for aqueous activity during the Late Hesperian to Amazonian. Planet Space Sci 87:130–145
65. Bouley S, Ansan V, Mangold N, Masson P, Neukum G (2009) Fluvial morphology of Naktong Vallis, Mars: a late activity with multiple processes. Planet Space Sci 57:982–999
66. Buhler PB, Fassett CI, Head JW, Lamb MP (2011) Evidence for paleolakes in Erythraea Fossa, Mars: implications for an ancient hydrological cycle. Icarus 213:104–115
67. Butcher FEG, Conway SJ, Arnold NS (2016) Are the Dorsa Argentea on Mars eskers? Icarus 275:65–84
68. Butcher FEG, Balme MR, Gallagher C, Arnold NS, Conway SJ, Hagermann A, Lewis SR (2017) Recent basal melting of a mid-latitude glacier on Mars. J Geophys Res Planets 122:2445–2468
69. Carr MH, Head JW (2003) Basal melting of snow on early Mars: a possible origin of some valley networks. Geophys Res Lett, 30: 2245, doi:10.1029/2003GL018575
70. Craddock RA, Lorenz RD (2017) The changing nature of rainfall during the early history of Mars. Icarus 293:172–179
71. Fairén AG, Stokes CR, Davies NS et al (2014) A cold hydrological system in Gale crater, Mars. Planet Space Sci 93–94:101–118
72. Orosei R, Lauro SE, Pettinelli E et al (2018) Radar evidence of subglacial liquid water on Mars. Science 361:490–493
73. Feldman WC, Mellon MT, Gasnault O, Maurice S, Prettyman TH (2008) Volatiles on Mars: scientific results from the Mars Odyssey Neutron Spectrometer. In: Bell J (ed) Martian Surf. Cambridge University Press, Cambridge, pp 125–148

74. Russell JK, Stasiuk MV (1997) Characterization of volcanic deposits with ground-penetrating radar. Bull Volcanol 58:515–527
75. Neugebauer G, Miinch G, Kieffer H, Chase SCJ, Miner E (1971) Mariner 1969 infrared radiometer results: temperatures and thermal properties of the Martian surface. Astron J 76:719
76. Cutts JA, Blasius KR (1981) Origin of Martian outflow channels: the Eolian hypothesis. J Geophys Res Solid Earth 86:5075–5102
77. Smith PH, Tamppari LK, Arvidson RE et al (2009) H_2O at the Phoenix landing site. Science 325:58–61
78. Harrison KP, Grimm RE (2004) Tharsis recharge: a source of groundwater for Martian outflow channels. Geophys Res Lett 31:L14703
79. Head JW, Weiss DK (2014) Preservation of ancient ice at Pavonis and Arsia Mons: tropical mountain glacier deposits on Mars. Planet Space Sci 103:331–338
80. Kadish SJ, Head JW, Fastook JL, Marchant DR (2014) Middle to late Amazonian tropical mountain glaciers on Mars: the ages of the Tharsis Montes fan-shaped deposits. Planet Space Sci 91:52–59
81. Scanlon KE, Head JW, Wilson L, Marchant DR (2014) Volcano–ice interactions in the Arsia Mons tropical mountain glacier deposits. Icarus 237:315–339
82. Pedersen GBM, Head JW (2010) Evidence of widespread degraded Amazonian-aged ice-rich deposits in the transition between Elysium Rise and Utopia Planitia, Mars: guidelines for the recognition of degraded ice-rich materials. Planet Space Sci 58:1953–1970
83. Moyer EJ, Irion FW, Yung YL, Gunson MR (1996) ATMOS stratospheric deuterated water and implications for troposphere-stratosphere transport. Geophys Res Lett 23:2385–2388
84. Dundas CM, Byrne S, McEwen AS, Mellon MT, Kennedy MR, Daubar IJ, Saper L (2014) HiRISE observations of new impact craters exposing Martian ground ice. J Geophys Res Planets 119:109–127
85. Lange MA, Ahrens TJ (1982) Impact induced dehydration of serpentine and the evolution of planetary atmospheres. J Geophys Res 87:A451
86. Gillmann C, Lognonné P, Moreira M (2011) Volatiles in the atmosphere of Mars: the effects of volcanism and escape constrained by isotopic data. Earth Planet Sci Lett 303:299–309
87. Lammer H, Chassefière E, Karatekin Ö et al (2013) Outgassing history and escape of the Martian atmosphere and water inventory. Space Sci Rev 174:113–154
88. Kurokawa H, Sato M, Ushioda M, Matsuyama T, Moriwaki R, Dohm JM, Usui T (2014) Evolution of water reservoirs on Mars: constraints from hydrogen isotopes in Martian meteorites. Earth Planet Sci Lett. 394:179–185
89. Krasnopolsky VA (2015) Variations of the HDO/H_2O ratio in the martian atmosphere and loss of water from Mars. Icarus 257:377–386
90. Villanueva GL, Mumma MJ, Novak RE, Käufl HU, Hartogh P, Encrenaz T, Tokunaga A, Khayat A, Smith MD (2015) Strong water isotopic anomalies in the Martian atmosphere: probing current and ancient reservoirs. Science 348(6231):218-221.
91. Ehlmann BL, Mustard JF, Murchie SL (2010) Geologic setting of serpentine deposits on Mars. Geophys Res Lett 37:L06201
92. Stopar JD, Jeffrey Taylor G, Hamilton VE, Browning L (2006) Kinetic model of olivine dissolution and extent of aqueous alteration on mars. Geochim Cosmochim Acta 70:6136–6152
93. Oze C, Sharma M (2007) Serpentinization and the inorganic synthesis of H_2 in planetary surfaces. Icarus 186:557–561
94. Carr MH (1974) The role of lava erosion in the formation of lunar rilles and Martian channels. Icarus 22:1–23
95. Ferlito C, Siewert J (2006) Lava channel formation during the 2001 Eruption on Mount Etna: evidence for Mechanical Erosion. Phys Rev Lett 96:028501
96. Greeley R, Fagents SA, Harris RS, Kadel SD, Williams DA, Guest JE (1998) Erosion by flowing lava: field evidence. J Geophys Res Solid Earth 103:27325–27345
97. Hopper JP, Leverington DW (2014) Formation of Hrad Vallis (Mars) by low viscosity lava flows. Geomorphology 207:96–113
98. Hurwitz DM, Fassett CI, Head JW, Wilson L (2010) Formation of an eroded lava channel within an Elysium Planitia impact crater: distinguishing between a mechanical and thermal origin. Icarus 210:626–634
99. Williams DA, Greeley R, Lesher CM, Clynne MA, Kadel SD (2004) Erosion by flowing lava: geochemical evidence in the Cave Basalt, Mount St. Helens, Washington. Bull Volcanol 66:168–181
100. Williams DA, Greeley R, Hauber E, Gwinner K, Neukum G (2005) Erosion by flowing Martian lava: new insights for Hecates Tholus from Mars express and MER data. J Geophys Res 110:E05006
101. Leone G (2014) A network of lava tubes as the origin of Labyrinthus Noctis and Valles Marineris on Mars. J Volcanol Geotherm Res 277:1–8
102. Leone G (2016) Alignments of volcanic features in the southern hemisphere of Mars produced by migrating mantle plumes. J Volcanol Geotherm Res 309:78–95
103. Leone G (2019) Comments to "A composite mineralogical map of Ganges Chasma and surroundings, Valles Marineris, Mars" by Selby Cull-Hearth and M. Caroline Clark (Planetary and Space Science 142, 1–8). Planet Space Sci 174:56–57

104. Leverington DW, Maxwell TA (2004) An igneous origin for features of a candidate crater-lake system in western Memnonia, Mars. J Geophys Res 109: E06006
105. Leverington DW (2009) Reconciling channel formation processes with the nature of elevated outflow systems at Ophir and Aurorae Plana, Mars. J Geophys Res Planet 114:E10005
106. Sharp RP (1973) Mars: fretted and chaotic terrains. J Geophys Res. 78(20): 4073– 4083
107. McSween HY, Wyatt MB, Gellert R, et al (2006) Characterization and petrologic interpretation of olivine-rich basalts at Gusev Crater, Mars. J Geophys Res Planet 111:E02S10
108. Grotzinger JP, Gupta S, Malin MC, et al (2015) Deposition, exhumation, and paleoclimate of an ancient lake deposit, Gale crater, Mars. Sci 350 (80):172–176
109. Madden MEE, Bodnar RJ, Rimstidt JD (2004) Jarosite as an indicator of water-limited chemical weathering on Mars. Nature 431:821–823
110. Niles PB, Catling DC, Berger G, Chassefière E, Ehlmann BL, Michalski JR, Morris R, Ruff SW, Sutter B (2013) Geochemistry of carbonates on Mars: implications for climate history and nature of aqueous environments. Space Sci Rev 174:301–328
111. Bargery AS, Wilson L (2010) Dynamics of the ascent and eruption of water containing dissolved CO_2 on Mars. J Geophys Res 115:5008
112. Parker TJ, Stephen Saunders R, Schneeberger DM (1989) Transitional morphology in West Deuteronilus Mensae, Mars: implications for modification of the lowland/upland boundary. Icarus 82:111–145
113. Sagan C, Toon OB, Gierasch PJ (1973) Climatic change on Mars. Science 181:1045–1049
114. Plescia JB (1990) Recent flood lavas in the Elysium region of Mars. Icarus 88:465–490
115. Burr DM, McEwen AS, Sakimoto SEH (2002) Recent aqueous floods from the Cerberus Fossae, Mars. Geophys Res Lett 29:1013
116. Levy J, Head J, Marchant D (2009) Thermal contraction crack polygons on Mars: classification, distribution, and climate implications from HiRISE observations. J Geophys Res 114:E01007
117. Carr MH, Schaber GG (1977) Martian permafrost features. J Geophys Res 82:4039–4054
118. Squyres SW (1979) The distribution of lobate debris aprons and similar flows on Mars. J Geophys Res 84:8087
119. Di Achille G, Hynek BM (2010) Ancient ocean on Mars supported by global distribution of deltas and valleys. Nat Geosci 3:459–463
120. Bibring J-P, Langevin Y, Mustard JF, Ois Poulet F, Arvidson R, Gendrin A, Gondet B, Mangold N, Pinet P, Forget F (2006) Global mineralogical and aqueous Mars history derived from OMEGA/Mars express data. Sci 312(80):400–404
121. Watters TR, Campbell B, Carter L et al (2007) Radar sounding of the Medusae Fossae Formation Mars: equatorial ice or dry, low-density deposits? Science 318:1125–1128
122. Karlsson NB, Schmidt LS, Hvidberg CS (2015) Volume of Martian midlatitude glaciers from radar observations and ice flow modeling. Geophys Res Lett 42:2627–2633
123. Erkeling G, Leone G, Ákos K (2015) Paleoshoreline. In: Encycl. Planet. Landforms. Springer New York, New York, NY, pp 1–9
124. Malin MC, Edgett KS (1999). Oceans or seas in the Martian northern lowlands: high resolution imaging tests of proposed coastlines. Geophys. Res. Lett.26 (19):3049-3052
125. Sholes SF, Montgomery DR, Catling DC (2019) Quantitative high-resolution reexamination of a hypothesized ocean shoreline in Cydonia Mensae on Mars. J Geophys Res Planets 124:316–336
126. Carr MH, Head JW (2003) Oceans on Mars: an assessment of the observational evidence and possible fate. J Geophys Res 108:5042
127. Perron JT, Mitrovica JX, Manga M, Matsuyama I, Richards MA (2007) Evidence for an ancient martian ocean in the topography of deformed shorelines. Nature 447:840–843
128. Ivanov MA, Erkeling G, Hiesinger H, Bernhardt H, Reiss D (2017) Topography of the Deuteronilus contact on Mars: evidence for an ancient water/mud ocean and long-wavelength topographic readjustments. Planet Space Sci 144:49–70
129. Ghatan GJ, Zimbelman JR (2006) Paucity of candidate coastal constructional landforms along proposed shorelines on Mars: implications for a northern lowlands-filling ocean. Icarus 185:171–196
130. Turbet M, Forget F (2019) The paradoxes of the Late Hesperian Mars ocean. Sci Rep 9:5717
131. Gulick VC, Baker VR (1989) Fluvial valleys and martian palaeoclimates. Nature 341:514–516
132. Gulick VC (2001) Origin of the valley networks on Mars: a hydrological perspective. Geomorphology 37:241–268
133. Baker VR (2001) Water and the martian landscape. Nature 412:228–236
134. Bargery AS, Wilson L (2011) Erosive flood events on the surface of Mars: application to Mangala and Athabasca Valles. Icarus 212:520–540
135. Laskar J, Correia ACM, Gastineau M, Joutel F, Levrard B, Robutel P (2004) Long term evolution and chaotic diffusion of the insolation quantities of Mars. Icarus 170(2):343-364.
136. Carr MH (2012) The fluvial history of Mars. Philos Trans R Soc A Math Phys Eng Sci 370:2193–2215
137. Brakenridge GR, Newsom HE, Baker VR (1985) Ancient hot springs on Mars: origins and paleoenvironmental significance of small Martian valleys. Geology 13:859
138. Brakenridge GR (1990) The origin of fluvial valleys and early geologic history, Aeolis Quadrangle, Mars. J Geophys Res 95:17289

139. Irwin RP, Craddock RA, Howard AD, Flemming HL (2011) Topographic influences on development of Martian valley networks. J Geophys Res 116:E02005
140. Grant JA, Irwin RP, Wilson SA, Buczkowski D, Siebach K (2011) A lake in Uzboi Vallis and implications for Late Noachian-Early Hesperian climate on Mars. Icarus 212(1):110-122.
141. Forget F, Wordsworth R, Millour E, Madeleine J-B, Kerber L, Leconte J, Marcq E, Haberle RM (2013) 3D modelling of the early martian climate under a denser CO_2 atmosphere: temperatures and CO_2 ice clouds. Icarus 222:81–99
142. Wordsworth R, Forget F, Millour E, Head JW, Madeleine J-B, Charnay B (2013) Global modelling of the early martian climate under a denser CO_2 atmosphere: water cycle and ice evolution. Icarus 222:1–19
143. Wordsworth RD, Kerber L, Pierrehumbert RT, Forget F, Head JW (2015) Comparison of "warm and wet" and "cold and icy" scenarios for early Mars in a 3-D climate model. J Geophys Res Planets 120:1201–1219
144. Carr MH (2002) Elevations of water-worn features on Mars: Implications for circulation of groundwater. J Geophys Res Planets 107:5131
145. Mellon MT, Phillips RJ (2001) Recent gullies on Mars and the source of liquid water. J Geophys Res Planets 106:23165–23179
146. Hartmann WK (2001) Martian seeps and their relation to youthful geothermal activity. Springer, Dordrecht, pp 405–410
147. Gaidos EJ (2001) Cryovolcanism and the recent flow of liquid water on Mars. Icarus 153:218–223
148. Wynn-Williams DD, Cabrol NA, Grin EA, Haberle RM, Stoker CR (2001) Brines in seepage channels as eluants for subsurface relict biomolecules on Mars? Astrobiology 1:165–184
149. Knauth LP, Burt DM (2002) Eutectic brines on Mars: origin and possible relation to young seepage features. Icarus 158:267–271
150. Musselwhite DS, Swindle TD, Lunine JI (2001) Liquid CO_2 breakout and the formation of recent small gullies on Mars. Geophys Res Lett 28:1283–1285
151. Treiman AH (2003) Geologic settings of Martian gullies: implications for their origins. J Geophys Res 108:8031
152. Plescia JB (2003) Cerberus Fossae, Elysium, Mars: a source for lava and water. Icarus 164:79–95
153. Keszthelyi LP, Jaeger WL, Dundas CM, Martínez-Alonso S, McEwen AS, Milazzo MP (2010) Hydrovolcanic features on Mars: preliminary observations from the first Mars year of HiRISE imaging. Icarus 205:211–229
154. Brož P, Hauber E (2013) Hydrovolcanic tuff rings and cones as indicators for phreatomagmatic explosive eruptions on Mars. J Geophys Res Planets 118:1656–1675
155. Lanagan PD, McEwen AS, Keszthelyi LP, Thordarson T (2001) Rootless cones on Mars indicating the presence of shallow equatorial ground ice in recent times. Geophys Res Lett 28:2365–2367
156. Fagents SA, Lanagan P, Greeley R (2002) Rootless cones on Mars: a consequence of lava-ground ice interaction. Geol Soc London, Spec Publ 202:295–317
157. Bruno BC, Fagents SA, Thordarson T, Baloga SM, Pilger E (2004) Clustering within rootless cone groups on Iceland and Mars: effect of nonrandom processes. J Geophys Res 109:E07009
158. Baloga SM, Glaze LS, Bruno BC (2007) Nearest-neighbor analysis of small features on Mars: applications to tumuli and rootless cones. J Geophys Res 112:E03002
159. Jurado-Chichay Z, Rowland SK, Walker GPL (1996) The formation of circular littoral cones from tube-fed p?hoehoe: Mauna Loa, Hawai'i. Bull Volcanol 57:471–482
160. Bruno BC, Fagents SA, Hamilton CW, Burr DM, Baloga SM (2006) Identification of volcanic rootless cones, ice mounds, and impact craters on Earth and Mars: using spatial distribution as a remote sensing tool. J Geophys Res 111:E06017
161. Frey H, Lowry BL, Chase SA (1979) Pseudocraters on Mars. J Geophys Res 84:8075
162. Greeley R, Fagents SA (2001) Icelandic pseudocraters as analogs to some volcanic cones on Mars. J Geophys Res Planets 106:20527–20546
163. Hamilton CW, Fagents SA, Thordarson T (2011) Lava–ground ice interactions in Elysium Planitia, Mars: geomorphological and geospatial analysis of the Tartarus Colles cone groups. J Geophys Res 116:E03004
164. Farrand WH, Gaddis LR, Keszthelyi L (2005) Pitted cones and domes on Mars: observations in Acidalia Planitia and Cydonia Mensae using MOC, THEMIS, and TES data. J Geophys Res 110:E05005
165. Brož P, Hauber E (2012) A unique volcanic field in Tharsis, Mars: pyroclastic cones as evidence for explosive eruptions. Icarus 218:88–99
166. Mouginis-Mark PJ (1982) Late-stage summit activity of Martian shield volcanoes. Lunar Planet Sci Conf 12th, Houston, TX, March 16–20, 1981, Proceedings Sect 2 (A82–31677 15–91) New York Oxford, Pergamon Press, 12:1431–1447
167. El Maarry MR, Markiewicz WJ, Mellon MT, Goetz W, Dohm JM, Pack A (2010) Crater floor polygons: desiccation patterns of ancient lakes on Mars? J Geophys Res 115:E10006
168. Mutch TA, Grenander SU, Jones KL et al (1976) The surface of Mars: the view from the viking 2 lander. Science 194:1277–1283
169. Masursky H, Crabill NL (1976) Search for the viking 2 landing site. Science 194:62–68
170. McSween HY, Ruff SW, Morris RV, et al (2006) Alkaline volcanic rocks from the Columbia Hills, Gusev crater, Mars. J Geophys Res 111:E09S91

171. Jaeger JC (1961) The cooling of irregularly shaped igneous bodies. Am J Sci 259:721–734
172. Plaut JJ, Safaeinili A, Holt JW, Phillips RJ, Head JW, Seu R, Putzig NE, Frigeri A (2009) Radar evidence for ice in lobate debris aprons in the mid-northern latitudes of Mars. Geophys Res Lett 36:L02203
173. Hauber E, van Gasselt S, Chapman MG, Neukum G (2008) Geomorphic evidence for former lobate debris aprons at low latitudes on Mars: indicators of the Martian paleoclimate. J Geophys Res 113: E02007
174. Baker VR, Strom RG, Gulick VC, Kargel JS, Komatsu G, Kale VS (1991) Ancient oceans, ice sheets and the hydrological cycle on Mars. Nature 352:589–594
175. Bishop JL, Dobrea EZN, McKeown NK et al (2008) Phyllosilicate diversity and past aqueous activity revealed at Mawrth Vallis, Mars. Science 321:830–833
176. Michalski J, Poulet F, Bibring JP, Mangold N (2010) Analysis of phyllosilicate deposits in the Nili Fossae region of Mars: comparison of TES and OMEGA data. Icarus 206:269–289
177. Carter J, Poulet F, Bibring J-P, Mangold N, Murchie S (2013). Hydrous minerals on Mars as seen by the CRISM and OMEGA imaging spectrometers: Updated global view. J. Geophys. Res. Planets 118: 831– 858.
178. Michalski JR, Cuadros J, Bishop JL, Darby Dyar M, Dekov V, Fiore S (2015) Constraints on the crystal-chemistry of Fe/Mg-rich smectitic clays on Mars and links to global alteration trends. Earth Planet Sci Lett 427:215–225
179. Ehlmann BL, Mustard JF, Clark RN, Swayze GA, Murchie SL (2011) Evidence for low-grade metamorphism, hydrothermal alteration, and diagenesis on Mars from phyllosilicate mineral assemblages. Clays Clay Miner 59:359–377
180. Ehlmann BL, Berger G, Mangold N, Michalski JR, Catling DC, Ruff SW, Chassefière E, Niles PB, Chevrier V, Poulet F (2013) Geochemical consequences of widespread clay mineral formation in Mars' ancient crust. Space Sci Rev 174:329–364
181. Che C, Glotch TD (2014) Thermal alteration: a possible reason for the inconsistency between OMEGA/CRISM and TES detections of phyllosilicates on Mars? Geophys Res Lett 41:321–327
182. Christensen PR, Morris RV, Lane MD, Bandfield JL, Malin MC (2001) Global mapping of Martian hematite mineral deposits: remnants of water-driven processes on early Mars. J Geophys Res Planets 106:23873–23885
183. Christensen PR, Bandfield JL, Clark RN et al (2000) Detection of crystalline hematite mineralization on Mars by the Thermal Emission Spectrometer: evidence for near-surface water. J Geophys Res 105:9623–9642
184. Hynek BM, Arvidson RE, Phillips RJ (2002) Geologic setting and origin of Terra Meridiani hematite deposit on Mars. J Geophys Res 107:5088
185. Arvidson RE, Poulet F, Morris RV et al (2006) Nature and origin of the hematite-bearing plains of Terra Meridiani based on analyses of orbital and Mars Exploration rover data sets. J Geophys Res Planets 111:E12S08
186. Blaney DL, McCord TB (1989) An observational search for carbonates on Mars. J Geophys Res Solid Earth 94:10159–10166
187. Pollack JB, Roush T, Witteborn F, Bregman J, Wooden D, Stoker C, Toon OB, Rank D, Dalton B, Freedman R (1990) Thermal emission spectra of Mars (5.4–10.5 μm): evidence for sulfates, carbonates, and hydrates. J Geophys Res 95:14595
188. Calvin WM, King TVV, Clark RN (1994) Hydrous carbonates on Mars?: Evidence from Mariner 6/7 infrared spectrometer and ground-based telescopic spectra. J Geophys Res 99:14659
189. Wagner C, Schade U (1996) Measurements and calculations for estimating the spectrometric detection limit for carbonates in Martian soil. Icarus 123:256–268
190. Christensen PR, Bandfield JL, Hamilton VE et al (2001) Mars global surveyor thermal emission spectrometer experiment: investigation description and surface science results. J Geophys Res Planets 106:23823–23871
191. Bandfield JL, Glotch TD, Christensen PR (2003) Spectroscopic identification of carbonate minerals in the martian dust. Science 301:1084–1087
192. Yen AS, Morris R V., Clark BC, et al (2008) Hydrothermal processes at Gusev Crater: an evaluation of Paso Robles class soils. J Geophys Res 113:E06S10
193. Morris RV, Ruff SW, Gellert R et al (2010) Identification of carbonate-rich outcrops on Mars by the Spirit rover. Science 329:421–424
194. Boynton WV, Ming DW, Kounaves SP et al (2009) Evidence for calcium carbonate at the Mars Phoenix landing site. Science 325:61–64
195. Ehlmann BL, Mustard JF, Murchie SL et al (2008) Orbital identification of carbonate-bearing rocks on Mars. Science 322:1828–1832
196. Halevy I, Schrag DP (2009) Sulfur dioxide inhibits calcium carbonate precipitation: implications for early Mars and Earth. Geophys Res Lett 36:L23201
197. Christensen PR (2006) Water at the poles and in permafrost regions of Mars. Elements 2:151–155
198. Kerber L, Head JW, Madeleine JB, Forget F, Wilson L (2011) The dispersal of pyroclasts from Apollinaris Patera, Mars: implications for the origin of the Medusae Fossae Formation. Icarus 216:212–220
199. Leone G (2014) A network of lava tubes as the origin of Labyrinthus Noctis and Valles Marineris on Mars. J Volcanol Geotherm Res. 277:1-8.

200. McSween HY, Grove TL, Wyatt MB (2003) Constraints on the composition and petrogenesis of the Martian crust. J Geophys Res Planets. 108: 5135
201. Wyatt MB, McSween HY, Tanaka KL, Head JW (2004) Global geologic context for rock types and surface alteration on Mars. Geology 32:645–648
202. Grotzinger JP, Sumner DY, Kah LC, et al (2014) A habitable fluvio-lacustrine environment at Yellowknife Bay, Gale crater, Mars. Science 343 (80):1242777
203. Fernando J, Schmidt F, Pilorget C, Pinet P, Ceamanos X, Douté S, Daydou Y, Costard F (2015) Characterization and mapping of surface physical properties of Mars from CRISM multi-angular data: application to Gusev Crater and Meridiani Planum. Icarus 253:271–295
204. Ming DW, Gellert R, Morris R V., et al (2008) Geochemical properties of rocks and soils in Gusev Crater, Mars: results of the alpha particle X-ray spectrometer from Cumberland ridge to home plate. J Geophys Res 113:E12S39
205. Morris RV, Vaniman DT, Blake DF et al (2016) Silicic volcanism on Mars evidenced by tridymite in high-SiO_2 sedimentary rock at Gale crater. Proc Natl Acad Sci 113:7071–7076
206. Dietterich HR, Cashman KV (2014) Channel networks within lava flows: formation, evolution, and implications for flow behavior. J Geophys Res Earth Surf 119:1704–1724
207. Leverington DW (2007) Was the Mangala Valles system incised by volcanic flows? J Geophys Res 112:E11005
208. Baird AK, Clark BC (1984) Did komatiitic lavas erode channels on Mars? Nature 311:18–18
209. Leverington DW (2004) Volcanic rilles, streamlined islands, and the origin of outflow channels on Mars. J Geophys Res 109:E10011
210. Keszthelyi L, Jaeger W, Milazzo M, Radebaugh J, Davies AG, Mitchell KL (2007) New estimates for Io eruption temperatures: implications for the interior. Icarus 192:491–502
211. Ghatan GJ, Head JW, Wilson L (2005) Mangala Valles, Mars: assessment of early stages of flooding and downstream flood evolution. Earth Moon Planets 96:1–57
212. Lunine JI, Chambers J, Morbidelli A, Leshin LA (2003) The origin of water on Mars. Icarus 165:1–8
213. Marty B, Avice G, Sano Y, Altwegg K, Balsiger H, Hässig M, Morbidelli A, Mousis O, Rubin M (2016) Origins of volatile elements (H, C, N, noble gases) on Earth and Mars in light of recent results from the ROSETTA cometary mission. Earth Planet Sci Lett 441:91–102
214. Quintana SN, Schultz PH (2019) Observational constraints on the cometary impact flux on Mars and Earth. Icarus 326:176–185
215. JeongAhn (Chung) Y (2015) Orbital distribution of minor planets in the inner solar system and their impact fluxes on the Earth, the Moon and Mars. The University of Arizona
216. Laufer D, Notesco G, Bar-Nun A, Owen T (1999) From the interstellar medium to Earth's oceans via comets—an isotopic study of HDO/H_2O. Icarus 140:446–450
217. Hagemann R, Nief G, Roth E (1970) Absolute isotopic scale for deuterium analysis of natural waters. Absolute D/H ratio for SMOW. Tellus 22:712–715
218. Craig H (1961) Standard for reporting concentrations of deuterium and oxygen-18 in natural waters. Science 133(80):1833–1834
219. Fedorova A, Korablev O, Vandaele A-C, et al (2008) HDO and H_2O vertical distributions and isotopic ratio in the Venus mesosphere by Solar Occultation at Infrared spectrometer on board Venus Express. J Geophys Res 113:E00B22
220. Bertaux JL, Carr M, Des Marais DJ, Gaidos E (2007) Conversations on the habitability of worlds: the importance of volatiles. Space Sci Rev 129:123–165
221. De Bergh C, Bézard B, Owen T, Crisp D, Maillard JP, Lutz BL (1991) Deuterium on Venus: observations from Earth. Science 251(80):547–549
222. Owen T, Tobias (1992) The composition and early history of the atmosphere of Mars. In: Mars (A93-27852 09–91). pp 818–834
223. Webster CR, Mahaffy PR, Flesch GJ et al (2013) Isotope ratios of H, C, and O in CO_2 and H_2O of the martian atmosphere. Science 341:260–263
224. Leone G, Tackley PJ, Gerya TV, May DA, Zhu G (2014) Three-dimensional simulations of the southern polar giant impact hypothesis for the origin of the martian dichotomy. Geophys Res Lett 41:8736–8743
225. Filiberto J, Treiman AH (2009) Martian magmas contained abundant chlorine, but little water. Geology 37:1087–1090
226. Grott M, Morschhauser A, Breuer D, Hauber E (2011) Volcanic outgassing of CO_2 and H_2O on Mars. Earth Planet Sci Lett 308:391–400
227. Balta JB, McSween HY (2013) Water and the compo sition of Martian magmas. Geology 41:1115–1118
228. Ahrens TJ (1993) Impact erosion of terrestrial planetary atmospheres. Annu Rev Earth Planet Sci 21:525–555
229. Ahrens TJ, Shen AH, Ni S (2004) Giant impact induced atmospheric blow-off. AIP Conference Proceedings 706:1419.
230. Melosh HJ, Vickery AM (1989) Impact erosion of the primordial atmosphere of Mars. Nature 338:487–489
231. Frey H, Schultz RA (1988) Large impact basins and the mega-impact origin for the crustal dichotomy on Mars. Geophys Res Lett 15:229–232

232. Wilhelms DE, Squyres SW (1984) The martian hemispheric dichotomy may be due to a giant impact. Nature 309:138–140
233. Marinova MM, Aharonson O, Asphaug E (2008) Mega-impact formation of the Mars hemispheric dichotomy. Nature 453:1216–1219
234. Andrews-Hanna JC, Zuber MT, Banerdt WB (2008) The Borealis basin and the origin of the martian crustal dichotomy. Nature 453:1212–1215
235. Reese CC, Orth CP, Solomatov VS (2011) Impact megadomes and the origin of the martian crustal dichotomy. Icarus 213:433–442
236. Golabek GJ, Keller T, Gerya TV, Zhu G, Tackley PJ, Connolly JAD (2011) Origin of the martian dichotomy and Tharsis from a giant impact causing massive magmatism. Icarus 215(1):346-357.
237. Segura TL, Toon OB, Colaprete A, Zahnle K (2002) Environmental effects of large impacts on Mars. Science 298:1977–1980
238. West MD, Clarke JDA (2010) Potential martian mineral resources: mechanisms and terrestrial analogues. Planet Space Sci 58:574–582
239. Barabash S, Fedorov A, Lundin R, Sauvaud J-A (2007) Martian atmospheric erosion rates. Science 315:501–503
240. Brain DA, Jakosky BM (1998) Atmospheric loss since the onset of the Martian geologic record: combined role of impact erosion and sputtering. J Geophys Res Planets 103:22689–22694
241. Wordsworth RD (2016) The climate of early Mars. Annu Rev Earth Planet Sci 44:381–408
242. Tanaka KL, Robbins SJ, Fortezzo CM, Skinner JA, Hare TM (2014) The digital global geologic map of Mars: chronostratigraphic ages, topographic and crater morphologic characteristics, and updated resurfacing history. Planet Space Sci 95:11–24

The Formation of Mars and the Origin of Its Volcanic Provinces

Giovanni Leone

Abstract

This chapter explores the various hypotheses put forward for the formation of the Martian dichotomy, the topographic difference between the northern lowlands and the southern highlands of the planet. Among the various hypotheses, the Southern Polar Giant Impact (SPGI) was the only one validated with the discovery of twelve volcanic alignments as predicted by the 3D model. Furthermore, this hypothesis is the only one that matches both astronomical and geophysical data available in the literature and all the requirements to be seriously considered as a theory. The various volcanic centres of Mars, including the vast majority being part of the alignments, are described with full detail in this chapter.

1 The Formation of Mars

The accretion and the differentiation history of Mars was derived by the study of the Shergottites-Nakhlites-Chassignites (SNC) meteorites and from the first studies of the Martian soils during the Viking missions [1–3]. The SNC meteorites are rocks of mafic or ultramafic composition, they were stripped away from the Martian crust or possibly even from the mantle by impact processes; they are essentially of basaltic composition [2] with low-alkali and with silica contents ranging from 37.4 wt% of the chassignite to the 51.3 wt% of the shergottite [4]. The Martian soils, although it would be more appropriate to define them mainly lava fields with some mantling of pyroclastic deposits, showed a more varied igneous composition than the SNC ranging from picrite to rhyolite [5–15]. A study of the isotopes of neodymium in SNC meteorites suggested a rapid accretion and an early differentiation of Mars before 4539 million years ago [3]. The later stages of accretion were then characterized by giant impacts naturally occurring when only a number of large objects remains after the hierarchical coagulation of planetesimals, the main implication was that Mars did not experience giant impacts later than 27 million years after the origin of the solar system [3, 16]. The central star does not play a key role in this phase of planetary formation [17]. The origin of the solar system is thought starting from the formation of the calcium-aluminium inclusions (CAI) in the CV chondrite Efremovka, the lead isotopic age of this event is 4567.2 ± 0.6 million years ago [18]. The time of differentiation into metallic core and silicate mantle, derived from the decay of ^{182}Hf to ^{182}W in Nakhla, is about 15 Ma after CAI with a mean accretionary time of 5 Ma for Mars [19, 20]. The

G. Leone (✉)
Instituto de Investigación en Astronomía y Ciencias Planetarias, Universidad de Atacama, Copiapó, Chile
e-mail: giovanni.leone@uda.cl

G. Leone (ed.), *Mars: A Volcanic World*,
https://doi.org/10.1007/978-3-030-84103-4_2

mean differentiation time derived from the shergottites is of 3.3 Ma after CAI [21]. However, it is worth noticing that the important heating effect of the radiogenic elements ^{26}Al and ^{60}Fe would have kept Mars in a molten state and thus prevented from differentiation any other formed planetesimal until 2.5–3 Ma after CAI [22, 23]. Thus, according to the above-mentioned constraints, any giant impact event that might have perturbed the normal differentiation of the metallic core from the silicate mantle should have occurred within a window of time between 3 and 15 Ma after CAI [24]. Other studies, taking into account uncertainties in the Hf-W timing of core formation, placed a further constrain between ~1 Ma during runaway growth [25] and 10 Ma after CAI [26]. Conservative estimates for the 3D simulations of the Southern Polar Giant Impact (SPGI) were assumed at 4 Ma after CAI [24]. This was a step forward, which kept into account the early effect of the short-lived radiogenic elements ^{26}Al and ^{60}Fe, with respect to a previous estimate of 5 Ma made for 2D simulations [27].

Hundreds of Ma (up to 1 Ga) of time are needed to start modern-style plate tectonics from the early stagnant-lid phase of a planet, numerical models suggest initiation of mobile lid tectonics within ~100 Ma from magma ocean solidification [28, 29], provided there is enough water to develop it afterwards [30, 31]. The absence of water on Mars [32–35] may explain why plate tectonics could not develop there. Mars lost much of its water from 4.5 to 4.0 Ga [36–39]. This water loss event is coeval to the strongest phase of volcanism that Mars ever experienced after the SPGI with a Moon-sized impactor, which likely occurred between 4.563 and 4.548 Ga [24]. This early loss of water may be the reason why plate tectonics never developed on Mars [40] despite earlier speculations of a putative Martian plate tectonics [41].

2 The Hypotheses of Formation of the Martian Dichotomy and the Origin of Volcanism

An important implication of the giant impacts is the formation of the Martian dichotomy, the topographic difference between the southern highlands and the northern lowlands [24, 42, 43]. Various hypotheses have been proposed for the formation of this most ancient geologic feature of the planet [44]:

- Degree-1 mantle convection [45–48].
- Noachian-Hesperian crustal thinning [49, 50].
- Multiple impacts shaping the lowlands [51].
- The Northern Polar Giant Impact (NPGI) [52–55].
- The Southern Polar Giant Impact (SPGI) [24, 27, 56–58].

In order to be taken in serious consideration as a theory, the proposed hypotheses must verify, include, and respect all of the three following requirements [44, 50]:

1. It must account for the shape of the lowlands and the highlands
2. It must be physically consistent
3. It must be compatible with all the available geological, geochemical, and geophysical data.

2.1 Degree-1 Mantle Convection Hypothesis

The degree-1 mantle convection hypothesis is based on the upwelling of a single large plume, which may produce the Martian dichotomy, but only in a particular condition of mid-mantle jump in viscosity that it is not so easily met [46]. In earlier simulations of phase transitions, aimed at studying the Martian mantle convection and the evolution of its volcanism, the viscosity of the mantle was assumed depth-dependent with a viscous lid of 170 km of thickness at the top, with an increase of the viscosity of a factor 1000 from the upper mantle to the surface, and a factor of 3 with depth in the rest of the mantle [59]. However, the high internal heating rate leads to an increase of temperature into the mantle that affects the convection through the decrease of the T-dependent viscosity [46]. Furthermore, either iso-viscous or T-dependent viscosity do not produce a degree-1 convection in the mantle

[47]. The analysis of the Rayleigh–Taylor instabilities showed how an asthenosphere $\sim$100 times weaker than the underlying mantle in an early Mars is the requirement to obtain a self-consistent degree-1 mantle convection [47] or, alternatively, a mid-mantle jump in viscosity linked to the Martian mineralogical transition zone below 1000 km of depth [46]. The timing of the degree-1 process is quite long, up to 2×10^9 years [47], even though it was suggested that it is possible to obtain a dichotomy between 150 and 120 Ma [46] or even in less than 100 Ma [48]. However, such timescales are still much above the upper constraint of 15 Ma suggested for the core-mantle differentiation [24]. Also, these timescales are still above the upper limit of 30–50 Ma suggested for crustal formation upon solidification of a magma ocean [60, 61]. Probably the degree-1 mantle convection might explain the formation of Tharsis [47], although there was already a previous attempt to explain the whole volcanism of Mars through this hypothesis [62], but surely does not explain the alignments observed among the volcanoes of Mars [43]. Therefore, in the case of the degree-1 mantle convection hypothesis, the third requirement is missing.

2.2 Noachian-Hesperian Crustal Thinning Hypothesis

The Noachian-Hesperian crustal thinning hypothesis suggests that the difference in crustal thickness between the highlands and the lowlands was caused by extensional thinning and delamination above the rising half of a convection cell or mantle plume [49, 50]. This hypothesis also suggests that the extensive fracturing along the dichotomy boundary, particularly in the sector of Ismenius Lacus, would be the evidence of a long-lived extensional stress regime [49]. However, except for the sectors from Deuteronilus Mensae to Nilosyrtis Mensae, there is no more trace along the dichotomy boundary of something resembling by far the putative extensional fracturing suggested at Ismenius Lacus [42]. The extensional stress proposed by the crustal thinning hypothesis cannot occur in a stagnant lid regime characterized by a thick crust, the mantle convection is unable to break the lithosphere into plates and thus produce any fracturing [63]. Furthermore, the higher resolution images released by the Mars Reconnaissance Orbiter (MRO) revealed that what seemed fractures in previous lower resolution images are in reality lava channels, being the knobby terrain located downhill the result of extensive erosion of previous lava fields by subsequent lava flooding [43]. The crustal thinning hypothesis is not physically consistent and not compatible with the observations. Thus, in the crustal thinning hypothesis the second and the third requirements are missing.

2.3 Multiple Impacts Hypothesis

The hypothesis of multiple impacts for the formation of the Martian dichotomy proposed the combination of the various impact basins that formed the lowlands, as an alternative to the single impact proposed earlier for the formation of the Borealis basin [52], because "there was little observational evidence in support of a single impact" [51]. Although the multiple impact hypothesis may nicely explain some aspects of the topography of the lowlands, it shows some problems in explaining the lowlands themselves and it is clear that the three main impact basins (Utopia, Elysium, Chryse) cannot have formed the whole lowlands [49]. In the hypothesis of formation of the lowlands through the Borealis basin there is a variable number of "missing" basins that clearly depends on the size of the impactors [51]. Furthermore, it is statistically unlikely that multiple impacts occurred exactly and only in the northern hemisphere to shape the lowlands of Mars [24]. In fact, other large impact basins (i.e. Argyre and Hellas) almost coeval to Utopia and Chryse [64] are superposed to the highlands [51] thus clearly suggesting that the dichotomy was already formed when they were already emplaced. As well as in the case of the crustal thinning hypothesis, the multiple impact hypothesis does not match the second and the third requirements even assuming that might meet the first.

2.4 The Northern Polar Giant Impact (NPGI) Hypothesis

The NPGI hypothesis suggests that the Martian dichotomy was attained through excavation of the lowlands by an oblique impact, occurred at a velocity of 24 km s^{-1} with a body of ~600 km of diameter and a density of 3.0 g cm^{-3}, or 950 km at 12 km s^{-1} of velocity, which would have generated a kinetic energy of 10^{36} erg [52]. A later elaboration of this hypothesis used an anorthositic impactor of 320 km in radius at a vertical speed of 14 km s^{-1} producing a kinetic energy of 4×10^{28}–10^{29} J [55] suggesting that the origin of the northern lowlands was different from that of the southern highlands, being the latter the result of magma ocean solidification occurred a few tens of millions of years after the accretion of Mars. Both versions of the same hypothesis never mentioned anything about the extensive volcanism that a so energetic impact should have produced in the northern hemisphere. Actually, it was even stated that the energy generated during such an impact would have not produced "too much melt" and that the chronology of the NPGI was assumed almost coeval to the Moon-forming impact from 30 to 50 Ma after CAI [55]. For sufficiently large impact energies, like the Moon forming impact (10^{31} J), retention of an impact basin was deemed unlikely because a global magma ocean would have formed [58]. However, it was suggested that "moderately oblique" NPGI may have produced the lowlands with no much melt [55]. Assuming that the words "moderately oblique" mean an impact at an angle of 45°, the volume of impact melt would decrease by at most 20% [65]. Considering the remaining 80% of the total impact energy, the production of melt would have been sufficient to produce a widespread volcanism in the lowlands instead. Obviously, given that such a widespread volcanism is not present in the lowlands, exception made for the extensive lava flooding coming from the highlands [66, 67] or from Elysium [68], the NPGI would be ruled out in consideration of both the second and the third requirements.

2.5 The Southern Polar Giant Impact (SPGI) Hypothesis

The alternative hypothesis of the SPGI was based on the totally opposite view of the NPGI, that is, the giant impact occurred in the southern hemisphere of Mars to form a magma ocean that became the southern highlands upon cooling [24, 27, 58]. Impactors of 200–250 km in radius can form impact basins of the size of Utopia or Hellas [56] but the formation of magma oceans requires impactors of mass about one tenth of the target's mass [69]. Earlier two-dimensional versions of the SPGI used impactors of 700–1500 km in radius at a speed of 7 km s^{-1} [56] or 500 km at a speed of 10 km s^{-1} [27] and effectively achieved a hemispherical magma ocean. However, these 2D versions were unable to model the complete migration of the plumes due to the lack of the third dimension [44]. The three-dimensional (3D) version of the SPGI was more elaborated in terms of parametrization with a wide range of larger impactors (from 1000 to 2000 km in radius), impact speeds (from 5 to 10 km s^{-1}), and impactor compositions (from 50 to 80% of iron content in radius) [24]. The Martian dichotomy was obtained only in two specific best cases [24, 43]: a siderite (80% of iron in radius) of 1600 km of radius and a mesosiderite (50% of iron in radius) of 2000 km of radius, both at an impact speed of 5 km s^{-1}. The timing of the impact is placed between 4 and 15 Ma after CAI [24, 27], that is, after the end of the heating effect of the short-lived radiogenic elements ^{26}Al and ^{60}Fe [22, 23] and before that the core completed its formation as evidenced by studies of ^{182}Hf and ^{182}W [20, 21, 26, 70, 71]. In the 3D version of the SPGI the core was formed after ~5 Ma after CAI, that is, ~1 Ma after the lower time limit of the impact, and the first pockets of partially molten silicate appeared under a cooled crust after ~1.3 Ma after CAI [24]. Such a giant impact effectively generated a magma ocean all over the southern hemisphere, triggered its magnetic field, and started to form mantle plumes from four points located 90° of longitude apart along the sub-tropical regions

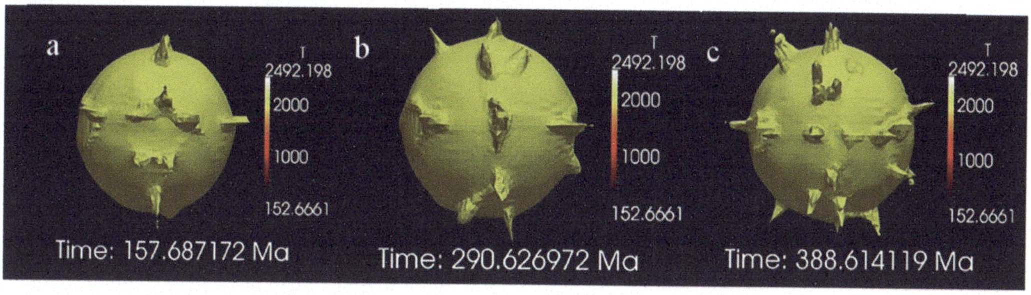

Fig. 1 **a** Time 157.68 Ma after the giant impact (AGI), the first mantle plumes nearly reached the South Pole whilst the second wave of plumes completed its formation along the subtropical regions of Mars; **b** time 290.62 Ma AGI, the first wave of plumes reached the South Pole, the second wave is on its way to the South Pole, a third wave is under formation in the subtropical regions; **c** time 358.61 Ma AGI, the second wave is fading on its way towards the South Pole whilst secondary plumes are on their way to cross the trajectories of previous plumes [24]

(Fig. 1a), on average 30° of latitude north, which then migrated towards the South Pole of Mars [24]. The first plumes appeared at an approximate mid-latitude of 35–40° N around 54.15 Ma after the giant impact (AGI), heading to the South Pole, spaced 90° of longitude among them (Fig. 2). The migration of these mantle plumes under the 100–150 km thick lithosphere of Mars formed twelve volcanic alignments with loxodromic trajectories starting from the four original points spaced 90° of longitude apart, these points correspond to Alba Patera–Tempe Terra, Hecates Tholus–Elysium Mons, Arabia Terra–Mawrth Vallis, and Syrtis Major (Fig. 3a) [43]. The discovery of these volcanic alignments was the evidence required to validate the hypothesis of the SPGI. Furthermore, the phase of plume migration coincided with the total time of duration of the magnetic field, which was related to the evolution of the core-mantle boundary (CMB) heat flux, and was estimated in ~500 Ma after CAI [24]. The end of the magnetic field of Mars, inferred from the study of the magnetic anomalies found on the surface of the planet by the Mars Global Surveyor, was estimated between 4.2–4 Ga [72–75]. The age of 4.06 Ga obtained from the fall of the CMB heat flux, under a threshold located between 50–20 mW m^{-2} calculated through the SPGI model, is perfectly consistent and within the range of ages estimated through the magnetic anomalies [24]. Another consistency is the timing of formation of Tyrrhenus Mons which was linked to the K/Ar age [76] of its lava flows deposited on the floor of Gale crater [77]. The SPGI explains the shape of the Martian dichotomy, it is physically

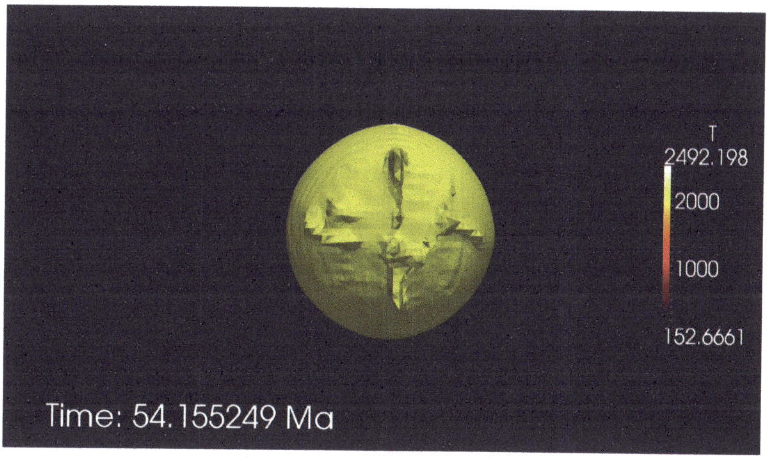

Fig. 2 Time 54.15 Ma AGI, the appearance of the first mantle plumes on their way to the South Pole of Mars [24]

consistent, and it is compatible with all the geological, geochemical, and geophysical data available.

3 The Volcanic Alignments of Mars

As already mentioned at the end of the previous section, the SPGI formed a magma ocean in the southern hemisphere of Mars. Then, ~50 Ma later the first mantle plumes formed at mid-latitude and then started their migration below the lithosphere as far as the South Pole of Mars (Fig. 3a). These plumes formed volcanic alignments [43] tracing loxodromic trajectories on the surface of Mars that can be seen in Fig. 3b.

The reason why the plumes moved along loxodromic trajectories is still unknown. The alignments started from a subtropical region included between 40° and 20°, on average 30°, of latitude north (Fig. 3b). According to the current axial tilt of Mars, 25.1° [78] with an historical average estimated into 24.4° [79], the northern tropic should be currently located along the latitude of 25.1° N. The total number of the alignments discovered on the surface of Mars is twelve, divided in primary and secondary alignments [43]: the primary alignments are those reaching directly the South Pole from the northern tropical regions; the secondary alignments are those not reaching the South Pole and crossing the primary ones. Regardless whether primary or secondary, the alignments' numbering is in order of discovery. This means that Alignment 1 was the first discovered, Alignment 2 was the second, and so on. A detailed description of the volcanic centres forming the various alignments is given in the following sections.

3.1 Alignment 1

Alignment 1 was tracked from Olympus Mons through the volcanic centres of Biblis Tholus, Phoenicis Lacus, Claritas Fossae, Warrego Valles, Thaumasia Fossae, Aonia Mons, Aonia Tholus, and Cavi Angusti as far as the South Pole. The size of the volcanic centres along the alignment is the largest at Olympus Mons and strongly decreases from Biblis Tholus onwards. Olympus Mons is regarded as a volcano with a long eruptive history because the latest flows appear young [80]. However, the age of its first emplacement of the bulk material might date back to Hesperian or even to Noachian times [81]. Consistent with such older ages is also Biblis Tholus that has a crater count age spanning from 3.68 Ga (Werner 2009) to 2.8 Ga [82], 3.35 Ga [83], and 3.65 Ga [84], likely postdating the formation of Olympus Mons.

After Biblis Tholus the Alignment 1 headed towards the Warrego Rise through Phoenicis Lacus coasting the Claritas Fossae to the east. The Warrego Rise was previously interpreted as a centre of magmatic-driven uplift, with a formation that dates back from the Noachian to the Early Hesperian [85].

Just out of the distal Tharsis lava flow fields there are the next volcanic centres: heads of valleys perpendicular to (and cut by) the Thaumasia Fossae, starting from a topographically high unnamed crater located just west of the crater Slipher; and next to them Aonia Mons and Aonia Tholus in a region of Noachian-Hesperian age. The Aonia volcanic centres were already mentioned, although not specifically described, in the orogenic activity postulated at a time (4 Ga) when the magnetic field of Mars was already over [86]. The alignment ends with the interesting structures of Cavi Angusti [87] and the confining Angustus Labyrinthus, also known as "Inca City" [88]. These features were interpreted as mobergs [89], the result of dike emplacement [90], or the result of ice melting by volcanic activity [87]. The volcanic hypothesis is the most plausible also for the formation of the confining larger volcanic plains of Parva Planum and Argentea Planum, both characterized by the volcano-ice interactions occurred during the Hesperian (3.7 Ga) in the Dorsa Argentea formation [91, 92]. The dark albedo patches found at Inca City show seasonal variations as a result of defrosting of the CO_2 polar cap that uncovers underlying dark material [93], likely volcanic material.

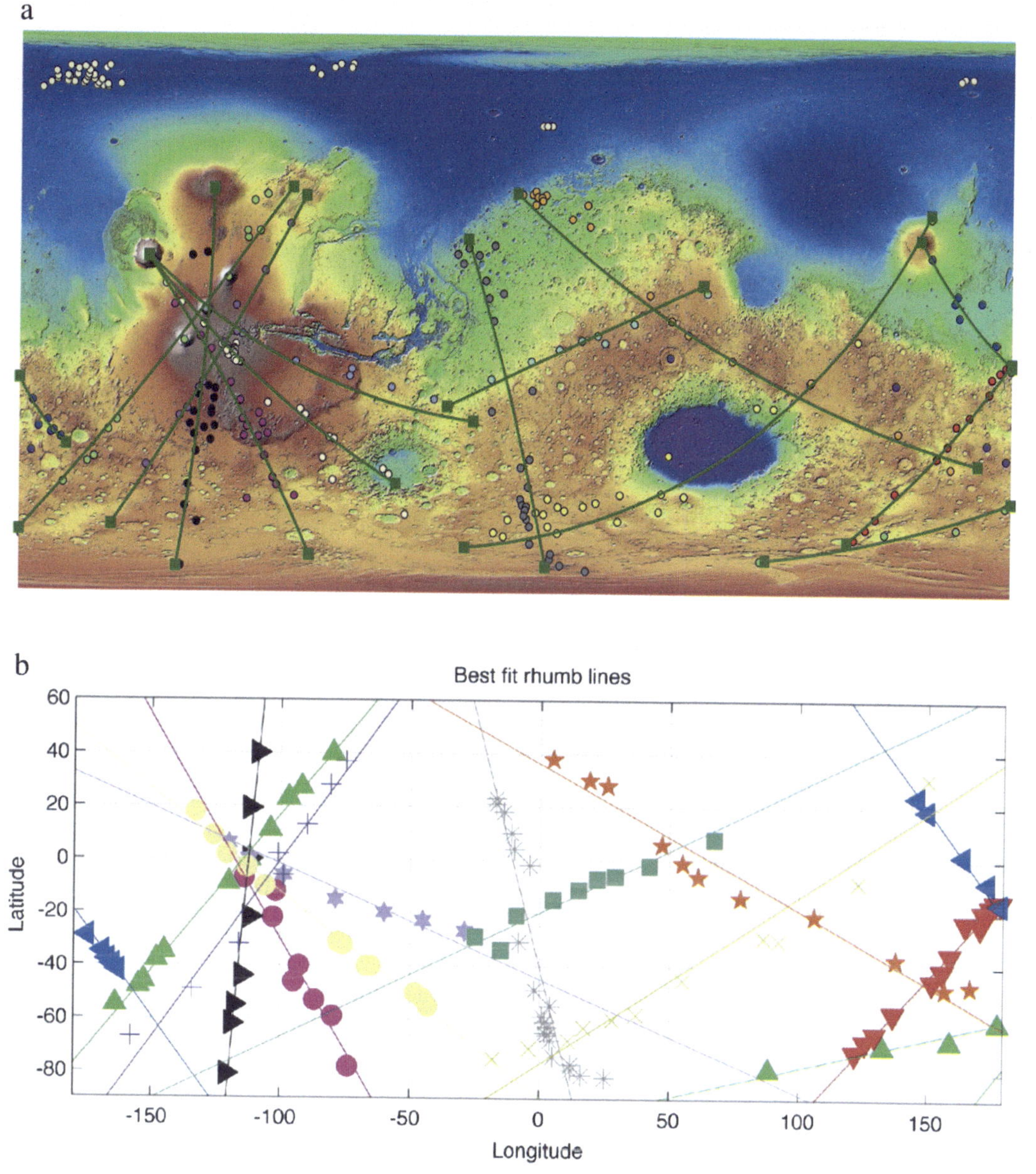

Fig. 3 **a** Map of the alignments of Mars obtained through the best fit instrument of ArcGIS; **b** map of the alignments of Mars obtained through a Matlab algorithm of best fit of data points [43]

3.2 Alignment 2

The Alignment 2 was tracked through Hellas as far as the Australe–Sisyphi Montes starting from the large volcanoes of Hecates Tholus and Elysium Mons. The volcanic domes located northeast of the Phlegra Montes were originally interpreted as erosional remnants of volcanic fields [94, 95] and thus not included in the alignment. Hecates Tholus and Albor Tholus, compared to the larger Elysium Mons, may suggest the alternative hypothesis that they could also be secondary cones but their mutual positions along the alignment suggest that Hecates Tholus formed before Elysium Mons and thus cannot be a secondary cone. If the respective positions of Hecates Tholus within the Alignment 2 and Albor Tholus within Alignment 6 were due to random magma rise from secondary (or stand-alone) conduits of Elysium, their respective and aligned positions along Alignment 2 and Alignment 6 would just be an amazing coincidence. Another argument in support to the hypothesis of the alignments is the older formation of Hecates Tholus and the younger of Albor Tholus with respect to Elysium Mons. Crater count ages of 3.8 Ga estimated for Hecates [96], 3.7 for Elysium, and 3.53 for Albor Tholus [97], suggest a decreasing pattern consistent with the one calculated with the SPGI [43]. According to the southern polar giant impact (SPGI) model [24], the plume volcanism that formed the alignments lasted only 500 Ma, from 4.456 to 3.950 Ga [43]. The crater count ages, or K–Ar ages where available, may help to identify a formation age for Elysium Mons within the range between 4.452 and 3.932 Ga, the latter age being consistent with the 3.9 Ga [98] and within the uncertainty derived by the method of the crater counts.

The subsequent volcanic centre after Elysium is an unnamed crater of ~60 km of diameter dated around Middle Noachian age [99], located on a topographic high northwest from the Herschel crater and thus unreachable by the lava flows coming from Tyrrhenum Mons [43]. The lava overflows from this unnamed crater joined the lava flows coming from Tyrrhenus Mons and together filled Gale crater [77].

The next volcanic feature along the Alignment 2 is Tyrrhenum Mons, a large volcano of Late Noachian—Early Hesperian age (3.7–3.4 Ga) as inferred by the crater counts [83, 97, 100]. However, the K/Ar measurements of its lava flows reaching Gale crater revealed an age of 4.21 Ga [76, 77]. From Tyrrhenum Mons onwards along the Alignment 2 the volcanic features seem to divide into two parallel branches, a northern one formed by massifs and a southern one formed by (pateras) calderas [43]. The northern branch starts from Anseris Mons and is followed by a massif, recently named Peneus Mensa (!), located in south-western Hellas after the Alpheus Colles and followed by a series of massifs as far as the end of the Sisyphi Montes. The southern branch starts from Hadriaca Patera, the central caldera of Hadriacus Mons, and is followed by Malea Patera, Peneus Patera, Amphitrites Patera, and Pityusa Patera on the other side of Hellas. The evidence of these branches does not necessarily mean that the original plume divided into two branches because the mutual distance between the two branches is still within the size of the plume [43]. Different mechanisms of formation have been suggested for the massifs and the calderas, mostly related to the different densities of magma rise in the lithosphere [100]. A mechanism of differentiation from the same kind of magma rising in a lithosphere characterised by inhomogeneity of density.

Anseris Mons is not genetically related to the impact that formed the Hellas and the Hellespontus Montes, although it is located in the rim of the Hellas basin as well as Hadriacus Mons. The combined THEMIS and CTX imagery showed volcanic channels on the slopes of Anseris Mons that were not observed along the Hellas Montes [43]. Furthermore, the key position of Peneus Mensa located on the southwestern side of Hellas' floor provided the only valuable connection between Anseris Mons and the series of aligned massifs heading to the Sisyphi Montes. The age of Peneus Mensa could be included between the Early Noachian formation of Hellas and the Middle Noachian Pityusa Patera, somewhere between 4.1 and 3.8 Ga as inferred by crater counts in the Circum-Hellas Volcanic Province (CHVP) [100]. The ages inferred from the crater counts for the CHVP calderas are quite heterogeneous, Hadriaca Patera (3.66 Ga), Amphitrites Patera (3.63 Ga), Malea Patera (3.81 Ga), Pityusa Patera (3.80 Ga) [100]. But the ages inferred from the SPGI model are older, around 3.95–4.05 Ga [43], more consistent with the 4.05–4.1 Ga for the previous formation of Hellas [64]. In this case, Peneus Mensa might be 4.05 Ga old. Just after Hellas there are massifs aligned with the Sisyphi Montes. The Sisyphi Montes are massifs already interpreted as of volcanic origin [101], which the crater counts place at Hesperian (3.7–3.5 Ga) age [102]. The

Sisyphi Montes form a crossing point [43]. The north–south direction of this crossing point coincides with the direction of the Alignment 8, described in Sect. 3.8, while the east–west direction is aligned with the massifs coming from Hellas. The migrating plume that formed the Alignment 2 ended its journey at the South Pole, the end of the alignment is Sisyphi Tholus.

3.3 Alignment 3

The Alignment 3 starts from the centre of Ascuris Planum, a Middle Noachian old lava-flooded plain [99, 103] crossed by the Tempe and Mareotis Fossae. The area is characterized by the 1815 m-high peak of Labeatis Mons, perfectly aligned with the other volcanic centres of Tharsis, and the (North, West and East) Mareotis Tholi. The Mareotis Tholi are offset of ~250–300 km but they are still within the average size of the plume (~600 km wide) and thus still belonging to the Alignment 3 [43]. The same criterion applies to the Issedon Paterae and Issedon Tholus, although located in the northwest Tempe Region [104], they still belong to the Alignment 3. In fact, although offset of ~560 km from the Alignment 3, Issedon Tholus belongs to it because it is still ~730 km away from Alba Mons, which is the beginning of the Alignment 7. Leone [43] has shown a system based on the mutual positions between the volcanoes and neighbour alignments to sort out eventual ambiguities.

The next important volcanic centres before reaching the Tharsis Montes are those forming the Uranius Group [105]: Uranius Mons, Ceraunius Tholus and Uranius Tholus. Different ages were inferred through crater counts among the Uranius Group and the next in line, Ascraeus Mons. Ceraunius Tholus, located between Uranius Mons and Ascraeus Mons, has an estimated age of 3.59 Ga [106] and thus still older than Ascraeus Mons but also younger than Uranius Mons; Uranius Tholus is located between Uranius Mons and Ceraunius Tholus, although shifted laterally about 105.84 km north from the alignment [43], and has an inferred age of 3.56–3.50 Ga [97, 106]. Ascraeus Mons has an estimated age of 3.3 Ga [84]. Although Uranius Mons may appear as one of the secondary cones of Ascraeus Mons, as well as the nearby Ceraunius and Uranius Tholus, the age inferred from the crater counts for Uranius Mons is included between 3.7 Ga [97] and 3.43–3.19 Ga [106], thus suggesting that Uranius Mons formed before Ascraeus Mons. The crater count ages here suggest that these features are not secondary with respect to Ascraeus Mons but rather primary and that follow the sequence of emplacement according to the direction of the Alignment 3. Pavonis Mons follows Ascraeus Mons along the alignment and has an estimated crater count average age of 3.0 Ga [84]; almost the same applies for Arsia Mons with its 3.2 Ga [84]. However, a descending trend of ages is given from Ascraeus Mons to Arsia Mons by the SPGI model [43].

After the volcanic fields of Arsia Mons, named Daedalia Planum, there are the aligned ancient highlands volcanoes of Sirenum Tholus and Sirenum Mons. These were considered older than the Tharsis volcanoes, the ages inferred for Sirenum Tholus and Sirenum Mons from the available crater counts were of 3.99 Ga and of 4.05 Ga respectively [107]. However, according to the SPGI model, the same descending trend of ages seen among the Tharsis Montes applies here too [43]. Sirenum Mons appears half-smashed by an impact crater while the edge of Sirenum Tholus appears shaped by several impact craters. East of Sirenum Tholus there are impact craters half-buried by the massive lava flow fields of Arsia Mons suggesting that there might be many others completely buried thus altering the crater counts [35, 43].

The next features along the alignment are Phaethontis and the nearby crater Li Fan, both of Late Noachian age (~3.8 Ga, [99]) from which spread the Tader Valles, and a group of unnamed calderas of Late Noachian age [99, 103] located south of the Copernicus crater [43]. Lava erupted from these calderas contributed to the infill of Copernicus and of the other surrounding impact craters. The last, but interesting, volcanic feature surrounded by several volcanic plains is

Chronius Mons, a Late Noachian [99] isolated massif located between Terra Cimmeria and Terra Sirenum. Its importance simply lays on the fact that it is the last volcanic centre providing a connection between Alignment 3 and the volcanic plains of the South Pole [43]. No other volcanic centre is seen around as far as one thousand km, other than those belonging to other alignments, thus excluding a widespread and random formation of volcanoes outside the alignments in this region. The alignment ends at the South Pole, after the Ulyxis and Thyles Rupes, where flow fronts and plains of likely volcanic origin [108] are recognizable south of the Promethei Rupes.

3.4 Alignment 4

The first secondary alignment discovered is the Alignment 4. The Alignment 4 was tracked from some unnamed massifs located to the east of Ma'adim Vallis, near the border between the highlands and the transition topography, through Apollinaris Tholus and Zephyria Tholus [43]. The alignment includes Draconis Mons, Cimmeria Tholi, Electris Mons, Eridania Mons, the crater Byrd, the Thyles Rupes, and the crater Burroughs to end into Alignment 3 immediately after the crater Rayleigh. Alignment 4 is the only alignment that does not start from any of the four main focal points where all the other alignments start. The initial unnamed massifs show volcanic channels starting directly from their slopes and continuing to the surrounding plains previously interpreted as volcanic plains [94]. Other volcanic channels contributed to the lava flooding that came from the chaos terrains located in the middle of the Ariadnes Colles and that are also at the origin of Ma'adim Vallis. Ma'adim Vallis is thought to be among the oldest features of Mars, dating back to 4.4–4.3 Ga [84]. The presence of unaltered olivine-phyric basalts at Gusev crater [10], where Ma'adim Vallis discharged its lava flows [109], suggests a pure volcanic origin. This fact is quite remarkable considering that Gusev was regarded as a place where liquid water should have ponded for long time [110].

The next centre along the alignment is Apollinaris Tholus, a volcanic centre with similar characteristics to Zephyria Tholus, a middle Noachian (3.9 Ga [103]) stratovolcano (or composite volcano) that originated the Durius Valles [111]. Following these volcanoes along the alignment, there is a couple of unnamed volcanic domes (tholi) in Terra Cimmeria that were obliterated but not completely destroyed by impact craters. These have been recently named Cimmeria Tholi. Their identification was possible thanks to the presence of pit chains, suggesting the presence of likely lava tubes with collapsed roofs, and of lava flows found on their flanks. These domes are quite large, between ~70 and 100 km, and are rich of volcanic channels along their slopes. Both domes contribute to the infill of the surrounding ridged volcanic plains providing a valuable connection between Zephyria Tholus and Electris and Eridania Montes [43]. The crater Byrd shows overflows and valleys on its western side [43], geomorphologically similar to others seen in confirmed volcanic centres. Also, a volcanic origin for the Thyles Rupes region has already been mentioned [112].

3.5 Alignment 5

The Alignment 5 is another secondary alignment and was tracked from the calderas of Arabia Terra, the geometric centre (also known as centroid) of the start of the alignment is located on the largest unnamed caldera located in the middle of the triangle formed by Ismenia Patera, Euphrates Patera, and Siloe Patera. Then the alignment follows through the heads of the Mamers Valles, the heads of the Locras Valles, the crater Schroeter, the head of the Naro Vallis, the 600–700 km long Huygens-Hellas system of ridges interpreted as a huge dyke [113], Tyrrhenum Mons (where the Alignment 5 crosses the Alignment 2), as far as the craters Huggins and Rossby located beyond the Alignment 3. The crater Huggins is taken as the nearest named reference point to an unnamed volcanic cone located nearby to the east, the Rossby crater is instead the source of the Drava Valles [43].

The Arabia Terra calderas, grouped in a cluster and elongated towards the south-east, are the track on the ground of the passage of the migrating plume and indicate the beginning of the Alignment 5. The next significant feature along the alignment is the source area of the Mamers Valles, located around the crater Cerulli, proposed as formed through the melting of subsurface ice by a warm blanket of ejecta [114]. However, the head of the eastern Mamers Valles, leading to Ismenius Lacus, originated from a point located 250 km northeast from Cerulli, well outside from its ejecta, and the channel connecting the western slope of Cerulli with the western Mamers Valles was formed on an area where Mars Odyssey data showed scarce presence of molecular water [115]. The ~300,000 km^3 of volcanic material removed from the Mamers Valles, a lower estimate not including the volcanic infill of the valleys, and from the connected Deuteronilus Mensa would require up to two or three orders of magnitude of water [33, 116, 117] that cannot simply be justified by the scarce amount in the limited area of the Cerulli impact. Even considering the whole ~ 40,000 km^2 of the area covered by Cerulli and its ejecta as pure water ice, which is already unrealistic, the column of water needed to carve the valleys would still be deep 7.5 km raising serious problems about the porosity of the crust that should host it [43]. Melting of water ice does not seem a viable hypothesis and thus an alternative volcanic source must be invoked. The volcanic hypothesis is instead supported by the presence of a series of pit chains spreading radially from Cerulli and by a widespread lava flooding. Not all the craters in the close area show incisions along their rims thus making the hypothesis of localized eruptive points much stronger. In such a case, the crater Cerulli could be a volcanic caldera. This hypothesis should not be surprising because of the presence of the nearby Arabia Terra calderas.

The next feature along the alignment is an unnamed topographic high, devastated by impact craters and characterized by the heads of the Locras Valles, from which the Indus Vallis and the Cusus Valles originated [43]. The area was already studied through Viking images for the dendritic development of several valley networks, with some ambiguity of interpretation on the nature of the upland plains (volcanic vs. sedimentary), concluding that the valleys formed in two distinct time intervals between 2–3 Ga ago after the release of groundwater or magmatic volatile that contributed to the destruction of the impact craters [118]. However, another study based on more recent THEMIS and MOLA data concluded that not even the valleys approaching the morphology of terrestrial networks (i.e. Indus Vallis), which are Noachian in age, satisfy the scaling relationships of perennial fluvial channels [119]. The analysis of the CTX images also revealed an intact platy-ridged valley infill, likely formed by lava flows, which sprouts from the dust also in the median course of the Locras Valles [43].

The crater Schroeter, from which the Tisia Valles originated with the contribution of the crater Huygens, is at the beginning of two major systems of broadly arcuate ridges subdivided in sectors and that can be traced for 600–700 km, perhaps extending farther in the subsurface, consistent with the emplacement of shallow dikes [113]. The direction of such a long system of ridges follows the same direction of the Alignment 5. The head of the Naro Vallis starts from three nested (and filled) craters located near the sector C of the ridge system. The sector F is superposed on smooth plains of Hesperian age and interpreted as composed of lava flows [113], these plains are then connected to the inter-crater plains of Oenotria Plana to the north and to the plains gradating towards Hellas to the south. The width and the geometry of these dikes, as well as the inferred effusion rate, are consistent with the volcanic flooding of the pre-existent Noachian topography [43, 113]. The Noachian age of such lava flooding is also consistent with the 3.9 Ga of the Locras Valles [120] and the 3.81 Ga of Naro Vallis [121].

Tyrrhenum Mons, already mentioned in the Alignment 2 subsection, is at the centre of an important volcanic province located between Terra Tyrrhena and Terra Promethei. A labyrinth and a valley network, Tyrrhenus Labyrinthus and

Vichada Valles respectively, are on its western slope while the CHVP system of outflow channels and valleys (Dao, Niger, Reull, Waikato, and Harmakhis Vallis) are on its southern slope [43]. The next volcanic centre, located beyond the inter-crater volcanic plains flooded by the eastern lava fields of Tyrrhenum Mons, is an unnamed domical mountain (tholus) surrounded by impact craters [43]. Its slopes are dissected with channels that terminate into a southern volcanic plain interpreted as ridged plains material unit Hr and characterized by extensive lava flows of low viscosity [94]. This unnamed tholus provided the connection towards the last volcanic centres of the alignment, an unnamed volcanic cone located east of the Huggins crater and the Rossby crater, from which originated the Drava Valles. The unnamed volcanic cone, the Drava Valles, and Electris Mons flooded the inter-crater plains located west of the Rossby crater [43].

3.6 Alignment 6

The Alignment 6 is another secondary alignment and was tracked from Elysium Mons, an important volcanic centre of Mars already mentioned in the Alignment 2 subsection. The alignment then follows through the volcanic centres of Albor Tholus, the low shields of the Cerberus Tholi, Apollinaris Mons, and a series of unnamed montes and tholi aligned through the Atlantis region ending into Alignment 3 [43]. The Tartarus Montes and the Hibes Montes were not included in the alignment because they could be the remnants of the lava fields erupted by Elysium, a similar situation of the Phlegra Montes described in the Alignment 2 section. The next in line is Albor Tholus, which is younger than Elysium Mons, 3.53 versus 3.7–3.9 Ga according to the crater counts, respectively [97, 98]. However, as already mentioned in Sect. 1, the older SPGI age of Elysium Mons, between 4.452 and 3.932 Ga, and the average velocity of $\sim$25 km/Ma^{-1} of the migrating plume calculated along the alignment 6 [43], could place the formation of Albor Tholus $\sim$17 Ma later than Elysium, between 4.435 and 3.915 Ga.

A few low shields located near the Hibes Montes and near the crater Tombaugh, the Cerberus Tholi [122], were instead included in the volcanic alignment. The ages of the Cerberus Tholi were initially set to 3.8–3.7 Ga in the available geologic maps of the region [103, 123], then the ages of the shields were set between 234 and 13.5 Ma [122], but the SPGI ages would be located between 4.405 and 3.885 Ga. The latter age would be almost close to the previous crater counts. Apollinaris Mons, already studied for the origin of Lucus Planum and the Medusae Fossae Formation [124] and for its possible role in the dispersal of ashes to long distance [125], is an important landmark for the direction of the Alignment 6. According to the velocity of migration of the plume along this alignment, the formation age would be between 4.350 and 3.830 Ga. Ages derived from crater counts ranged from Late Noachian to Early Hesperian (3.8–2.9 Ga) [126], which was also estimated as coeval to the lava flooding of Gusev crater, thus reflecting a period of intense volcanism in the region [109]. However, a subsequent analysis of the direction of lava flows showed that the lava flooding of Gusev crater was not associated to Apollinaris Mons but to Ma'adim Vallis, which is fed by other volcanic features located to the south [43, 127].

Between the Alignment 4 and the Alignment 6 there are several chaos terrains crossed by the Sirenum Fossae. These are Gorgonum Chaos, Atlantis Chaos, and Caralis Chaos ending towards the Ariadnes Colles. East of the chaos terrains there are Late Noachian ($\sim$3.7 Ga) volcanic plains [99] surrounding a series of unnamed and aligned tholi. The channels of the tholi give a contribution to the flooding of the chaos terrains coming prevalently from the south. The contribution from the Simois Colles, already interpreted as a volcanic feature [103, 128], seems not much significant instead. These tholi are separated by short distances ending the Alignment 6 into the Alignment 3 [43].

3.7 Alignment 7

The Alignment 7 is a primary alignment tracked from Alba Mons through the volcanic centres of Jovis Fossae, Pavonis Mons (crossing point with the Alignments 1 and 3), a cluster of massifs west of Claritas Fossae, an unnamed crater, an isolated and large massif along the Icaria Fossae, a couple of heads of channels, the (offset) crater Steno, and a cluster of calderas at the heads of the Chico Valles as far as the South Pole.

Alba Mons is a polygenic volcano probably formed 3.8 Ga ago according to the crater counts [84, 129], between 4.517 and 4.017 Ga according to the SPGI model [43], and it is crossed and surrounded at the same time by several systems of fossae. The fossae surrounding Alba Mons are the Tractus Fossae and the Cyane Fossae, these appear to originate from Tharsis [130, 131] and have been reinterpreted as an asymmetric rift zone resulting from plume volcanism [131]. Volcano-tectonic processes related to magma rise and dike emplacements were also invoked for Alba Mons [132, 133]. Other fossae crossing Alba Mons are found on its eastern slope, Phlegeton and Acheron Catenae for example, and were interpreted as the surface manifestation of shallow dikes [117, 129, 134]. There are also other systems of fossae originating directly from Alba Mons, these are the Alba Fossae and the Tantalus Fossae, explained as formed by the late stage volcanic loading of the summit and the solidification of the magma reservoir [135].

However, many of these fossae might just be lava channels, which should not be surprising being located on the flanks of a volcano, a new dedicated topographic study might shed light on the real tectonic situation of Alba Mons. The map of the Tharsis region shows several alignments of vents. Aligned fields of vents were indeed published in works regarding Pavonis Mons [136] and Syria Planum [137]. The aligned fields of vents located south of Pavonis Mons and Ascraeus Mons, and even inside Arsia Mons [136], follow the directions of the hemispherical Alignments 3 and 7 respectively [43]. The field of vents south of Pavonis Mons follows the direction of Alignment 7 towards Claritas Fossae. The field of vents located south of Ascraeus Mons, as well as the field of vents located inside the lava fields of Arsia Mons, is aligned with the direction of the Alignment 3. A chain of vents is visible north of Labyrinthus Noctis and another one within Syria Planum ending at Phoenicis Lacus along the direction of Alignment 11. Fields of vents are not observed along the direction of Alignment 3 towards Daedalia Planum and Sirenum Tholus probably because they have been covered by the massive emplacement of lava flows along the slope of Arsia Mons. Flows along Daedalia Planum were so abundant that many impact craters were completely buried [35].

The next feature after Pavonis Mons along this alignment is a cluster of volcanic massifs located west of Claritas Fossae [107]. The centre of the cluster, which is the selected landmark on the basis of geometrical criteria, is an unnamed massif located at the coordinates 22° 0.39′ S–113° 0.27′ W [43]. The cluster is characterized by several lines of massifs oriented in direction north–south and embayed by the massive lava flows erupted by Arsia Mons. The cluster ends at the extreme limit of the Arsia Mons flow fields after which the inter-crater volcanic plains of Icaria Planum are crossed by the Icaria Fossae. For these volcanic plains a formation age of 3.59 Ga was determined through crater counts [138]. Considering the high uncertainty in the crater counts methods, this is an estimate essentially consistent with another one that placed the volcanic plains (their Hnf unit) of Icaria Planum in the Early Hesperian (3.7 Ga) [139].

West from Icaria Planum there is an unnamed crater that shows several channels on its southern slope, these channels end into the volcanic inter-crater plains where wide lava flow fronts spread concentrically from the channels [43]. This unnamed crater shows overflows almost all over its rim and a flooded inner crater with a central pit where is placed the centroid, a filled crater is also visible on its south-eastern slope. A close view of the southern rim shows that the channels were formed directly from the overflows. Both the unnamed crater and the related volcanic plains are located within the Npl2 unit, which is aged Late Noachian (3.8–3.7 Ga) [139].

The next eruptive points along the alignment are located south of the crater Brashear and south-east of the crater Dokuchaev, where ghost craters also appear from the volcanic plains and where the valley networks end [43]. The CTX coverage of the area is not complete yet so THEMIS mosaics showed the location of the channels. The channels south of Brashear formed over a 3–6 km topographic high that is unreachable from the lava flows eventually coming from the Icaria fossae. This means that the topographic high is an eruptive centre belonging to the alignment. The channels spreading from this topographic high contributed to the lava filling of the volcanic plains located at its foothills. In fact, these plains were already interpreted as smooth Hpl3 unit formed by thick inter-bedded lava flows of Early Hesperian age (3.7–3.6 Ga) covered by a veneer of aeolian deposits [103], so it could not be excluded that the channels that contributed to their formation might have their same age. The eruptive point located southeast of the crater Dokuchaev does not seem related to the nearby Icaria Fossae as well [43].

The last eruptive points of Alignment 7 were found around the southern polar region: one, although the position is still uncertain, was inferred from a topographic high obliterated by the 104.31 km offset impact crater Steno; the other was found on a chain of massifs and domes of probable volcanic origin located just north-west from the Australe Scopuli [43].

3.8 Alignment 8

This primary alignment is the one that starts from the fourth focal point formed by the SPGI, Tharsis is the first, Elysium is the second, and the Arabia Terra calderas are the third. The Alignment 8 was tracked through the head of Mawrth Vallis, the heads of the Coogoon Valles, including several other eruptive points in Western Arabia Terra like the crater Becquerel, a filled crater near Thymiamata, a possibly buried vent located near the crater Vernal, the crater Crommelin, a floor-fractured crater attached to the Miyamoto crater (although some other vent cannot be excluded in Meridiani planum), the heads of Marikh and Bashkaus Valles, a floor-fractured crater near the crater Maunder, ending with a series of aligned massifs in continuity with the Sisyphi–Australe Montes.

Mawrth Vallis originated from an unnamed crater located north of the crater Trouvelot [43] and it was always considered a fluvial valley of astrobiological and climatological significance because of the presence of phyllosilicates and their putative formation in aqueous environment [140–144]. The age of formation inferred from the crater counts for the main layered unit is located between 4.0 and 3.9 Ga, followed by putative fluvial activity 3.8 Ga ago, ending between 3.7 and 3.6 Ga ago [143]. Mawrth Vallis today shows evident lava flows fitting the width of the bed-floor along the direction of the stream, nearby flows (coming probably from Becquerel) cover only part of its initial bed-floor instead [43]. The lava flows are also confirmed by the presence of unaltered olivine unequivocally found along the bed-floor of Mawrth Vallis [145]. Furthermore, the presence of unaltered olivine indicates a dry climate since the original emplacement of the lava flows occurred in the Noachian [34]. A dry climate is also confirmed by the presence of jarosite [140], a mineral that becomes rapidly altered in presence of humidity in the atmosphere [146]. Other eruptive points feeding lava tributaries were found along Mawrth Vallis and some others at the heads of the Coogon Valles [43]. The eruptive point of Becquerel was inferred from the directions of the lava flow fronts but also from the presence of pit chains in the area. A channel originating from Becquerel crosses a nearby lava filled crater. No later fluvial erosion is observed, but just transport of lava that filled the crater [43].

The next eruptive points were tracked southward from the Mawrth Vallis source area in a filled crater located west of Thymiamata, in the crater Crommelin, and in a likely buried vent located near the crater Vernal, these locations were found thanks to the good resolution of the CTX images [43]. The presence of well visible lava flows favoured the tracking of their eruptive source, an unnamed filled crater showing also lava overflows located near Thymiamata. These lava flows, dated about 3.8 Ga ago, filled other pre-existing impact craters and were already observed since the post-Mariner age [147]. Also, the eruptive source of Vernal crater was inferred from the geometric centre of radially spreading lava flow fronts in the area, some of them heading to the west towards Oxia Palus [43]. Volcanic processes have already been studied in Oxia Palus, a region located between Ares Vallis and Meridiani Planum where floor-fractured and filled craters were ascribed to volcanic processes [148]. The next volcanic centre along the alignment, an unnamed crater close to the Miyamoto crater, was not easy to find because of the widespread presence of stratifications of platy-ridged lava flows on Meridiani Planum that also filled the crater Miyamoto. Fluvial processes were usually proposed to explain its morphology [149, 150] but, as usual, without a detailed explanation of the potential sources of water in the area. The presence of an unnamed and externally isolated crater of about 50 km on its western rim indicates the possible presence of underground igneous activity [150]. This crater appears geomorphologically similar to the floor-fractured craters seen on the Moon [151], the fractures look channels at THEMIS resolution, a morphology that will also be seen within Maunder and nearby craters.

A thorough check between Meridiani Planum and Noachis Terra has shown that there is a large drainage basin previously interpreted as a fluvial network formed by precipitation-recharged groundwater sapping, where water was believed to have ponded in a warmer and wetter paleoclimate from the Late Noachian (3.8–3.7 Ga ago) to the Early Hesperian (3.7–3.6 Ga ago) [152]. There are now several doubts about the fluvial origin [34]. In fact, many floodings have the unequivocal morphology of the lava flows already at the source, along the course, and at the mouth of the valleys [33, 43, 153]. And now it is also well known that lava can form its own drainage networks resembling those formed by water [154]. The ages of these volcanic terrains, dated between Noachian and Hesperian, were essentially confirmed by a test on different age-dating techniques on the nearby Parana Valles [155]. The Alignment 8 in this region shows only two source points: one at the head of Marikh Vallis and one at the head of the Bashkaus Valles. The nearby Samara Valles share the same eruptive point with the Bashkaus Valles and then discharge into the Morava Valles (where also the Parana and the Loire Valles discharge) that become Ares Vallis after passage through Margaritifer and Iani Chaos [43]. Unaltered olivine was found along Ares Vallis and at its sources [145] suggesting that lava had no contact with water for more than 10 k years [156, 157] or never [34]. It is worth noticing that the valleys formed on or directly from the lava flows present in the region. Evros Vallis and Marikh Vallis have segments of their courses characterized by sheet lava flows on the surface before being channelized again by the local topography [43]. The source of Evros Vallis is crossed by the Alignment 10 that will be described in detail in Sect. 3.10 ahead.

Further to the south a cluster of several apparently floor-fractured craters is suggestive of volcanic activity as observed on the Moon [151], the higher resolution of the CTX images showed that the fractures are in reality lava channels. The cluster of these craters is almost centred between the degraded craters Asimov and Maunder at the centre of several inter-crater plains [43]. The degraded geomorphology of the crater Asimov was explained as the result of basal ice melting during a period of favourable obliquity estimated in >8 Ma ago [158] but there are problems with this hypothesis. First, it is difficult to understand why such a climate-induced basal melting of ice was limited to this crater and not extended to the whole Noachis Terra, or to nearby craters like Kaiser or Proctor, for example. Second, it is difficult to understand why the global buried cryosphere should only be limited to the area of Asimov, Maunder, and to the two other nearby unnamed craters. An alternative explanation is given by the passage of the plume that formed Alignment 8 exactly below the cluster of Asimov and Maunder. The directions of the lava flow fronts around Maunder crater show how the eruptive points following Alignment 8 should be centred right at Maunder, Asimov, and at the unnamed crater located to the east of Maunder and to the south-east of Asimov, where several channels and overflows can be seen [43]. Furthermore, the observation of the Asimov's moat shows a striking geomorphological similarity with incipient chasmata (i.e. Coprates and Candor Chasma; see [117]), a morphology also partially observed within Maunder and nearby craters. South of Maunder at 381 km of distance Alignment 8 continues along a chain of massifs heading to the Sisyphi Montes. The Sisyphi Montes have already been studied in the past, several mechanisms for their formation have been evaluated with the conclusion that a volcanic origin best explains their morphology and origin [101]. It is interesting to note how these volcanic massifs are aligned with considerable continuity with the Australe Montes to the south and also how they cross the alignment of massifs of Alignment 2 coming from the east. The Australe Montes have been classified as knobby material, unit k in the geologic map of the Mare Australe area of Mars, interpreted as knobs either formed by ejecta material or rims of older crater material [159]. There are several elements that favour an alternative interpretation of a volcanic origin: (a) the continuity of alignment with the volcanic Sysiphi Montes, including the average distance of about 100 km (see Table 1 in [43]); (b) the presence of Sisyphi Cavi between the Sisyphi Montes and the Australe Montes, a volcanic origin was proposed for the morphologically similar Cavi Angusti [87]; (c) if the hypothesis of the remains of an old impact basin rim was true, the >2000 km length of the Sysiphi–Australe alignment would make it one of the largest impact basins of Mars, even larger than Hellas, but there is no topographic evidence for such a structure; (d) the average height of 1500 m of the Australe Montes, provided by MOLA data, is consistent with the average height of 1000–1500 m [101] of the Sisyphi Montes suggesting a similar origin by the same migrating plume.

3.9 Alignment 9

The Alignment 9 begins from Olympus Mons, as well as Alignment 1, and was tracked through Pavonis Mons, Noctis Labyrinthus, a unnamed volcanic massif in Sinai Planum, a couple of unnamed volcanic massifs in Thaumasia Planum, an unnamed massif (heavily destroyed by impact craters) near the head of Nirgal Vallis, and the head of the Erythraea Fossa [43]. The tracking was facilitated by the previous identification of volcanic centres in Thaumasia Planum [107]. Of course, no need of particular interpretation to identify the volcanic centres located at the beginning of the alignment, particularly for Olympus and Pavonis Montes that were already included in Alignment 1 and Alignment 3 respectively.

The volcanic massifs in Sinai Planum and Thaumasia Planum as far as the unnamed massif located north of the head of Nirgal Vallis were already mentioned in a previous work [107]. This latter massif is located on a local topographic high and its valleys are tributaries of Her Desher Vallis. A closer look through available CTX and THEMIS images revealed once more the presence of lava flows at the heads of Her Desher Vallis and Nirgal Vallis and that these valleys

Table 1 Areal extent of the known volcanic provinces of Mars [100, 185–187]

Province	Area (km^2)	Ranking
[a]Tharsis	7.25×10^6	1
[b]Elysium	4.08×10^6	2
[c]Circum-Hellas	2.10×10^6	3
[d]Syrtis Major	$1.6–3.2 \times 10^5$	4

Source of data [a]Schaber et al. (1978); [b]Hall et al. (1986); [c]Williams et al. (2009); [d]Hiesinger and Head (2004)

formed directly from lava flows [43]. The Her Desher Vallis tributaries coming from the unnamed massif also formed from lava flows (Fig. A.17e in [43]). The presence of lava was observed even inside one of the craters located on top of the unnamed massif. Both Her Desher and Nirgal Vallis, although the latter more evidently, are tributaries of Uzboi Vallis, which in turn ends up into Holden crater to the north and into Argyre to the south. The presence of unaltered olivine was already confirmed along Nirgal Vallis [145] and inside Holden crater [160] thus excluding a fluvial origin or fluvial processes [34].

The next eruptive point is located at the head of the Erythraea Fossa. The analysis of the topography showed that the Ladon Valles received a contribution from the Erythraea Fossa but also that this contribution is too small to justify the volume of the Ladon Valles [43]. Erythraea Fossa is roughly 600 km^3 (considering its maximum values of width in the lower course) while Ladon Valles is ~4000 km^3 (considering both its minimum depth and width), the Holden crater does not show any evident distributary, so it would be reasonable to think that the Ladon Valles received contributions directly from the Mare Erythraeum (and thus from the Tharsis volcanoes) through the lava flow fields of Thaumasia Planum. On this regard, the lava flows coming from Tharsis must have crossed the Eberswalde crater and indeed there are channels conveying lava into it [43]. The delta of Eberswalde was interpreted as formed by alluvial deposits [161] but the origin of the water remained speculative without any credible explanation [34]. The inverted relief of the alluvial channels was then explained with aeolian deflation [162], although an alternative explanation could have included the typical shape of lava tubes visible on the lava fields. Although the volcanic option was never considered for this feature in the literature, it is considered here. Lava can form deltas as already observed at Gale crater [77]. Lava flows were also observed at the head of the Erythraea Fossa, interpreted as an Hesperian fluvial system [163], which ends up into the Ladon Valles [43]. The Ladon Valles, an important channel located downhill of the Holden crater and of Nirgal Vallis, were dated with crater counts from 4.0–3.7 Ga [84] to 3.7 Ga [164]. The last feature of the alignment is the region of Vulcani Pelagus, several lava channels formed on the slopes of a topographic rise located northwest of the crater Vogel [43].

3.10 Alignment 10

The tracking of this alignment was facilitated by the previous identification of volcanic centres in Terra Sabaea [107] but also by the decreasing age of the valley networks from Terra Sabaea to Margaritifer Sinus [165]. Starting from Meroe and Nili Paterae in the Syrtis Major region of Mars, before the crossing point with Alignment 5 in Terra Sabaea, the alignment was tracked through the heads of Naktong Vallis, an unnamed massif at the head of Verde Vallis, an unnamed massif in Sinus Sabaeus at the head of the Brazos Valles, an unnamed (and heavily bombarded) massif at the head of Evros Vallis, an unnamed crater contributing to Marikh Vallis, the crater Peta at the head of the Erythraeum Chaos and the Paranà Valles, and an unnamed crater located on a topographic high in the Margaritifer region of Mars [43].

The activity of the Syrtis Major calderas was dated with crater counts from Late Noachian to Early Hesperian (3.9–3.7 Ga) age [166] and is characterized by a 2 km deep central depression perhaps caused by the collapse of a shallow magma reservoir located into the mega-regolith of Mars [167]. Nili Patera in particular has shown traces of differentiated (felsic) lava [168, 169] that might support such a view. The observation of the volcanic centres along this alignment has led to the surprising discovery that the heads of several fluvial valleys, most of them located near highlands volcanoes, are aligned with the Syrtis Major calderas through Terra Sabaea [43]. It is interesting to note how the formation and the ages of these valleys, estimated from crater counts [165], are decreasing in space and in time according to the direction of the alignment. The ages show a decreasing order from the Syrtis Major as far as the heads of the valley networks located in Margaritifer Sinus, a time span of 210–260 Ma is observed between the formation of the valleys in Terra Sabaea and the valleys in Margaritifer Sinus [165]. The presence of the volcanic massifs [107] provided a good point of reference for the probable volcanic origin of Verde Vallis, Nakhtong Vallis and other unnamed valleys in Terra Sabaea. The head of Evros Vallis is located on the southern slope of an unnamed massif located south of the crater Schiaparelli. The course of Evros Vallis was followed as far as its mouth in Meridiani Planum and revealed a surface lava flow limited to the segment located to the east of the crater Mädler [43]. Furthermore, the same massif from which Evros Vallis originated was also a contributing source for another drainage basin located just west of the lava-flooded crater Schiaparelli. The lava flows coming from this basin directly formed two unnamed valleys that discharged into Meridiani Planum [43].

The next point along the alignment is a crater located on a topographic high that contributed to Marikh Vallis with some channels and to the lava filling of the crater Mädler through a chain of filled craters, a morphological similarity already seen along Alignment 8 where another topographic high formed Marikh Vallis [43]. The Paraná Valles originated from the crater Peta to drain into Erythraeum Chaos. It is possible to see on any topographic map of Mars that Erythraeum Chaos discharged to Morava Valles through the Loire Valles. As already mentioned in the Alignment 8 section, the Morava Valles become Ares Vallis where unaltered olivine was found. South-east of the Erythraea Fossa there is the last eruptive point of the alignment, at the centre of several heads of volcanic channels which contribute to the lava filling of several large inter-crater plains extending as far as Argyre to the south and to the valley networks of Oltis and Himera Valles to the north [43].

3.11 Alignment 11

The Alignment 11 starts with Labeatis Mons, a Noachian massif located on Western Tempe Terra [99], surrounded by the curvilinear Tempe Fossae. The next feature is an elongated vent of about 20 km of diameter located ~100 km north from the head of the Labeatis Fossae, basically at the head of Tractus Albus, which shows overflows all along its southern perimeter [43]. The next feature, Tharsis Tholus, has an inferred crater count age for the shield between 3.54 Ga [97] and 3.24 Ga [83], and it is located at a distance of 531.29 km from the Alignment 3. Another minor feature along the alignment is an unnamed vent, at coordinates 2° 07′ 12.99″ N and 101° 20′ 09.04″ W, located north of Noctis Fossae and west of Fortuna Fossae, with the same morphological characteristics of several nearby other vents studied by Hauber et al. [170]. These low shields do not occur sparsely everywhere on Tharsis as they are distributed among the alignments crossing Tharsis [43].

The Alignment 11 crosses regions with volcanic centres already seen in other alignments, Noctis Labyrinthus, Syria Planum, and the western Claritas massifs before reaching the volcanic plains extending as far as the Icaria Fossae. The origin of these volcanic plains might be ascribed both to the Icaria Fossae and to an unnamed massif of about 150 km of diameter heavily bombarded by impact craters [43]. Several channels spread from this massif in almost all the directions and their morphology is similar to many others observed along other alignments. A closer view on the Icaria Fossae has shown possible skylights and volcanic channel breakouts, basically features that have nothing to do with tectonics, suggesting that these fossae might be remains of lava tubes or erosional lava channels [43].

The last feature of the Alignment 11 is the source of the system of channels of the Chico Valles that feeds two main volcanic inter-crater plains: one east of the crater Charlier and one south-west of the crater Stoney. The sources of the channels were located from the radial spreading of the channels and following back the directions of the main lava flow fronts observed in the region [43]. The Icaria Fossae seem not to be involved in the formation of the northern Chico Valles but rather are encountering their foothills from the north–east. The lava flows going to the south–west from the foothills of the Chico Valles formed several ghost craters as far as the crater Richardson. The same happened in the inter-crater plains south–west of the crater Stoney. The alignment ends at the South Pole towards the Ultimi Scopuli where the presence of the southern polar cap makes difficult to observe further volcanic features there [43].

3.12 Alignment 12

As well as the Alignment 1 and the Alignment 9, the Alignment 12 begins from Olympus Mons. The next eruptive point along the Alignment 12 is Ulysses Patera, tracked along some aligned vents on both sides of the Ulysses Fossae [43]. The crater count age of Ulysses Patera ranges from 3.92–3.73 Ga [97], 3.75 Ga [84], to 3.4 Ga for the flanks [82] while the youngest age reported is 1.92 Ga [83]. A chain of vents crossing Ulysses Patera is also visible along Alignment 12, crossing the Alignment 3 between Pavonis and Arsia Mons [43].

The next eruptive point of this alignment is found along the region of Syria Planum, characterized by the southern expansion of the lava tubes coming from Labyrinthus Noctis [117], in a field of vents aligned mainly along the direction of Solis Planum towards the Coracis Fossae [43]. Some of the vents are located on top of low shields indicating three volcanic episodes for the total duration of ~900 Ma from 3.7 Ga to 2.8 Ga ago [137]. Along Solis Planum the Alignment 12 does not show vents or any other low shield likely due to a thicker cover of lava flows.

The next eruptive points along the Alignment 12 were found in two volcanic massifs near the Coracis Fossae previously listed in the map of Xiao et al. [107]. These massifs, together with the Coracis Fossae, have contributed to the lava mantling of large parts of Ogygis Regio, Phrixi Regio and Bosporus Planum. However, this is not the only contribution that these regions received. A few craters located within Ogygis and Phrixi Regio contributed to the lava mantling and formed some lava channels networks [43].

Other eruptive points were located inside Argyre Planitia suggesting that this large impact basin was already formed before that the migration of the Alignment 12 plume took place. The most unambiguous volcanic centre is Argyre Mons, which shows a domical shape with a central caldera of 10–20 km of diameter and a main channel starting from its south-western rim [43, 171]. Considering that Argyre Mons is aligned with the other volcanoes of the Alignment 12 and considering that the plume stage following the SPGI is located between 4.456 and 3.950 Ga [43], the ages of the Argyre basin and of the Argyre Mons should be older than those estimated with the crater counts, 3.93 and ~3 Ga respectively [171, 172]. However, the crater count ages might have some consistency with the lower limit of the SPGI ages. Although the Nereidum and the Charitum Montes are associated to the formation of the Argyre impact basin in Noachian age [173], the mountains located inside Argyre Planitia (Argyre Mons, Octantis Mons, and Oceanidum Mons) do not seem to be related to the same event but to a likely volcanic origin occurred after the formation of the impact basin [43, 171]. Other eruptive sources inside the Argyre impact basin include low relief features: the cavi located north of Argyre Mons [171]; the filled crater Mari, from which the sinuous pit chain Octantis Cavi is directly formed; and the crater Lodwar, from which departs a 100 m wide channel of possible volcanic origin [43]. The various dorsa features located near Octantis and Oceanidum Montes were interpreted as eskers formed by putative glacial activity 3.6 Ga ago [174], although they are morphologically more similar to the lava tubes present on Chryse Planitia [117]. Other authors gave different interpretations, lava flows or wrinkle ridges [108], exhumed igneous dikes

[175], glacial moraines [173], but the prevailing view remained that of the eskers [173, 176, 177]. The interpretation and the formation of the eskers relies on the putative past presence of water brought into the basin by the valleys located along the Argyre's rim: Surius Vallis, Dzigai Vallis, Doanus Vallis, and Pallacopas Vallis. An hypothesis that is not certainly supported by the widespread presence of unaltered olivine all over the Argyre basin [178], see also Ehlmann et al. [145] for unaltered olivine in Uzboi Vallis, and by the observation that these valleys, once more, start from volcanic areas.

Pallacopas Vallis and Doanus Vallis start from Sisyphi Planum, at the foothills of the Sisyphi Montes; Surius Vallis from the volcanic Argentea Planum; and Dzigai Vallis from an unnamed volcanic centre located northeast of the crater Phillips, exactly where the Alignment 12 ends [43]. The length of the valleys and their origin from the volcanic plains of Argentea Planum and Sisyphi Planum and the radial spreading of the main dorsa (in this case, lava tubes) from Surius Vallis show that the valleys indeed contributed to the volcanic infill of Argyre Planitia and post-dated the infill produced by the volcanic centres inside Argyre. However, the infill brought by the valleys was the lava flooding of the volcanic plains of Argentea Planum and Sisyphi Planum and thus not formed by water at all. This hypothesis is also confirmed by the observation of lava tubes like Pasithea Dorsum embaying and surrounding Oceanidum Mons and of the volcanic infill embaying also the Chalce Montes [43]. Furthermore, a closer view of the Auxo Dorsum and Charis Dorsum already at CTX resolution showed that the morphology of these dorsa features is more compatible with lava tubes rather than eskers, collapsed sections in proximity of a breakout and terminal sections as channels are clearly visible in the available images (Fig. A.20, panel K) published in Leone [43].

At last, also the Alignment 12 has features that mark a lateral distance that can be useful to infer the size of the plume, these are: (1) an unnamed crater located west of Ogygis Regio from which several channels spread all along the south–west and north directions; (2) an unnamed crater located southeast of Douglass from which many channels spread in almost all the directions; an unnamed massif of the same size of those observed along the Sisyphi Montes located northeast of the crater Phillips.

4 The Volcanic Provinces of Mars

First, it is important to define a volcanic (or igneous) province. On Earth, a large igneous province (LIP) was defined as an area larger than 100,000 km^2 that includes erupted igneous rocks [179]. On the basis of this definition, all the ignimbrites deposited among Amazonis Planitia, Memnonis and Aeolis quadrangles, also known as the Medusae Fossae Formation (MFF) [180, 181] including only Apollinaris Mons as a significant volcanic edifice, would form a volcanic province of 2.1–2.2 $\times$ 10^6 km^2 [180, 182]. Thus, according to this definition, there might be large igneous provinces in some regions of the lowlands of Mars where volcanoes are not present but only their basaltic products. It is the case of the basaltic lavas filling Chryse and Acidalia Planitia [66]. It is quite clear that these basaltic lavas came from Tharsis [117], but according to the above definition they could be considered a different volcanic province. Some main volcanic provinces (Olympus-Tharsis, Tempe Terra plateau, Southern Highlands) have already been described by Scott [183], although with no clear criteria defined except for the recognition of volcanic features. The same occurred for the description of the Tempe volcanic province [184] where no clear criteria were established to define a volcanic province. In fact, many of the volcanic products that reached Tempe Terra came from Tharsis [43], which is in theory another different volcanic province according to Scott [183]. Furthermore, always according to the above mentioned definition of LIP, many other provinces can be defined according to arbitrary boundaries all over the surface of Mars, given the interplay or superposition that some provinces may have among each other. More recent papers, with respect to those published in the 80′s, at least better describe, although not yet clearly define, the boundary of a LIP according to a visual continuity between the volcanic centres and their generated or related igneous products [100, 185]. Therefore, it is clear that a new and better definition of LIP must be established to avoid confusion. I here propose a new definition in which a LIP is defined as "*an area larger than* 100,000 km^2 *containing volcanic edifices and their related volcanic products in continuity among them*". The main known volcanic provinces, which were already recognized by other authors [100, 185–187], could be resumed into four ones that include the main volcanic centres and their erupted products. These are by ranking order (Table 1): Tharsis, Elysium, Circum-Hellas, Syrtis Major. This is an important variation with respect to the original main provinces described by Scott [183]. According to the observation of the volcanic alignments of Mars, Tharsis is crossed by 6 of the 12 alignments [43] and it is considered the largest province of Mars [183] with its 7.25 × 10^6 km^2 [186]. Elysium, with its 4.08 × 10^6 km^2 is the second largest province of Mars [187], although it is crossed by only 2 alignments [43]. The third is Circum-Hellas with 2.10 × 10^6 [100] and 3 alignments. The fourth is Syrtis Major with 1.6–3.2 × 10^5 km^2 [185] and 2 alignments.

If all the ignimbrites of the Amazonis Planitia, Memnonis and Aeolis quadrangle could be ascribed to Apollinaris Mons as a possible source [125, 188], the MFF could be the fourth volcanic province in the ranking of the known provinces listed in Table 1. However, according to the above mentioned definition and, again, on the basis of the observation of the volcanic alignments [43], other much larger provinces can be distinguished on Mars (Table 2): Arabia Terra–Meridiani Planum, approximately 12.05 × 10^6 km^2, crossed by 3 alignments; Cimmeria–Sirenum, approximately 8.61 × 10^6 km^2, crossed by 4 alignments; the Southern Polar province, approximately 5.11 × 10^6 km^2, where all the primary alignments end; and the Northern Polar province, which includes all the units described by Tanaka et al. [189], approximately 2.91 × 10^6 km^2. These numbers were obtained circling the volcanic areas, including source centres and erupted products, with the polygon tool available on Google Earth/Mars. Such an operation was facilitated by the availability and ease of accessibility of THEMIS, HiRISE, and CTX data on Google Earth. Clearly, although these numbers are quite rough and approximate, new and accurate digital elevation models (DEM) would be required to have improved numbers, but at least they give a global view and a general idea of the areal extent of the volcanic provinces of Mars, including new ones that were never described as such in the literature.

In consideration of these new provinces, and the above defined known provinces, a new total ranking could be described in Table 3 as follows.

From the new ranking shown in Table 3 it is possible to see that Tharsis would not be anymore the largest volcanic province of Mars but just the third. However, considering a conservative estimate of the areal extent of the whole Tharsis bulge, including Olympus Mons and its lava fields, Ascraeus, Pavonis, Arsia Mons and their lava fields as far as Mangala Valles to the west, Thaumasia Rise to the south, Acidalia Planitia to the east, Tempe Terra to the north-east and Alba Patera and its lava fields extending to the north, the new area of Tharsis obtained through the polygon tool of Google Earth/Mars would approximately be 15.74 × 10^6 km^2 and

Table 2 Areal extent of the new volcanic provinces [43]

Province	Area (km^2)	Ranking
Arabia–Meridiani	12.05 × 10^6	1
Cimmeria–Sirenum	8.61 × 10^6	2
Southern Polar	5.11 × 10^6	3
Northern Polar	2.91 × 10^6	4

Source of data this work and Leone (2016)

Table 3 Areal extent of the volcanic provinces of Mars [43, 100, 180, 181, 185–187]

Province	Area (km^2)	Ranking
Arabia–Meridiani	12.05×10^6	1
Cimmeria–Sirenum	8.61×10^6	2
[a]Tharsis	7.25×10^6	3
Southern Polar	5.11×10^6	4
[b]Elysium	4.08×10^6	5
Northern Polar	2.91×10^6	6
[c,d]MFF	$2.1\text{–}2.2 \times 10^6$	7
[e]Circum-Hellas	2.10×10^6	8
[f]Syrtis Major	$1.6\text{–}3.2 \times 10^5$	9

Source of data this work, Leone (2016) and [a]Schaber et al. (1978); [b]Hall et al. (1986); [c]Kerber and Head (2010); [d]Harrison et al. (2010); [e]Williams et al. (2009); [f]Hiesinger and Head (2004)

thus more than double than previously estimated. This new areal extent would bring back Tharsis to the rank of volcanic province number 1 of Mars.

5 The Magnetic Field of Mars: Origin and Demise

Mars today does not possess an active magnetic field, an estimate of the upper limit of the current Mars dipole corresponds to an average surface equatorial field of <0.5 nT [190], which is well below the 3 nT set for the noise measurement in current combined MGS and MAVEN data [191]. Mars Global Surveyor (MGS) measurements revealed magnetic crustal anomalies up to ±200–1000 nT on the surface of Mars [192–194], depending on the altitude from which the measurements are done [191, 194]. Patches of magnetite and pyrrhotite observed in the ALH84001 meteorite revealed a stable natural remanent magnetization (NRM) up to ±100 nT [194, 195]. Numbers consistent with a surface equatorial field of 25–125 nT [196]. The magnetization of the crust appears today mainly concentrated on the ancient heavily cratered highlands, consistent with a reversing dynamo that ceased to exist early in Martian history [72, 197]. From these data it was possible to infer that the Red Planet had an active dynamo producing a magnetic field comparable to that of the Earth today, which varies from 30,000 to 60,000 nT [73, 190]. Although the age of the magnetic field was debated between post-4 Ga [198] and pre-4 Ga [199], it is still thought that the magnetic field lasted for the first 500 Ma of the Martian history [200] because of the evident lack of magnetic anomalies in the large impact basins formed after 4.1 Ga ago [72, 199, 201, 202]. From the magnetized patches present in the ALH84001 meteorite it was alternatively inferred that the magnetic dynamo of Mars formed from 450 to 650 Ma after the formation of the planet, corresponding from Early to Middle Noachian (4.1–3.9 Ga), an age consistent with the model of inner core solidification of Schubert et al. [198], and that lasted indefinitely afterwards [195]. Alternative models of dynamo formation confirmed a magnetic field lasting 500 Ma after the initial planetary formation through a mechanism of an initial core hotter 150 °K than the mantle [198]. The core of Mars did not experienced solidification with time, otherwise a longer dynamo (>1 Ga) would have survived due to the consequent release of latent heat and gravitational energy powering the magnetic field [198, 200]. Also, giant impacts can alternatively stop [203] or trigger [204] a core dynamo [191]. The pattern of the core-mantle boundary (CMB) heat flux all over the Martian history suggests that a transient magnetic field of Mars might have been triggered by the SPGI to shut down when the heat flux dropped and then stabilised to a today value of 22–14 mWm^{-2} around 4.1 Ga ago [24]. This age is consistent with the age of the magnetic anomalies detected on Mars and estimated in 4.09 Ga [72], which are absent in younger (4 Ga) impact basins like Hellas [199, 202]. On the basis of the various arguments explained so far the most likely scenario is that the magnetic field of Mars had a fast decrease around 4.1 Ga, due to the shut-down of the core dynamo, and a definitive cessation in a range of time located between 4.1 and 4.0 Ga to never restart again [73]. From this time onwards the planet was vulnerable to the radiation coming from space.

6 Effect of Volcanism on the Climate of Mars

Although the Mariner and the Viking missions have shown since the years 60s and 70s how Mars is a dry planet (see also the detailed discussion in the Introduction Chapter of this book), there has been in the 80s the belief of a regolith-atmosphere exchange of water and carbon dioxide based on climatic changes on Mars; the role of the regolith was inferred on the basis of the observation that the polar caps were not enough extensive to buffer the atmospheric CO_2 [205, 206]. However, the available studies are not concurring on any specific kind of climate for Mars [207, 208]. There are those that put forward a warm and wet climate [209–211] and those that sustain a cold climate with little difference from the one observed today [212]. Some authors did not see any problem for water flowing on the surface of Mars even in a cold climate and thin atmosphere [213, 214] although this is physically impossible even in the presence of salts because of the low pressure [35]. In this debated scenario what could be the effect of volcanism on the climate of Mars? Volcanoes contribute with ashes and volatiles to the atmosphere and the lower atmospheric pressure of Mars should in principle facilitate the development of explosive volcanism [215]. There is clear evidence of extended pyroclastic deposits even of a certain significance like the Medusae Fossae Formation [124, 125, 181] and the ignimbrites of Amazonis Planitia [182] or smaller deposits like on Arsia Mons [216] or Hecates Tholus [217], although the traces of explosive volcanism are not widespread at global level as it should be expected in the low atmospheric pressure of Mars [218–220]. Past degassing of water was inferred from isotopic studies [36–39] and this coincided with the strongest phase of volcanism in the history of Mars [35, 43]. The point is understanding the volume of such a contribution and whether this might have somehow affected the climate of Mars. The volumes of the pyroclastic deposits found on Mars are large enough to have certainly obscured the Martian sky for some periods of time, likely in the order of years when compared to the eruptions occurred on Earth [221–223], significantly attenuating the solar radiation and thus decreasing the global temperature of the planet. Additional complementary studies aimed at estimating the contribution of Alba Patera to the global volatile inventory suggested a total contribution ranging from 5.1 to 51.7 mm of global equivalent layer (GEL), which is a small fraction of the 1.47 m GEL calculated for all the volcanic centres of Mars [224]. Compared to the various global contributions seen in the previous Introduction chapter, the contribution of all the volcanic centres of Mars accounts for only the 0.05–0.2% of the total. Nearly the 50% of the total has been degassed away from the atmosphere already in the pre-Noachian while the remainder left the planet during the whole Martian history [38]. From these numbers it is clear that the contribution of the volcanoes to the water budget of the planet was really low if not insignificant. Yet, there were authors who inferred climate changes on Mars from putative ice deposits 1–10 m deep corresponding to 10–40 cm GEL [225] or from putative ice contents of 1.53 m GEL contained in polar cavi units considered one of the largest water ice reservoirs on Mars [226]. Of course, these amounts are even less than those estimated for the volcanoes in the first case or comparable in the second case, far from being the largest water reservoirs of Mars. Their contributions are unlikely to have any effect on the climate of Mars, as well as the periodic variations in obliquity [79, 227], any past warmer and wetter climate $\geq 10^5$ yr [228] would have left evident mineralogical traces of alteration in the olivine [156, 157] that have not been found even in the oldest units of Mars [35]. Objectively, it can be said that the strong eruptions during the peak of volcanism occurred in the first 500 Ma of Martian history [24] might have contributed to lower the whole temperature of the planet thus contributing to a colder climate than expected.

References

1. Wänke H, Dreibus G (1994) Chemistry and accretion history of Mars. Philos Trans R Soc London Ser A Phys Eng Sci 349(1690):285–293.
2. Halliday AN, Wänke H, Birck JL, Clayton RN (2001) The accretion, composition and early differentiation of Mars. Space Sci Rev 96:197–230
3. Harper CL, Nyquist LE, Bansal B, Wiesmann H, Shih CY (1995) Rapid accretion and early differentiation of Mars indicated by 142Nd/144Nd in SNC meteorites. Science 267(5195):213–217.
4. Lodders K (1998) A survey of shergottite, nakhlite and chassigny meteorites whole-rock compositions. Meteorit Planet Sci 33(S4):A183–A190.
5. Sautter V, Fabre C, Forni O, Toplis MJ, Cousin A, Ollila AM et al (2014) Igneous mineralogy at Bradbury rise: The first ChemCam campaign at Gale crater. J Geophys Res Planets 119(1):30–46.
6. Sautter V, Toplis MJ, Beck P, Mangold N, Wiens R, Pinet P et al (2016) Magmatic complexity on early Mars as seen through a combination of orbital, in-situ and meteorite data. Lithos 254–255:36–52
7. Payré V, Fabre C, Cousin A, Sautter V, Wiens RC, Forni O et al (2017) Alkali trace elements in Gale crater, Mars, with ChemCam: calibration update and geological implications. J Geophys Res Planets [Internet] [cited 2019 Mar 27] 122(3):650–679. http://doi.wiley.com/https://doi.org/10.1002/2016JE005201
8. Sautter V, Toplis MJ, Wiens RC, Cousin A, Fabre C, Gasnault O et al (2015) In situ evidence for continental crust on early Mars. Nat Geosci [Internet] [cited 2018 Nov 20] 8(8):605–609. http://www.nature.com/articles/ngeo2474
9. Schmidt ME, Campbell JL, Gellert R, Perrett GM, Treiman AH, Blaney DL et al (2014) Geochemical diversity in first rocks examined by the Curiosity Rover in Gale Crater: Evidence for and significance of an alkali and volatile-rich igneous source. J Geophys Res Planets [Internet] [cited 2018 Oct 18] 119(1):64–81. http://doi.wiley.com/https://doi.org/10.1002/2013JE004481
10. McSween HY, Wyatt MB, Gellert R, Bell JF, Morris RV, Herkenhoff KE et al (2006) Characterization and petrologic interpretation of olivine-rich basalts at Gusev Crater, Mars. J Geophys Res Planets 111: E02S10
11. Rieder R, Economou T, Wänke H, Turkevich A, Crisp J, Brückner J et al (1997) The chemical composition of Martian soil and rocks returned by the mobile alpha proton X-ray spectrometer: preliminary results from the X-ray mode. Science [Internet] [cited 2019 Aug 14] 278(5344):1771–1774. http://www.ncbi.nlm.nih.gov/pubmed/9388173
12. Baird AK, Toulmin P, Clark BC, Rose HJ, Keil K, Christian RP et al (1976) Mineralogic and petrologic implications of viking geochemical results from Mars: interim report. Science [Internet] [cited 2019 Apr 5] 194(4271):1288–1293. http://www.ncbi.nlm.nih.gov/pubmed/17797085
13. Cousin A, Sautter V, Payré V, Forni O, Mangold N, Gasnault O et al (2017) Classification of igneous rocks analyzed by ChemCam at Gale crater, Mars. Icarus 288:265-283
14. Stolper EM, Baker MB, Newcombe ME, Schmidt ME, Treiman AH, Cousin A et al (2013) The petrochemistry of Jake_M: a Martian Mugearite. Science (80-)341(6153):1239463
15. Ollila AM, Newsom HE, Clark B, Wiens RC, Cousin A, Blank JG et al (2014) Trace element geochemistry (Li, Ba, Sr, and Rb) using Curiosity's ChemCam: early results for Gale crater from Bradbury landing site to rocknest. J Geophys Res E Planets. 119:255–285
16. Foley CN, Wadhwa M, Borg LE, Janney PE, Hines R, Grove TL (2005) The early differentiation history of Mars from 182 W–142 Nd isotope systematics in the SNC meteorites. Geochim Cosmochim Acta 69(18):4557–4571
17. Apai D, Pascucci I, Bouwman J, Natta A, Henning T, Dullemond CP (2005) The onset of planet formation in brown dwarf disks. Science 310 (5749):834–836.
18. Amelin Y, Krot AN, Hutcheon ID, Ulyanov AA (2002) Lead isotopic ages of chondrules and calcium-aluminum-rich inclusions. Science [Internet] [cited 2019 Oct 11] 297(5587):1678–1683. http://www.ncbi.nlm.nih.gov/pubmed/12215641
19. Lee D-C, Halliday AN (1997) Core formation on Mars and differentiated asteroids. Nature [Internet] 388(6645):854–857. http://www.nature.com/articles/42206
20. Righter K, Shearer CK (2003) Magmatic fractionation of Hf and W: constraints on the timing of core formation and differentiation in the Moon and Mars. Geochim Cosmochim Acta [Internet] [cited 2019 Oct 12] 67(13):2497–2507. https://www.sciencedirect.com/science/article/pii/S0016703702013492
21. Jacobsen SB (2005) The Hf-W isotopic system and the origin of the Earth and Moon. Annu Rev Earth Planet Sci [Internet] [cited 2019 Oct 12] 33(1):531–

570. http://www.annualreviews.org/doi/https://doi.org/10.1146/annurev.earth.33.092203.122614
22. Šrámek O, Milelli L, Ricard Y, Labrosse S (2012) Thermal evolution and differentiation of planetesimals and planetary embryos. Icarus [Internet] [cited 2019 Oct 12] 217(1):339–354. https://www.sciencedirect.com/science/article/pii/S0019103511004489#b0350
23. Yoshino T, Walter MJ, Katsura T (2003) Core formation in planetesimals triggered by permeable flow. Nature [Internet] [cited 2019 Oct 12] 422 (6928):154–157. http://www.nature.com/articles/nature01459
24. Leone G, Tackley PJ, Gerya TV, May DA, Zhu G (2014) Three-dimensional simulations of the southern polar giant impact hypothesis for the origin of the Martian dichotomy. Geophys Res Lett 41 (24):8736–8743
25. Chambers JE, Wetherill GW (1998) Making the terrestrial planets: N-body integrations of planetary embryos in three dimensions. Icarus [Internet] [cited 2019 Oct 12] 136(2):304–327. https://www.sciencedirect.com/science/article/abs/pii/S0019103598960079
26. Nimmo F, Kleine T (2007) How rapidly did Mars accrete? Uncertainties in the Hf-W timing of core formation. Icarus 191(2):497–504
27. Golabek GJ, Keller T, Gerya T V., Zhu G, Tackley PJ, Connolly JAD (2011) Origin of the Martian dichotomy and Tharsis from a giant impact causing massive Magmatism. Icarus 215(1):346-357.
28. Bercovici D, Ricard Y (2014) Plate tectonics, damage and inheritance. Nature 508(7497):513–516
29. Foley BJ, Bercovici D, Elkins-Tanton LT (2014) Initiation of plate tectonics from post-magma ocean thermochemical convection. J Geophys Res Solid Earth [Internet] [cited 2020 Jan 17] 119(11):8538–8561. http://doi.wiley.com/https://doi.org/10.1002/2014JB011121
30. Korenaga J (2013) Initiation and evolution of plate tectonics on earth: theories and observations. Annu Rev Earth Planet Sci [Internet] [cited 2020 Jan 17] 41(1):117–151. http://www.annualreviews.org/doi/https://doi.org/10.1146/annurev-earth-050212-124208
31. Maruyama S, Santosh M, Azuma S (2018) Initiation of plate tectonics in the Hadean: eclogitization triggered by the ABEL bombardment. Geosci Front 9(4):1033–1048
32. de Pater I, Davies AG, McGregor A, Trujillo C, Ádámkovics M, Veeder GJ et al (2014) Global near-IR maps from Gemini-N and Keck in 2010, with a special focus on Janus Patera and Kanehekili Fluctus. Icarus [Internet] [cited 2019 Jan 19] 242:379–395. https://www.sciencedirect.com/science/article/pii/S0019103514003303
33. Leverington DW (2011) A volcanic origin for the outflow channels of Mars: Key evidence and major implications. Geomorphology 132:51–75
34. Leone G (2020) The absence of an ocean and the fate of water all over the Martian history. Earth Sp Sci [Internet] 7(4):1–16. https://onlinelibrary.wiley.com/doi/abs/https://doi.org/10.1029/2019EA001031
35. Leone G (2017) Mangala Valles, Mars: a reassessment of formation processes based on a new geomorphological and stratigraphic analysis of the geological units. J Volcanol Geotherm Res 337:62–80
36. Gillmann C, Lognonné P, Moreira M (2011) Volatiles in the atmosphere of Mars: the effects of volcanism and escape constrained by isotopic data. Earth Planet Sci Lett [Internet] [cited 2019 Apr 5] 303(3–4):299–309. https://www.sciencedirect.com/science/article/pii/S0012821X11000239
37. Krasnopolsky VA (2015) Variations of the HDO/H_2O ratio in the Martian atmosphere and loss of water from Mars. Icarus [Internet] 257:377–386. http://dx.doi.org/https://doi.org/10.1016/j.icarus.2015.05.021
38. Kurokawa H, Sato M, Ushioda M, Matsuyama T, Moriwaki R, Dohm JM et al (2014) Evolution of water reservoirs on Mars: constraints from hydrogen isotopes in Martian meteorites. Earth Planet Sci Lett 394:179–185
39. Villanueva GL, Mumma MJ, Novak RE, Käufl HU, Hartogh P, Encrenaz T et al (2015) Strong water isotopic anomalies in the Martian atmosphere: probing current and ancient reservoirs. Science 48 (6231):218-221
40. O'Rourke JG, Korenaga J (2012) Terrestrial planet evolution in the stagnant-lid regime: size effects and the formation of self-destabilizing crust. Icarus 221:1043–1060
41. Sleep NH (1994) Martian plate tectonics. J Geophys Res 99(E3):5639– 5655
42. Leone G (2015) Transition Topography (Mars). In: Encyclopedia of planetary landforms. Springer, New York, NY, pp. 1–6. https://doi.org/10.1007/978-1-4614-9213-9_650-1
43. Leone G (2016) Alignments of volcanic features in the southern hemisphere of Mars produced by migrating mantle plumes. J Volcanol Geotherm Res 309:78–95
44. Leone G (2016) The Southern polar giant impact hypothesis for the origin of the Martian dichotomy and the evolution of volcanism on Mars. PhDThesis ETH Zurich. https://doi.org/10.3929/ethz-a-010652097
45. Wise DU, Golombek MP, McGill GE (1979) Tectonic evolution of Mars. J Geophys Res [Internet] [cited 2019 Oct 13] 84(B14):7934. http://doi.wiley.com/https://doi.org/10.1029/JB084iB14p07934
46. Keller T, Tackley PJ (2009) Towards self-consistent modeling of the Martian dichotomy: the influence of one-ridge convection on crustal thickness distribution. Icarus [Internet] [cited 2019 Oct 13] 202

(2):429–443. https://www.sciencedirect.com/science/article/pii/S0019103509001377
47. Zhong S, Zuber MT (2001) Degree-1 mantle convection and the crustal dichotomy on Mars. Earth Planet Sci Lett 189:75–84
48. Roberts JH, Zhong S (2006) Degree-1 convection in the Martian mantle and the origin of the hemispheric dichotomy. J Geophys Res [Internet] [cited 2019 Oct 13] 111(E6):E06013. http://doi.wiley.com/https://doi.org/10.1029/2005JE002668
49. McGill GE, Dimitriou AM (2019) Origin of the Martian global dichotomy by crustal thinning in the Late Noachian or Early Hesperian. J Geophys Res [cited 2019 Oct 13] 95(B8):12595. http://doi.wiley.com/https://doi.org/10.1029/JB095iB08p12595
50. McGill GE, Squyres SW (1991) Origin of the Martian crustal dichotomy: evaluating hypotheses. Icarus 93(2):386–393
51. Frey H, Schultz RA (1988) Large impact basins and the mega-impact origin for the crustal dichotomy on Mars. Geophys Res Lett [Internet] [cited 2019 Jan 16] 15(3):229–232. http://doi.wiley.com/https://doi.org/10.1029/GL015i003p00229
52. Wilhelms DE, Squyres SW (1984) The Martian hemispheric dichotomy may be due to a giant impact. Nature [Internet] [cited 2019 Jan 16] 309 (5964):138–140. http://www.nature.com/articles/309138a0
53. Andrews-Hanna JC, Zuber MT, Banerdt WB (2008) The Borealis basin and the origin of the Martian crustal dichotomy. Nature 453:1212–1215
54. Marinova MM, Aharonson O, Asphaug E (2008) Mega-impact formation of the Mars hemispheric dichotomy. Nature 453:1216–1219
55. Nimmo F, Hart SD, Korycansky DG, Agnor CB (2008) Implications of an impact origin for the Martian hemispheric dichotomy. Nature 453 (7199):1220–1223
56. Reese CC, Solomatov VS (2006) Fluid dynamics of local Martian magma oceans. Icarus [Internet] [cited 2019 Jan 10] 184(1):102–120. https://www.sciencedirect.com/science/article/pii/S0019103506001308
57. Reese CC, Solomatov VS (2010) Early Martian dynamo generation due to giant impacts. Icarus [Internet] [cited 2019 Jan 10] 207(1):82–97. https://www.sciencedirect.com/science/article/pii/S0019103509004321
58. Reese CC, Orth CP, Solomatov VS (2011) Impact megadomes and the origin of the Martian crustal dichotomy. Icarus [Internet] [cited 2019 Jan 10] 213 (2):433–442. https://www.sciencedirect.com/science/article/pii/S0019103511001199
59. Breuer D, Yuen DA, Spohn T, Zhang S (1998) Three dimensional models of Martian mantle convection with phase transitions. Geophys Res Lett [Internet] [cited 2019 Nov 14] 25(3):229–232. http://doi.wiley.com/https://doi.org/10.1029/97GL03767
60. Elkins-Tanton LT, Hess PC, Parmentier EM (2005) Possible formation of ancient crust on Mars through magma ocean processes. J Geophys Res Planets 110:E12S01
61. Elkins-Tanton LT (2008) Linked magma ocean solidification and atmospheric growth for Earth and Mars. Earth Planet Sci Lett [Internet] [cited 2019 Oct 13] 271(1–4):181–91. https://www.sciencedirect.com/science/article/pii/S0012821X08002306
62. Harder H, Christensen UR (1996) A one-plume model of martian mantle convection. Nature 380 (6574):507–509
63. Schubert G, Bercovici D, Glatzmaier GA (1990) Mantle dynamics in Mars and Venus: influence of an immobile lithosphere on three-dimensional mantle convection. J Geophys Res [Internet] [cited 2019 Oct 14] 95(B9):14105. http://doi.wiley.com/https://doi.org/10.1029/JB095iB09p14105
64. Frey H (2008) Ages of very large impact basins on Mars: Implications for the late heavy bombardment in the inner solar system. Geophys Res Lett 35(13): L13203
65. Pierazzo E, Melosh HJ (2000) Melt production in oblique impacts. Icarus [Internet] [cited 2019 Oct 16] 145(1):252–261. https://www.sciencedirect.com/science/article/abs/pii/S0019103599963327
66. Salvatore MR, Mustard JF, Wyatt MB, Murchie SL (2010) Definitive evidence of Hesperian basalt in Acidalia and Chryse planitiae. J Geophys Res [Internet] [cited 2019 Aug 13] 115(E7):E07005. http://doi.wiley.com/https://doi.org/10.1029/2009JE003519
67. Salvatore MR, Christensen PR (2014) On the origin of the Vastitas Borealis formation in Chryse and Acidalia Planitiae, Mars. J Geophys Res Planets [Internet] [cited 2019 Jan 10] 119(12):2437–2456. http://doi.wiley.com/https://doi.org/10.1002/2014JE004682
68. Hopper JP, Leverington DW (2014) Formation of Hrad Vallis (Mars) by low viscosity lava flows. Geomorphology [Internet] [cited 2018 Oct 25] 207:96–113. https://www.sciencedirect.com/science/article/pii/S0169555X13005552
69. Tonks WB, Melosh HJ (1993) Magma ocean formation due to giant impacts. J Geophys Res Planets. 98(E3):5319–5333
70. Kleine T, Münker C, Mezger K, Palme H (2002) Rapid accretion and early core formation on asteroids and the terrestrial planets from Hf-W chronometry. Nature 418(6901):952–955
71. Kleine T, Mezger K, Münker C, Palme H, Bischoff A (2004) 182Hf-182W isotope systematics of chondrites, eucrites, and martian meteorites: chronology of core formation and early mantle differentiation in Vesta and Mars. Geochim Cosmochim Acta 68(13):2935–2946
72. Lillis RJ, Frey HV, Manga M, Mitchell DL, Lin RP, Acuña MH et al (2008) An improved crustal magnetic field map of Mars from electron

reflectometry: highland volcano magmatic history and the end of the Martian dynamo. Icarus 194 (2):575–596
73. Lillis RJ, Robbins S, Manga M, Halekas JS, Frey HV (2013) Time history of the Martian dynamo from crater magnetic field analysis. J Geophys Res Planets [Internet] [cited 2019 Aug 14] 118 (7):1488–1511. https://doi.org/10.1002/jgre.20105
74. Fairén AG, Ruiz J, Anguita F (2002) An origin for the linear magnetic anomalies on Mars through accretion of terranes: implications for dynamo timing. Icarus 160(1):220–223
75. Acuña MH, Connerney JEP, Wasilewski P, Lin RP, Mitchell D, Anderson KA et al (2001) Magnetic field of Mars: summary of results from the aerobraking and mapping orbits. J Geophys Res Planets [Internet] [cited 2019 Nov 16], 106 (E10):23403–23417. http://doi.wiley.com/https://doi.org/10.1029/2000JE001404
76. Farley KA, Malespin C, Mahaffy P, Grotzinger JP, Vasconcelos PM, Milliken RE et al (2014) In situ radiometric and exposure age dating of the Martian surface. Science (80-) [Internet] 343(6169):1247166-1–1247166-5. http://science.sciencemag.org/
77. Gasparri D, Leone G, Cataldo V, Punjabi V, Nandakumar S (2020) Lava filling of Gale crater from Tyrrhenus Mons on Mars. J Volcanol Geotherm Res 389:106743
78. Ward WR (1973) Large-scale variations in the obliquity of Mars. Science (80-) 181(4096):260–262
79. Rubincam DP (1990) Mars: change in axial tilt due to climate? Science (80-) 248(4956):720–721
80. Neukum G, Jaumann R, Hoffmann H, Hauber E, Head JW, Basilevsky AT et al (2004) Recent and episodic volcanic and glacial activity on Mars revealed by the high resolution stereo camera. Nature 432(7020):971–979
81. Isherwood RJ, Jozwiak LM, Jansen JC, Andrews-Hanna JC (2013) The volcanic history of Olympus Mons from paleo-topography and flexural modeling. Earth Planet Sci Lett 363:88-96
82. Plescia JB (1994) Geology of the Small Tharsis Volcanoes: Jovis Tholus, Ulysses Patera, Biblls Patera. Mars. Icarus. 111(1):246–269
83. Robbins SJ, Achille G Di, Hynek BM (2011) The volcanic history of Mars: high-resolution crater-based studies of the calderas of 20 volcanoes. Icarus [Internet] [cited 2018 Nov 13] 211(2):1179–203. https://www.sciencedirect.com/science/article/pii/S0019103510004318
84. Neukum G, Hiller K (1981) Martian ages. J Geophys Res Solid Earth [Internet] [cited 2018 Nov 13] 86(B4):3097–3121. http://doi.wiley.com/https://doi.org/10.1029/JB086iB04p03097
85. Dohm JM, Ferris JC, Baker VR, Anderson RC, Hare TM, Strom RG et al (2001) Ancient drainage basin of the Tharsis region, Mars: Potential source for outflow channel systems and putative oceans or paleolakes. J Geophys Res Planets [Internet] 106 (E12):32943–32958. http://doi.wiley.com/https://doi.org/10.1029/2000JE001468
86. Anguita F, Fernández C, Cordero G, Carrasquilla S, Anguita J, Núñez A et al (2006) Evidences for a Noachian-Hesperian orogeny in Mars. Icarus 185 (2):331–357
87. Ghatan GJ (2003) Cavi Angusti, Mars: Characterization and assessment of possible formation mechanisms. J Geophys Res 108(E5):5045–5063
88. Sharp RP (1973) Mars: South polar pits and etched terrain. J Geophys Res 78(20):4222–4230
89. Allen CC (1979) Volcano-ice interactions on Mars. J Geophys Res Solid Earth. 84(B14):8048–8059
90. Mutch T (1976) The geology of Mars [Internet], 1st edn. Princeton University Press, Princeton, NJ [cited 2019 Dec 6], 400 pp. https://www.abebooks.com/first-edition/GEOLOGY-MARS-Mutch-Thomas-A-Princeton/434425546/bd
91. Carr MH, Head JW (2010) Geologic history of Mars. Earth Planet Sci Lett 294(3):185–203
92. Tanaka KL, Kolb EJ (2001) Geologic history of the polar regions of Mars based on Mars Global survey data. I. Noachian and Hesperian Periods. Icarus. 154(1):3–21
93. Cantor B, Malin M, Edgett KS (2002) Multiyear Mars Orbiter Camera (MOC) observations of repeated Martian weather phenomena during the northern summer season. J Geophys Res E Planets. 107(E3):5014–5022
94. Greeley R, Guest JE (1987) Geologic map of the eastern equatorial region of Mars [Internet]. IMAP [cited 2019 Jul 27]. https://pubs.er.usgs.gov/publication/i1802B
95. Tanaka K, Skinner Jr J, Hare T (2005) Geologic map of the Northern Plains of Mars. Sci Invest Map 2888
96. De Pablo MA, Michael GG, Centeno JD (2013) Age and evolution of the lower NW flank of the Hecates Tholus volcano, Mars, based on crater size-frequency distribution on CTX images. Icarus 226 (1):455–469
97. Werner SC (2009) The global Martian volcanic evolutionary history. Icarus 201:44–68
98. Platz T, Michael G (2011) Eruption history of the Elysium Volcanic Province, Mars. Earth Planet Sci Lett [Internet] [cited 2018 Sep 23] 312(1–2):140–151. https://www.sciencedirect.com/science/article/pii/S0012821X11005887
99. Tanaka KL, Robbins SJ, Fortezzo CM, Skinner JA, Hare TM (2014) The digital global geologic map of Mars: chronostratigraphic ages, topographic and crater morphologic characteristics, and updated resurfacing history. Planet Space Sci [Internet] [cited 2019 Mar 26] 95:11–24. https://www.sciencedirect.com/science/article/pii/S0032063313000652
100. Williams DA, Greeley R, Fergason RL, Kuzmin R, McCord TB, Combe JP et al (2009) The Circum-Hellas volcanic province, Mars: overview. Planet Space Sci 57:895–916

101. Ghatan GJ, Head JW III (2002) Candidate subglacial volcanoes in the south polar region of Mars: morphology, morphometry, and eruption conditions. J Geophys Res–Planets. 107(E7):5048–5071
102. Fenton LK, Hayward RK (2010) Southern high latitude dune fields on Mars: Morphology, Aeolian inactivity, and climate change. Geomorphology 121 (1–2):98–121
103. Scott DH, Tanaka KL (1986) Geologic map of the western equatorial region of Mars [Internet]. IMAP 1802 [cited 2019 Apr 6]. https://pubs.er.usgs.gov/publication/i1802A
104. Zimbelman JR, Garry WB, Bleacher JE, Crown DA (2015) Volcanism on Mars. In: The encyclopedia of volcanoes [Internet]. Academic Press [cited 2019 Jul 21], pp 717–728. https://www.sciencedirect.com/science/article/pii/B9780123859389000419
105. Plescia JB (2000) Geology of the Uranius group volcanic constructs: Uranius Patera, Ceraunius Tholus, and Uranius Tholus. Icarus [Internet] [cited 2019 May 17] 143(2):376–396. https://www.sciencedirect.com/science/article/abs/pii/S0019103599962590
106. Robbins SJ, Di Achille G, Hynek BM (2011) The volcanic history of Mars: high-resolution crater-based studies of the calderas of 20 volcanoes. Icarus
107. Xiao L, Huang J, Christensen PR, Greeley R, Williams DA, Zhao J et al (2012) Ancient volcanism and its implication for thermal evolution of Mars. Earth Planet Sci Lett [Internet] [cited 2019 May 26] 323–324:9–18. https://www.sciencedirect.com/science/article/pii/S0012821X12000453
108. Tanaka KL, Scott DH (1987) Geologic map of the Polar Regions of Mars. USGS Miscellaneous Investigations Series Map I–1802–C
109. Greeley R, Foing BH, McSween HY, Neukum G, Pinet P, van Kan M et al (2005) Fluid lava flows in Gusev crater. Mars. J Geophys Res E Planets. 110: E05008
110. Cabrol NA, Grin EA, Carr MH, Sutter B, Moore JM, Farmer JD et al (2003) Exploring Gusev Crater with spirit: review of science objectives and testable hypotheses. J Geophys Res Planets [Internet] [cited 2019 Apr 5] 108(E12) http://doi.wiley.-com/https://doi.org/10.1029/2002JE002026
111. Stewart EM, Head JW (2001) Ancient Martian volcanoes in the Aeolis region: new evidence from MOLA data. J Geophys Res Planets [Internet] [cited 2019 May 17] 106(E8):17505–17513. https://doi.org/10.1029/2000JE001322
112. Ackiss SE, Wray JJ (2014) Occurrences of possible hydrated sulfates in the southern high latitudes of Mars. Icarus 5(243):311–324
113. Head JW, Wilson L, Dickson J, Neukum G (2006) The Huygens-Hellas giant dike system on Mars: implications for late Noachian-early Hesperian volcanic resurfacing and climatic evolution. Geology 34 (4): 285–288
114. Mangold N (2012) Fluvial landforms on fresh impact ejecta on Mars. Planet Space Sci 62(1):69–85
115. Christensen PR (2006) Water at the poles and in permafrost regions of Mars. Elements [Internet] [cited 2019 Apr 5] 2(3):151–155. https://pubs.geoscienceworld.org/elements/article/2/3/151-155/137698
116. Andrews-Hanna JC, Phillips RJ (2007) Hydrological modeling of outflow channels and chaos regions on Mars. J Geophys Res Planets [Internet] [cited 2019 Apr 5] 112(E8). http://doi.wiley.com/https://doi.org/10.1029/2006JE002881
117. Leone G (2014) A network of lava tubes as the origin of Labyrinthus Noctis and Valles Marineris on Mars. J Volcanol Geotherm Res 277:1–8
118. Grant JA, Schultz PH (1990) Gradational epochs on Mars: Evidence from West-Northwest of Isidis Basin and Electris. Icarus 84(1):166–195
119. Som SM, Montgomery DR, Greenberg HM (2009) Scaling relations for large Martian valleys. J Geophys Res [Internet] [cited 2019 Dec 7] 114(E2): E02005. http://doi.wiley.com/https://doi.org/10.1029/2008JE003132
120. Maxwell TA, Craddock RA (1995) Age relations of Martian highland drainage basins. J Geophys Res 100(E6):11765– 11780
121. Fassett CI, Head JW (2008) Valley network-fed, open-basin lakes on Mars: distribution and implications for Noachian surface and subsurface hydrology. Icarus 198(1):37–56
122. Vaucher J, Baratoux D, Mangold N, Pinet P, Kurita K, Grégoire M (2009) The volcanic history of central Elysium Planitia: implications for Martian magmatism. Icarus 204(2):418–442
123. Scott D, Tanaka K (1978) Geologic map of Mars [Internet]. Reston. https://pubs.er.usgs.gov/publication/i1083
124. Kerber L, Head JW, Madeleine JB, Forget F, Wilson L (2011) The dispersal of pyroclasts from Apollinaris Patera, Mars: implications for the origin of the Medusae Fossae Formation. Icarus 216 (1):212–220
125. Kerber L, Head JW, Madeleine J-B, Forget F, Wilson L (2012) The dispersal of pyroclasts from ancient explosive volcanoes on Mars: Implications for the friable layered deposits. Icarus [Internet] [cited 2019 Aug 13] 219(1):358–381. https://www.sciencedirect.com/science/article/pii/S0019103512001091
126. Robinson MS, Mouginis-Mark PJ, Zimbelman JR, Wu SSC, Ablin KK, Howington-Kraus AE (1993) Chronology, eruption duration, and atmospheric contribution of the Martian volcano Apollinaris Patera. Icarus 104(2):301–323
127. Lang NP, McSween HY, Tornabene LL, Hardgrove CJ, Christensen PR (2010) Reexamining the relationship between Apollinaris Patera and the basalts of the Gusev crater plains, Mars. J Geophys

Res [Internet] [cited 2018 Sep 22] 115(E4):E04006. http://doi.wiley.com/https://doi.org/10.1029/2009JE003397
128. Capitan RD, Van De Wiel M (2011) Landform hierarchy and evolution in Gorgonum and Atlantis basins, Mars. Icarus 211(1):366-388
129. Öhman T, McGovern PJ (2014) Circumferential graben and the structural evolution of Alba Mons. Mars. Icarus 1(233):114–125
130. Keszthelyi L, Jaeger W, McEwen A, Tornabene L, Beyer RA, Dundas C et al (2008) High resolution imaging science experiment (HiRISE) images of volcanic terrains from the first 6 months of the Mars reconnaissance orbiter primary science phase. J Geophys Res E Planets 113:E04005
131. Spagnuolo MG, Figueredo PH, Ramos VA (2008) Reinterpretation of Tractus Fossae region as an asymmetric rift system on Mars. Icarus 198(2):318–330
132. Wilson L, Head JW (2002) Tharsis-radial graben systems as the surface manifestation of plume-related dike intrusion complexes: models and implications. J Geophys Res E Planets. 107(8):1–1
133. Cailleau B, Walter TR, Janle P, Hauber E (2005) Unveiling the origin of radial grabens on Alba Patera volcano by finite element modelling. Icarus 176(1):44–56
134. Scott ED, Wilson L, Head JW (2002) Emplacement of giant radial dikes in the northern Tharsis region of Mars. J Geophys Res E Planets 25:107(4)
135. Ivanov MA, Head JW (2006) Alba Patera, Mars: topography, structure, and evolution of a unique late Hesperian-early Amazonian shield volcano. J Geophys Res E Planets. 111(E9):E09003
136. Bleacher JE, Glaze LS, Greeley R, Hauber E, Baloga SM, Sakimoto SHE et al (2009) Spatial and alignment analyses for a field of small volcanic vents south of Pavonis Mons and implications for the Tharsis province, Mars. J Volcanol Geotherm Res 185(1-2):96-102
137. Richardson JA, Bleacher JE, Glaze LS (2013) The volcanic history of Syria Planum, Mars. J Volcanol Geotherm Res 252:1-13
138. Platz T, Michael G, Tanaka KL, Skinner JA, Fortezzo CM (2013) Crater-based dating of geological units on Mars: methods and application for the new global geological map. Icarus 225(1):806–827
139. Dohm JM, Tanaka KL, Hare TM (2001) Geologic Map of the Thaumasia Region, Mars [Internet] Reston [cited 2019 Dec 8]. https://pubs.usgs.gov/imap/i2650/
140. Farrand WH, Glotch TD, Rice JW, Hurowitz JA, Swayze GA (2009) Discovery of Jarosite within the Mawrth Vallis region of Mars: Implications for the geologic history of the region. Icarus 204(2):478–488
141. McKeown NK, Bishop JL, Noe Dobrea EZ, Ehlmann BL, Parente M, Mustard JF et al (2009) Characterization of phyllosilicates observed in the central Mawrth Vallis region, Mars, their potential formational processes, and implications for past climate. J Geophys Res [Internet] [cited 2019 Dec 9] 114:E00D10. http://doi.wiley.com/https://doi.org/10.1029/2008JE003301
142. Michalski JR, Bibring J-P, Poulet F, Loizeau D, Mangold N, Dobrea EN et al (2010) The Mawrth Vallis region of Mars: a potential landing site for the Mars Science Laboratory (MSL) mission. Astrobiology [Internet] [cited 2019 Dec 9] 10(7):687–703. http://www.liebertpub.com/doi/https://doi.org/10.1089/ast.2010.0491
143. Loizeau D, Werner SC, Mangold N, Bibring JP, Vago JL (2012) Chronology of deposition and alteration in the Mawrth Vallis region, Mars. Planetary Space Science 72(1):31-43
144. Bishop JL, Loizeau D, McKeown NK, Saper L, Dyar MD, Des Marais DJ et al (2013) What the ancient phyllosilicates at Mawrth Vallis can tell us about possible habitability on early Mars. Planet Space Sci 15(86):130–149
145. Ehlmann BL, Mustard JF, Murchie SL (2010) Geologic setting of serpentine deposits on Mars. Geophys Res Lett 37:L06201
146. Madden MEE, Bodnar RJ, Rimstidt JD (2004) Jarosite as an indicator of water-limited chemical weathering on Mars. Nature 431:821–823
147. Wilhelms DE (1974) Comparison of Martian and lunar geologic provinces. J Geophys Res [Internet] [cited 2018 Nov 19] 79(26):3933–3941. http://doi.wiley.com/https://doi.org/10.1029/JB079i026p03933
148. Greeley R, Spudis PD (1978) Volcanism in the cratered terrain hemisphere of Mars. Geophys Res Lett [Internet] [cited 2018 Sep 23] 5(6):453–455. http://doi.wiley.com/https://doi.org/10.1029/GL005i006p00453
149. Andrews-Hanna JC, Zuber MT, Arvidson RE, Wiseman SM (2010) Early Mars hydrology: Meridiani playa deposits and the sedimentary record of Arabia Terra. J Geophys Res [Internet] [cited 2019 Dec 9] 115(E6):E06002. http://doi.wiley.com/https://doi.org/10.1029/2009JE003485
150. Newsom HE, Lanza NL, Ollila AM, Wiseman SM, Roush TL, Marzo GA et al (2010) Inverted channel deposits on the floor of Miyamoto crater. Mars. Icarus. 205(1):64–72
151. Schultz PH, Glicken H (1979) Impact crater and basin control of igneous processes on Mars. J Geophys Res [Internet] [cited 2019 Aug 5] 84(B14):8033. http://doi.wiley.com/https://doi.org/10.1029/JB084iB14p08033
152. Grant JA, Parker TJ (2002) Drainage evolution in the Margaritifer Sinus region. Mars. J Geophys Res E Planets. 107(9):4–1
153. Leverington DW (2006) Volcanic processes as alternative mechanisms of landform development at a candidate crater-lake site near Tyrrhena Patera, Mars. J Geophys Res E Planets. 111(E11):E11002
154. Dietterich HR, Cashman KV (2014) Channel networks within lava flows: Formation, evolution, and

implications for flow behavior. J Geophys Res Earth Surf 119:1704–1724
155. Bouley S, Craddock RA, Mangold N, Ansan V (2010) Characterization of fluvial activity in Parana Valles using different age-dating techniques. Icarus 207(2):686–698
156. Oze C, Sharma M (2007) Serpentinization and the inorganic synthesis of H2in planetary surfaces. Icarus 186:557–561
157. Stopar JD, Jeffrey Taylor G, Hamilton VE, Browning L (2006) Kinetic model of olivine dissolution and extent of aqueous alteration on mars. Geochim Cosmochim Acta 70:6136–6152
158. Morgan GA, Head JW, Marchant DR (2011) Preservation of Late Amazonian Mars ice and water-related deposits in a unique crater environment in Noachis Terra: Age relationships between lobate debris tongues and gullies. Icarus 211 (1):347–365
159. Condit C, Soderblom L (1978) Geologic map of the Mare Australe area of Mars [Internet]. Reston, VA. https://pubs.er.usgs.gov/publication/i1076
160. Rogers AD, Bandfield JL (2009) Mineralogical characterization of Mars Science Laboratory candidate landing sites from THEMIS and TES data. Icarus 203(2):437–453
161. Malin MC, Edgett KS (2003) Evidence for persistent flow and aqueous sedimentation on early Mars. Science (80-)12, 302(5652):1931–1934
162. Pain CF, Clarke JDA, Thomas M (2007) Inversion of relief on Mars. Icarus 190(2):478–491
163. Buhler PB, Fassett CI, Head JW, Lamb MP (2011) Evidence for paleolakes in Erythraea Fossa, Mars: implications for an ancient hydrological cycle. Icarus [Internet] [cited 2019 Jul 7] 213(1):104–15. https://www.sciencedirect.com/science/article/pii/S0019103511000881
164. Grant JA (2000) Valley formation in Margaritifer Sinus, Mars, by precipitation-recharged groundwater sapping. Geology 28(3):223–226
165. Hoke MRT, Hynek BM (2009) Roaming zones of precipitation on ancient Mars as recorded in valley networks. J Geophys Res Planets [Internet] [cited 2019 Dec 11] 114(E8):2008JE003247. https://onlinelibrary.wiley.com/doi/abs/https://doi.org/10.1029/2008JE003247
166. Michalski JR, Bleacher JE (2013) Supervolcanoes within an ancient volcanic province in Arabia Terra. Mars. Nature. 502(7469):47–52
167. Hiesinger H (2004) The Syrtis major volcanic province, Mars: synthesis from Mars global surveyor data. J Geophys Res 109:E01004
168. Christensen PR, McSween HY, Bandfield JL, Ruff SW, Rogers AD, Hamilton VE et al (2005) Evidence for magmatic evolution and diversity on Mars from infrared observations. Nature 436 (7050):504–509
169. Wray JJ, Hansen ST, Dufek J, Swayze GA, Murchie SL, Seelos FP et al (2013) Prolonged magmatic activity on Mars inferred from the detection of felsic rocks. Nat Geosci 6(12):1013–1017
170. Hauber E, Bleacher J, Gwinner K, Williams D, Greeley R (2009) The topography and morphology of low shields and associated landforms of plains volcanism in the Tharsis region of Mars. J Volcanol Geotherm Res 185(1-2):69-95
171. Williams J-P, Dohm JM, Soare RJ, Flahaut J, Lopes RMC, Pathare AV et al (2017) Long-lived volcanism within Argyre basin, Mars. Icarus [Internet] [cited 2019 Aug 13] 293:8–26. https://www.sciencedirect.com/science/article/pii/S0019103516307254
172. Robbins SJ, Hynek BM, Lillis RJ, Bottke WF (2013) Large impact crater histories of Mars: the effect of different model crater age techniques. Icarus [Internet] [cited 2018 Oct 18] 225(1):173–184. https://www.sciencedirect.com/science/article/pii/S0019103513001358
173. Hiesinger H, Head III JW (2002) Topography and morphology of the Argyre Basin, Mars: implications for its geologic and hydrologic history. Planet Space Sci [Internet] [cited 2019 Jul 7] 50(10–11):939–981. https://www.sciencedirect.com/science/article/pii/S0032063302000545
174. Bernhardt H, Hiesinger H, Reiss D, Ivanov M, Erkeling G (2013) Putative eskers and new insights into glacio-fluvial depositional settings in southern Argyre Planitia, Mars. Planet Space Sci [Internet] [cited 2019 Jul 7] 85:261–278. https://www.sciencedirect.com/science/article/pii/S0032063313001633
175. Carr M, Evans N (1980) Images of Mars: the Viking extended mission [Internet] [cited 2019 Dec 13]. https://ui.adsabs.harvard.edu/abs/1980NASSP.444.....C/abstract
176. Kargel JS, Strom RG (1992) Ancient glaciation on Mars. Geology 20(1):3–7
177. Banks ME, Lang NP, Kargel JS, McEwen AS, Baker VR, Grant JA et al (2009) An analysis of sinuous ridges in the southern Argyre Planitia, Mars using HiRISE and CTX images and MOLA data. J Geophys Res [Internet] [cited 2019 Jul 7] 114 (E9):E09003. http://doi.wiley.com/https://doi.org/10.1029/2008JE003244
178. Poulet F, Gomez C, Bibring J-P, Langevin Y, Gondet B, Pinet P et al (2007) Martian surface mineralogy from Observatoire pour la Minéralogie, l'Eau, les Glaces et l'Activité on board the Mars Express spacecraft (OMEGA/MEx): Global mineral maps. J Geophys Res Planets [Internet] [cited 2020 Jan 11] 112(E8). http://doi.wiley.com/https://doi.org/10.1029/2006JE002840
179. Coffin MF, Eldholm O (1992) Volcanism and continental break-up: a global compilation of large igneous provinces. Geol Soc Spec Publ 68:17–30
180. Kerber L, Head JW (2010) The age of the Medusae Fossae Formation: evidence of Hesperian emplacement from crater morphology, stratigraphy, and ancient lava contacts. Icarus 206(2):669–684

181. Harrison SK, Balme MR, Hagermann A, Murray JB, Muller JP (2010) Mapping Medusae Fossae formation materials in the southern highlands of Mars. Icarus 209(2):405–415
182. Scott DH, Tanaka KL (1982) Ignimbrites of Amazonis Planitia region of Mars. J Geophys Res Solid Earth [Internet] [cited 2019 Aug 8] 87(B2):1179–1190. http://doi.wiley.com/https://doi.org/10.1029/JB087iB02p01179
183. Scott DH (1982) Volcanoes and volcanic provinces: Martian western hemisphere. J Geophys Res [Internet] [cited 2019 Jul 27] 87(B12):9839. http://doi.wiley.com/https://doi.org/10.1029/JB087iB12p09839
184. Plescia JB (1981) The Tempe volcanic province of Mars and comparisons with the Snake River Plains of Idaho. Icarus [Internet] [cited 2018 Nov 6] 45 (3):586–601. https://www.sciencedirect.com/science/article/pii/0019103581900245
185. Hiesinger H, Head JW (2004) The Syrtis major volcanic province, Mars: synthesis from mars global surveyor data. J Geophys Res E Planets. 109(1):E01004
186. Schaber GG, Horstman KC, Dial AL (1978) Lava flow materials in the Tharsis region of Mars. In: Lunar and planetary science conference, 9th [Internet]. Houston, Tex., March 13–17, 1978: New York, Pergamon Press, Inc [cited 2020 Jan 16], pp 3433–3458. https://ui.adsabs.harvard.edu/abs/1978LPSC....9.3433S/abstract
187. Hall JL, Solomon SC, Head JW (1986) Elysium Region, Mars: Tests of lithospheric loading models for the formation of tectonic features. J Geophys Res 91(B11):11377
188. Kerber L, Forget F, Madeleine J-B, Wordsworth R, Head JW, Wilson L (2013) The effect of atmospheric pressure on the dispersal of Pyroclasts from Martian volcanoes. Icarus [Internet] [cited 2018 Sep 22] 223(1):149–156. https://www.sciencedirect.com/science/article/pii/S0019103512004915
189. Tanaka KL, Fortezzo CM, Hayward RK, Rodriguez JAP, Skinner JA (2011) History of plains resurfacing in the Scandia region of Mars. Planet Space Sci 59(11–12):1128–1142
190. Acuña MH (2003) The magnetic field of Mars. Lead Edge (Tulsa, OK). 22(8):769–771
191. Langlais B, Thébault E, Houliez A, Purucker ME, Lillis RJ (2019) A new model of the crustal magnetic field of mars using MGS and MAVEN. J Geophys Res Planets [Internet] [cited 2019 Aug 19] 124(6):2018JE005854. https://onlinelibrary.wiley.com/doi/abs/https://doi.org/10.1029/2018JE005854
192. Langlais B, Lesur V, Purucker ME, Connerney JEP, Mandea M (2010) Crustal magnetic fields of terrestrial planets. Space Sci Rev 152(1–4):223–249
193. Connerney JEP, Acuña MH, Ness NF, Kletetschka G, Mitchell DL, Lin RP et al (2005) Tectonic implications of Mars crustal magnetism. Proc Natl Acad Sci 102(42):14970–14975
194. Hood LL, Harrison KP, Langlais B, Lillis RJ, Poulet F, Williams DA (2010) Magnetic anomalies near Apollinaris Patera and the Medusae Fossae Formation in Lucus Planum, Mars. Icarus 208 (1):118-131
195. Weiss BP, Vali H, Baudenbacher FJ, Kirschvink JL, Stewart ST, Shuster DL (2002) Records of an ancient Martian magnetic field in ALH84001. Earth Planet Sci Lett [Internet] [cited 2019 Aug 9], 201 (3–4):449–463. https://www.sciencedirect.com/science/article/pii/S0012821X02007288
196. Riedler W, Möhlmann D, Oraevsky VN, Schwingenschuh K, Yeroshenko Y, Rustenbach J et al (1989) Magnetic fields near Mars: first results. Nature 341(6243):604–607
197. Connerney JEP, Acuña MH, Wasilewski PJ, Kletetschka G, Ness NF, Rème H et al (2001) The global magnetic field of Mars and implications for crustal evolution. Geophys Res Lett [Internet] [cited 2020 Mar 3] 28(21):4015–4018. http://doi.wiley.com/https://doi.org/10.1029/2001GL013619
198. Schubert G, Russell CT, Moore WB (2000) Timing of the Martian dynamo. Nature 408(6813):666–667
199. Acuña MH, Connerney JEP, Ness NF, Lin RP, Mitchell D, Carlson CW et al (1999) Global distribution of crustal magnetization discovered by the Mars Global Surveyor MAG/ER experiment. Science (80-)284(5415):790–793.
200. Williams JP, Nimmo F (2004) Thermal evolution of the Martian core: Implications for an early dynamo. Geology 32(2):97–100
201. Milbury C, Schubert G, Raymond CA, Smrekar SE, Langlais B (2012) The history of Mars' dynamo as revealed by modeling magnetic anomalies near Tyrrhenus Mons and Syrtis Major. J Geophys Res Planets [Internet] [cited 2019 Aug 24] 117(E10). http://doi.wiley.com/https://doi.org/10.1029/2012JE004099
202. Stevenson DJ (2001) Mars' core and magnetism. Nature [Internet] [cited 2019 Aug 9] 412 (6843):214–219. http://www.nature.com/articles/35084155
203. Monteux J, Amit H, Choblet G, Langlais B, Tobie G (2015) Giant impacts, heterogeneous mantle heating and a past hemispheric dynamo on Mars. Phys Earth Planet Inter [Internet] [cited 2019 Aug 14] 240:114–24. https://www.sciencedirect.com/science/article/pii/S0031920114002532
204. Roberts JH, Lillis RJ, Manga M (2009) Giant impacts on early Mars and the cessation of the Martian dynamo. J Geophys Res [Internet] [cited 2020 Jan 18] 114(E4):E04009. http://doi.wiley.com/https://doi.org/10.1029/2008JE003287
205. Fanale FP, Jakosky BM (1982) Regolith-atmosphere exchange of water and carbon dioxide on Mars: effects on atmospheric history and climate change. Planet Space Sci 30(8):819–831
206. Fanale FP, Salvail JR, Zent AP, Postawko SE (1986) Global distribution and migration of subsurface ice on mars. Icarus 67(1):1–18

207. Gulick VC, Baker VR (1989) Fluvial valleys and martian palaeoclimates. Nature [Internet] [cited 2019 May 26] 341(6242):514–516. http://www.nature.com/articles/341514a0
208. Wordsworth RD, Kerber L, Pierrehumbert RT, Forget F, Head JW (2015) Comparison of "warm and wet" and "cold and icy" scenarios for early Mars in a 3-D climate model. J Geophys Res Planets [Internet] [cited 2019 Apr 5] 120(6):1201–1219. http://doi.wiley.com/https://doi.org/10.1002/2015JE004787
209. Pollack JB, Kasting JF, Richardson SM, Poliakoff K (1987) The case for a wet, warm climate on early Mars. Icarus 71(2):203–224
210. Craddock RA, Howard AD (2002) The case for rainfall on a warm, wet early Mars. J Geophys Res E Planets [Internet] [cited 2020 Mar 4] 107 (11):21-1–21-36. http://doi.wiley.com/https://doi.org/10.1029/2001JE001505
211. Squyres SW, Kasting JF (1994) Early Mars: how warm and how wet? Science 265:744–749. American Association for the Advancement of Science
212. Haberle RM (1998) Early Mars climate models. J Geophys Res Planets [Internet] [cited 2020 Feb 10] 103(E12):28467–28479. http://doi.wiley.com/https://doi.org/10.1029/98JE01396
213. Gaidos E, Marion G (2003) Geological and geochemical legacy of a cold early Mars. J Geophys Res E Planets. 108(6):9–1
214. Fairén AG (2010) A cold and wet Mars. Icarus 208 (1):165–175
215. Wilson L, Head JW (1994) Mars: Review and analysis of volcanic eruption theory and relationships to observed landforms. Rev Geophys [Internet] [cited 2018 Sep 22] 32(3):221. http://doi.wiley.com/https://doi.org/10.1029/94RG01113
216. Mouginis-Mark PJ (2002) Prodigious ash deposits near the summit of Arsia Mons volcano, Mars. Geophys Res Lett [Internet] [cited 2019 Apr 5] 29 (16):15-1–15-4. http://doi.wiley.com/https://doi.org/10.1029/2002GL015296
217. Mouginis-Mark PJ, Wilson L, Head JW (2018) Explosive volcanism on Hecates Tholus, Mars: investigation of eruption conditions. J Geophys Res [Internet] [cited 2018 Nov 6] 87(B12):9890–904. http://doi.wiley.com/https://doi.org/10.1029/JB087iB12p09890
218. Brož P, Hauber E (2012) A unique volcanic field in Tharsis, Mars: Pyroclastic cones as evidence for explosive eruptions. Icarus [Internet] [cited 2018 Nov 6] 218(1):88–99. https://www.sciencedirect.com/science/article/pii/S001910351100457X
219. Brož P, Hauber E (2013) Hydrovolcanic tuff rings and cones as indicators for phreatomagmatic explosive eruptions on Mars. J Geophys Res Planets [cited 2019 Jul 22], 118(8):1656–75. http://doi.wiley.com/https://doi.org/10.1002/jgre.20120
220. Reimers CE, Komar PD (1979) Evidence for explosive volcanic density currents on certain Martian volcanoes. Icarus [Internet] [cited 2018 Sep 22] 39(1):88–110. https://www.sciencedirect.com/science/article/pii/0019103579901039
221. Lamb HH (1970) Volcanic, dust in, the atmosphere, with a chronology and assessment of its meteorological significance. Phil Trans Roy Soc London Ser A Math Phys Sci. 266(1178):425–533
222. Pueschel RF, Garcia CJ, Hansen RT, Pueschel RF, Garcia CJ, Hansen RT (1974) Solar radiation: effects of atmospheric water vapor and volcanic aerosols. Journal of Applied Meteorology and Climatology 13(3):397-401
223. Santer BD, Bonfils C, Painter JF, Zelinka MD, Mears C, Solomon S et al (2014) Volcanic contribution to decadal changes in tropospheric temperature. Nat Geosci 7(3):185–189
224. Wilson L, Mouginis-Mark PJ (1987) Volcanic input to the atmosphere from Alba Patera on Mars. Nature 330(6146):354–357
225. Mustard JF, Cooper CD, Rifkin MK (2001) Evidence for recent climate change on Mars from the identification of youthful near-surface ground ice. Nature 412(6845):411–414
226. Nerozzi S, Holt JW (2020) Buried ice and sand caps at the North pole of Mars: revealing a record of climate change in the Cavi unit with SHARAD. Geophys Res Lett [Internet] [cited 2020 Mar 3] 46 (13):7278–7286. https://doi.org/10.1029/2019GL082114
227. Laskar J, Correia ACM, Gastineau M, Joutel F, Levrard B, Robutel P (2004) Long term evolution and chaotic diffusion of the insolation quantities of Mars. Icarus 170(2):343-364
228. Kite ES (2019) Geologic constraints on early mars climate. Space Sci Rev 215:10

Petrologic Evolution of Martian Volcanism and Clues from Meteorites

Caitlin Ahrens and Giovanni Leone

Abstract

Our knowledge of the formation and evolution of the Martian surface, and thus the fate of its volcanic regions, are revealed in the mineralogy of the crust (as observed by orbiters and landers) and the chemistry hidden within Martian meteorites. The composition of the Martian crust has been estimated from SNC (shergottite—nakhlite—chassignite) meteorites, which are chemically similar to terrestrial basalts. However, these meteorites reveal a complexity in other mineralogies, such as the silicic implications for volcanism across Mars. Although the exact locations of such SNCs are difficult to discern, the variety of geochemical data from these meteorites gives us a clue as to the mineralogical regions, and thus probable varieties of volcanism on Mars. In this chapter, we will explore the Martian meteorites and their clues to the Martian volcanic regions and crustal components of the North–South dichotomy.

C. Ahrens (✉)
Universities Space Research Association, NASA Goddard Space Flight Center, Greenbelt, MD, USA
e-mail: caitlin.ahrens@nasa.gov

G. Leone
Instituto de Investigación en Astronomía Y Ciencias Planetarias, Universidad de Atacama, Copiapó, Chile

G. Leone (ed.), *Mars: A Volcanic World*,
https://doi.org/10.1007/978-3-030-84103-4_3

1 Introduction

The formation and evolution of the Martian surface, and ultimately the fate of its various volcanic regions, are revealed in the chemistry of Martian meteorites, and the mineralogy of the crust. The interpretation of the chemistry from remote sensing instrumentation from orbiters and landers/rovers has been strongly influenced by our knowledge of geochemical processes on Earth, the Moon, and other asteroidal, rocky bodies [1]. The composition, and thus evolution, of the Martian crust has been resultant from SNC (shergottite—nakhlite—chassignite) meteorites, which are the igneous counterpart of rocks of Martian origin [2–4]. These meteorites are chemically similar to terrestrial basalts and cumulates, with the exception of higher concentrations of iron and other volatile elements (i.e., phosphorous, chlorine, chalcophile) [5]. According to Tuff et al. [2], most Martian meteorites have relatively young crystallization ages ranging from 1.4 billion years to 180 million years ago [6]. These are considered to be derived from young, usually lightly-cratered volcanic regions, such as the Tharsis volcanic region [6, 7]. Though mainly basaltic in composition, surface rocks are richer in nickel and sulfur with lower manganese/iron ratios than Martian meteorites [2], which can be explained by the differences in oxygen fugacity during melting of the sulfur-rich mantle [2, 8]. From this difference, Martian meteorites are tied to surface rocks through an early (3.7 billion years ago) oxidation of the uppermost mantle, producing the more recent volcanic rocks [2]. The variety of compositions from basaltic, olivine volcanic rocks through pyroxene-cumulate nakhlites and peridotite chassignites, enable the hypothetical reconstructions of the silicic evolution of Mars similar to techniques used on Earth [2, 5, 9].

In this chapter, we will explore the basic Martian composition of the crust and mantle in regards to volcanic regions, the crustal components of the North–South dichotomy, and more detail of the SNC meteorites in relation to the evolution of the mantle for volcanism.

2 Temporal Evolution of Volcanism

Mars is not uniformly distributed as observed by a diversity of volcanic landforms such as shield volcanoes, tholi, paterae, and small domes, as well as vast volcanic plains with various textures and lava flow patterns, as noted throughout this book. Such a diversity of volcanism comes from different eruption styles and changes of eruption through time (see also chapter "Volcanic Eruptions on Mars, Lava Flow Morphology, and Thermodynamics"), though it is the localization of the various compositions and subsurface material (i.e., more silicic components could influence the overall ascension of mantle material to form different volcanic constructs and eruption style). The two topographically-dominating volcanic provinces are the Tharsis and Elysium regions, whose morphologies are analogous to basaltic landforms found on Earth [10]. Both are situated close to the dichotomy boundary between the cratered (older) highlands and the northern lowlands (see Sect. 2.2 of this chapter). The Tharsis region consisting of Olympus Mons, Ascraeus Mons, Pavonis Mons, and Arsia Mons share many characteristics with Hawaiian basaltic shield volcanoes [11], as with their multiple lobate flows of lava [12, 13] and complex nested summit calderas (chapter "Volcanic Eruptions on Mars, Lava Flow Morphology, and Thermodynamics"). Several authors [10, 14, 15] summarized the chronology and classification of Martian volcanic activity, grouping volcanic landforms into: (i) shield volcanoes; (ii) domes; (iii) highland paterae; (iv) volcano-tectonic features.

The temporal connection of extensive fluvial activity (e.g., valley networks) and volcanic episodes (highland paterae) is hypothesized from phreatomagmatic interactions (e.g., Tyrrhena Patera or western flanks of Elysium Mons) [10, 16]. Werner and Greeley and Spudis [10, 17] make note that the Hellas basin formation and the accumulation of highland paterae (at least of later-stage patera activities) were not triggered by the impact event itself. Back at the Tharsis region, there are notable differences in the structural content between Olympus Mons and Alba Patera, suggesting an environmental change [10]. Poulet et al. [18] deduced the geologic record of early Mars from a variety of features by: (i) widespread silicate-dominant formation and related alterations (e.g., phyllosilicates); (ii) existence of an active magnetic dynamo; (iii) erosion of extensive valley networks; (iv) large-basin scale impact cratering; and (v) construction of the Tharsis volcanic province. While an early Mars climate would have influenced these factors and subsequent features, volcanism is known to be a major factor in the long-term climate evolution of Mars, liberating volatile species from the interior to the atmosphere [19–21].

The massiveness of the Tharsis region has been implicated in a transition from a silicate-forming era (phyllosian) to a sulfate-forming era (theiikian) [18, 22, 23]. Poulet et al. [18] stated that the timing of Tharsis volcanism is critical to understanding whether this change in mineralogical environments is plausible. While the bulk of the Tharsis region was probably emplaced in the Noachian [20], the load and emplacement on the regional lithosphere may have influenced the orientation of Late Noachian/Early Hesperian valley networks. This is further implied by evidence from crater counts and tectonic mapping from remote sensing techniques [18, 24, 25]. However, the notion of the Tharsis region being of Noachian-age has been disputed by [21] by the lack of craters on much of Tharsis, implying that this region to be

Hesperian or Amazonian and requiring significant post-Noachian resurfacing. However, Poulet et al. [18] express a plausibility for these two contrasting views to be viable. From this, the majority of the Tharsis crust is constructed in the Noachian [26], but where extensive volcanic resurfacing persists through the Hesperian/Amazonian eras [27]. Parts of Tharsis from magnetization observations may also pre-date Hellas [18, 28]. The interpretation of the Tharsis construction near the end of the Noachian led to the silicate to sulfate transition may not be consistent with the bulk "old-age" of Tharsis [22, 23]. In the Hesperian, volcanic plain emplacement (particularly in the northern lowlands) resurfaced ~30% of Mars [29]. From Viking mapping, more than half of the volcanic resurfacing on Mars is early Hesperian or younger [30, 31].

2.1 Primordial Crust

Such volcanic transitions imply the changing of the Martian crust altogether, suggesting a need to differentiate the modern crust and primordial crust. The primordial crust would then be the crystallization of a magma ocean within the first 100 My after the formation of the Solar System [32]. In terms of volcanism, the first crust formed by early phases of intense partial melting of the mantle [33, 34], probably having a thickness of ~20–30 km [35]. However, the timing of the primordial crust is still under observation.

Isotopic anomalies preserved in SNC meteorites provide vital constraints about the timing of this pre-Noachian global scale differentiation within the deep Martian interior [36]. For example, the Hf-W-Th indicates that core formation occurred in the first 10–15 Myr [37, 38], though impacting to create large basins or even the dichotomy may accelerate this process (see chapters "History of Scientific Studies and Current Views of Mars and The Formation of Mars and the Origin of Its Volcanic Provinces"). U-Th-Pb and Sm–Nd measurements show that the initial differentiation between the Martian mantle and primordial crust occur in the first 30–50 My [32, 39]. This supports the hypothesis where fractionation producing the core and the crust occurred sometime within the first 30 Myr [38]. The Sm–Nd petrology of shergottites provide a clear separation of at least two distinct Martian mantle reservoirs (depleted versus enriched in Rare Earth Elements—REE) in the first 10–100 My of Martian geologic history [32, 40]. An early magmatic build-up from the Martian highland crust required a rapid release of volatiles from the interior, forming an early atmosphere of Mars [41, 42].

The formation processes of the primordial crust remain debatable. The first hypothesis is the crystallization of a magma ocean (like that of the Moon), which would have been formed a resultant anorthositic crust [43]. However, this case seems unlikely given the larger size of Mars and its volatile content and anorthositic detections by CRISM (Compact Reconnaissance Imaging Spectrometer for Mars) would rely on a different mineralogical process [44]. The second hypothesis is the crystallization of a magma ocean as that for an early Earth [5], where plagioclase crystallization is delayed and solidification proceeds with denser Fe-rich cumulate overlying Mg-rich material, which creates a gravitational instability and mantle overturn [45, 46]. Thus, a basaltic crust would form by partial melting from adiabatic decompression of the rising Mg-rich cumulate. The third hypothesis is a partial melting of a fertile primitive mantle composition. This comes from the REE analysis of the NWA 7533 meteorite breccia, where low degree partial melting (~4%) of a primitive Martian mantle is sufficient to explain the formation of an enriched primordial crust and a magma ocean would no longer be necessary [47]. However, this would ultimately form a uniformly thick global layer [48].

The reshaping and ultimate burying (or eradication) of the primordial crust is also controversial. From [40], we know that there is no crustal recycling within the mantle as observed from the Sm–Nd signatures of shergottites. There are three potential processes that may have contributed to the processes of such reshaping of the primordial crust: (i) the formation of the North–South dichotomy dating back from the early Noachian, may have been synchronic with the primordial crust [42]. However, the nature of the dichotomy and its origin is still relatively unknown; (ii) the early Noachian period produced an intense volcanic activity time, leading to abundant crustal formation and partly burying the ancient primordial crust [33, 49]; and (iii) differentiation-fractional crystallization occurred within a thick impact melt formed by a gigantic impact ~4.43 Ga ago, leading to alkali-rich clasts as observed in the meteorite breccia NWA 7533 [47].

2.2 Dichotomy

The North–South dichotomy is one of the most distinct textural differences observed on the surface of Mars. The southern hemisphere appears older due to the highly-cratered surface (~40%) and higher in elevation (by ~6 km) than the northern hemisphere (Fig. 1). This dichotomy dates back between 4.5 and 4 Gyr [50], where age is determined from density crater measurements [51–53]. The bulk crust is basaltic, implying that a thick crust in the southern hemisphere should compensate for its relatively higher elevation [35]. However, there is a discrepancy regarding the density of such a thick crust compared to the geoid-topography ratio from [3-old] and may lead to lower crustal flow [54, 55]. A lighter crustal component comprised of feldspathic/anorthositic materials in relation to a thinner crust might be buried under subsequent basaltic volcanic products in the southern highlands [56], matching current geophysical data measured by orbital instruments. Crustal density values assumed from gravity and topography data are significantly lower (2,700–3,100 kg/m^3) than those of a purely homogeneous basaltic crust (3,100–3,700 k/m^3) inferred from meteorites and mineralogical observations from orbiters [56].

In geologic detail, the majority of the southern highlands is dominated by brecciated terrains that are mid-Noachian in age with some localized areas of early Noachian age. These zones of early-Noachian surfaces form isolated highland massif, whereas the mid-Noachian regions are included in the Circum-Hellas Volcanic Province (CHVP), and the late Noachian terrains are found in Syrtis major [61]. Later Hesperian terrains (3.7–3.1 Ga) are found in the large volcanic provinces of Hesperia Planum.

The igneous rocks, as revealed by orbital spectroscopy techniques, are dominantly basaltic with a relatively high LCP/HCP (low Ca-pyroxene—high Ca-pyroxene) ratio and olivine phenocrysts [62, 63]. The scattered Noachian terrains are rather phyllosilicate-rich [64, 65]. Any remnants of an ancient crust should be located in the deepest places of Mars, or likely exposed at

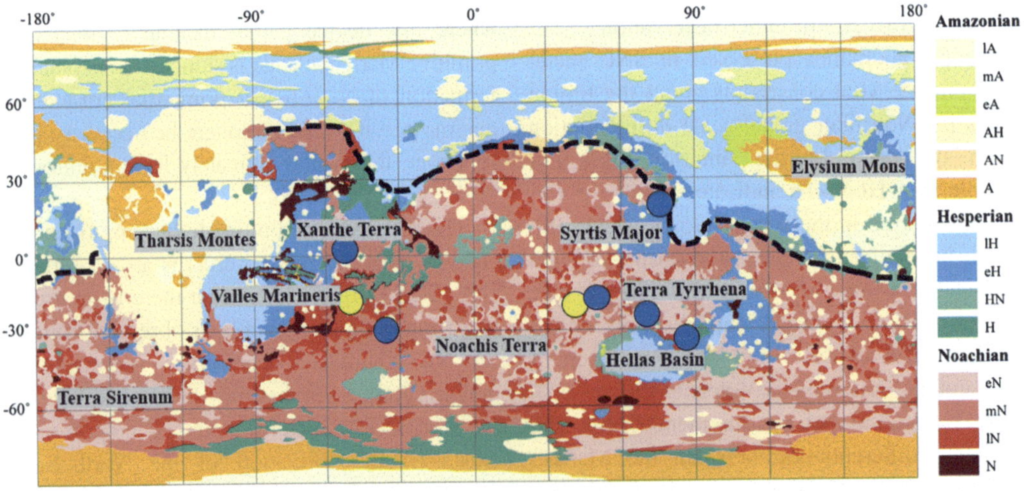

Fig. 1 Latest chronostratigraphic map of Mars with noted surface ages. Names throughout this chapter are noted. Black dashed line indicates the major dichotomy. Blue circles are feldspar-rich zones [44, 57, 58]. Yellow circles indicate plagioclase-rich zones [59, 60]. Modified and adapted from [61]

central uplifts of large impact craters and deep canyons, such as Valles Marineris [66, 67]. Resurfacing processes from constant volcanic activities likely complicate the identification of most ancient crust. Early Noachian terrains do show evidence of volcanic lava flows and pyroclastic deposits [68–70]; see Chaps. History of Scientific Studies and Current Views of Mars, Volcanic Eruptions on Mars, Lava Flow Morphology, and Thermodynamics and The Origin and Evolution of Volcanism at Martian Highland Paterae: A Review of the Current State of Knowledge].

In the Aeolis region along the dichotomy boundary at Gusev crater, where the Spirit rover landed in 2004, is the highly cratered Noachian southern highland and south of the Medusae fossae formation. The crater floor is covered by a Hesperian basaltic flow [71] containing olivine phenocrysts comparable to olivine-rich shergottites [72, 73]. This volcanic flow area encloses elevated terrains of the Columbia hills as the first alkali-basalt lithology identified on Mars [74]. However, the stratigraphic sequence is still unclear, so this could either represent a central uplift of older Noachian material or post-date the crater by infilling of Hesperian olivine-basalt lava flows [75]. Another rover at the dichotomy boundary, Curiosity at Gale crater in 2012, was the first mission to probe such ancient terrains (older than Late Noachian) [76]. From several analyses with the rover's suite of spectroscopic instruments, mafic and light-toned felsic igneous rocks have been analyzed [57, 77–80]. This included magmatic suites of: basalts, gabbros, norites, trachy-andesites, trachytes (alkaline-trend); and diorites and quartzo-diorites (sub-alkaline trends) [57, 77]. Such a diversity in the chemistry can lead to a variety of speculations in the magma processes, ascension dynamics, and overall volcanic structure (discussed in Chaps. History of Scientific Studies and Current Views of Mars and Volcanic Eruptions on Mars, Lava Flow Morphology, and Thermodynamics).

These observations raise several key questions, such as: *Are these specific areas corresponding to localized plutons or an extensive ancient, differentiated crust? Are these felsic outcrops part of an excavated layer by huge impacts and younger basalt lava flows? Might such an evolved layer be formed by deep melting of a thickened basaltic crust preceded by crystallization of a magma ocean, or rather differentiated magma derived from low-degree partial melting of a primitive mantle?*

3 Mineralogy in Volcanic Regions

The composition of magmas, produced by partial melting of planetary interiors, has the probable preservation of pressure–temperature conditions of melting throughout geologic time. For Mars, surface chemistry data must be obtained, such as thermal spectrometers or the gamma-ray spectrometer onboard Mars Odyssey [81, 82], although this method is limited to more extensive regions that show clear morphological evidence of volcanism. Orbital spectroscopy has provided a complementary view on the mineralogy of magmatic rocks, showing that Martian igneous rocks have a mineralogy similar to that of Earth: dominated by plagioclase [83], pyroxenes [84–86], and olivine (sometimes lack thereof) [87]. The OMEGA (Observatoire pour la Minéralogie, l'Eau, les Glaces, et l'Activité) and CRISM imaging spectrometers provide evidence for a higher proportion of LCP in the Noachian highlands relative to the Hesperian-dated Syrtis Major [62, 84].

Baratoux et al. [33] stated there is a clear transition in the nature of pyroxenes observed in magmatic rocks and a possible increase in olivine at the surface of Mars over geologic time. This transition is most prominent near the Noachian/Hesperian boundary [84, 85]. LCP-rich outcrops have been suggested to be associated with higher degrees of partial melting, or trace the crystallization of the magma, followed by HCP-rich volcanism in the Hesperian [62]. Olivine-rich young volcanism and the evolution of the LCP + HCP ratios observed at the Hesperian/Noachian transition would be a natural sequence (and consequence) of primary melt compositions over geologic time, assuming that primitive magmas are a dominant component of the Martian crust [33, 72, 82]. Modeling of crystallization products of liquids produced by partial melting of a possible Martian mantle from conditions of the earliest Noachian era to the most recent Amazonian times have been attempted [33, 88, 89]. Such a transition is interpreted to be the consequence of the thermal evolution of the mantle. In that case, according to Baratoux et al. [33], Noachian rocks exposed at the surface are indeed linked to volcanism rather than being associated with mantle overturn following the crystallization of a magma ocean.

On Mars, the primary silicate paragenesis is of basaltic to andesitic-basaltic composition, as observed by several orbital observations such as Mars Global Surveyor (MGS), Mars Odyssey, and Mars Express [59, 90–93], in addition to in-situ measurements by the Viking landers and rovers [94] as well as SNC meteorite analyses [95]. Main silicate types observed on the surface are olivine pyroxene and plagioclase, although silica-rich feldspathic minerals [96] indicate magmatic differentiation. Mars Global Surveyor (MGS) Thermal Emission Spectrometer (TES) has also detected plagioclase on Mars [92, 97]. Although silicate mineralogical data alone cannot support any one specific Martian magmatism model over another, data does support the basic igneous reservoirs proposed for Mars and used to constrain specific petrogenetic models [98].

Iron phases are also presumed on the surface of Mars [99]. The Martian regolith contains ~20 wt.% Fe^{3+} mostly under the form of iron oxides [100, 101]. Such iron phases have been confirmed by in-situ measurements from Viking 1 and 2 at Chryse Planitia and Utopia Planitia), Mars Pathfinder at Ares Vallis, and Spirit (at Gusev Crater) and Opportunity (Meridiani Planum) [92]. Columbia Hills, while having an age uncertainty compared to the 3.6 Gy-Hesperian olivine-basalt mineralogy of the Gusev crater floor [71, 75, 102], it has been noted that the tephrites are comparable in alkali constituents to some of the effusive alkaline lavas (e.g., basalt, trachyandesite, trachyte—Fig. 2) described in Gale crater [72, 75]. However, these alkali-rich rocks have not been observed (so far) in younger Amazonian shergottites, which are essentially olivine-basalts and detected in lava flows throughout the southern hemisphere [103].

Low-albedo (dark) regions on Mars provide the best opportunity for determining the composition of newer, pristine crustal lithologies [63]. Many of these regions are associated with volcanic provinces (e.g., Syrtis Major volcanic plateau) and analyses of their spectral properties from orbiters show a mafic/ultramafic composition [105–108]. Such compositional analyses can reveal the eruption and emplacement of magma, even in understanding the crystallization processes that may occur [63, 109]. Besides being ultimately mafic and basaltic in nature, there are also altered materials on the surface of Mars.

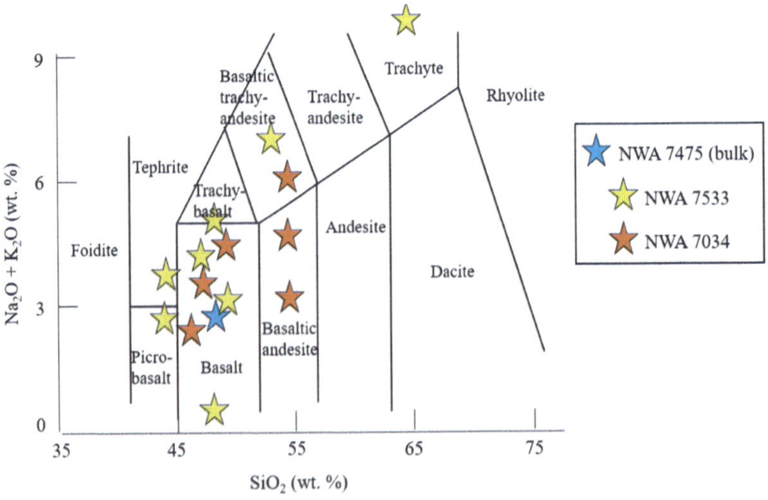

Fig. 2 Total alkali content versus silica content adapted from [95], including a geochemical classification of Martian meteorites analyzed. Examples in this figure include geochemical data [47, 104] of NWA 7475, 7533, 7034

Interestingly, Mustard et al. [63] observed that relatively unaltered surface materials were found in volcanic terrains of Syrtis Major, Valles Marineris, Ophir Planum, and Sinus Meridiani. As stated early regarding the observation of two-pyroxene mineralogy consistent with SNC meteorites indicates that (i) the SNC mineralogies are representative of large volcanic regions of Mars as old as 2–3 Ga; (ii) the mantle was depleted in aluminum; and (iii) there has been little evolution in the composition of mantle sources [63].

LCP-rich outcrops have been identified in the Noachian walls of Valles Marieneris [110], while higher ratios of HCP/LCP have been found in the Hesperian volcanic deposits of the Circum-Hellas Volcanic province [111] and Chryse and Acidalia Planitia of the northern plains [112]. Although dust is an extensive issue for properly characterizing the younger volcanic regions, there have been observations of how LCP concentrations at certain locations, such as the Amazonis Planitia [113], Noctis Labyrinthus, Echus Chasma [114], and volcanic domes in the northern plain [115], all strongly dominated by HCP and plagioclase. Amazonis Planitia is noted to also be rich in olivine, as typified by the effusive volcanism in the area [113].

At Terra Tyrrhena, dated from early to mid-Noachian (see chapter "The Origin and Evolution of Volcanism at Martian Highland Paterae: A Review of the Current State of Knowledge"), CTX and HiRISE images observed bright rocks, where the CRISM instrument detected >95% of feldspars in a nearby crater, which have been interpreted as anorthosites [44]. However, Rogers and Nekvasil [116] stated that sub-millimeter feldspars detected in the near-infrared spectra (such as CRISM) contain a mixture of at least <50 vol% mafic minerals. The co-existence of Al-rich clays (i.e., kaolin) are likely formed by the alteration of highly felsic parent rocks [44], which have been detected at Xanthe Terra. Additionally, northeast Noachis Terra has been observed to have felsic polygonal light-toned outcrops. South of Xanthe Terra, light-toned massive bedrocks have been observed by HiRISE [60] and are dominated by LCP by CRISM data. Flahaut et al. [60] suggested that these LCP-rich rocks are potentially a Noachian pristine crust exposure. Fractional crystallization at low pressure of a basaltic melt can explain these felsic detections. CRISM data revealed that these terrains with feldspar-rich materials along with Al-silicates likely formed by the alteration of the parent felsic rocks [57]. Either way, the occurrence of feldspar-rich rocks points out early evolved magmatism on Mars. The occurrence of such felsic outcrops within large mafic terrains may be implied as buried ancient evolved crust or localized plutons.

Terra Sirenum and Terra Cimmeria, dating from early to late Noachian, display the strongest crustal remnant magnetism ranging from −100 to +100 nT/deg measured by Mars Global Surveyor [117]. A geomorphological study by Xiao et al. [49] has revealed these could be the early Noachian volcanic edifice remains are likely shield-like volcanoes with 50–100 km diameter and 2–3 km high. Terra Sirenum was also observed to have Fe/Mg phyllosilicates, hypothesized to be the alteration of an ancient Noachian crust rich in volcanic and impact-generated mafic glasses [118, 119], although CRISM detections of Al-silicates likely reflect the alteration of felsic parent rocks [120].

Few mineralogical data exist for plain-style volcanism with which to constrain their composition independently from the rheological evidence of broad lava flows [114, 121] suggesting a high fluidity (see chapter "Volcanic Eruptions on Mars, Lava Flow Morphology, and Thermodynamics"). East of the large volcanoes of Tharsis is Echus Chasma, the source area of the Kasei Valles outflow channels [114]. The Echus-Kasei valley system is partially filled by lava flows originating from the main volcanic centers from the west [122], though no vent is observed [114], suggesting that localized volcanic fissures exist [123] or that Kasei Valles branched out from Valles Marineris [70] and references therein]. Echus Chasma and Eastern Noctis Chasma have platy-ridged textures, as observed by Mangold et al. [114]. Both areas display a mineralogy of basaltic lava flows with predominant plagioclases and pyroxenes. Models of platy-ridged lava flows generally show a viscosity in the range of 100–1000 Pa s [124]. Such low viscosities and yield strengths less than 200 Pa are usually found in the central Elysium Planitia [125], where extensive platy flows have also been reported [114, 124]. While Mangold et al. [114] argue that all platy-ridge flows may not be magmatic in origin, the textures at Echus and Noctis Chasma may indeed be associated with basaltic lava flows. Low viscosity and low yield strengths for silicate magma may be related to different parameters (see chapter "Volcanic Eruptions on Mars, Lava Flow Morphology, and Thermodynamics"), including increased temperatures or low proportions of silicates.

4 Meteoritic Evidence for Volcanism

4.1 SNCs

Crater-forming impact events on Mars can generate sufficient energy to eject fragments of the crust through the atmosphere and into space (with an escape velocity ~5 km/s) [126] through near-surface bombardment [29]. Fragments as a result of these ejection events represent the currently recovered Martian meteorites that have been found on Earth in places such as Antarctica, Morocco, Libya, Egypt, Tunisia, France, Chile, USA, India, Nigeria, Mali, Mauritania, Oman, and Brazil [127].

Martian meteorites are divided into three main groups: shergottites (basalts of pyroxene and plagioclase, ~82% of known samples), nakhlites (Ca-rich clinopyroxenes with minor olivine, ~14%), and chassignites (greater amounts of olivine, ~3%), all known collectively as the Shergottite-Nakhlite-Chassignite (SNC) [33, 98, 128, 129]. Shergottites also have different subtypes, such as poikilitic, gabbroic, basaltic, and olivine-phyric. Additional rock types from Mars includes Lherzolites (coarse-grained ultramafic rocks of olivine and pyroxenes) and Orthopyroxenite (cumulate rock almost entirely of orthopyroxene) [98]. The readers are encouraged to refer to (https://imca.cc/mars/martian-meteorites-list.htm) for the most up-to-date list of Martian meteorite samples. These have been verified to be of Martian origin based on their distinct oxygen isotope signatures as identified by the techniques of Clayton and Mayeda [130]. While Martian meteorites offer an opportunity to study a number of surface and subsurface localities, cosmic ray exposures [6, 131] relate the ejection age of the rocks to be limited to a few locations on Mars. Nonetheless, they provide petrological, mineralogical, and geochemical data, which could otherwise be difficult to ascertain by landers and orbital spacecraft.

SNC meteorites being labeled as specimens from Mars was observed from their young crystallization ages [6] and fractionated rare-earth patterns [132, 133], both indicators that a sample came from a planet-sized rocky body, in addition to the presence of trapped Martian atmosphere in shock-produced glass and basaltic matrix [134, 135]. Most Martian meteorites are geologically young, with shergottites chiefly being mid- to late-Amazonian in age (<716 Myr crystallization ages), whereas nakhlites and chassignites dated at ~1.3 Gyr [6, 114, 129, 136–142]. This is consistent with the HCP-rich and olivine mineral assemblages predicted to crystallize from primary melts [33]. This is also comparable to the youngest volcanic areas on Mars.

However, the extensive dust overlying young volcanic regions and basaltic shergottites do not hold a very clear link [90, 114]. Most basaltic shergottites are richer in LCP (ratios from 1:1 to 18:1) [143–145] when compared to the proportions derived from OMEGA data [114]. From [114] there are young (Amazonian-aged) volcanic areas that show LCP/HCP ratios from 1:4 to 1:6. Considering the chemistry of these Martian meteorites, REE abundances and isotopic data suggest a rather heterogeneous source region for the shergottites [146, 147].

All and all, Martian meteorites (along with orbital and in-situ results) revealed the occurrence of felsic rocks dating from early Noachian [104, 148]. Such a discovery leads to some interesting questions regarding the nature of felsic igneous materials possibly remaining in the primary crust. A fundamental question that arises concerns the fate of the aluminum. As noted early, it is unlikely that Mars had an ancient anorthositic crust similar to the Moon. Treiman and Lindstrom [149] suggest that a basaltic crust 120–150 km thick could in fact account for the depletion of Al, which is consistent with the alteration products in Martian dust, but not of anorthosite [150]. Various felsic and alkaline rock compositions were analyzed at the surface by the Spirit and Curiosity rovers [72, 76–78, 151, 152] and fractional crystallization (or integration-assimilation) could be probable processes that formed these rocks [127, 152, 153].

Due to dating constraints, these meteorites do not directly reveal the composition of the nearly Noachian crust, but rather younger surfaces. Several authors [6, 47, 154] however note that two specimens are aged from Noachian that could potentially tell us more about the nature of a primitive Martian crust: ALH8400 ("Allan Hills" orthopyroxene sample) [155] and the regolith breccias of NWA7034 (with several paired stone such as 7533/7445/7475/7907 etc.) [154]. ALH8400 has been dated to about 4.5 to 4.071 Gyr by [6, 33, 129, 156] obtained by Lu-Hf and Pb isotope data. Because of its cumulate nature and homogenized mineralogy by diffusive processes, this meteorite serves as a complex sample for the reconstruction of magmatic processes on early Mars.

Sub-alkaline magmatic constituents analyzed within Martian breccia samples (e.g., NWA 7034) and in-situ measurements from Pathfinder and Curiosity reveal the putative andesitic-rich protolith [148]. The Curiosity rover in particular detected quartz-dioritic rocks [78] and likely rhyolitic protolith in Gale crater [157]. While the spatial and temporal relationships of the various rocks are continuously being researched, these findings within the Martian southern hemisphere are rather linked temporally, pointing out a common early crust. The combination of Pathfinder and MGS-TES data indicate the presence of andesites [94, 158], and with the zircon inclusion studies from the NWA 7034 meteorite sample [148], the 4.54 Gyr proto-crust could be andesitic. Recently, the InSight lander and its seismograph instrument at Elysium Planitia in the northern hemisphere also supports an andesitic crust [159]. The formation of the andesitic proto-crust either being primarily extracted from a primitive mantle or through magmatic processes affecting an even earlier primary crust remains unclear.

The second type of felsic lithology is alkaline. The Curiosity rover at Gale crater analyzed a variety of rocks ranging from basalt to trachyte, including anorthoclase and sanidine throughout the Bradbury and Kimberly formations [78, 80, 160–162]. These alkali igneous rocks echo within the NWA 7533 regolith breccia meteorite [47, 163]. While the alkaline rocks within Gale crater have likely been formed by fractional crystallization from a basaltic melt [164], the alkali lithologies in the meteorite sample likely formed 4.43 Gyr by differentiation and fractional crystallization within an impact melt sheet, likely reworking the andesitic proto-crust 4.54 Gy described earlier [47, 148].

Interestingly, the chemistry of plagioclase is considered a great record-keeper of the chemical (and physical) processes of basaltic volcanism, especially utilizing different elements that segregate into olivine and pyroxene [98]. Papike et al. [98] noted that there is no plagioclase detection in meteorite samples of Yamato-980459 or MIL-00346.

NWA 7475 is considered to be the first catalogued indurated-polymict-breccia from Mars that could be representative of the anciently cratered southern hemisphere [165], providing constraints on the nature of the primitive crust and possibly early magmatism. From several studies on this sample [47, 154, 163, 165], this sample has a mixture of effusive and extrusive igneous polymineralic coarse-grained clasts including both felsic and mafic constituents. A unique feature of this meteorite is the presence of leucocratic felsic igneous clasts, in which this identification is the first Noachian felsic clasts that have been identified as trachy-andesites, trachytes, and monzonites [47, 154, 163, 165]. The mafic clasts have been identified as norites and basaltic andesites, mostly formed by pyroxenes (including orthopyroxenes and augites), plagioclases with andesines and oligoclases, and trace amounts of Fe-Ti–oxides [47, 154, 165]. Interestingly, the orthopyroxenes are similar to the cumulate nature analyzed in the ALH84001 sample [163]. Although basaltic clast compositions are typically characteristic of mafic rocks (with SiO_2 <49.5 wt.%) [166, 167], the sample's Na_2O contents are higher than the average crustal composition, while the MgO concentrations are lower. Moreover, the exsolution features in coarse crystals (whether in noritic and monzonitic fragments or in mineral clast forms) point to slower cooling, which in turn means deep-seated plutonic origins. This breccia is dated at 4.43 Gyr, which could be representative of a reworked Noachian surface regolith, possibly excavated from several deep impact melts [163]. This early Noachian Martian breccia provides two interesting insights to the early stages of crustal formation: (i) a buildup of an andesitic proto-crust no more than 20 Myr after Solar System formation, and (ii) several impacts on this primordial crust weathered 100 Myr later to form deep melts where differentiation occurred, which formed more alkali-rich rocks in the upper portion.

4.2 Mantle Reservoirs

Within the framework of a possible magmatic ocean on early Mars in that the volcanic products of mantle overturned, the igneous mineralogy of surface rocks does not require a peak in volcanism around the Noachian/Hesperian boundary [33]. Volcanic gas composition strongly depends on the atmospheric pressure [168], namely sulfur-dioxide rich volcanic gases being produced. A decrease of atmospheric pressure at the Noachian/Hesperian boundary would therefore solidify the hypothesis that Hesperian volcanic activity is responsible for the formation of massive sulfate deposits [22, 33, 169, 170]. However, it remains unclear if shergottites and Martian magmas are sulfide-saturated (or undersaturated) and therefore how much of this sulfur can contribute via volcanism [171]. Experimental results from Righter et al. [171] have found that primitive shergottite magmas are capable of dissolving ~4000 ppm S at sulfide saturation. With the melting of the Martian mantle and consequent volcanism, this provides more than enough sulfur for the sulfate-rich soil and deposits observed on the surface of Mars.

Several authors [33, 49, 172] concluded that the Early Noachian was the prime time of widespread volcanism associated with planetary cooling and most rocks exposed at the surface should be related to this period of intense volcanism. Martian meteorites have provided a means to modeling the Martian mantle and basaltic magmatism. During the main shield period on Mars and a large eruption volume of shergottite lavas, load is emplaced (unevenly) on the underlying lithosphere, leading to flexure and consequently the development of a flexural bulge [127]. Flexural bulges and moats are observed on Earth, most prominently at the Hawaiian chain volcanoes, and based on gravity, on Mars in the Tharsis volcanic province [173, 174]. As with terrestrial rejuvenated lavas, a previously depleted mantle has to be chemically altered in order to induce partial melting through decompression during lithospheric flexure. On Mars, this is required for the source of nakhlites based on their Sr/Nd isotopic ratios.

Shergottites from enriched, intermediate, and depleted sources have been calculated to originate from mantle sources with a range of potential temperatures, such as ~1750 °C compared to Noachian rocks at Gale crater (~1450 °C), representative of products from a hot mantle plume [175, 176]. Such geochemical and isotopic characteristics suggest the occurrence of at least four silicate reservoirs that formed during the primordial differentiation of Mars [98, 142].

Wadhwa and Borg [142] described them as: (i) Depleted Mantle Reservoir 1—source of light REE (LREE) depleted shergottites, such as Yamato-980459, QUE94201, Dar al Gani 476, Sayh al Uhaymir 005. This reservoir also has distinctive Hf/W isotopic ratios; (ii) Depleted Mantle Reservoir 2—source of all clinopyroxenites (Nakhlites). While also LREE-depleted, this reservoir type has characteristic Nd and

Hf/W isotopic signatures; (iii) Depleted Mantle Reservoir 3—source of ancient orthopyroxenite crust, representative of the ALH-84001 meteorite. This mantle reservoir type has notable Sm/Nd isotopic signatures compared to the other types; and (iv) Enriched Mantle Reservoir—either an oxidized crustal or mantle component that contributed to the LREE enrichment of shergottites, such as Shergotty, Zagami, Los Angeles, and NWA 1110.

The identification of such distinct mantle reservoirs (depleted versus enriched) from which the shergottites were derived has given rise to two general models [98]: (i) assimilation of crustal material by mantled basaltic magmas (Fig. 3a); and (ii) derivation of basaltic magmas from multiple reservoirs (Fig. 3b, c).

Figure 3a proposes that the Enriched Reservoir is located within the Martian crust and Depleted Reservoir 1 is the shergottite mantle source [177–181]. Both of these reservoirs would have been produced during the initial stages of differentiation through crystallization of a magma ocean [98]. In this particular scenario, there are varying degrees of crustal adaptation by mantle-derived basalts that which produces the isotropic signatures observed in shergottites [137, 142, 181–183]. This particular scenario has the advantage of explaining temporal separation and long-term isolation of the two reservoirs (crust versus mantle emplacement). Furthermore, this model provides a means in which enriched and depleted components could mix during basaltic generation and consequent eruption [98].

For Fig. 3b, c, the Depleted Reservoir 1 and the Enriched Reservoir are located in the Martian mantle [137, 142, 183–185]. These multi-reservoir models must be isolated within the mantle for at least 4000 Myr and only mixed at the time of melting to produce relevant shergottite isotopic characteristics [98]. A fundamental aspect of these multi-reservoir models is that they must have been produced by the crystallization of a deep Martian magma ocean [186]. Symes et al. [184] explains a two-stage model of the depleted, intermediate, and enriched sources (Fig. 3b). Heat-producing elements (e.g., K, Th, U, etc.) are concentrated in the late-stage crystallization of the putative magma ocean and can be mixed by convection processes [98]. Figure 3b illustrates where melts rise with no (or little) crystallization until they enter magma chambers in the Martian

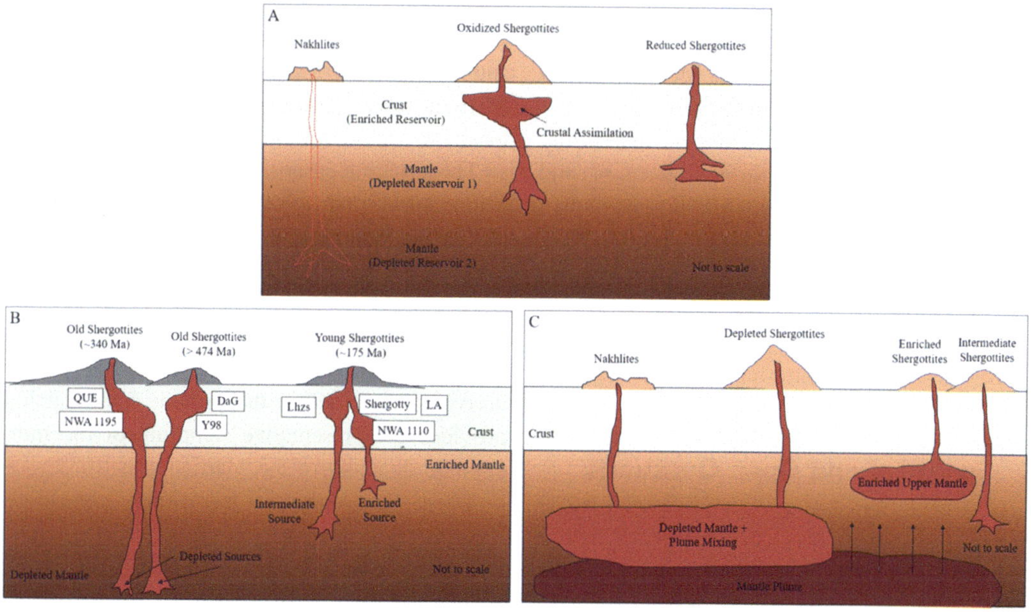

Fig. 3 Models of Martian magmatism as hypothesized by shergottite geochemistry. **a** Crustal assimilation model [180]. **b** Multiple mantle source [184]. **c** Multiple mantle source [185]. See text for more details for each scenario. Adapted from Papike et al. [98]

crust. Crystallization processes would then commence, where various mixtures can be derived by fractional crystallization [181, 184] before being transported to the Martian surface. Shearer et al. and Symes et al. [181, 184] proposed that this process links the olivine-basalts (e.g., Yamato-980459) to pyroxene-basalts (e.g., QUE-94201) within each shergottite end-member (depleted versus enriched).

Blinova and Herd [185] proposed a more complex, three-stage model to account for the REE systematics of shergottites (Fig. 3c). In this scenario, the source of the depleted shergottites is a mixture of upper mantle cumulates (nakhlite parental magmas) at 1.3 Gyr and a deep mantle lithology that was located near the mantle-core boundary [98]. These two distinctive sources progressed in Nd isotopic systematics due to their different Nd/Sm ratios [98]. The deep mantle source was then transported to the shallower mantle by a plume and experienced melting over a variety of pressure regimes. Such a mixture of melts of the plume and mantle residue produces the REE patterns of the depleted shergottites. The enriched reservoir is rather a late-stage Martian magma ocean cumulation emplaced in the upper mantle, but the lower mantle plumes could also provide a heat source to initiate melting for the enriched (E.g., Zagami, Los Angeles) and intermediate Martian basalts (e.g., LEW-88516, NWA 4797).

Although these illustrative models do not fully explain the timing of reservoir isolation and mixing processes, each do impose their own requirements on the mantle processes [98, 184]. With the isolation of the reservoirs, the reservoirs must have thermal boundaries that allow for the transport of heat, though not the exchange of mass, while the mixing must initiate melting [98].

5 Summary and Open Questions

The multiple observations of a variety of mineralogy from orbiters, landers, and SNC meteorites reveal a complex history of Martian crustal evolution. The detections of early Noachian-aged felsic outcrops throughout the southern hemisphere suggests that there are remnant exposures of an extensive ancient andesitic crust deeply impacted early in Martian geologic history. Concerning the paradox of the crustal average density, a fully-basaltic crust would be considered too high [35, 187], therefore Wieczorek and Zuber [35] proposed the occurrence of large crustal proportions composed of felsic materials. This is also evident in Martian meteorites.

One of the major knowledge gaps for Martian meteorites is the location of their excavated origin. Comparison of meteorite ages with crater chronology of the surface can show distinct "ancient" versus "young" [188], but a better understanding of the ejection physics of the meteorites (and consideration of shock effects and isotopic compositions) strengthens the conclusions of possible origins [29, 189, 190]. Hartmann and Neukum [52] noted that less than 10% of the Martian surface is younger than 1 Ga, including Tharsis, Amazonis Planitia, and Elysium.

Another prominent feature on Mars is the hemispheric dichotomy from early Noachian time, when the earliest crust of Mars being more likely basaltic than anorthositic. Long-lived mantle upwelling related to high heat production produced a thickened felsic alkaline crust localized in the southern hemisphere. Consistent and widespread volcanism have likely curried the Martian felsic crust, only to be observed on localized scales from impact crater excavation.

The SNC meteorites provide the most detailed geochemical and mineralogical data for Mars [63, 128]. However, it should be noted that the implications for a global understanding of planetary evolution and widespread volcanism are very much limited by the inherent uncertainties in the exact locations from which the SNCs were derived, and whether the current SNC catalog provides a representative sampling of the mineralogic diversity of the Martian crust [63].

There are several remaining questions regarding Martian meteorites and the geological evolution of Mars, such as: *How compositionally variable is the Martian interior and surface? How did the magma ocean crystallize? How have the diversity of mantle and crustal sources changed over time? How has magmatic behavior, from fractional crystallization to accumulation, evolved over geologic time? Is there a way to resolve the contrast between meteorite samples and remote-sensing data?* Nonetheless, the diverse mineralogy that such Martian meteorites can provide show us just how complex volcanism can be dependent on the minerals present throughout Martian geologic history.

References

1. Newsom H (2007) Geochemical analogs and Martian meteorites. In: Chapman M (ed) The geology of mars. Cambridge Planetary Science, Cambridge University Press
2. Tuff J, Wade J, Wood B (2013) Volcanism on Mars controlled by early oxidation of the upper mantle. Nat Lett 498:342–346
3. McSween HY et al (1979) Petrogenetic relationship between Allan Hills 77005 and other achondrites. Earth Planet Sci Lett 45:275–284
4. Walker D, Stolper EM, Hays JF (1979) Basaltic volcanism: the importance of planet size. Proc Lunar Planet Sci Conf 10:1995–2015
5. Dreibus G, Wanke H (1985) Mars, a volatile-rich planet. Meteoritics 20:367–381
6. Nyquist LE et al (2001) Ages and geologic histories ofMartianmeteorites. Space Sci Rev 96:105–164
7. Treiman AH (2005) The nakhlite meteorites: augite-rich igneous rocks from Mars. Chemie Erde Geochem 65:203–270
8. Lodders K, Fegley B Jr (1997) An oxygen isotope model for the composition of Mars. Icarus 126 (2):373–394
9. McDonough WF, Sun S-S (1995) The composition of the Earth. Chem Geol 120:223–253
10. Werner S (2009) The global martian volcanic evolutionary history. Icarus 201:44–68
11. Carr MH (1973) Volcanism on Mars. J Geophys Res 78(17):4049–4062
12. Bleacher JE, Greeley R, Williams DA, Cave SR, Neukum G (2007) Trends in effusive style at the Tharsis Montes, Mars, and implications for the development of the Tharsis province. J Geophys Res 112:E09005
13. Hiesinger H, Head JW, Neukum G (2007) Young lava flows on the eastern flank of Ascraeus Mons: rheological properties derived from High Resolution Stereo Camera (HRSC) images and Mars Orbiter Laser Altimeter (MOLA) data. J Geophys Res 112:5011
14. Plescia JB, Saunders RS (1979) The chronology of the martian volcanoes. Proceedings of the 10th Lunar Planet Sci: 2841–2859
15. Neukum G, Hiller K (1981) Martian ages. J Geophys Res 86(15):3097–3121
16. Wilson L, Mouginis-Mark PJ (2003) Phreatomagmatic explosive origin of Hrad Vallis, Mars. J Geophys Res 108:5082
17. Greeley R, Spudis PD (1981) Volcanism on Mars. Rev Geophys Space Phys 19:13–41
18. Poulet F, Mangold N, Loizeau D, Bibring JP, Langevin Y, Michalski J, Gondet B (2008) Abundance of minerals in the phyllosilicate-rich units on Mars. Astron Astrophys 487(2):L41–L44
19. Jakosky B, Phillips R (2001) Mars' volatile and climate history. Nature 412(6843):237–244
20. Phillips RJ, Zuber MT, Solomon SC, Golombek MP, Jakosky BM et al (2001) Ancient geodynamics and global-scale hydrology on Mars. Science 291(5513):2587–2591
21. Craddock R, Greeley R (2009) Minimum estimates of the amount and timing of gases released into the martian atmosphere from volcanic eruptions. Icarus 204(2):512–526
22. Bibring JP, Langevin Y, Mustard JF, Poulet F, Arvidson R et al (2006) Global mineralogical and aqueous Mars history derived from OMEGA/Mars Express data. Science 312(5772):400–404
23. Mustard JF, Murchie SL, Pelkey SM, Ehlmann BL, Milliken RE et al (2008) Hydrated silicate minerals on Mars observed by the Mars Reconnaissance Orbiter CRISM instrument. Nature 454(7202):305–309
24. Plescia JB, Saunders RS (1982) Tectonic history of the Tharsis region, Mars. J Geophys Res Solid Earth 87(B12):9775–9791
25. Anderson RC, Dohm JM, Golombek MP, Haldemann AF, Franklin BJ et al (2001) Primary centers and secondary concentrations of tectonic activity through time in the western hemisphere of Mars. J Geophys Res Planets 106(E9):20563–20585
26. Neumann GA, Zuber MT, Wieczorek MA, McGovern PJ, Lemoine FG, Smith DE (2004) Crustal structure of Mars from gravity and topography. J Geophys Res Planets 109(E8):E08002
27. Solomon SC, Head JW (1982) Evolution of the Tharsis province of Mars: the importance of heterogeneous lithospheric thickness and volcanic construction. J Geophys Res Solid Earth 87 (B12):9755–9774
28. Johnson CL, Phillips RJ (2005) Evolution of the Tharsis region of Mars: insights from magnetic field observations. Earth Planet Sci Lett 230(3–4):241–254
29. Head JW III, Kreslavsky MA, Pratt S (2002) Northern lowlands of Mars: evidence for widespread volcanic flooding and tectonic deformation in the Hesperian period. J Geophys Res Planets 107(E1):1–3
30. Tanaka K, Scott D (1987) Geological map of the polar regions of Mars. USGS Miscellaneous Investigations Series Map I–1802–C
31. Greeley R, Schneid BD (1991) Magma generation on Mars: amounts, rates, and comparisons with Earth, Moon, and Venus. Science 254(5034):996–998

32. Debaille V, Brandon AD, Yin QZ, Jacobsen B (2007) Coupled 142Nd–143Nd evidence for a protracted magma ocean in Mars. Nature 450 (7169):525–528
33. Baratoux D, Toplis MJ, Monnereau M, Sautter V (2013) The petrological expression of early Mars volcanism. J Geophys Res Planets 118(1):59–64
34. Zuber MT (2001) The crust and mantle of Mars. Nature 412(6843):220–227
35. Wieczorek MA, Zuber MT (2004) Thickness of the Martian crust: improved constraints from geoid-to-topography ratios. J Geophys Res Planets 109(E1): E01009
36. Mezger K, Debaille V, Kleine T (2013) Core formation and mantle differentiation on Mars. Space Sci Rev 174:27–48
37. Dauphas N, Pourmand A (2011) Hf–W–Th evidence for rapid growth of Mars and its status as a planetary embryo. Nature 473(7348):489–492
38. Lee D-C, Halliday AN (1997) Core formation on Mars and differentiated asteroids. Nature 388 (6645):854
39. Debaille V, Brandon AD, O'Neill C, Yin Q-Z, Jacobsen B (2009) Early martian mantle overturn inferred from isotopic composition of nakhlite meteorites. Nat Geosci 2(8):548–552
40. Debaille V, Yin Q-Z, Brandon AD, Jacobsen B (2008) Martian mantle mineralogy investigated by the 176Lu–176Hf and 147Sm–143Nd systematics of shergottites. Earth Planet Sci Lett 269(1): 186–199
41. Marty B, Marti K (2002) Signatures of early differentiation of Mars. Earth Planet Sci Lett 196 (3):251–263
42. Solomon SC, Aharonson O, Aurnou JM, Banerdt WB, Carr MH, Dombard AJ et al (2005) New perspectives on ancient Mars. Science 307 (5713):1214–1220
43. Wood JA, Dickey JS Jr, Marvin UB, Powell BN (1970) Lunar anorthosites and a geophysical model of the moon. Geochim Cosmochim Acta Suppl 1:965
44. Carter J, Poulet F (2013) Ancient plutonic processes on Mars inferred from the detection of possible anorthositic terrains. Nat Geosci 6(12):1008–1012
45. Elkins-Tanton LT, Hess PC, Parmentier EM (2005) Possible formation of ancient crust on Mars through magma ocean processes. J Geophys Res Planets 110 (E12):E12S01
46. Elkins-Tanton LT (2012) Magma oceans in the inner solar system. Annu Rev Earth Planet Sci 40 (1):113–139
47. Humayun M, Nemchin A, Zanda B, Hewins RH, Grange M, Kennedy A et al (2013) Origin and age of the earliest Martian crust from meteorite NWA 7533. Nature 503(7477):513–516
48. Zuber MT, Solomon SC, Phillips RJ, Smith DE, Tyler GL, Aharonson O et al (2000) Internal structure and early thermal evolution of Mars from Mars global surveyor topography and gravity. Science 287(5459):1788–1793
49. Xiao L, Huang J, Christensen PR, Greeley R, Williams DA, Zhao J et al (2012) Ancient volcanism and its implication for thermal evolution of Mars. Earth Planet Sci Lett 323:9–18
50. Bottke WF, Andrews-Hanna JC (2017) A post-accretionary lull in large impacts on early Mars. Nat Geosci 10(5):344–348
51. Carr MH, Head JW (2010) Geologic history of Mars. Earth Planet Sci Lett 294(3):185–203
52. Hartmann WK, Neukum G (2001) Cratering chronology and the evolution of Mars. Chronology and evolution of Mars. Springer, pp 165–194
53. Wyatt MB, McSween HY, Tanaka KL, Head JW (2004) Global geologic context for rock types and surface alteration on Mars. Geology 32(8):645–648
54. Nimmo F, Stevenson DJ (2001) Estimates of Martian crustal thickness from viscous relaxation of topography. J Geophys Res 106:5085–5098
55. Bertka CM, Fei Y (1998) Density profile of an SNC model Martian interior and the moment-of-inertia factor of Mars. Earth Planet Sci Lett 157(1):79–88
56. Baratoux D, Samuel H, Michaut C, Toplis MJ, Monnereau M, Wieczorek M et al (2014) Petrological constraints on the density of the Martian crust. J Geophys Res Planets 119(7):1707–1727
57. Wray JJ, Hansen ST, Dufek J, Swayze GA, Murchie SL, Seelos FP et al (2013) Prolonged magmatic activity on Mars inferred from the detection of felsic rocks. Nat Geosci 6(12):1013–1017
58. Bandfield JL (2006) Extended surface exposures of granitoid compositions in Syrtis Major, Mars. Geophys Res Lett 33(6):L06203
59. Bandfield JL (2002) Global mineral distributions on Mars. J Geophys Res Planets 107(E6):1–9
60. Flahaut J, Quantin C, Clenet H, Allemand P, Mustard JF, Thomas P (2012) Pristine Noachian crust and key geologic transitions in the lower walls of Valles Marineris: insights into early igneous processes on Mars. Icarus 221(1):420–435
61. Tanaka KL, Robbins SJ, Fortezzo CM, Skinner JA Jr, Hare TM (2014) The digital global geologic map of Mars: chronostratigraphic ages, topographic and crater morphologic characteristics, and updated resurfacing history. Planet Space Sci 95:11–24
62. Poulet F, Mangold N, Platevoet B, Bardintzeff J-M, Sautter V, Mustard JF et al (2009) Quantitative compositional analysis of martian mafic regions using the MEx/OMEGA reflectance data: 2. Petrological implications. Icarus 201(1):84–101
63. Mustard J, Murchie S, Erard S, Sunshine J (1997) In situ compositions of Martian volcanics: inmplications for the mantle. JGR 102(E11):25605–25615
64. Carter J, Poulet F, Bibring J-P, Mangold N, Murchie S (2013) Hydrous minerals on Mars as seen by the CRISM and OMEGA imaging spectrometers: updated global view. J Geophys Res Planets 118(4):831–858

65. Ehlmann BL, Mustard JF, Murchie SL, Bibring J-P, Meunier A, Fraeman AA et al (2011) Subsurface water and clay mineral formation during the early history of Mars. Nature 479(7371):53–60
66. Skok JR, Mustard JF, Tornabene LL, Pan C, Rogers D, Murchie SL. (2012), A spectroscopic analysis of Martian crater central peaks: formation of the ancient crust. J Geophys Res Planets 117 (E11):E00J18
67. Taylor SR, McLennan S (2009) Planetary crusts: their composition, origin and evolution, vol 10. Cambridge University Press, Cambridge
68. Michalski JR, Bleacher JE (2013) Supervolcanoes within an ancient volcanic province in Arabia Terra, Mars. Nature 502(7469):47–52
69. Grott M, Baratoux D, Hauber E, Sautter V, Mustard J, Gasnault O et al (2013) Long-term evolution of the Martian crust-mantle system. Space Sci Rev 174(1–4):49–111
70. Leone G (2016) Alignments of volcanic features in the southern hemisphere of Mars produced by migrating mantle plumes. J Volcanol Geotherm Res 309:78–95
71. Greeley R, Foing BH, McSween HY, Neukum G, Pinet P, van Kan M et al (2005) Fluid lava flows in Gusev crater, Mars. J Geophys Res E Planets 110: E05008
72. McSween HY, Ruff SW, Morris RV, Bell JF, Herkenhoff K, Gellert R et al (2006) Alkaline volcanic rocks from the Columbia Hills, Gusev crater, Mars. J Geophys Res 111(E9):E09S91
73. Filiberto J, Dasgupta R (2011) Fe2+–Mg partitioning between olivine and basaltic melts: applications to genesis of olivine-phyric shergottites and conditions of melting in the Martian interior. Earth Planet Sci Lett 304(3–4):527–537
74. Squyres SW, Arvidson RE, Blaney DL, Clark BC, Crumpler L, Farrand WH et al (2006) Rocks of the Columbia Hills. J Geophys Res Planets 111(E2): E02S11
75. Arvidson RE, Squyres SW, Anderson RC, Bell JF, Blaney D, Brückner J et al (2006) Overview of the spirit Mars exploration Rover Mission to Gusev Crater: landing site to Backstay Rock in the Columbia Hills. J Geophys Res Planets 111(E2):E02S01
76. Cousin A, Sautter V, Payré V, Forni O, Mangold N, Gasnault O et al (2017) Classification of igneous rocks analyzed by ChemCam at Gale crater, Mars. Icarus 288:265–283
77. Sallé B, Lacour JL, Mauchien P, Fichet P, Maurice S, Manhes G (2006) Comparative study of different methodologies for quantitative rock analysis by laser-induced breakdown spectroscopy in a simulated Martian atmosphere. Spectrochim Acta Part B 61(3):301–313
78. Sautter V, Toplis MJ, Wiens RC, Cousin A, Fabre C, Gasnault O et al (2015) In situ evidence for continental crust on early Mars. Nat Geosci 8 (8):605–609
79. Sautter V, Fabre C, Forni O, Toplis MJ, Cousin A, Ollila AM et al (2014) Igneous mineralogy at Bradbury Rise: the first ChemCam campaign at Gale crater. J Geophys Res Planets 119(1):30–46
80. Schmidt ME, Campbell JL, Gellert R, Perrett GM, Treiman AH, Blaney DL et al (2014) Geochemical diversity in first rocks examined by the Curiosity Rover in Gale Crater: evidence for and significance of an alkali and volatile-rich igneous source. J Geophys Res Planets 119(1):64–81
81. Boynton WV, Feldman WC, Mitrofanov IG, Evans LG, Reedy RC et al (2004) The Mars Odyssey gamma-ray spectrometer instrument suite. In: 2001 Mars Odyssey. Springer, Dordrecht, pp 37–83
82. Baratoux D, Toplis MJ, Monnereau M, Gasnault O (2011) Thermal history of Mars inferred from orbital geochemistry of volcanic provinces. Nature 472(7343):338–341
83. Rogers AD, Bandfield JL, Christensen PR (2007) Global spectral classification of Martian low-albedo regions with Mars Global Surveyor Thermal Emission Spectrometer (MGS-TES) data. J Geophys Res Planets 112(E2):E02004
84. Mustard JF, Poulet F, Gendrin A, Bibring JP, Langevin Y et al (2005) Olivine and pyroxene diversity in the crust of Mars. Science 307 (5715):1594–1597
85. Pinet PC, Heuripeau F, Clenet H, Chevrel S, Daydou Y et al (2007) Mafic mineralogy variations across Syrtis Major shield and surroundings as inferred from visible-near-infrared spectroscopy by OMEGA/Mars Express. In: Seventh international Mars conference, Pasadena, CA, Abstract, vol 3146
86. Cloutis EA, Gaffey MJ (1991) Pyroxene spectroscopy revisited: spectral-compositional correlations and relationship to geothermometry. J Geophys Res Planets 96(E5):22809–22826
87. Koeppen WC, Hamilton VE (2008) Global distribution, composition, and abundance of olivine on the surface of Mars from thermal infrared data. J Geophys Res Planets 113(E5):E05001
88. Ding N, Bray VJ, McEwen AS, Mattson SS, Okubo CH, Chojnacki M, Tornabene LL (2015) The central uplift of Ritchey crater, Mars. Icarus 252:255–270
89. McGetchin TR, Smith JR (1978) The mantle of Mars: Some possible geological implications of its high density. Icarus 34(3):512–536
90. Hamilton VE, Christensen PR, McSween HY, Bandfield JL (2003) Searching for the source regions of martian meteorites using MGS TES: integrating Martian meteorites into the global distribution of igneous materials on Mars. Meteorit Planet Sci 38(6):871–885
91. Wyatt MB, McSween HY (2002) Spectral evidence for weathered basalt as an alternative to andesite in the northern lowlands of Mars. Nature 417 (6886):263–266

92. Chevrier V, Mathé PE (2007) Mineralogy and evolution of the surface of Mars: a review. Planet Space Sci 55(3):289–314
93. Bibring JP, Langevin Y, Gendrin A, Gondet B, Poulet F, Berthé M et al (2005) Mars surface diversity as revealed by the OMEGA/Mars express observations. Science 307(5715):1576–1581
94. McSween HY, Murchie SL, Crisp JA, Bridges NT, Anderson RC, Bell JF et al (1999) Chemical, multispectral, and textural constraints on the composition and origin of rocks at the Mars Pathfinder landing site. J Geophys Res Planets 104(E4):8679–8715
95. McSween HY Jr (2002) The rocks of Mars, from far and near. Meteorit Planet Sci 37(1):7–25
96. Bandfield JL, Hamilton VE, Christensen PR, McSween Jr HY (2004) Identification of quartzofeldspathic materials on Mars. J Geophys Res Planets 109(E10):E10009
97. Hamilton VE, Wyatt MB, McSween Jr HY, Christensen PR (2001) Analysis of terrestrial and Martian volcanic compositions using thermal emission spectroscopy: 2. Application to Martian surface spectra from the Mars Global Surveyor Thermal Emission Spectrometer. J Geophys Res Planets 106 (E7):14733–14746
98. Papike J, Karner J, Shearer C, Burger P (2009) Silicate mineralogy of martian meteorites. Geochim Et Cosmochem Acta 73:7443–7485
99. Schwertmann U (1958) The effect of pedogenic environments on iron oxide minerals. Advances in soil science. Springer, New York, NY, pp 171–200
100. Bell JF, McSween HY, Crisp JA, Morris RV, Murchie SL et al (2000) Mineralogic and compositional properties of Martian soil and dust: Results from Mars Pathfinder. J Geophys Res Planets 105 (E1):1721–1755
101. Foley CN, Economou T, Clayton RN (2003). Final chemical results from the Mars Pathfinder alpha proton X-ray spectrometer. J Geophys Res Planets 108(E12):8096
102. McCoy TJ, Sims M, Schmidt ME, Edwards L, Tornabene LL, Crumpler LS et al (2008) Structure, stratigraphy, and origin of Husband Hill, Columbia Hills, Gusev Crater, Mars. J Geophys Res Planets 113(E6):E06S03
103. Ody A, Poulet F, Bibring JP, Loizeau D, Carter J, Gondet B et al (2013) Global investigation of olivine on Mars: insights into crust and mantle compositions. J Geophys Res Planets 118(2):234–262
104. Sautter V, Toplis MJ, Beck P, Mangold N, Wiens R, Pinet P et al (2016), Magmatic complexity on early Mars as seen through a combination of orbital, in-situ and meteorite data. Lithos 254–255
105. Adams J, McCord T (1969) Mars: interpretation of spectral reflectivity of light and dark regions. JGR 74(20):4851–4856
106. McCord T, Clark R, Singer R (1982) Mars: near-infrared spectral reflectance of surface regions and compositional implications. JGR Solid Earth 87 (B4):3021–3032
107. Reyes D, Christensen P (1994) Evidence for Komatiite-type lavas on Mars from Phobos ISM data and other observations. GRL 21(10):887–890
108. Mustard J, Sunshine J (1995) Seeing through the dust: Martian crustal heterogeneity and links to the SNC meteorites. Science 267(5204):1623–1626
109. Wilson L, Head J III (1994) Mars: review and analysis of volcanic eruption theory and relationships to observed landforms. Rev Geophys 32 (3):221–263
110. Flahaut J (2011) Minéralogie de Valles Marineris (Mars) par imagerie hyperspectrale: histoire magmatique et sédimentaire de la region. Doctoral dissertation
111. Williams DA, Greeley R, Werner SC, Michael G, Crown DA, Neukum G, Raitala J (2008) Tyrrhena Patera: geologic history derived from Mars express high resolution stereo camera. J Geophys Res Planets 113(E11):E11005
112. Salvatore MR, Mustard JF, Wyatt MB, Murchie SL (2010) Definitive evidence of Hesperian basalt in Acidalia and Chryse planitiae. J Geophys Rese Planets 115(E7):E07005
113. Stockstill-Cahill KR, Scott AF, Hamilton VE (2008) A study of low-albedo deposits within Amazonis Planitia craters: evidence for locally derived ultramafic to mafic materials. J Geophys Res 113:E07008
114. Mangold N, Loizeau D, Poulet F, Ansan V et al (2010) Mineralogy of recent volcanic plains in the Tharsis region, Mars, and implications for platy-ridged flow composition. Earth Planet Sci Lett 294:440–450
115. Farrand WH, Lane MD, Edwards BR, Yingst RA (2011) Spectral evidence of volcanic cryptodomes on the northern plains of Mars. Icarus 211(1):139–156
116. Rogers AD, Nekvasil H (2015) Feldspathic rocks on Mars: compositional constraints from infrared spectroscopy and possible formation mechanisms. Geophys Res Lett 42(8):2619–2626
117. Connerney JEP, Acuña MH, Ness NF, Kletetschka G, Mitchell DL, Lin RP et al (2005) Tectonic implications of Mars crustal magnetism. Proc Natl Acad Sci 102(42):14970–14975
118. Glotch TD, Bandfield JL, Tornabene LL, Jensen HB, Seelos FP (2010) Distribution and formation of chlorides and phyllosilicates in Terra Sirenum, Mars. Geophys Res Lett 37(16)
119. Osterloo MM, Hamilton VE, Bandfield JL, Glotch TD, Baldridge AM, Christensen PR et al (2008) Chloride-bearing materials in the southern highlands of Mars. Science 319(5870):1651–1654
120. Wray JJ, Milliken RE, Dundas CM, Swayze GA, Andrews-Hanna JC, Baldridge AM et al (2011) Columbus crater and other possible groundwater-fed paleolakes of Terra Sirenum, Mars. J Geophys Res 116(E1):E01001

121. Hodges CA, Moore HJ (1994) Atlas of volcanic landforms on Mars. USGS Prof. paper, p 1534. Washington
122. Scott DH, Tanaka KL (1986) Geologic map of western equatorial region of Mars. U.S. Geol. Surv. Misc. Invest Ser., Map I-1802-A
123. Chapman MG, Neukum G, Dumke A, Michael G, van Gasselt S, Kneissl T, Zuschneid W, Hauber E, Mangold N (2010) Amazonian geologic history of the Echus Chasma and Kasei Valles system on Mars: nsew data and interpretations. Earth Planet Sci Lett 294(3–4):238–255
124. Keszthelyi L, McEwen A, Thordarson T (2000) Terrestrial analogs and thermal models for Martian flood lavas. J Geophys Res 105(E6):15027–15049
125. Vaucher JD, Baratoux TMJ, Pinet P, Mangold N, Kurita K (2009) The morphologies of volcanic landforms at Central Elysium Planitia: evidence for recent and fluid lavas on Mars. Icarus 200:39–51
126. Fritz J, Artemieva N, Greshake A (2005) Ejection of Martian meteorites. Meteorit Planet Sci 40(9–10):1393–1411
127. Udry A, Howarth G, Herd C, Day J, Lapen T, Filiberto J (2020) What martian meteorites reveal about the interior and surface of Mars. JGR Planets 125(12):e2020JE006523
128. McSween HY (1994) What we have learned about Mars from SNC meteorites. Meteoritics 29(6)
129. Bouvier A, Blichert-Toft J, Albarède F (2009) Martian meteorite chronology and the evolution of the interior of Mars. Earth Planet Sci Lett 280(1–4):285–295
130. Clayton RN, Mayeda TK (1983) Oxygen isotopes in eucrites, shergottites, nakhlites, and chassignites. Earth Planet Sci Lett 62:106
131. Eugster O, Herzog GF, Marti K, Caffee MW (2006) Irradiation records, cosmic-ray exposure ages, and transfer times of meteorites. Meteorit Early Solar Syst II:829–851
132. McSween HY (1984) SNC meteorites—are they martian rocks? Geology 12:3–6
133. Shih CY et al (1982) Chronology and petrogenesis of young achondrites, Shergotty, Zagami, and Alha77005—late magmatism on a geologically active planet. Geochim Cosmochim Acta 46:2323–2344
134. Becker RH, Pepin RO (1984) The case for a martian origin of the shergottites: nitrogen and noble gases in EETA-79001. Earth Planet Sci Lett 69:225–242
135. Bogard DD, Johnson P (1983) Martian gases in an Antarctic meteorite. Science 221:651–654
136. Borg LE, Nyquist LE, Wiesmann H, Reese Y (2002) Constraints on the petrogenesis of Martian meteorites from the Rb-Sr and Sm-Nd isotopic systematics of the lherzolitic shergottites ALH77005 and LEW88516. Geochim Cosmochim Acta 66(11):2037–2053
137. Borg LE, Nyquist LE, Wiesmann G, Shih CY, Reese Y (2003) The age of Dar al Gani 476 and the differentiation history of the martian meteorites inferred from their radiogenic isotopic systematics. Geochim Cosmochim Acta 67(18):3519–3536
138. Brennecka GA, Borg LE, Wadhwa M (2014) Insights into the Martian mantle: the age and isotopics of the meteorite fall Tissint. Meteoritics & Planetary Sciences 49(3):412–418
139. Cohen BE, Mark DF, Cassata WS, Lee MR, Tomkinson T, Smith CL (2017) Taking the pulse of Mars via dating of a plume-fed volcano. Nat Commun 8(1):640
140. Lapen TJ, Righter M, Andreasen R, Irving AJ, Satkoski AM, Beard BL et al (2017) Two billion years of magmatism recorded from a single Mars meteorite ejection site. Sci Adv 3(2):e1600922.
141. Harper CL, Nyquist LE, Bansal B, Wiesmann H, Shih CY (1995) Rapid accretion and early differentiation of Mars indicated by 142Nd/144Nd in SNC meteorites. Science 267(5195):213–217
142. Wadhwa M, Borg LE (2006) Trace element and 142nd systematics in the nakhlite MIL03346 and the orthopyroxenite ALH84001: implications for the martian mantle. Lunar Planet Sci XXXVII, Houston #2045
143. McSween HY, Jarosewich E (1983) Petrogenesis of the Elephant Moraine A79001 meteorite multiple magma pulses on the shergottite parent body. Geochim Cosmochim Acta 47:1501–1513
144. Barrat JA, Jambon A, Bohn M, Gillet P, Sautter V, Gopel C, Lesourd M, Keller F (2002) Petrology and chemistry of the pricritic shergottite NorthWest Africa 1068 (NWA 1068). Geochim Cosmochim Acta 66:3505–3518
145. Taylor LA et al (2002) Martian meteorite Dhofar 9: a new shergottite. Meteorit Planet Sci 37:1107–1128
146. Peslier AH, Hnatyshin D, Herd CDK, Walton EL, Brandon AD, Lapen TJ, Shafer JT (2010) Crystallization, melt inclusion, and redox history of a Martian meteorite: olivine-phyric shergottite Larkman Nunatak 06319. Geochim Cosmochim Acta 74 (15):4543–4576
147. Shafer JT, Brandon AD, Lapen TJ, Righter M, Peslier AH, Beard BL (2010) Trace element systematics and 147Sm-143Nd and 176Lu-176Hf ages of Larkman Nunatak 06319: closed-system fractional crystallization of an enriched shergottite magma. Geochim Cosmochim Acta 74(24):7307–7328
148. Bouvier LC, Costa MM, Connelly JN, Jensen NK, Wielandt D, Storey M et al (2018) Evidence for extremely rapid magma ocean crystallization and crust formation on Mars. Nature 558(7711):586
149. Treiman AH, Lindstrom DJ (1997) Trace element geochemistry of Martian iddingsite in the Lafayette meteorite. J Geophys Res Planets 102(E4):9153–9163
150. Gooding JL (1992) Soil mineralogy and chemistry on Mars: Possible clues from salts and clays in SNC meteorites. Icarus 99(1):28–41
151. Edwards PH, Bridges JC, Wiens R, Anderson R, Dyar D, Fisk M et al (2017) Basalt-trachybasalt

samples in Gale Crater, Mars. Meteorit Planet Sci 20:1–20
152. Payré V, Siebach KL, Dasgupta R, Udry A, Rampe EB, Morrison SM (2020). Constraining ancient magmatic evolution on Mars using crystal chemistry of detrital igneous minerals in the sedimentary Bradbury group, Gale crater, Mars. J Geophys Res Planets 125(8):e2020JE006467
153. Ostwald AM, Udry A, Gazel E, Payré V (2020) Assimilation-fractional crystallization on Mars as a formation process for felsic rocks. Paper presented at 51st Lunar and Planetary Institute science conference, Abstract #2200
154. Agee CB, Wilson NV, McCubbin FM, Ziegler K, Polyak VJ, Sharp ZD et al (2013) Unique meteorite from early Amazonian Mars: water-rich basaltic breccia Northwest Africa 7034. Science 339 (6121):780–785
155. Mittlefehldt DW (1994) ALH84001, a cumulate orthopyroxenite member of the martian meteorite clan. Meteoritics 29(2):214–221
156. Lapen TJ, Righter M, Brandon AD, Debaille V, Beard BL, Shafer JT et al (2010) A younger age for ALH84001 and its geochemical link to shergottite sources in Mars. Science 328(5976):347–351
157. Morris RV, Vaniman DT, Blake DF, Gellert R, Chipera SJ et al (2016) Silicic volcanism on Mars evidenced by tridymite in high-SiO2 sedimentary rock at Gale crater. Proc Natl Acad Sci 113 (26):7071–7076
158. McSween HY, Grove TL, Wyatt MB (2003) Constraints on the composition and petrogenesis of the Martian crust. J Geophys Res Planets 108(E12):5135
159. Deng S, Levander A (2020) Autocorrelation reflectivity of mars. Geophys Res Lett 28:47(16)
160. Schmidt ME, Perrett GM, Bray SL, Bradley NJ, Lee RE, Berger JA et al (2018) Dusty rocks in gale crater: assessing areal coverage and separating dust and rock contributions in APXS analyses. J Geophys Res Planets 123(7):1649–1673
161. Stolper EM, Baker MB, Newcombe ME, Schmidt ME, Treiman AH, Cousin A et al (2013) The Petrochemistry of Jake M: A Martian Mugearite. Science 341(6153):1239463
162. Treiman AH, Bish DL, Vaniman DT, Chipera SJ, Blake DF, Ming DW et al (2016) Mineralogy, provenance, and diagenesis of a potassic basaltic sandstone on Mars: CheMin X-ray diffraction of the Windjana sample (Kimberley area, Gale Crater). J Geophys Res Planets 121(1):75–106
163. Hewins RH, Zanda B, Humayun M, Nemchin A, Lorand J-P, Pont S et al (2017) Regolith breccia Northwest Africa 7533: Mineralogy and petrology with implications for early Mars. Meteorit Planet Sci 52(1):89–124
164. Udry A, Gazel E, McSween HY (2018) Formation of Evolved Rocks at Gale Crater by Crystal Fractionation and Implications for Mars Crustal Composition. J Geophys Res Planets 123(6):1525–1540
165. Wittmann A, Korotev RL, Jolliff BL, Irving AJ, Moser DE, Barker I et al (2015) Petrography and composition of Martian regolith breccia meteorite Northwest Africa 7475. Meteorit Planet Sci 50 (2):326–352
166. Weiss BP, Vali H, Baudenbacher FJ, Kirschvink JL, Stewart ST, Shuster DL (2002) Records of an ancient Martian magnetic field in ALH84001. Earth Planet Sci Lett 201(3–4):449–463
167. Santos AR, Agee CB, McCubbin FM, Shearer CK, Burger PV, Tartèse R, Anand M (2015) Petrology of igneous clasts in Northwest Africa 7034: implications for the petrologic diversity of the martian crust. Geochim Cosmochim Acta 157:56–85
168. Gaillard F, Scaillet B (2009) The sulfur content of volcanic gases on Mars. Earth Planet Sci Lett 279 (1–2):34–43
169. Farquhar J, Savarino J, Jackson TL, Thiemens MH (2000) Evidence of atmospheric sulphur in the martian regolith from sulphur isotopes in meteorites. Nature 404(6773):50–52
170. Halevy I, Zuber MT, Schrag DP (2007) A sulfur dioxide climate feedback on early Mars. Science 318(5858):1903–1907
171. Righter K, Pando K, Danielson LR (2009) Experimental evidence for sulfur-rich martian magmas: implications for volcanism and surficial sulfur sources. Earth Planet Sci Lett 288(1–2):235–243
172. O'Neill C, Lenardic A, Jellinek AM, Kiefer WS (2007). Melt propagation and volcanism in mantle convection simulations, with applications for Martian volcanic and atmospheric evolution. J Geophys Res Planets 112(E7):E07003
173. Di Genova D, Hess KU, Chevrel MO, Dingwell DB (2016) Models for the estimation of Fe3+/Fetot ratio in terrestrial and extraterrestrial alkali-and iron-rich silicate glasses using Raman spectroscopy. Am Miner 101(4):943–952
174. Sandwell DT, Müller RD, Smith WHF, Garcia E, Francis R (2014) New global marine gravity model from CryoSat-2 and Ja-son-1 reveals buried tectonic structure. Science 346:65–67
175. Filiberto J (2017) Geochemistry of Martian basalts with constraints on magma genesis. Chem Geol 466:1–14
176. Day JMD, Tait KT, Udry A, Moynier F, Liu Y, Neal CR (2018) Martian magmatism from plume metasomatized mantle. Nat Commun 9(1):4799
177. Jones JH (1986) A discussion of isotopic systematics and mineral zoning in the shergottites: evidence for a 180 m.y. igneous crystallization age. Geochimica et Cosmochimica Acta 50(6):969–977
178. Jones JH (1989) Isotopic relationships among the shergottites, the nakhlites and Chassigny. Proceedings of 19th Lunar and planetary science conference:465-474
179. Wadhwa M (2001) Redox state of Mars' upper mantle and crust from Eu anomalies in shergottite pyroxenes. Science 291(5508):1527–1530

180. Herd CDK, Borg LE, Jones JH, Papike JJ (2002) Oxygen fugacity and geochemical variations in the martian basalts: implications for martian basalt petrogenesis and the oxidation state of the upper mantle of Mars. Geochim Cosmochim Acta 66 (11):2025–2036
181. Shearer CK, Burger PV, Papike JJ, Borg LE, Irving AJ, Herd C (2008) Petrogenetic linkages among martian basalts. Implications based on trace element chemistry of olivine. Meteorit Planet Sci 43:1241–1258
182. Jones JH (2003) Constraints on the structure of the martian interior determined from the chemical and isotopic systematics of the SNC meteorites. Meteorit Planet Sci 38:1807–1814
183. Borg LE, Drake MJ (2005) A review of meteorite evidence for the timing of magmatism and of surface or near-surface liquid water on Mars. J Geophys Res 110(E12), E12S03
184. Symes SJK, Borg LE, Shearer CK, Irving AJ (2008) The age of the martian meteorite Northwest Africa 1195 and the differentiation history of the shergottites. Geochim Cosmochim Acta 72(6):1696–1710
185. Blinova A, Herd CDK (2009) Experimental study of polybaric REE partitioning between olivine, pyroxene and melt of the Y 980459 composition: insights into the petrogenesis of depleted shergottites. Geochim Cosmochim Acta 73:3471–3492
186. Borg LE, Draper DS (2003) A petrogenetic model for the origin and compositional variation of the martian basaltic meteorites. Meteorit Planet Sci 38 (12):1713–1731
187. Leone G, Tackley PJ, Gerya TV, May DA, Zhu G (2014) Three-dimensional simulations of the southern polar giant impact hypothesis for the origin of the martian dichotomy. Geophys Res Lett 41 (24):8736–8743
188. Nyquist LE, Borg LE, Shih CY (1998) The Shergottite age paradox and the relative probabilities for Martian meteorites of differing ages. J Geophys Res 103(E13):31445
189. Walton EL, Kelley SP, Herd CDK (2008) Isotopic and petrographic evidence for young Martian basalts. Geochim Cosmochim Acta 72(23):5819–5837
190. Treiman AH (1995) S-NC: Multiple source areas for Martian meteorites. J Geophys Res 100(E3):5329

Volcanic Eruptions on Mars, Lava Flow Morphology, and Thermodynamics

Caitlin Ahrens, Vincenzo Cataldo,
and Giovanni Leone

Abstract

Mars displays a variety of volcanic structures, thus giving us insight into the variety of volcanic eruption processes and lava flow morphologies. These eruption processes, from plumbing systems, supply rates, and magma mineralogy can all affect the eruption style. Understanding these different types of eruption styles represents a distinct geophysical process on Mars that in some cases resemble volcanic systems of Earth. The similarities are in the gas exsolution from magmas at shallow depth, but then depends on factors such as composition, volatile content, and temperature. The ascent of the magma can form dikes and sills, which may undergo chemical and physical evolution, which can then lead to a diversification of volcanic activity and lava flow viscosity. This chapter explores our current understanding of eruptive styles, magma reservoirs, and flow textures across Mars with a look at the thermodynamical perspectives of lava flow morphology.

C. Ahrens (✉)
Universities Space Research Association; NASA Goddard Space Flight Center, Greenbelt, MD, USA
e-mail: caitlin.ahrens@nasa.gov

V. Cataldo
School of Earth and Space Exploration, Arizona State University, Phoenix, USA

G. Leone
Instituto de Investigación en Astronomía Y Ciencias Planetarias, Universidad de Atacama, Copiapó, Chile

G. Leone (ed.), *Mars: A Volcanic World*,
https://doi.org/10.1007/978-3-030-84103-4_4

1 Introduction

It is almost impossible to think of a volcano without contemplating the power displayed during an eruption. Mars has a variety of volcanism as displayed in this book, and therefore reflect the varied processes that shaped them, and with a wide range of plumbing systems, supply rates, magma types, and eruption styles. Understanding eruptive styles and hypotheses on such mechanisms represents a distinct geophysical process on the Martian surface and subsurface. On the surface as imaged by numerous spacecraft, such as Mariner 9, Viking, and most recently the Mars Reconnaissance Orbiter (MRO), are diversities in the ancient lava flows, leading to discussions on the pyroclastic activity [1–9].

Whenever magma is produced by partial melting at depth, it will have a tendency to ascend to the surface as long as its density contrast with the surrounding materials is to essentially produce a positive buoyancy [10, 11]. Like Earth's volcanism, the ascent may occur in stages, with the magma pausing to form intrusive bodies (i.e., dikes, sills) which may undergo chemical and physical evolution. These evolutions may then lead to a diversification of eruptions and product compositions [1, 10]. One approach to understanding the styles of Martian volcanic activity has been to compare the morphologies with those of terrestrial analogs [10, 12, 13]. While useful for basic generalization, this method may overlook potentially dramatic effects between terrestrial and Martian environments on eruptive processes [14–16]. Wilson and Head [10] suggest that considering the initial processes of ascent and eruption of magma on Mars would lead to predictive analyses on the types of volcanic landforms. On the evaluation of the crustal configuration of Mars and a theoretical perspective of the ascent and eruption of magma through this crust in the Martian gravitational and atmospheric conditions, the Martian environmental conditions operate to modulate the various eruptive styles (this chapter) and morphology of the resulting landforms (see chapters "History of Scientific Studies and Current Views of Mars" and "The Origin and Evolution of Volcanism at Martian Highland Paterae: A Review of the Current State of Knowledge").

The objective of this chapter is to outline our current understanding of the eruption styles on Mars. This includes the different types of eruptions (from gas-free flows to eruptive pyroclastic activity) that may have been present on Mars, intrusive mechanisms, and the periodic activity related to magma reservoirs and the presence of calderas. We also explore the thermodynamics of lava flow morphology.

2 Eruption Styles

The range of eruption styles similar to terrestrial volcanoes have a common denominator—gas exsolution from magmas at shallow depth, which depends on factors such as composition, volatile content, and temperature (which controls magma rheology), and environmental settings (i.e., atmospheric pressure and gravity) [14, 17, 18]. The terminology for Martian eruption types is essentially and conveniently identical to eruptions on Earth [1, 10, 19]. However, we emphasize that the classifications are based on physical processes, not necessarily on the eruption deposits or landforms produced [14, 17, 20]. Wilson and Head [10] state that there are two events that may occur as a volatile-bearing magma nears the surface and the total pressure decreases. The first critical event is gas nucleation, where gas bubbles begin to nucleate if the amount of the volatile present is enough to ensure some small degree of supersaturation before the magma reaches the surface. Martian lavas have a typical volatile content of ~0.5 wt % water, ~0.7 wt% carbon dioxide, and ~0.14 wt% sulfur dioxide, and several other constituents [10, 21, 22]. The second critical event is magma disruption, where the magma may disrupt into a mixture of released gas and pyroclastics if the volatile content is relatively greater [see 14, 23]. In essence, whenever a steadily erupting magma contains adequate volatiles to ensure that it disrupts into pyroclasts, the kinetic energy produced by expansion of the released gas warrants pyroclasts and gas emergence from the vent at eruption speeds of at least a few tens of meters per second [10, 15, 24].

In steady explosive eruptions, the style is determined by a combination of the size distribution of pyroclasts and the mass fraction of released volatiles in the eruption products. These factors control the degree of coupling between the motions of clasts and gas. Large clasts have a larger terminal fall velocity through the gas than smaller, same-density clasts, and higher-density clasts have a larger terminal velocity than lower-density clasts. As a result, [25] states that there is a maximum size of clast of a given density which can be transported out of the vent.

2.1 Gas-Free

There is little disruption of the magma and its exit velocity is small (<10 ms^{-1}; <1ms^{-1} for mafic magmas) comparable to the velocity it would have had due to buoyancy or excess pressure forces [21, 26] if it had been volatile-free. In such cases, magma overflows quietly from the vent, forming a low, domelike fountain, most possibly only a few meters high, and feeds directly into a lava flow (Fig. 1).

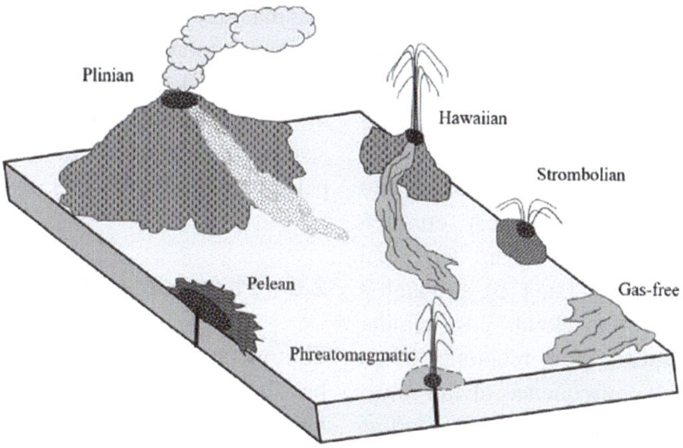

Fig. 1 The swath of physical processes occurring in the vicintity of a vent on Mars can result in a wide array of eruption styles, as shown. (*Counterclockwise*): Gas-free magma erupts to produce low, fluid feeding flows; Strombolian eruptions are due to gases exceeding magma ascension speeds, causing bubbles to grow and coalesce and lava to spatter; Hawaiian eruptions are steady explosive eruptions, creating a pyroclastic fire fountain; Plinian eruptions are due to fragmented magma locked into a gas stream, contacting a convected atmosphere and spread laterally, or pyroclastic flows (ignimbrites) stream down slopes; Pelean eruptions occur when low magma rise speed causes cooling and forms a viscous dome with pyroclastic flows; Phreatomagmatic eruptions are the solidification of magma and gas buildup in a localized conduit, causing an explosive disruption. Adapted from [10]

2.2 Strombolian

This is considered a special case when a gas/volatile-poor magma rises at a low speed and the coalescence of gas bubbles to occur in the rising magma, which can vary from passive degassing to violent fire fountaining [27, 28]. When large bubbles rise buoyantly faster than smaller bubbles, a runaway process can occur in which large bubbles emerge intermittently from the vent [10, 29], through a "puffing" action [29, 30] to more eruptive processes [31, 32]. This sweeping of a vertical column of magma nearly clean of other bubbles describes the strombolian eruption style (Fig. 1). The magma then erupts quietly during the intervals (one to tens to seconds) between the explosive emergence of the large gas bubbles [26, 27, 33].

2.3 Hawaiian

Hawaiian eruptions tend to occur in mafic magmas when most of the magma is disrupted into clots not much smaller than the maximum eruptible size, which decouples quickly from the gas to form a fountain of material and fall relatively near the vent (Fig. 1). Through this fountain process, the accumulating clasts depend on their sizes and passage out of the fountain [20, 26]. Three possible clast formations may occur: (i) cold enough to form brittle scoria deposits; (ii) hot enough to form spatter deposits; (iii) uncooled to coalesce into a lava pond which can feed one (or more) lava flows.

2.4 Plinian

Unlike Hawaiian eruptions, Plinian eruptions can occur in any magma type (Fig. 1). However, the stipulation is that most of the magma that disrupts into clots must be small that they are effectively locked by drag forces to the venting gas [10]. Fundamentally all of the eruption products ascend into an eruption cloud over the vent, provided the bulk density of such eruptions products become less than the density of the atmosphere (as a result of the heating of atmospheric gases) [10, 14, 21]. Interesting to note from [25] is that small pyroclasts can transfer heat efficiently to the atmospheric gases entrained into the eruption cloud and thus maximize buoyant convection of the eruptive cloud into the atmosphere.

Differences in the atmospheric pressure and temperature structure cause Plinian eruption clouds to rise $\sim 5 \times$ higher than terrestrial clouds for the same eruption rate [10], although it is speculated that eruption clouds pose similar cloud-height and deposit-width relationships on Mars as on Earth. A consequence of the lower Martian atmospheric pressure is that Martian Plinian deposits of any magma composition will be systematically finer-grained than those on Earth (by a factor of ~ 100 in sub-cm size) [10, 17]. Large-scale Plinian deposits, though mainly of basaltic composition [17], may be linked to the enhanced fragmentation of basaltic magma (rather than silicic) in the Martian environment, or to the interaction of basaltic magma with groundwater. However, silicic-type Plinian eruption styles must have a sufficiently large magma volatile content enough for thorough distribution of the magma around the vent. Thus if the magma had both low viscosity and sufficiently low rise velocity through the lithosphere, then gas bubble coalescence could be significant enough to lead to a more intermittent explosive style, consisting of a more coarse pyroclastic grain distribution [34].

2.5 Pelean

The main characteristic of pelean explosions (Fig. 1). is that the maximum excess pressure in the trapped gas is limited (≤ 10 MPa) by the tensile strength of the retaining cooled exterior [10, 24]. For magmas of intermediate-basaltic to silicic composition, it is common that the rise speed of the magma nearing the surface is relatively small. This leads to extensive cooling of the magma in contact with the fissure walls in the form of an extrusive dome. These factors, along with the efficient retention of gases released from the decompressing magma, can lead to the slow accumulation in the interior of a near-surface magma body (or extrusive dome) of a substantial pressure (more so than the local hydrostatic pressure) [10]. The eventual failure in tension of the cooled retention exterior leads to the explosive decompression of the gas, which drives the pelean eruption style [35].

2.6 Phreatomagmatic

The interaction between magma and ground ice has been supported to explain various (and notably unusual) landforms of possible volcanic origin on Mars, including: massive pyroclastic deposits; large-scale ground collapse and chaotic terrain; major outflow channels; lahars; sub-ice-sheet eruptions; pseudocraters; landslides on volcanic edifice flanks; and hydrothermal sites [22, 36–39].

Phreatomagmatic explosions appear to be the consequence of the interaction of juvenile material with near-surface layers of ice volatiles [10, 17, 40, 41]. These types of eruptions (Fig. 1) differ from pelean mainly in that coarse, non-juvenile, lithic material is incorporated into the ejecta. Phreatomagmatic volcanism is highly expected on Mars, particularly at the Hrad Vallis, Granicus Valles, and Hebrus Valles regions, due to the presence of significant amounts of near-surface ices in many regions [10, 42–48] whereas pelean activity would occur if/when high-viscosity magmas approach the surface.

3 Effusive Activity

The term *effusive* in volcanology refers to the non-explosive extrusion of magma at the surface, which includes all eruptions of lava flows, coulees, and domes. As previously discussed, a number of factors must be taken into account for the wide variety of effusive activity, such as environmental temperatures, pressures, and gravity; composition; volatility; porosity and other rheological characteristics. Several studies in recent years have involved the relationship between lava flow morphology to both rock chemistry and eruption mechanism through rheological physics [7, 49–51]; see Sect. 4.7 of this chapter. Martian samples through landers have found that samples from Chryse Planitia [6, 52] and Utopia Planitia [6, 52] show ~44 wt.% SiO2, ~9 wt.% Fe2O3, and lesser amounts of Al2O3, MgO, and SO3. It should be noted that the amount of SiO2 is critical in the general thickness, discharge rate, and ultimately flow of the lava, where relatively higher levels (>55 wt% SiO2) would typically be andesitic or dactite composition, much larger thickness and slower discharge rate [53]. These types are comparable to the terrestrial analogues at Trident, Alaska and Hibok-Hibok, Philippines [53]. Lower levels of SiO2 are basaltic/basaltic andesitic, with smaller thicknesses and faster discharge rate, and comparable to terrestrial sites like Etna and Vesuvius, Italy [53]. From a photogeological standpoint, a number of lava flow morphologies have been identified Fig. 2 [6, 10, 51]. This section will review the variety of volcanic features with associated lava flow morphology and lava flow morphology across volcanic plains.

3.1 Central Volcanoes

Past images from the Mariner 9 orbiter showed a variety of volcanic and volcanic-related features described primarily on the basis of their topographic profiles and include: (i) tholii (domes with steep slopes); (ii) shields; (iii) patera; and (iv) a host of small (~1 km) positive relief features considered to be cinder cones. Shields were compared to shield volcanoes on Earth, and thus interpreted to be composed essentially of basaltic lavas; tholii were then representative of more viscous, possibly silicic lavas.

Alba Patera does not fit into this category and tends to be its own classification [54, 55]. However, Alba Patera displays a diversity of lava flows that make up the structure. Four main lava flows are identified [3, 54, 55]: tube-fed and tube-channel flows; sheet flows; and undifferentiated flows (Fig. 3).

Tube-fed and tube-channel flows are characterized by radial-centric ridges to the center of Alba Patera (Fig. 3a,b). Lava channels and partially-collapsed tubes can be outlined down the axes of the flows that were considered to be the main conduits feeding the advancing flow [54]. Lava tube construction in comparison with terrestrial and lunar environments have basaltic rheological properties. More silicic lavas are too viscous, or extremely fluid lavas also do not seem to develop tubes [6, 56–58]. The width of the tubular flows around Alba Patera are only representative of the exposed part of the flow. From terrestrial counterparts, lava channels and

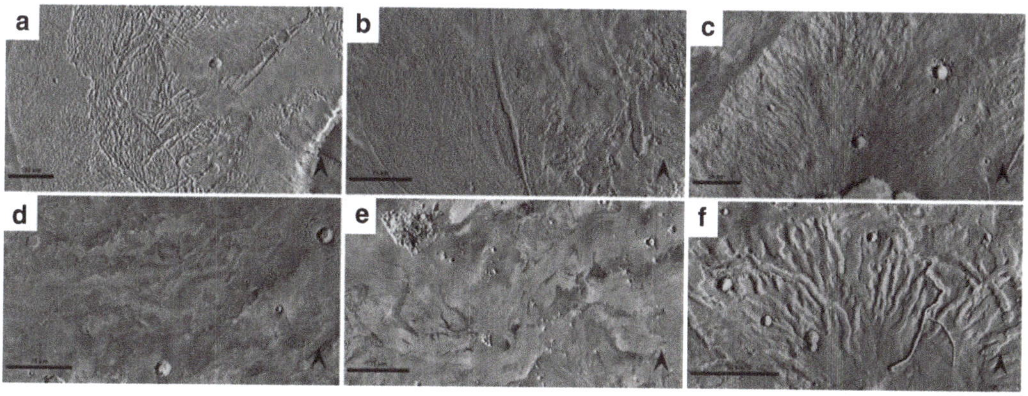

Fig. 2 Sampling of lava flow morphologies across various volcanic constructs on Mars as imaged by the THEMIS camera. **a** Olympus Mons (scale bar = 100 km); **b** Arsia Mons (scale bar = 75 km); **c** Uranius Tholus (scale bar = 25 km); **d** Daedalia Planum (scale bar = 75 km); **e** Elysium Mons (scale bar = 75 km); **f** Tyrrhena Patera (scale bar 50 km)

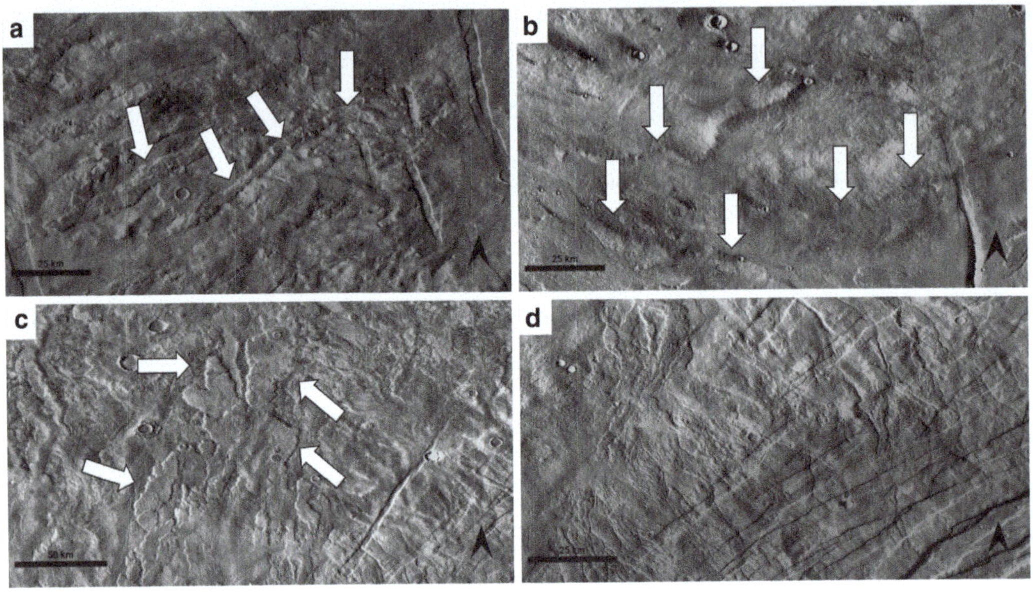

Fig. 3 THEMIS observations of various Alba Patera flows at its western and northwestern flanks. **a** tube-channel flows (scale bar = 25 km); **b** Tube-channel flows (scale bar = 25 km); **c** Sheet flows (scale bar = 50 km); **d** Undifferentiated flows (scale bar = 25 km)

tubes form constructional arches along their axes, principally by overflow and ultimately act as topographic barriers [54, 59], with the potential of subsequent flows burying the lower flanks of the arch or not able to breach the arch. Because of this, it is possible to determine the actual width of the tube-fed flows. However, the incidence angle plus the exposed portion of the flows as a minimum can estimate a flow volume [54]. Comparing such volumes to typical effusion rates of basaltic lavas on Earth gives some consideration on the curiousness on Alba. It should be noted that many flows appear to have operated essentially concurrently, either with extremely high rates of effusion or the eruptions were of very long duration. [54] favors the latter interpretation due to the lack of lava tube formation from high effusion rates (flood eruptions) on Earth, therefore Alba tube-fed flows may have developed though the eruption of basaltic lavas of moderate rates of effusion over many years of activity [55, 60, 61]. Other features that may be present in tube-fed flows are domes at the terminus of the flow, which may be the result of lava flowing within the tube system from the vent under hydrostatic loading. At the lower end of the flow, the lava may rupture the tube roof (forming a pseudo-vent) and resulting in local sprouting of flows to build up into domes [62].

Sheet flows are the most noticeable type of lava flows on Alba Patera (Fig. 3c). The actual vents for the sheet flows are not visible, where they may have been buried by their own effusive products, though appear to originate near the ring fracture. Typically sheet flows lack flow surface structures, such as tubes and channels, but form multiple, overlapping lobes of material. The relationship between the rate of effusion and length of the lava flows suggests that the sheet flows involve high rates of effusion, possibly comparable to terrestrial flood eruptions [54].

The fourth type of flow observed at Alba Patera are the undifferentiated flows (Fig. 3d). Although many of these flows are initiated by tubes and channels, they differ by having surfaces of rugged relief and indistinct flow margins [54]. It is also apparent that from the superposition relations and their levels of degradation that the undifferentiated flows at the distal ends of Alba Patera may be the oldest flows [54, 55].

Yet, the undifferentiated flows within the ring fractures are fresh appearing with well-defined channels with levees, though some cut by caldera-originated fractures, which implies a contemporaneous eruption-caldera formation sequence. The combinations of these flow types all over the vicinity of Alba Patera suggest that this volcanic region involved both flood-type eruptions (from a central region) and Hawaiian styles over long durations [54, 61].

3.2 Highland Pateras

This sub-category of central volcano includes Tyrrhena Patera, Hadriaca Patera, and the Amphitrites Patera complex [5; see chapter "Infrasound Propagation on Mars Atmosphere"]. These features are all distinguished by their low profile, radial channels, and complex central calderas. It has been proposed by several studies [63] that some of the channels on Tyrrhena Patera are primarily formed by ash eruptions along with some levels of erosion by wind, subsurface volatiles, and mass wasting [Fig. 2f; see Chapter "Infrasound Propagation on Mars Atmosphere"]. Flood lavas of the nearby and greater Hesperia Planum encroached and partly buried the flanks of Tyrrhena Patera, as observed by flow contacts through erosional channels and other depressions. From previous studies [6, 64], the last stage of volcanism at Tyrrhena Patera is the eruption of a small volume of lava in the central volcanic region, producing upper shield materials and lava channels to supply plains materials from the Tyrrhena summit [6, 54].

3.3 Shield Volcanoes

The shield volcanoes of the Tharsis region provided the first clear evidence from orbital data that volcanism was quintessential in the evolution of Mars. This category of volcano is characterized by central vent(s) often in the form of complex calderas, and composed of numerous overlapping flows (and flow units), many of which are fed by lava tubes and channels. The main Martian shield volcanoes occur in the Tharsis region (Fig. 2a–d), Elysium region (Fig. 2e), and southeast of Elysium (Apollinaris Patera region) [see chapters "History of Scientific Studies and Current Views of Mars", "The Origin and Evolution of Volcanism at Martian Highland Paterae: A Review of the Current State of Knowledge" and "Infrasound Propagation on Mars Atmosphere"]. Many of the flows appear to originate from the shields, where some cases the flanks are buried by lava flows from younger volcanoes (or unidentified sources) [54]. Although the details of the flows (and sequences thereof) vary from shield to shield, all have a complex history and share the same general morphology. The steep slopes of shield volcanoes could result from several factors, including: (i) increased proportion of pyroclastics in relation to lavas; (ii) decreased rate of effusion, producing relatively shorter lava flows; and (iii) more silicic compositions that produce more viscous flows [54, 57].

3.4 Dome Volcanoes

Domes, or tholii, on Mars are characterized by relatively steeper slopes than the aforementioned shield volcanoes, usually exceeding $\sim 8°$ in slope. The notable steepness of the domes is attributed to more viscous lavas, higher proportion of pyroclastics, lower rates of effusion, or a combination of these factors. The surface texture of these dome structures range from relatively featureless and smooth (Tharsis Tholus) to higher relief lava flows and channels (Ceraunius Tholus). However, these features are usually buried (or partly buried) by younger lava flows erupted from adjacent lava plains vents or mantled material of non-volcanic origin, which suggests an older geological age of these domes [54].

Along with the more viscous lavas, this leads to a more felsic-based composition. This was a geomorphologic case for felsic volcanic domes in western Arcadia Planitia (eastern Phlegra Montes) [65, 66], and felsic spectral signatures across volcanic features at the Nili Fossae caldera and Terra Sirenum [67–69]. These felsic domes are morphologically similar to terrestrial felsic volcanic domes by the presence of concentric structures, lava injections, viscous flows, and summit fissures [66]. These domes have the potential to form in the subsurface as *crypto-domes*. While domes are associated with lobate aprons and aureoles, [65] suggests that "pancake"-shaped domes (similar to the terrestrial "torta" domes found in the Andes [70]) were extrusive or partially extrusive, and that extrusive boulders were formed by autobrecciation—the process whereby volatile-bearing extruded magmas fragment when exposed to the lower pressure surface.

3.5 Volcanic Plains

These plains make up for roughly 60% of the surface area of Mars. Many of these smooth regions are of volcanic origin, though several criteria are used to classify these as actual volcanic lava flows. These include surface morphology, such as: lava flow fronts; embayment relations suggesting fluid emplacement; and mare-type wrinkle ridges [1, 2]. The volcanic plains are categorized into four broad groups based on more detailed and distinct morphological differences.

The first classification is the simple flow plains. Simple flow plains display low relief, and numerous mare-type wrinkle ridges, typically found in Hesperia Planum. This unit also appears to consist of a thick sequence of lavas [6, 54]. Simple flows may have been emplaced as flood lavas, extruded at extremely high rates though long fissure vent systems that spread rapidly, thus resurfacing extensive regions [54, 55, 61]. Therefore, this process is primarily important as basin-filloing and highland-resurfacing. Type examples are the Hellas, Argyre, and Isidis basins that are filled with simple flows.

The second type of volcanic plain are complex flows. These are particularly found in the vicinity of the Tharsis bulge, displaying a complex of overlapping flow lobes and lack wrinkle ridges. This type also includes some of the youngest lavas on Mars, though there is indicators of flow-lobed lava plains to be found among the oldest plains on Mars [71]. This suggests that the presence of flow lobes is an indicator of emplacement rather than a result of erosion [54]. To distinguish from the simple flows, complex flow plains result from a style of volcanism involving more protracted, sporadic, and lower rates of effusion. Although relatively higher flows in comparison to Hawaiian-type, shield-building rates, the complex flows are composed of compound, overlapping flow units of comparable thickness [72].

The third type of volcanic plain are the undifferentiated flow plains. These are classified by a lack of widespread features with the exception of a few, scattered features such as flow lobes. [54] state this classification is loosely defined and may include aeolian or periglacial relations, particularly in the northern latitudes [3, 73]. This type of volcanic plain is rather heavily based on their proximity to well-defined volcanic regions, such as the Tharsis region [74, 75].

Werner [54] includes a possible fourth classification of volcanic plains labelled the "*questionable*" plains. These plains are extensively modified by fracturing, mass wasting, and erosional processes. Some of these peripheral to the Tharsis bulge are heavily fractured and almost certainly older that have been modified by such faulting. Other regions of interest with this type of volcanic plain are at the aureoles of the Tharsis volcanoes, where origins are uncertain, but could include volcano-clastic debris, such as mud and ash flows [76, 77]. There are also numerous areas of questionable plains surrounding the Elysium complex, modified by complex tectonic and channeling processes [6, 78, 79].

In summary, volcanic plains being the majority of the Martian surface leads to implications of the phreatic activity on Mars, given the presence of volatiles within the regolith [21, 80, 81].

3.6 Leveed Lava Flows

Leveed lava flows have an interesting emplacement history, as observed by several previous studies [82–84]. Such flows on Mars (also on the Moon and Earth), deposit a significant amount of active lava volume into stationary components (e.g., levees or margins). Terrestrial analogues [85–88] and experimental experiments [89] indicate that levees are established as the front flow passes, and that additional products to the levees continue long after the frontal lobe has passed. The flow front leaves a stationary deposit along the margins of the flow to establish essentially a channel, yet subsequent clogs and breakouts occur upstream, adding additional material to the levee [82]. On Mars specifically, it is difficult to tell from morphology alone to distinguish these processes dominating levee formation. However, this flow emplacement process can lead to differences in volumetric flow rate, duration of emplacement, and certainly the rheology of the lava over time. Such examples of leveed lava flows can be observed at the Tharsis volcanoes, Arsia (Fig. 2b), Pavonis, and Ascraeus Mons [82].

4 Intrusive Activity

As we have explored eruption styles and lava flow types across the Martian surface, the conditions for flow and eruption physics also extend to the subsurface where the movement of magma ascension through the Martian rock layers can also depend on gravity/isostacy, composition, density, viscosity, and rheology. Because of the Martian lower gravity environment, fluid convective motions and settling processes driven by buoyancy forces, as well as overall diapiric ascent rates, will be slower on Mars than on Earth [10, 90]. This ascension process, along with the lower gravity on Mars, causes cooling-limited flows to be longer (nearly 6 × longer on Mars), and also favors an increase in dike widths on Mars (by a factor of 2) and subsequently higher effusion rates (by a factor of 5) [10]. With the lower Martian gravity and lower atmospheric pressure, this ensures that both nucleation and disruption of magma occur at greater depths on Mars than on Earth, although lava flow cooling rates and surface textures would be relatively the same. Heat is conducted out of the magma into the dike walls during dike emplacement. In the subsurface aquifer system, heated volatiles are incorporated into the magma as vapor bubbles [91, 92]. Conservatively, the estimate of the rate at which heat penetrates laterally is found by assuming that no convection occurs, therefore heat transfer is then limited by the thermal diffusivity of the dike wall material [91].

The propagation of magma-filled dikes and cracks from magma source regions often concludes in the dike reaching the surface to cause eruption of the magma and production of lava flows. Commonly, dikes (or portions thereof) approach the surface but do not reach it, setting up an extensional near-surface stress field that does produce consequential surficial fractures and graben [93]. Long graben systems are then evidence for lateral dike swarms radiating from major volcanic centers, where the upper part of the dikes is trapped at shallow depths by the regional stress field and magma density [93]. Wilson and Parfitt [94] and Wilson and Head [95] addressed to investigation of the influence of gravity on dike widths and eruption rates. It was found that dike propagation is limited by the dike tip stresses must overcome the effective fracture toughness of the country rocks. This implies that the mean width of a dike should be inversely proportional to the cube root of the acceleration due to gravity, and its length perpendicular to the direction of magma motion should be inversely relational to the gravity raised to the 2/3 power [10]. Flow velocities of laminar-motion-oriented magmas in dikes are proportional to the total pressure gradient and square of the dike width; whereas in more turbulent motion, the velocity is proportional to the square root of the product between the pressure gradient and dike width [10]. From this, the upper limit on magma ascent speed is set by the mean width of the widest dike that can be formed in a given stress field [10].

Density, as well as other factors, can exert controls on the upward migration of plumes and magma material. Ascending diapirs may stall upon encountering rheological barriers of cooler—more viscous—surroundings, such as the outer thermal boundary layer [10]. If the magma is still buoyant, instabilities may develop in the source region, or the viscous lid and small diapirs may ascend. Depending on the rheological properties of the country rock, diapirs may propagate from the stalled diapir or deep magma reservoir [10, 96].

There are several clues of intrusive activity across the surface of Mars. [46] observed intrusive-generated surface flows in the region around Hrad Vallis, and dike formations around Olympus Mons [97] and Cerberus Fossae [93]. In the NW area of Elysium Planitia, [46] observed some clues to dike-fed sill intrusions around ridged mountains, similar to hyaloclastic ridges in Iceland. It was postulated that these ridges formed as a consequence of a dike rising close to the surface and intruding a sill, though no explosive decompression of the sill took place, quite possibly due to a lower substrate volatile content. With the observance of additional narrow ridges running along the long axis of the mountains suggests that a dike approached close to the surface before the mountain-building production [46].

The emplacement of lateral dikes extending to significant distances (>2000 km) from Olympus Mons may be linked to the latest caldera collapsing processes, and that the last phase of summit inflation was insufficient to initiate further dike intrusions [97, 98]. From this possible connection, there may be a minimum-volume-limited threshold for dike emplacement and caldera collapse. At the smaller Tharsis region volcanoes, the volumes of the lateral dikes are greater than that of large paterae, probably due to the magma being "buffered" within the edifice [98, 100–102] or that any magma chamber was only partially emptied [97]. These dike intrusions could occur at the end-life of a magma chamber when the creation of a large dike rising from the mantle inflates the magma chamber to the point where multiple ruptures can happen [97]. Therefore, not only is a lateral dike initiated, the stresses on the overlying rocks cause caldera collapsing to begin [100].

The presence of subsurface ice (along with more felsic, silicic materials) could also influence intrusive volcanic interactions and form cryptodomes and "brain terrain" [66, 103], most notably in western Arcadia Planitia [104, 105]. Volcano-ice interactions have also been described on the northern plains of Mars [38, 39] and in the Sysiphi Montes region in the southern hemisphere [106, 107]. As noted by Farrand et al. [108], the past presence of ice in the Arcadia Planitia region is indicated by the brain terrain surface texture, a texture similar to those found in mid-latitude lobate debris aprons (LDAs) [109–112]. More discussion on this issue will be provided in chapter "The Origin and Evolution of Volcanism at Martian Highland Paterae: A Review of the Current State of Knowledge".

5 Magma Reservoirs

The lithospheric structure of Mars has been well studied by a variety of techniques [10, 16, 39, 61, 113]. [113] inferred a mean global thickness of (basaltic) crust of 34–40 km, with maximum thicknesses under Tharsis (61–77 km) and minimum thicknesses under Hellas (8–10 km). This thickness from the upper mantle strongly suggests upper mantle source regions may have an evolutionary trend of compositions as a function of time and degree of upper mantle depletion [10]. Systematic thickening of the Martian lithosphere as a function of time from thermal history models [114] estimates that lithospheric thicknesses from the tectonic response to individual loads show spatial and temporal variations in elastic thickness, with distinctly lower values beneath major volcanic provinces [115].

As stated earlier, the lower gravity on Mars plus positive/negative buoyancy forces will consequently have slower diapiric ascension rates. The differences in both gravity and surface atmospheric pressure, in addition to the differences in lithospheric bulk density, then result magma reservoirs to be deeper on Mars than on Earth by a factor of $\sim$4 [10, 116]. Thus although lithospheric thickness generally increases as a function of time, thus influencing the position of the brittle-ductile transition and level of diapiric rise versus dike emplacement, prolonged regional volcanic phases can ultimately affect the regional thermal gradient [117]. Wilson and

Head [10] also assume that the basic processes controlling partial melting of mantle materials (and magma segregation] at depth on Mars is similar to processes on Earth, occurring as a result of adiabatic decompression of the ascending mass, and gradients in density, compaction, or deviatoric stresses [118, 119]. Such thermal inhomogeneities cause localized regions of the mantle to ascend diapirically, and the sizes, shapes, and spacings of these regions (and their ascent rates) are determined by the overall advective heat flux, along with the physical properties of the volcanic bodies and environmental material [120–122]. Wilson et al. [123] argue that any one magma reservoir can survive as a thermal entity so long as it is supplied with source-region magma at a sufficient rate that its contents remain largely molten (not a high accumulation of crystallization). Once the fluid cools to ~25% crystals, it will cease to act as a Newtonian fluid and behave rheologically as a more solidified one [124].

Shallow magma chambers are extended by the injection of melt from below. The melt bodies rise buoyantly from deeper zones of partial melting as diapirs or elongated dikes. Each newly implemented melt body added to the existing reservoir brings a finite volume and internal pressure in excess of the ambient lithospheric pressure. Therefore, the reservoir reflects a constant flux (readjusting) between its internal volume and stress state of the surrounding rocks [125]. For substantial cooling to occur in an existing reservoir (of any given size), there must be a significant reduction of magma supply from the mantle or relatively long period of time. This in itself implies periodicity in the magma supply rate from the mantle [123]. As the reservoir contents cool and solidify, denser crystals sink to the bottom, and the overall density of the contents increases due to thermal contraction. The reservoir will then adjust to a new gravitational equilibrium by sinking to a deeper level within the density-stratified edifice to be neutrally buoyant [126], increasing the vertical and radial stress component beneath the lower portion of the magma reservoir and decreasing the stress around the upper part of the reservoir [127].

Relatively deep magma reservoirs can form in several different mantle environmental settings, which we will review here. Some result from the fusing of a rising diapir (or hot spot) as the adiabatic temperature reaches the melting point curve, as well as voluminous pressure release melting occurs, producing a large, partially molten magma source region [10]. Deep reservoirs are also concurrent with rift zones on Earth as a result of crustal thinning, localized mantle upwelling, and pressure release melting. Rift zones that on Mars do not exist due to the lack of plate tectonics. Commonly, magma reaches shallow levels where it achieves a neutral buoyancy with the country rock due to the decreased density of the shallow crustal materials [10].

If sufficient material reaches this level, a secondary reservoir is possible, as is observed at the Kilauea volcano on Earth [128]. These secondary magma reservoirs evolve with their centres at neutral buoyancy levels [123, 128, 129], where at depths of the magma density is equal to the density of the country rocks. Generally, silicate compositions (and thus phases) expand on melting, and so positive buoyancy of the magma in the lower portion of a chamber is expected [57]. Negative buoyancy may be due to several factors, including shallow layers in the country rock to likely consist of ancient lava flows, ash deposits, and other low-density material [10].

6 Caldera Types and Associated Processes

Calderas are large, sometimes irregular depressions, the diameter of which is many times greater than that of included subsurface vents. The basic formation of calderas are by way of a roof collapse over an underlying shallow magma reservoir, though their shape and morphology vary depending on several factors, including composition, flow rate of the subsiding magma, depth of the magma chamber, roof geometries, and environmental setting. Generally, the caldera diameter increases in proportion to volume of the associated eruption.

Buoyant rise of the magma body may cause broad uplifting and therefore production of ring and radial fractures. Volcanic leaks from the evolving magma chamber commonly produce multiple clusters of vents over the magma chamber, at times in linear or arcuate patterns [130]. However, some of this pre-collapse morphology may sometimes be covered during the caldera-collapse event.

There are a variety of tell-tale signs of a caldera collapse. These structural geomorphologies serve as a general basis of a caldera, though more complex subsidence and multi-collapse events may highly morph the overall structure. The topographic rim is simply the top of the escarpment that bounds the caldera, enclosing both the subsidence area and the area of scarp retreat due to rock falls and mass wasting [130]. Next is the inner topographic wall, which is typically the steepest in its upper portions and tends to have a concave profile that flattens downslope. Material removed by mass wasting and scarp retreat defines a collapse collar. Arcuate bounding faults (arcuate ring faults) are exposed at some deeply eroded volcanic areas (>5 km in diameter). These can also accommodate uplift as well as subsidence [130]. Most observed ring faults are near-vertical or dip steeply inward. Such the geometry and placement of the ring faults may also reflect sequential change, from an early stress field related to magma withdrawal to later stresses related to slumping. Some sizeable calderas can lack bounded arcuate ring faults, but limited erosion depths preclude unambiguous interpretation. The next geomorphic structure typical of calderas is the intracaldera fill. This provides key evidence of caldera processes because collapse occurs during associated eruption events, and ignimbrites and inter-leaved caldera breccias accumulate to multikilometre thicknesses within the subsided area. The distribution of such accumulations can influence the timing and geometry of subsidence [130]. The caldera floor geometry and overall shape can give a lot of evidence on the eruptive and post-eruptive processes. We will review the different types of caldera subsidences and their diversity of geometry.

6.1 Subsidence Processes

The complexity of Martian calderas gives rise to the variety of types of processes for such collapse events to take place. Size of the eruption and geometry of the subsurface magma chamber are also valuable factors for differing subsidence developments. We briefly review the types of subsidence that may occur for different caldera-forming processes [see also 130].

Many large calderas dominantly involve *plate* (or *piston*) collapse, which consists of a coherent floor bounded by steeply dipping faults. These are mainly from a single, large volume eruption. *Trapdoor* subsidence, bounded by an incomplete arcuate fault and by a hinged segment, reflecting early or incomplete downsagging and plate collapse. In the subsurface, this type may also include an asymmetric intrusive system or magma chamber. Simple *downsag* subsidence is fairly rare, containing inward-tilted blocks along their margins. These also require a smaller volume, though much deeper intrusive dike or magma chamber system. Terrestrial volcanoes, such as Kilauea and Mauna Loa, are notable downsag calderas that have become muti-cyclic. A component of the *piecemeal-fault* disruption of caldera floors is common, although such structures so not usually accommodate the bulk of subsidence in larger calderas. However, the caldera floors are not usually simple and smooth, more likely relating to a multi-cyclic nature [130]. Brecciated *chaotic* disruption of subsiding caldera floor and the *funnel* type can be identified to smaller (<3–5 km in diameter) calderas. Chaotic subsidence has been proposed to generate low-density material within calderas that can account for the observed negative gravity, and to generate lithic breccias by collapse of the rood over a depressurizing magma chamber. Funnel-types are associated with explosive eruptions from a central vent, lacking a bounding ring fault (or subsiding blocks), and of more mafic magma to form tuff rings and maars.

6.2 Martian Caldera Types

6.2.1 Shields

The prominent shield volcanoes of the Tharsis region have at least one summit caldera. Several nested calderas are visible on the summits of Olympus Mons (Fig. 4a), Ascraeus Mons (Fig. 4f), and Alba Patera [16, 123, 131], and all show evidence for at least one caldera collapse event. This variety of summit caldera morphologies have been classified by several previous studies [123, 132–134].

Both Ascraeus and Olympus Mons have multi-sets of summit calderas, consisting of a central caldera with smaller ones located around its margin. Arsia and Pavonis Mons (Fig. 4b), the two major southerly Tharsis volcanoes do not show this complexity in multi-caldera summit morphology. Instead, Arsia Mons has a single large caldera, and Pavonis Mons has a sag-like depression with small collapse pitting. These calderas infer the presence of shallow magma reservoirs, fed by batches of small rising magma from the mantle, within which magma resided for some time before expulsion to form either an intrusion, an extrusive lava flow, or pyroclastic flow [128]. The existence of a single caldera collapse, let alone multiple collapse depressions, implies that at least once (and then possibly several times) during the lifetime of each reservoir, a sufficiently large volume of magma was withdrawn in that the stresses of the roof exceeded the country rock strength, allowing subsidence to take place [123, 135].

Olympus Mons is by far the most morphologically complex caldera as observed on Mars. It has been noted that as much as a 2.5 km of collapse took place within the 80 × 65 km diameter caldera [136], with an elevation difference of about 2.0 km (low elevations around the oldest collapse depressions). Tectonic features on the caldera floor indicate an extensional (graben) environment around the perimeter to transition to a more compressional (ridge) formation. This transition occurs at a radial distance of ~7 km from the caldera center, which is important to note considering that this may infer the magma chamber to be relatively shallow, nearly ≤6 km beneath the caldera floor [137]. There is also evidence found within the Olympus Mons caldera for solidified lava lakes more than 30 km in width and for the localization of withdrawal of lava within the lakes [136]. Mouginis-Mark [138] identified both compressional and extensional features on the floor of the Olympus Mons complex (see chapter "The Origin and Evolution of Volcanism at Martian Highland Paterae: A Review of the Current State of Knowledge").

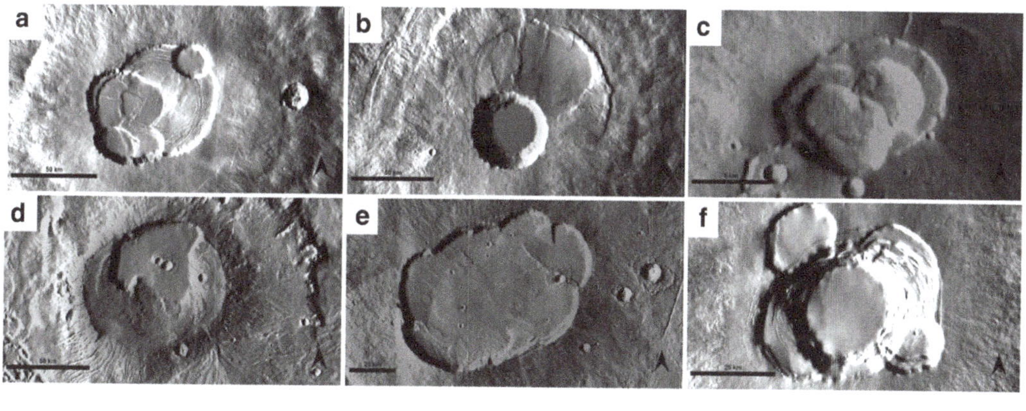

Fig. 4 Caldera complexities across the surface of Mars as imaged by THEMIS. **a** Olympus Mons nested caldera complex (scale bar 50 km); **b** Pavonis Mons simple caldera (scale bar 50 km); **c** Hecates Tholus nested complex (scale bar 5 km); **d** Apollinaris Patera downsag or multicyclic caldera (scale bar 50 km); **e** Uranius Tholus smooth-floored, scalloped-rim caldera (scale bar 25 km); **f** Ascraeus Mons with multi-arcuate ring fractures (scale bar 25 km)

These include compressional ridges (1–3 km wide) that are morphologically similar to lunar wrinkle ridges, and narrower (<1 km wide) linear ridges approximately circumferential to the perimeter of the caldera crater.

Mouginis-Mark and Robinson [136] have remarked that there is an eight-stage evolutionary sequence for the Olympus Mons caldera subsidence and found that it was a long-term process rather than near-instantaneous events from comparable terrestrial examples. The proposed sequence of deformation have been well-estimated by multiple image data collected by a number of Mars missions, including the Viking 1 Orbiter, Mars Express, Mars Global Surveyor, and Mars Reconnaissance Orbiter.

The following observations of the Olympus Mons complex have been inferred in the following stages:

(1) The first preserved summit event was the catastrophic collapse and subsequential partial infilling, which essentially defines the caldera boundary. Post-collapse talus is interpreted to be presently buried beneath younger materials.
(2) Continuation of subsidence of the central portion of the caldera crater, which created an extensional environment close to the caldera wall and a compressional environment closer to the caldera centre. Concentric graben and ridges formed at this stage as well.
(3) New collapse events occurred at the western part of the caldera, forming a 1700-m deep crater. After the solidification of the floor of this newly formed crater, an episode of circumferential graben formed close to the caldera wall, presumably due to the subsidence of the floor centre.
(4) To the south of the caldera centre, additional new subsidence events formed. These events postdate the formation of the main caldera due to the cross-cutting relationships of the graben and scarp boundaries.
(5) Resurfacing (burial of earlier tectonic features and the creation of a smooth caldera floor) by a large lava lake during (or just after) its formation. The linear ridges are interpreted to be compressional, produced by local convergence (or rafting) of solidified lava plates on the surface of the lava lake.
(6) This stage consists of the partial drainage of the lava lake, producing a topographic relief "bench."
(7) Continued subsidence of the central portion of the summit area produce a compressional stress regime [137] that promoted the formation of the larger (> 3 km wide) wrinkle ridges. The numerous wrinkle ridges formed from the partial drainage in Stage 6 were thus preserved on top of these larger wrinkle ridges.
(8) The final collapse event. Unlike the preservation of events from Stage 5, there is no evidence of surface features (i.e., linear ridges and wrinkle ridges of large-scale subsidence or deformation) that could be associated with upending of the lake on the floor.

However, there are alternative hypotheses to explain the observed the ~2 km variation in rim topography of the Olympus Mons nested caldera complex and the younger segments were built to higher elevations. Mouginis-Mark and Robinson [136] suggest that the original caldera formed off-centre, similarly to what has occurred at Mauna Loa, Hawaii. Another option is that the intrusive events associated with continued activity raised the rim crest where activity continued the longest, although extensional graben would be expected, but no such features are observed. Alternatively, one side of the summit area may have subsided relative to the other, while younger lava flows bury bounding faults, like that at Kilauea volcano, Hawaii [13, 139]. The commonality with all these hypothetical caldera-forming sequences and partial caldera infilling demonstrates that Olympus Mons underwent similar summit activity to Hawaiian volcanoes [127].

These collapse events at Olympus Mons may have spanned thousands to a few millions of years [99, 140], but there are still insufficient data to constrain the exact timing and duration of this complex activity. All of the volcanic events are implied to occur in a relatively short time period compared to the entire geologic history of Mars (over 4 billion years) [140].

6.2.2 Paterae and Tholi

Alba Patera has been considered a patera type volcanic construct [3], but also considered a shield [5] or a unique volcanic type all on its own category [6]. Although its edifice shape is certainly peculiar (although much wider and lower than any Martian shield), the caldera is essentially shield-like, similar to Uranius Tholus (Fig. 4e). There is little topographic relief, and the highest point is not the centre but part of the fractured perimeter [141, 142], reminiscent of Mauna Loa, Hawaii.

Highland paterae have very low relief with complex calderas, with incised radial channels at their flanks [5, 6; see chapter "Infrasound Propagation on Mars Atmosphere"]. The lack of radial channels at volcanoes with obvious lava flows at shield and domes suggests that the paterae are composed of pyroclastic materials. The Apollinaris Patera edifice stands about 5–6 km in relief and 200 km in diameter at its base [143]. The summit of Apollinaris has at least two separate calderas (Fig. 4d). The outer-seeming caldera is larger than the nested calderas of the Tharsis calderas at Olympus (Fig. 4a), Ascraeus (Fig. 4f), and Pavonis (Fig. 4b). Episodic eruptions and caldera formation could have taken place throughout the growth of the edifice [144]. The size of a caldera, as discussed earlier, increases with the volume of eruption materials, though the link between eruption volumes and caldera diameters are not that straight-forward due to infilling and erosion [144–146]. Apollinaris is thought to have been active during the early to mid-Hesperian [147], composed mainly of low-viscosity lavas or pyroclastic deposits [148–150].

Hecates Tholus (Fig. 4c) is another example of complex caldera formation of paterae and tholi structures [34, 151] and is the northernmost of the three volcanic constructs in the Elysium Planitia region. Lack of impact craters indicates a very young age for this eruption, possibly even postdating the most recent collapse episode at the Olympus Mons caldera complex $\sim$300 My [34, 152]. The summit of Hecates Tholus is characterized by a nested caldera nearly 11.3 $\times$ 9.1 km in extent. At least four distinct caldera collapse events are evident, with the maximum caldera depth to be $\sim$470 m. Unlike the larger Tharsis shields that show evidence of lobate lava flows on their flanks [3, 7], Hecates has no flow features of this type. However, the sinuous rilles are similar to those at Uranius and Ceraunius tholi [63], possibly by-products of erosion from volcanic density currents [63, 153]. Craters observed on the flanks of Hecates gives an approximate age of 1680 $\pm$ 326–3600 $\pm$ 38 My [5, 34]. Mouginis-Mark et al. [34] analysed Hecates Tholus and suggested that the Hecates deposit was produced by a relatively steady, Plinian explosive eruption. As mentioned earlier in this chapter, the Plinian eruption style is associated with silicic magmas with a sufficiently large magma volatile content enough for thorough distribution of the magma in and below the vent.

7 Thermal Properties and Erosional Power of Lava

Several studies have involved efforts to relate lava flow morphology to rock chemistry and eruption mechanisms through consideration of the rheological properties of the lava itself [6, 7, 154–156]. This section defines the rationale for a three-dimensional model of thermal erosion mechanisms and rheological mechanisms by lava flows.

7.1 Rationale for Creation of a Three-Dimensional Model of Thermal Erosion by Lava

Cataldo et al. [157] created a 3-D model of thermal erosion by turbulently flowing lava to assess how channel bank erosion relates to erosion at the channel bed, something that is not addressable by one- or two-dimensional models. When applied to sinuous rilles or curved channels in general, a 3-D model has the unique advantage of revealing the details of flow circulation and how flow velocity, viscosity and pressure at meander bends relate to the same flow parameters at the channel centerline and bed. *How does flow circulation at meander bends vary when varying temperature distribution within the lava? Will the details of flow circulations be the same for crusted versus uncrusted flows?* The lava that flowed at sinuous flow sites might have either been crusted over or might have lost heat by radiation. Alternatively, it might have flowed insulated and this likely caused it to reach greater downstream distances from the lava source. These different emplacement environments compel us to describe a range of possible scenarios and explore how different vertical distributions of lava temperature for each scenario might have related to the details of flow circulation at the channel bed and banks. The other advantage of a three-dimensional model lies in its ability to reveal spatial variations of erosion rates with changing flow rate, flow temperature, and meander/channel geometry.

7.2 Open Source Field Operation and Manipulation (OpenFOAM)

The new model uses the resources made available by OpenFOAM, a C + + library of applications called solvers and utilities. Solvers enable simulation of specific problems of Computational Fluid Dynamics (CFD), whereas utilities are designed to perform simple pre-and post-processing tasks like those involved in mesh generation. The OpenFOAM finite volume software is extremely versatile because users can create their own solvers and utilities or modify existing ones. Among the solvers that are used in turbulence modeling and are available in OpenFOAM, "simpleFoam" was chosen for the purpose of simulating steady-state flow scenarios. The software allows for grid design that can be tailored to the scenario of interest. Several meshes of curved channels of different length and width and with meanders of different wavelength and amplitude were generated to simulate flow.

7.3 The Finite-Volume Method

A thorough description of the method is found in [158]. The term finite volume refers to the small volume surrounding each node point in a mesh. Analogous to other numerical methods developed for the simulation of fluid flow, the finite volume method transforms a chosen set of partial differential equations into a system of linear algebraic equations. The discretization process involves two basic steps: in the first step, the partial differential equations are integrated and transformed into conservation equations over an individual volume or cell. In the second step, interpolation profiles are chosen to approximate the variation of the variables within the cell and relate the surface values of the variables to their cell values. This later step leads to the algebraic equations previously mentioned.

A key advantage of this methodology is the fact that the flux entering a given volume is identical to that leaving the same volume, i.e. the method is conservative. Furthermore, a key ingredient in the implementation of the methodology is setting up the geometrical support framework for the problem at hand, which also allows for unstructured meshes.

7.4 Turbulence Modelling

The motion of a turbulent lava flow—an incompressible fluid—can be described by the Navier–Stokes momentum equations in the x, y, and z direction. To reduce the large computational cost associated with a direct solution of the Navier–Stokes equations, statistical analyses can be used to simplify the resolution of turbulent flows. Specifically, the time-dependent nature of turbulence together with its wide range of time scales suggest that statistical averaging techniques can be applied to approximate fluctuations. The process of time-averaging introduces two new unknowns (Reynolds stresses) into the system of equations, and is applicable to Netwonian fluids only because Reynolds stresses are assumed to be a linear function of the mean velocity gradients [158]. The process of calculating Reynolds stresses is referred to as turbulence modeling and consists of finding two additional transport equations. Two among the most widely used turbulence models are the k-epsilon and SST k-omega models. Each of the two models contains two additional transport equations. The standard "k-ε" model [159, 160] is a high Reynolds number model that does not perform well in cases of adverse pressure gradients and flow separation. The k expression quantifies the turbulent kinetic energy that is produced, convected, and diffused within the flow, and the ε equation approximates turbulent dissipation. The "SST k-ω" model [161] switches to a "k-ε" behavior in the free stream, and can be used as a low Reynolds number model all the way down to the wall, thanks to the Shear Stress Transport (SST) formulation. Omega is the rate of dissipation of the turbulent eddies or, in other words, the rate at which turbulence kinetic energy is converted into internal thermal energy per unit volume and time [158]. Thanks to SST formulation, the SST k-ω model is very well-suited to determine erosion rates into the lava substrate and is especially effective in cases of adverse pressure gradients and flow separation. In this paper, we will present results obtained with the SST k-ω model.

7.4.1 Modifying SimpleFoam—The OpenFOAM Steady-State Solver for Turbulent Flow

Although our model adopts an initial value of flow velocity at the lava source (inlet value in the three directions, U_x, U_y, and U_z), the velocity field within the flowing lava is not known and has to be computed by solving the set of Navier–Stokes equations. For incompressible flows this task is complicated by the strong coupling that exists between pressure and velocity and by the fact that pressure does not appear as a primary variable in either the momentum or continuity equations [158]. The Semi Implicit Method for Pressure Linked Equations (SIMPLE) algorithm was specifically designed to reformulate the Navier–Stokes equations in terms of a momentum and a pressure equation, which are then discretized and solved sequentially. The sequence is repeated until the velocity and pressure fields satisfy both mass and momentum conservation [162]. The simpleFoam solver that is made available in the OpenFOAM library is well-suited for the purpose of modeling turbulent, incompressible flow. The key input parameters of the flow are velocity and pressure, and flow is investigated in steady conditions. However, no initial temperature is specified in the solver, a severe limitation when modeling turbulently flowing lava.

To solve the temperature limitation inherent in the solver, a 3-D temperature model was coupled to simpleFoam. The first step was to choose an energy expression that could be applicable to a fixed control volume and fit a scenario of turbulently flowing lava. An energy equation can be written in terms of specific internal energy, specific enthalpy, specific total enthalpy, or temperature. An energy expression written in terms of temperature assumes that the flow is Newtonian because the specific enthalpy of the flow is only a function of pressure and temperature variations [158]. Such an assumption is consistent with a scenario of turbulently flowing lava for which lava temperatures are assumed to be above at or near the liquidus value. Finally, our steady-state energy expression as a function of temperature can be written as:

$$\nabla\left[\rho c_p \mathbf{v} T\right] = \nabla[k \nabla T] \qquad (1)$$

in which ρ is density of the liquid lava, c_p is lava specific heat capacity, **v** is the velocity vector ($\mathbf{v} = v_x, v_y, v_z$), T is lava temperature, and k is lava thermal conductivity. Because we are dealing with an incompressible fluid, lava density is taken as a constant. If we normalize Eq. (1) by density, we obtain:

$$\nabla\left[c_p \mathbf{v} T\right] = \nabla\left[\frac{k}{\rho} \nabla T\right] \qquad (2)$$

Conveniently, the lava *effective thermal diffusivity* α_{eff} can be expressed as a function of lava thermal conductivity, k, as follows:

$$\alpha_{eff} = \frac{k}{\rho c_p} \quad (3)$$

If we solve for k and then substitute Eq. (3) into Eq. (2), we obtain: Eq. (3).

$$\nabla\left[c_p \mathbf{v} T\right] = \nabla\left[c_p \alpha_{eff} \nabla T\right] \quad (4a)$$

or

$$\nabla[\mathbf{v} T] = \nabla[\alpha_{eff} \nabla T] \quad (4b)$$

The left-hand side term in Eq. (4a) is a convection term and the right-hand side one is a diffusion term and they are both normalized by density. Equation (4b) represents the temperature model that was coupled to the SimpleFoam solver, and the suffix "eff" stands for effective. The *effective thermal diffusivity* accounts for the laminar and turbulent components of diffusivity, and is given by:

$$\alpha_{eff} = \frac{\upsilon}{Pr + \alpha_t} \quad (5)$$

in which υ is the kinematic viscosity of the lava, Pr is the molecular (laminar) Prandtl number and α_t is the kinematic turbulent thermal diffusivity that is expressed as:

$$\alpha_t = \frac{\upsilon_t}{Pr_t} \quad (6)$$

7.5 Deriving an Expression for Erosion Rate of the Lava Substrate

The one-dimensional expression that allows determination of the erosion rate into the lava substrate, u_m, is obtained by adopting the approach taken by [163–166], as follows:

$$u_m = \frac{h_T(T_l - T_m)}{\rho_g\left[c_g(T_m - T_0) + L_g\right]} \quad (7)$$

in which the numerator represents the convective heat flux to the base of the flow and the denominator defines the energy required to melt the ground. Specifically, h_T is the convective heat transfer coefficient, T_l is the temperature of the liquid lava, T_m is the temperature of melting of the ground, ρ_g is the density of the lava substrate, c_g is the substrate specific heat capacity, T_0 is substrate temperature away from the lava/substrate interface, and L_g is latent heat of fusion of the ground. Equation (7) calculates the steady-state rate of advance of the ground/lava interface, when thermal equilibrium is assumed at the lava/ground interface.

To find an expression that can be used in our model, we must equate the numerator of Eq. (7) to a heat flux expression that contains thermal diffusivity and temperature gradient of the lava in a direction perpendicular to the lava/substrate interface, as follows:

$$h_T(T_l - T_m) = -\rho c_p \alpha_{eff} \frac{\partial T}{\partial n} \quad (8)$$

in which the left-hand side term represents the convective heat flux to the base of the flow and $\frac{\partial T}{\partial n}$ is the temperature gradient of the lava in a direction perpendicular to the channel bed and banks. The convective heat transfer coefficient [167] can be expressed as:

$$h_T = \frac{0.027\, k\, Re^{0.8} Pr^{0.33}}{h} \left(\frac{\mu_b}{\mu_g}\right) 0.14 \quad (9)$$

in which Re is the Reynolds number of the flow, expressed as:

$$Re = \frac{UL}{\upsilon} \quad (10)$$

in which U is flow velocity, L is the characteristic length of the flow (here the hydraulic diameter of the channel) and υ is the kinematic viscosity of the lava. In Eq. (9), h is lava thickness μ_b is the bulk dynamic viscosity of the flow and μ_g is the dynamic viscosity of the lava substrate. Equation (9) shows how the value of the heat transfer coefficient is influenced by flow thickness, and also accounts for the viscosity of the substrate. By combining Eqs. (8) and (7), we obtain:

$$u_m = \frac{-\rho c_p \alpha_{eff} \frac{\partial T}{\partial n}}{\rho_g [c_g (T_m - T_0) + L_g]} \quad (11)$$

which is used to calculate erosion rates into the channel bed and banks. In the following section, we will describe the key input parameters and assumptions of the model. This will help identify the flow parameters in Eq. (9) that are most effective in determining variations in erosion rate Eq. (11).

7.6 Model Input Parameters and Assumptions

The input parameters of the model can be divided into physical, thermo-physical and turbulence parameters of the flowing lava. Among the physical parameters, the model requires initial flow velocities or velocities at the channel inlet. The temperature of the lava at time of eruption is also another key input parameter. The model adopts the Dragoni et al. [168] temperature-dependent viscosity that—though it was originally used for the purpose of investigating flow behavior in the Bingham laminar regime—can be used to model the evolution of lava temperature and viscosity as long as flow rate is fixed (a scenario for which flow height and velocity are both independent of lava yield stress and flow regime is Newtonian [168]. The vertical distribution of temperatures within the flow is similar to that modelled by previous authors [169, 170], with higher temperature values at the flow top and core and lower values at the channel bed. Also, we assume for lava temperature and viscosity at the channel banks to be the same as those at the channel bed. The actual temperature values at the channel bed and banks were chosen based on our understanding of how convective heat flux is transported to the base of the flow. Holding the initial lava temperature and viscosity equal, the higher the flow thickness the higher the convective heat flux to flow boundaries, i.e., the value of the left-hand side term of Eq. (8). As a result, the temperature gap occurring between the flow top (and core) and the boundaries will be greater for a thicker flow. As lava temperature decreases, viscosity increases at both the channel bed and banks.

The model assumes steady-state flow conditions, an assumption justified by the likely long-lasting duration of eruptive events associated with the formation of sinuous rilles [171]. The lava substrate is modeled as consolidated and unconsolidated, depending on available information at individual flow sites.

8 Summary

In this chapter we summarized the present state of knowledge on Martian magma reservoirs, consequent eruption styles, lava flows and caldera processes, and the rheological implications of lava flow dynamics using current modeling techniques. These processes are essential to understanding the geologic foundations of Martian volcanism and evolution of varying volcanic constructs across the Martian surface. Despite over many years of analysis, many outstanding questions regarding the construction of these various volcanoes (including paterae and tholii), such as the transition from explosive to effusive activity. The model also presented in this chapter should also help identify lava flow thermophysical properties and flow parameters for erosion rates. We hope that this chapter will provide a guide for eruption, magma and volcanic construction, and lava flow types.

References

1. Carr M (1973) Volcanism on Mars. J Geophys Res 78(20):4049–4062
2. Carr M, Clow G (1981) Martian channels and valleys: their characteristics, distribution, and age. Icarus 48(1):91–117
3. Carr M, Greeley R, Blasius K, Guest J, Murray J (1977) Some Martian volcanic features as viewed from the Viking orbiters. JGR 82(28):3985–4015
4. Greeley R (1973) Mariner 9 photographs of small volcanic structures on Mars. Geology 1(4):175–180

5. Plescia J, Saunders R, (1979) The chronology of the martian volcanoes. In: 10th lunar and planetary science conference proceedings, vol 3. pp 2841–2859
6. Greeley R, Spudis P (1981) Volcanism on Mars. Rev Geophys Space Phys 19(1):13–41
7. Schaber G, Horstman K, Dial A (1978) Lava flow materials in the Tharsis region of Mars. In: 9th lunar and planetary science conference proceedings, vol 3. pp 3433–3458
8. Schaber G (1982) Syrtis Major: a low-relief volcanic shield. J Geophys Res Solid Earth 87 (B12):9852–9866
9. Hodges C, Moore H, (1994) Atlas of volcanic landforms on Mars. US Government Printing Office (No. 1534)
10. Wilson L, Head J (1994) Mars: review and analysis of volcanic eruption theory and relationships to observed landforms. Rev Geophys 32(3):221–263
11. Black B, Manga M (2016) The eruptibility of magmas at Tharsis and Syrtis Major on Mars. JGR Planets 121(6):944–964
12. Gulick V, Baker V (1990) Origin and evolution of valleys on Martian volcanoes. JGR Solid Earth 95 (B9):14325–14344
13. Seelos K, Arvidson R, Jolliff B, Chemtob S, Morris R, Ming D, Swayze G (2010) Silica in a Mars analog environment: Ka'u Desert, Kilauea Volcano, Hawaii. J Geophys Res Planets 115(E4):E00D15
14. Wilson L, Head J (1983) A comparison of volcanic eruption processes on Earth, Moon, Mars, Io and Venus. Nature 302:663–669
15. Wilson L (1984) The influences of planetary environments on the eruption styles of volcanoes. Vistas Astron 27:333–360
16. Mouginis-Mark P, Wilson L, Zuber M (1992) The physical volcanology of Mars; In: Kieffer HH, Jakosky BM, Snyder CW, Matthews MS (eds) Mars. University of Arizona Press, Tuscon, pp 424–452
17. Wall K, Rowe M, Ellis B, Schmidt M, Eccles J (2014) Determining volcanic eruption styles on Earth and Mars from crystallinity measurements. Nat Commun 5(5090):1–8
18. Houghton BF et al (2004) The influence of conduit processes on changes in style of basaltic Plinian eruptions: Tarawera 1886 and Etna 122 BC. J Volcanol Geotherm Res 137:1–14
19. Fisher R, Schmincke H (1984) Pyroclastic rocks. p 472, Springer-Verlag, New York
20. Head J, Wilson L (1986) Volcanic processes and landforms on Venus: theory predictions and observations. J Geophys Res 91:9407–9446
21. Craddock R, Greeley R (2009) Minimum estimates of the amount and timing of gases released into the martian atmosphere from volcanic eruptions. Icarus 204(2):512–526
22. Kusanagi T, Matsui T (2000) The change of eruption styles of Martian volcanoes and estimates of the water content of the Martian mantle. Phys Earth Planet Inter 117(1–4):437–447
23. Sparks R (1978) The dynamics of bubble formation and growth magmas: a review and analysis. J Volcanol Geotherm Res 3:1–37
24. Wilson L (1980) Relationships between pressure, volatile content and ejecta velocity in three types of volcanic explosion. J Volcanol Geotherm Res 8:297–313
25. Wilson L (1976) Explosive volcanic eruptions, III, Plinian eruption columns. Geophys J R Astron Soc 45:543–556
26. Wilson L, Head J (1981) Ascent and eruption of basaltic magma on the Earth and Moon. J Geophys Res 86:2971–3001
27. James M, Lane S, Wilson L, Corder S (2009) Degassing at low magma-viscosity volcanoes: quantifying the transition between passive bubble-burst and Strombolian eruption. J Volcanol Geotherm Res 180(2–4):81–88
28. Parfitt E, Wilson L (1995) Explosive volcanic eruptions—IX. The transition between Hawaiian-style lava fountaining and Strombolian explosive activity. Geophys J Int 121(1):226–232
29. Harris A, Ripepe M (2007) Temperature and dynamics of degassing at Stromboli. J Geophys Res Solid Earth 112(B3):B03205
30. Ripepe M, Harris A, Carniel R (2002) Thermal, seismic and infrasonic evidences of variable degassing rates at Stromboli volcano. J Volcanol Geoth Res 118(3–4):285–297
31. Chouet B, Hamisevicz N, McGetchin T (1974) Photoballistics of volcanic jet activity at Stromboli Italy. J Geophys Res 79(32):4961–4976
32. Blackburn E, Wilson L, Sparks R (1976) Mechanisms and dynamics of strombolian activity. J Geol Soc London 132:429–440
33. Parfitt E (2004) A discussion of the mechanisms of explosive basaltic eruptions. J Volcanol Geoth Res 134(1–2):77–107
34. Mouginis-Mark P, Wilson L, Head J (1982) Explosive volcanism at Hecates Tholus, Mars: investigation of eruption conditions. J Geophys Res 87:9890–9904
35. Eichelberger J, Hayes D (1982) Magmatic model for the Mount St. Helens blast of May 18, 1980. J Geophys Res 87:7727–7738
36. Head, J. W., and Wilson, L. (2002). Mars: a review and synthesis of general environments and geological settings of magma-H2O interactions, vol 202 (1). Geological Society, London, Special Publications, pp 27–57
37. Schulze-Makuch D, Dohm J, Fan C, Fairén A, Rodriguez J et al (2007) Exploration of hydrothermal targets on Mars. Icarus 189(2):308–324
38. Allen CC (1979) Volcano-ice interactions on Mars. J Geophys Res Solid Earth 84(B14) 8048–8059
39. Chapman M, Allen C, Gudmundsson M, Gulick V, Jakobsson S et al (2000) Volcanism and ice

interactions on Earth and Mars; environmental effects on volcanic eruptions. Springer, Boston, MA, pp 39–73
40. Self S, Wilson L, Nairn I (1979) Vulcanian eruption mechanisms. Nature 277:440–443
41. Brož P, Hauber E (2013) Hydrovolcanic tuff rings and cones as indicators for phreatomagmatic explosive eruptions on Mars. J Geophys Res Planets 118 (8):1656–1675
42. Clifford SM (1981) A pore volume estimate of the Martian megaregolith based on a lunar analog. In: 3rd international colloquium on Mars, vol 441. p 46
43. Clifford SM (1984) A model for the climatic behavior of water on Mars. UMass dissertations, Scholarworks. https://scholarworks.umass.edu/cgi/viewcontent.cgi?article=2782&context=dissertations_1
44. Clifford SM (1993) A model for the hydrologic and climatic behavior of water on Mars. J Geophys Res Planets 98(E6):10973–11016
45. Squyres SW, Cliffor SM, Kuzmin RO, Zimbelman JR, Costard F (1992) 16: Ice in the Martian Megaregolith. In: Mars. pp 523–554
46. Wilson L, Mouginis-Mark PJ (2003) Phreatomagmatic explosive origin of Hrad Vallis, Mars. J Geophys Res Planets 108(E8):5082
47. De Hon RA (1992) Martian lake basins and lacustrine plains. Earth Moon Planet 56(2):95–122
48. Mouginis-Mark PJ (1985) Volcano/ground ice interactions in Elysium Planitia Mars. Icarus 64 (2):265–284
49. Baloga S, Mouginis-Mark PJ, Glaze LS (2003) Rheology of a long lava flow at Pavonis Mons, Mars. J Geophys Res Planets 108(E7):5066
50. Bleacher JE, Greeley R, Williams DA, Cave SR, Neukum G (2007) Trends in effusive style at the Tharsis Montes, Mars, and implications for the development of the Tharsis province. J Geophys Res Planets 112(E9):E09005
51. Crown D, Ramsey M (2017) Morphologic and thermophysical characteristics of lava flows southwest of Arsia Mons Mars. J Volcanol Geotherm Res 342:13–28
52. Toulmin P, Baird A, Clark B, Keil K, Rose H et al (1977) Geochemical and mineralogical interpretation of the Viking inorganic chemical results. J Geophys Res 82:4625–4634
53. Killburn C (2000) Lava flows and flow fields. In: Sigurdsson H et al (eds) Encyclopedia of volcanoes. Academic Press, pp 291–307
54. Werner SC (2009) The global martian volcanic evolutionary history. Icarus 201(1):44–68
55. Cattermole P (1990) Volcanic flow development at Alba Patera Mars. Icarus 83(2):453–493
56. Greeley R (1977) Basaltic 'plains' volcanism, volcanism of the eastern snake river plain, Idaho, NASA Contract Rep CR-154621, 23–43
57. Skok J, Mustard J, Ehlmann B, Milliken R, Murchie S (2010) Silica deposits in the Nili Patera caldera on the Syrtis Major volcanic complex on Mars. Nat Geosci 3:838–841
58. Francis P, Wood C (1982) Absence of silicic volcanism on Mars: implications for crustal composition and volatile abundance. J Geophys Res 87:9881–9889
59. Greeley R (1971) Observations of actively forming lava tubes and associated structures, Hawaii. Mod Geol 2:207–223
60. Mouginis-Mark PJ, Wilson L, Zimbelman JR (1988) Polygenic eruptions on Alba Patera Mars. Bull volcanol 50(6):361–379
61. Lopes RM, Kilburn CR (1990) Emplacement of lava flow fields: application of terrestrial studies to Alba Patera, Mars. J Geophys Res Solid Earth 95 (B9):14383–14397
62. Greeley R, Hyde J (1972) Lava tubes of the Cave basalt, Mount St. Helens, Washington. Geol Soc Am Bull 83:2397–2418
63. Riemers C, Komar P (1979) Evidence for explosive volcanic density currents on certain Martian volcanoes. Icarus 39:88–110
64. Greeley R, Crown D (1990) Volcanic geology of Tyrrhena Patera, Mars. J Geophys Res 95:7133–7149
65. Rampey ML, Milam KA, McSween Jr HY, Moersch JE, Christensen PR (2007) Identity and emplacement of domical structures in the western Arcadia Planitia, Mars. J Geophys Res Planets 112 (E6):E06011
66. Farrand WH, Rice JW, Chuang FC, Rogers AD (2020) Spectral and geological analyses of domes in western Arcadia Planitia, Mars: evidence for intrusive alkali-rich volcanism and ice-associated surface features. Icarus 357:114111
67. Brož P, Hauber E, Platz T, Balme M (2015) Evidence for Amazonian highly viscous lavas in the southern highlands on Mars. Earth Planet Sci Lett 415:200–212
68. Christensen PR, McSween HY, Bandfield JL, Ruff SW, Rogers AD et al (2005) Evidence for magmatic evolution and diversity on Mars from infrared observations. Nature 436(7050):504–509
69. Wray JJ, Hansen ST, Dufek J, Swayze GA, Murchie SL et al (2013) Prolonged magmatic activity on Mars inferred from the detection of felsic rocks. Nat Geosci 6(12):1013–1017
70. Calder ES, Lavallée Y, Kendrick JE, Bernstein M (2015) Lava dome eruptions. In: The encyclopedia of volcanoes. Academic Press, pp 343–362
71. Greeley R, Spudis PD (1978) Volcanism in the cratered terrain hemisphere of Mars. Geophys Res Lett 5(6):453–455
72. Walker G (1972) Compound and simple lava flows and flood basalts. Bull Volcanol 35:579–590
73. Carr M, Schaber G (1977) Martian permafrost features. J Geophys Res 82:4039–4054
74. Guest JE, Butterworth PS, Greeley R (1977) Geological observations in the Cydonia region of Mars from Viking. J Geophys Res 82(28):4111–4120

75. Scott D (1979) Geologic problems in the northern plains of Mars. In: 10th proceeding lunar planet science conference, 3039–3054
76. Hodges C, Moore H (1979) The subglacial birth of Olympus Mons and its aureoles. J Geophys Res 84:8061–8074
77. Lopes RM, Guest JE, Wilson CJ (1980) Origin of the Olympus Mons aureole and perimeter scarp. Moon Planets 22(2):221–234
78. Platz T, Michael G (2011) Eruption history of the Elysium volcanic province Mars. Earth Planet Sci Lett 312(1–2):140–151
79. Plescia JB (1990) Recent flood lavas in the Elysium region of Mars. Icarus 88(2):465–490
80. King JS, Riehle JR (1974) A proposed origin of the Olympus Mons escarpment. Icarus 23(2):300–317
81. Squyres SW, Wilhelms DE, Moosman AC (1987) Large-scale volcano-ground ice interactions on Mars. Icarus 70(3):385–408
82. Glaze LS, Baloga SM, Garry WB, Fagents SA, Parcheta C (2009) A hybrid model for leveed lava flows: implications for eruption styles on Mars. J Geophys Res Planets 114(E7):E07001
83. Hulme G (1974) The interpretation of lava flow morphology. Geophys J R Astron Soc 39:361–383
84. Carr MH (1974) The role of lava erosion in the formation of lunar rilles and Martian channels. Icarus 22:1–23
85. Linneman SR, Borgia A (1993) The blocky andesitic lava flows of Arenal volcano, Costa Rica. In: Kilburn RJ, Luongo G (eds) Active lavas. Routledge, London, pp 25–72
86. Lipman PW, Banks NG (1987) A flow dynamics, Mauna Loa 1984, in Volcanism in Hawaii. In: Decker RW et al (ed) Geological survey professional paper, vol 1350. pp 1527–1567
87. Moore, H. J. (1987), Preliminary estimates of the rheological properties of 1984 Mauna Loa lava, in Volcanism in Hawaii. In: Decker RW et al (ed) Geological survey professional paper, vol 1350. U.S. Geological Survey, pp 1569–1588
88. Guest JE, Spudis P, Greeley R, Taylor G, Baloga S (1995) Emplacement of xenolith nodules in the Kaupulehu lava flow, Hualalai Volcano, Hawaii. Bull Volcanol 57:179–184
89. Garry WB (2006) The emplacement of channeled flows in subaerial, submarine, simulated, and extraterrestrial environments, Ph.D. dissertation, State University of New York at Buffalo, p 240
90. Carrigan C (2000) Plumbing systems. In: Sigurdsson H et al (eds) Encyclopedia of volcanoes. Academic Press, pp 219–237
91. Wilson L, Head III JW (2004) Evidence for a massive phreatomagmatic eruption in the initial stages of formation of the Mangala Valles outflow channel, Mars; Geophys Res Lett 31(15):L15701
92. Bonafede M, Mazzanti M (1997) Hot fluid migration in compressible saturated porous media. Geophys J Int 128(2):383–398
93. Head JW, Wilson L, Mitchell K (2003) Generation of recent massive water floods at Cerberus Fossae, Mars by dike emplacement, cryospheric cracking, and confined aquifer groundwater release. Geophys Res Lett 30(11):1577
94. Wilson L, Parfitt EA (1990) Widths of dikes on Earth and Mars. In: Mars: evolution of volcanism, Tectonics, and Volatiles, p 308
95. Wilson L, Head JW (1988) The influence of gravity on planetary volcanic eruption rates. Proceedings of the 19th Lunar and planetary science conference: Abstract #1283
96. Rubin A (1993) Dikes vs. diapirs in viscoelastic rock. Earth Planet. Sci Lett 119(4):641–659
97. Mouginis-Mark PJ, Wilson L (2019) Late-stage intrusive activity at Olympus Mons, Mars: summit inflation and giant dike formation. Icarus 319:459–469
98. Scott ED, Wilson L, Head J (2002) Emplacement of giant radial dikes in the northern Tharsis region of Mars. J Geophys Res Planets 107(E4):3–1
99. Neukum G, Jaumann R, Hoffmann H, Hauber E, Head J et al (2004) Recent and episodic volcanic and glacial activity on Mars revealed by the High Resolution Stereo Camera. Nature 432(7020):971–979
100. Robbins SJ, Di Achille G, Hynek BM (2011) The volcanic history of Mars: high-resolution crater-based studies of the calderas of 20 volcanoes. Icarus 211(2):1179–1203
101. Parfitt EA, Head JW (1993) Buffered and unbuffered dike emplacement events on the Earth and Venus: implications for the magma reservoir size, depth, and rate of magma replenishment. Earth Moon Planet 61(3):249–281
102. Scott ED, Wilson L (1999) Evidence for a sill emplacement event on the upper flanks of the Ascraeus Mons shield volcano Mars. J Geophys Res Planets 104(E11):27079–27089
103. Levy JS, Head JW, Marchant DR (2009) Concentric crater fill in Utopia Planitia: history and interaction between glacial "brain terrain" and periglacial mantle processes. Icarus 202(2):462–476
104. Bramson AM, Byrne S, Putzig NE, Sutton S, Plaut JJ et al (2015) Widespread excess ice in Arcadia Planitia Mars. Geophys Res Lett 42 (16):6566–6574
105. Piqueux S, Buz J, Edwards CS, Bandfield JL, Kleinböhl A et al (2019) Widespread shallow water ice on Mars at Highlatitudes and Midlatitudes. Geophys Res Lett 46(24):14290–14298
106. Farrand WH, Lane MD, Edwards BR (2008) North and south: possible tuyas and hyaloclastite hills on the northern plains and in the southern dorsa argentea region of Mars. LPI 1391:1761
107. Ackiss S, Horgan B, Seelos F, Farrand W, Wray J (2018) Mineralogic evidence for subglacial volcanism in the Sisyphi Montes region of Mars. Icarus 311:357–370

108. Farrand WH, Lane MD, Edwards BR, Yingst RA (2011) Spectral evidence of volcanic cryptodomes on the northern plains of Mars. Icarus 211(1):139–156
109. Squyres SW (1979) The distribution of lobate debris aprons and similar flows on Mars. J Geophys Res Solid Earth 84(B14):8087–8096
110. Mangold N (2003) Geomorphic analysis of lobate debris aprons on Mars at Mars Orbiter Camera scale: evidence for ice sublimation initiated by fractures. J Geophys Res Planets 108(E4):8021
111. Head JW, Nahm AL, Marchant DR, Neukum G (2006) Modification of the dichotomy boundary on Mars by Amazonian mid-latitude regional glaciation. Geophys Res Lett 33(8):L08S03
112. Holt JW, Safaeinili A, Plaut JJ, Young DA, Head JW et al (2008) Radar sounding evidence for ice within lobate debris aprons near Hellas Basin, mid-southern latitudes of Mars. LPI 1391:2441
113. Bills BG, Ferrari AJ (1978) Mars topography harmonics and geophysical implications. J Geophys Res Solid Earth 83(B7):3497–3508
114. Stevenson DJ, Spohn T, Schubert G (1983) Magnetism and thermal evolution of the terrestrial planets. Icarus 54(3):466–489
115. Comer R, Solomon SC, Head JW (1985) Mars: thickness of the lithosphere from the tectonic response to volcanic loads. Rev Geophys 23 (1):61–92
116. Elder JW (1987) The structure of the planets. Academic, San Diego, California, p 210
117. Solomon SC, Head JW (1990) Heterogeneities in the thickness of the elastic lithosphere of Mars: constraints on heat flow and internal dynamics. J Geophys Res 95:11073–11083
118. Oxburgh ER (1980) Heat flow and magma genesis. In: Hargraves RB (ed) Physics of magmatic processes. Princeton University Press, Princeton, N.J., pp 161–199
119. Sleep N (1988) Tapping of melt by veins and dikes. J Geophys Res 93:10255–10272
120. Marsh B, Carmichael I (1974) Benioff zone magmatism. J Geophys Res 79:1196–1206
121. Marsh BD, Kantha L (1978) On the heat and mass transfer from an ascending magma, Earth Planet. Sci Lett 39:435–443
122. Spera FJ (1980) Aspects of magma transport. In: Hargraves RB (ed) Physics of magmatic processes. Princeton University Press, Princeton, N.J., pp 265–323
123. Wilson L, Scott ED, Head JW (2001) Evidence for episodicity in the magma supply to the large Tharsis volcanoes. J Geophys Res Planets 106(E1):1423–1433
124. Marsh B (1984) Mechanics and energetics of magma formation and ascension. Studies in Geophysics. Explosive volcanism: inception, evolution, and hazards 67–83
125. Blake S (1981) Volcanism and the dynamics of open magma chambers. Nature 289:783–785
126. Walker GPL, Self S, Wilson L (1984) Tarawera, 1886, New Zealand—A basaltic plinian fissure eruption. J Volcanol Geotherm Res 21:61–78
127. Gudmundsson A (1990) Emplacement of dikes, sills and crustal magma chambers at divergent plate boundaries. Tectonophysics 176(3–4):257–275
128. Ryan MP (1987) Neutral buoyancy and the mechanical evolution of magmatic systems, in magmatic processes: physiochemical Principles. In: Mysen BO (ed) Spec Publ Geochem Soc 1:259–287
129. Rubin AM, Pollard DD (1987) Origins of blade-like dikes in volcanic rift zones. US Geol Surv Prof Pap 1350(2):1449–1470
130. Lipman P (2000) Calderas. In: Sigurdsson H et al (eds) Encyclopedia of volcanoes Academic Press, pp 643–663
131. Zimbelman JR, Edgett KS (1992) The Tharsis Montes, Mars: comparison of volcanic and modified landforms. In: proceedings of lunar and planetary science, vol 22. Lunar and Planetary Institute
132. Crumpler LS, Head JW, Aubele JC (1994) Calderas on Mars: classification, characteristics, and processes related to mechanisms of formation. In: Lunar and planetary science conference, vol 25. p 305
133. Crumpler LS, Head JW, Aubele JC (1996) Calderas on Mars: Characteristics, structure, and associated flank deformation. Geological Society, London, Special Publications 110(1):307–348
134. Zimbelman J (2000) Volcanism on Mars. In: Sigurdsson H et al (eds) Encyclopedia of volcanoes. Academic Press, pp 771–783
135. Marti J, Ablay GJ, Redshaw LT, Sparks RS (1994) Experimental studies of collapse calderas. J Geol soc 151(6):919–929
136. Mouginis-Mark PJ, Robinson MS (1992) Evolution of the Olympus Mons Caldera Mars. Bulletin Volcanol 54(5):347–360
137. Zuber MT, Mouginis-Mark PJ (1992) Caldera subsidence and magma chamber depth of the Olympus Mons volcano. Mars. J Geophys Res Planets 97(E11):18295–18307
138. Mouginis-Mark PJ (1990) Recent water release in the Tharsis region of Mars. Icarus 84(2):362–373
139. Holcomb, R. T. (1987), Eruptive history and long-term behavior of Kilauea Volcano, in Volcanism in Hawaii. In: Decker RW et al (ed) Geological survey professional paper, vol 1350. U.S. Geological Survey, pp 261–350
140. Scott D, Carr M (1978) Geologic map of Mars, scale 1:25,000,000. U.S. Geol. Surv. Misc. Invest. Ser., Map I-1083
141. Carr MH (1976) The volcanoes of Mars. Sci Am 234(1):32–43
142. Wood CA (1984) Calderas: a planetary perspective. J Geophys Res Solid Earth 89(B10):8391–8406

143. Plescia JB (2004) Morphometric properties of Martian volcanoes. J Geophys Res 109:E03003
144. Kerber L, Head JW, Madeleine JB, Forget F, Wilson L (2011) The dispersal of pyroclasts from Apollinaris Patera, Mars: implications for the origin of the Medusae Fossae Formation. Icarus 216 (1):212–220
145. Lipman P (1997) Subsidence of ash-flow calderas: relation to caldera size and magma-chamber geometry. Bull Volcanol 59(3):198–218
146. Mason BG, Pyle DM, Oppenheimer C (2004) The size and frequency of the largest explosive eruptions on Earth. Bull Volcanol 66(8):735–748
147. Scott DH, Dohm JM, Applebee DJ (1993) Geologic map of science study area 8, Apollinaris Patera region of Mars. USGS 2351
148. Robinson MS, Mouginis-Mark PJ, Zimbelman JR, Wu SS, Ablin KK, Howingtonkraus AE (1993) Chronology, eruption duration, and atmospheric contribution of the Martian Volcano Apollinaris-Patera. Icarus 104:301–323
149. Farrell AK, Lang NP (2010) Distribution of explosive and effusive volcanic deposits at Apollinaris Patera. Mars. LPI 1533:2072
150. Crown DA, Greeley R (1989) The martian highland Paterae: evidence for explosive volcanism on Mars. In: MEVTV workshop on early tectonic and volcanic evolution of Mars, 29–31
151. Hauber E, van Gasselt S, Ivanov B, Werner S, Head JW et al (2005) Discovery of a flank caldera and very young glacial activity at Hecates Tholus, Mars. Nature 434(7031):356–361
152. Neukum G, Hiller K (1981) Martian ages. J Geophys Res Solid Earth 86(B4):3097–3121
153. Malin MC, Dzurisin D (1977) Landform degradation on Mercury, the Moon, and Mars: evidence from crater depth/diameter relationships. J Geophys Res 82(2):376–388
154. Walker G (1973) Lengths of lava flows. Philos Trans R Soc London. Ser A 274:107–118
155. Hulme G (1976) The determination of the rheological properties and effusion rate of an Olympus Mons lava. Icarus 27(2):207–213
156. McGetchin TR, Smith JR (1978) The mantle of Mars: some possible geological implications of its high density. Icarus 34(3):512–536
157. Cataldo V, Williams DA, Schmeeckle MW, Leone G (2019). Vallis Schröteri, Moon: Results of first 3 -D model of thermal erosion by turbulently flowing lava reveal how erosion likely shaped the inner rille. 50th Lunar and Planetary Science Conference, abstract #2297
158. Moukalled F, Mangani L, Darwish M, (2015) The finite volume method in computational fluid dynamics: an advanced introduction with OpenFOAM and Matlab, 1st edition. Springer Internatational Publish
159. Jones WP, Launder BE (1972) The prediction of laminarization with a two-equation model of turbulence. Int J Heat Mass Trans 15:301–314. Kakaç S, Shah RK, Aung W (1987) Handbook of single-phase convective heat transfer. John Wiley & Sons, New York
160. Launder BE, Sharma BI (1974) Application of the energy dissipation model of turbulence to the calculation of flow near a spinning disk. Lett Heat Mass Trans 1(2):131–138
161. Menter FR (1993) Zonal two-equation k-ω turbulence models for aerodynamic flow. AIAA paper 93–2906
162. Patankar SV, Spalding DB (1972) A calculation procedure for heat, mass and momentum transfer in three-dimensional parabolic flows. Int J Heat Mass Transf 15(10):1787–1806
163. Hulme G (1973) Turbulent lava flow and the formation of lunar sinuous rilles Mod. Geol 4:107–117
164. Huppert HE, Sparks RSJ (1985) Komatiites, I, Eruption and flow. J Petrol 26:694–725
165. Williams DA, Kerr RC, Lesher CM (1998) Emplacement and erosion by Archean komatiite lava flows at Kambalda: revisited. J Geophys Res Solid Earth 103(B11):27533–27549
166. Williams DA, Fagents SA, Greeley RA (2000) A reassessment of the emplacement and erosional potential of turbulent, low-viscosity lavas on the Moon. J Geophys Res 10520(E8):189–205
167. Kakac S, Yener Y, Pramuanjaroenkij A (2013) Convective heat transfer. CRC Press
168. Dragoni M, Bonafede M, Boschi E (1986) Downslope flow models of a Bingham liquid: implications for a lava flow. J Volcanol Geotherm Res 30:305–325. In: Hulme G (1973) Turbulent lava flows and the formation of lunar sinuous rilles. Mod Geol 4:107–117
169. Kerr RC (2001) Thermal erosion by laminar lava flows. J Geophys Res 106(B11):26,453–26,465
170. Siewert J, Ferlito C (2008) Mechanical erosion by flowing lava. Contemp Phys 49:43–54. https://doi.org/10.1080/00107510802077388
171. Wilson L, Head JW (2017) Generation, ascent and eruption of magma on the Moon: new insights into source depths, magma supply, intrusions and effusive/explosive eruptions (Part 1: Theory). Icarus 283:146–175

Volcanic Channels and Volcanic Features on Mars

Henrik Hargitai and Giovanni Leone

Abstract

Volcanic channels are common features on the terrestrial planets. They have diverse morphologies including sinuous rilles, channelized flows, and potentially collapsed lava tubes. Many of these channels are a part of a typical igneous assemblage of landforms. In addition, some channels have an ambiguous origin, where a volcanic interpretation explains many of the observed features in some settings. This chapter describes volcanically formed sinuous channels, pitted channels, channelized flows, streamlined forms, platy terrain, and discusses a potential volcanic context for outflow channels and some valley networks on Mars.

1 Introduction

Mars, like other terrestrial planets and the Moon, shows a variety of linear morphologies that are potentially produced by lava. These include sinuous, lava channels, also called rilles, and channelized flows with levees. These features concentrate on volcanic rises and are often accompanied by fault zones and pitted graben. In this chapter we outline the types of these features, and the efforts to characterize their conditions of origin. We also review the debate on many of the channel's origin: these features, and also outflow channels, have been proposed to have formed by fluvial/diluvial, volcanic, and a combination of these processes. In this paper we focus our attention to the volcanic interpretations of these features.

2 History of Research

This section further explores the literature already discussed in chapter "Volcanic Eruptions on Mars, Lava Flow Morphology, and Thermodynamics", the history of volcanism on Mars from Lowell's canals to the dawn of the space age. Channels, or 'canals', drawn initially by G. Schiaparelli on his maps [1] as straight lines of dark albedo markings, later populated the planet in P. Lowell's Mars cartography [2]. However, these channels were never interpreted as lava channels, in the absence of any potential volcanic sources for the naked-eye Mars observer, although today the term canali is used for describing lava channels on Venus. In 1950, E. J. Öpik calculated that surface gravity on Mars allows mountains 2.8 times higher than on Earth, however, "the mighty volcanic cones such as Kilimanjaro, Aconcagua, etc. on Earth, are definitely absent on Mars". Remarkably, the calculation's result, 24.6 km, is exactly the height of

H. Hargitai (✉)
Eötvös Loránd University, Budapest, Hungary
e-mail: hargitai.henrik@btk.elte.hu

G. Leone
Instituto de Investigación en Astronomía Y Ciencias Planetarias, Universidad de Atacama, Copiapó, Chile

G. Leone (ed.), *Mars: A Volcanic World*,
https://doi.org/10.1007/978-3-030-84103-4_5

Olympus Mons. The article was published in response to an observation of a 900-miles-long cloud, though to be potentially formed in a volcanic explosion and consisted of ashes, or by asteroid impact [3]. Dark albedo features with shifting boundaries, especially those horne-shaped, were attributed to volcanic ash deposition and therefore, active volcanism, comparable in volume to those in terrestrial eruptions. One prominent example was Syrtis Major, which later turned out to be indeed a large volcanic center and the dark materials basaltic. Up to the 1940s, the only other theory explained shifting-boundary dark markings as growing vegetation [4]. Sinuous valleys, or rilles were first observed on the Moon and were not seen on Mars until the orbiting spacecraft visited it. Julius Schmidt catalogued rilles on the Moon in 1866, listing 425 such features [5]. The first formation models hypothesized fluvial origin. Their origin was heavily debated after the first detailed images were received from the Lunar Orbiters. Urey [6] pointed out that Lunar Orbiter images provided "overwhelming evidence" supporting the view that lunar rilles were carved by water, although many others suggested that rilles were produced by lava or dust-gas. Even after the first lunar samples were returned, Schubert et al. [7] argued that "the amount of water ultimately available on the lunar surface is two orders of magnitude greater than that required to erode all the sinuous rilles" and concluded that rilles were carved by water. Volcanic shields were first detected in Mariner 9 images in 1971 [8, 9] and sinuous channels with rille morphologies on or near volcanic constructs visible on Viking Orbiter images were initially interpreted as lava channels [10] (see a pictorial review in [11]). The debate over the fluid agent involved in the formation of sinuous channels was revived when channels on the flanks of Martian volcanoes, and elsewhere, were mapped in detail. The cause for these debates is that the origin of some channels (Fig. 1) cannot be determined from morphology alone, as constructional-leveed channels and incised channels can be both resulted from fluvial or volcanic processes [12]. Braided and sinuous channels can also result from both processes, although some reaches may provide evidence for one or the other theory [13]. Outflow channels are generally interpreted as fluvial features on Mars, however, an alternative, volcanic interpretation is also proposed [14–17].

3 Modes of Channel Formation

This section shows the visual results of thermal erosion, mechanical erosion, drainage and construction leading to channel formation already discussed in chapter "Volcanic Eruptions on Mars, Lava Flow Morphology, and Thermodynamics". A lava channel is "A channel on the upper surface of a partly or completely solidified body of lava through which liquid lava has flowed." [18, 19]. Lava channels are contained within marginal zones of solid and stationary lava, or raised rims called lava levees (see review by [20]). Lava channels on Earth are typically *constructional features* flanked by a pair of parallel, narrow levees (e.g., [21]). Lava channels may also be *incised into* the substrate by thermal, mechanical or thermo-mechanical erosion [22]. Unconsolidated material, such as porous regolith, is more likely to be eroded mechanically [23]. Thermal erosion occurs when the flowing fluid's temperature exceeds the melting temperature of the underlying material and partially melts the substrate before carrying melted substrate down-slope [23, 24]. Canali on Venus may have formed by a further type of erosion, upward erosion by magmatic stoping that thermally or thermomechanically erodes the *overlying* materials [25]. While the largest sinuous rille on Mars is 50 km long, maximum 2 km wide and ~220 m deep [26], canali on Mars can reach several thousand km [15, 16, 27]. The maximum distance the lava in a channel can travel depends on eruption rate, flow temperatures, heat loss by radiation, lava cooling, crystallization, and subsequent rheological changes [28]. The type of erosion of rilles is still debated today [29], despite Apollo 15 mission returned samples from the walls of Hadley Rille (a sinuous rille) [30]. Another open question is whether the lava was transported through stable lava

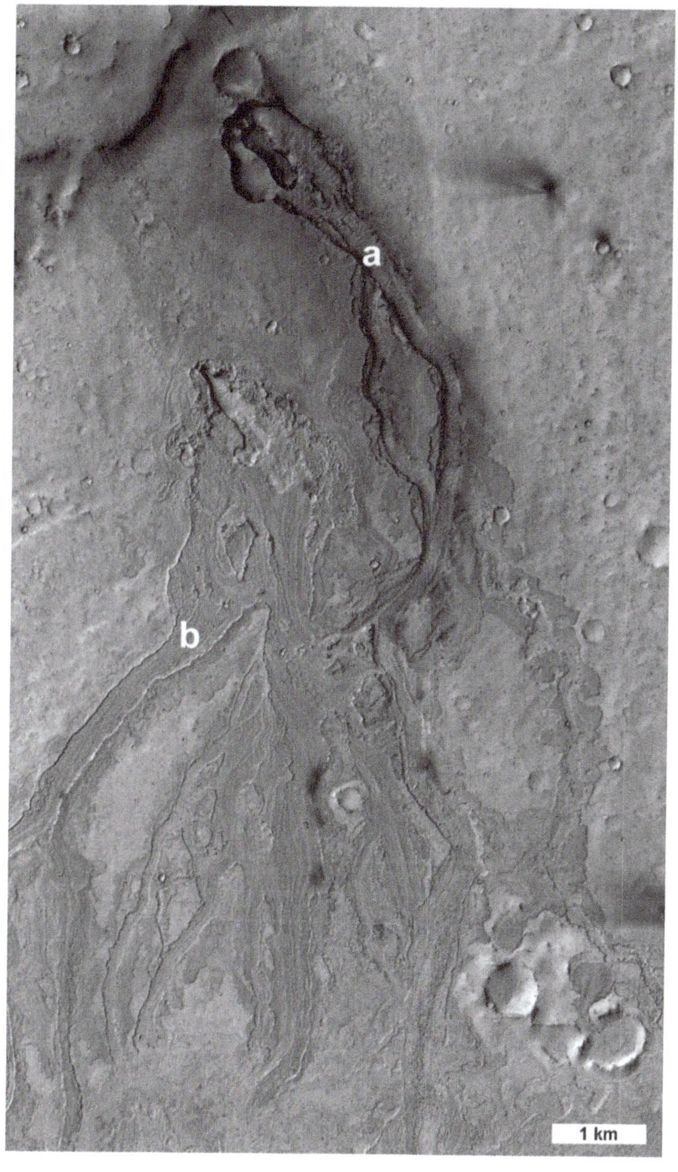

Fig. 1 Incised (erosional) and constructional (leveed) channels both associated with source vents, interpreted as lava channels [26]. Thomas [12] argues that the origin of these channels cannot be determined from morphology alone. HiRISE ESP_016361_1870 at 7.1°N 156.9°E. (NASA/JPL/UA)

tubes or partially roofed lava channels [31]. Leone [16] showed the channel evolution from lava tube to chasma in Valles Marineris.

4 Channel Types

4.1 Sinuous Channels, Rilles

This section describes morphologies and then inferred parameters, interpretations and models.

Narrow, incised, non-rimmed, sinuous channels on volcanic terrain on Mars are similar in morphology to channels on the Moon or Venus, however, also similar to fluvial features on Earth (for review, see [32]). Sinuous rilles are commonly found on volcanic terrain on Mars, such as on the flanks of Pavonis Mons [31], in Tempe volcanic province and on Syria Planum [26]. Sinuous channels without levees are also called rilles, from similar features on the Moon where they were first named [7]. They are rimless,

Fig. 2 Deeply carved outflow channel showing scour marks typical of outflow channels elsewhere. Streamlined islands, however, are on the surface level and a new flow fills the channels displaying curved pressure ridges. A sinuous channel is incised later into this material, or, alternatively, the channel may have formed as lava channel or lava tube during flow emplacement. Olympica Fossae. CTX mosaic image

winding, narrow channels. Sinuous channels typically emerge from a circular to irregular depression [33]. They may be composed of one channel or several channel segments [34]. Nested channels that show valley terraces within a larger valley may form either by incision or by the emplacement of volcanic valley deposits [14]. Sinuous rilles may be carved directly into the surface or may be developed within levées (without levée, Fig. 2). The volcanic formation theories of rilles range from ash flow erosion [35], lava tube collapse [36], and lava flow emplacement where lava channels formed by thermomechanical erosion during high-flux eruption of turbulent lava floods on the Moon [10, 33, 34, 37–39] and Venus [40–43]. Interestingly, fluvial origin was also proposed for the lunar rilles [7, 44] and for Martian rilles (e.g.,

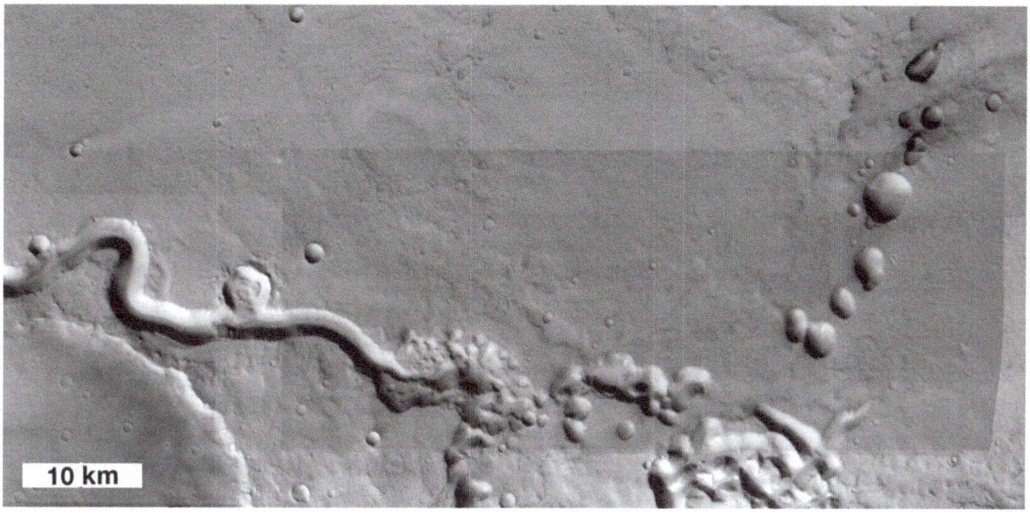

Fig. 3 The pitted origin of Tinto Vallis that flowed into Palos crater. CTX mosaic centered 4°S 111°W, north is to left

[45]), with theories including formation by surface flow to and subsurface geothermally heated water outflow from volcanic fractures [46, 47] and mixed volcanic and fluvial origin [48]. In a comparative study, Byrne et al. [49] suggested that the morphology of sinuous rilles are so similar in volcanic terrain to those in potentially fluvial sediments that all rilles should have formed by one single process. However, the historical instability of water on the surface of Mars [50] suggests that only lava processes may have formed the rilles [14].

4.2 Pitted Channels

Pit troughs can be linear, typically straight features that contain pits (rimless depressions) or a row of pits may form beyond their ends, or curvilinear [16]. Pits that have sharp rims and almost vertical walls are called atypical pit craters (APCs) that are thought to be entry points to subsurface caves [51], both linear and curvilinear can also be depressions formed by the collapsing roof of a lava tube [16, 52].

Various other origins have been suggested for the pit chains: dilational faulting [53], dike intrusion [54], and sapping/piping (see review in [55]). However, the origin by collapsing lava tubes seems the strongest explanation so far [16]. Tinto Vallis is a sinuous crater outflow channel or valley that emerges from a pit-rich region and is considered to be formed by igneous processes [56]. Similar forms are identified near Olympica Fossae where an outflow channel, lobate lava flows, straight troughs or fissures, sinuous channels and pit rows are concentrated [57] (Fig. 3).

Sinuous pitted troughs have been identified in the lower flanks of several volcanic structures and were initially suggested as formed by subsurface water flow, e.g., Ascraeus Mons [49] and Hadriacus Mons [58, 59]. However, as explained in the Introductory Chapter, the very low atmospheric pressure at the stratospheric height of the Tharsis volcanoes like Ascraeus would sublimate any water thus preventing erosion.

Pits and pitted troughs are important for the future human exploration of Mars because at least a part of these features are lava tubes that potentially provide stable shelter for the explorers. The issue will be discussed with more detail in chapter "Future perspectives of Mars after the latest studies".

4.3 Channelized Flows

Channelized flows are lobate lava flows where a wide channel forms along its centre, many of these channels emerge from the surroundings

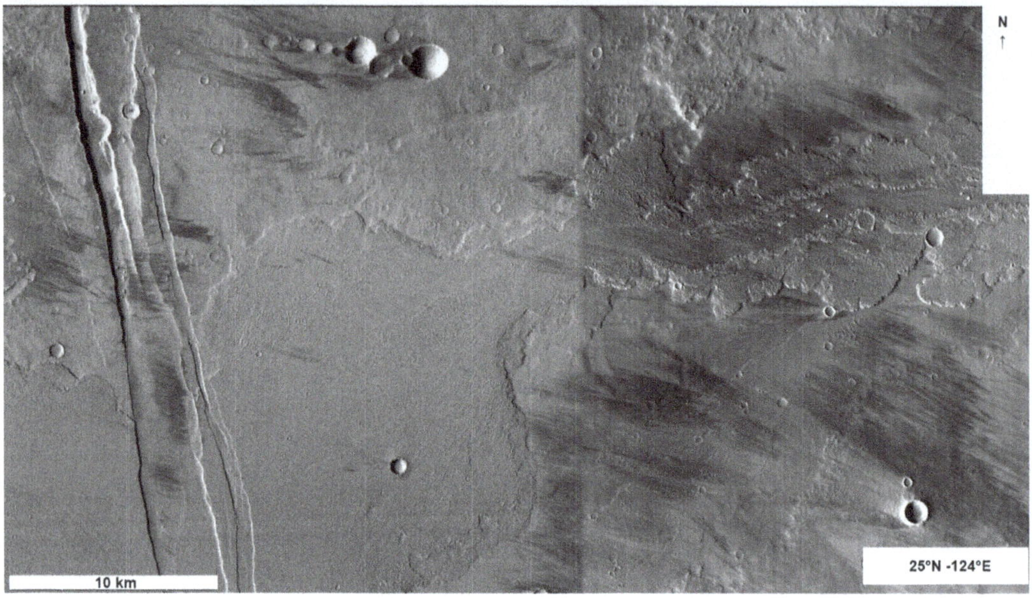

Fig. 4 Lava delta at the terminus of a lobate channelized flow in the Olympica Fossa region. CTX mosaic. A graben crosses the delta. The graben formed more recently, as shown by the continuation of the delta on the other (left) side of the graben [57]

[60]. Such perched lava channels form where pahoehoe lava flows are creating a raised bank or levee and the central flow becomes elevated over the surrounding terrain. Perched channels are constructional forms, formed when the channel walls are built from overspilled and solidified levee materials (e.g., [61]). The central channel may produce a lava delta (Fig. 4). In the East Tharsis region, these channels are 1–2 km wide (Fig. 5). Morphologies in these lava flows include a central, perched channel, may include few round islands, but are dominated by irregular islands; also, they may include slab terrain, lava flow with ogive pressure ridges, negative channel features, ghost islands, and rubbled-knobby channel margin instead of a rounded-smooth channel wall. Perched lava channels may form anastomosing channel belts [57, 62].

Narrow, leveed lava channels appear as double lineaments on relatively low-resolution images. Typical examples are in the Tyrrhena Mons lava flows [57, 58] and on the flanks of Olympus Mons where overlapping flows form the surface (Fig. 6).

4.4 Serpentes

Serpentes are linear features where the same channel exhibits both positive-relief (elevated) and negative (channel-like) segments (Fig. 7). Serpens morphology was first described from channel segments interpreted as inverted-fluvial in origin [63, 64]. However, an alternative interpretation of the serpentes as volcanic features, where the formative process was not geomorphologic inversion, but lava inflation in the elevated segments of the lava channel was later proposed [65]. Other serpens-like features start as raised elevations (perched channels) that are continued as collapsed channel that becomes discontinuous [66].

4.5 Outflow Channels

At the beginning of the Martian exploration era, the outflow channels were interpreted as of volcanic origin [9] but alternative hypotheses of catastrophic floods originating from putative

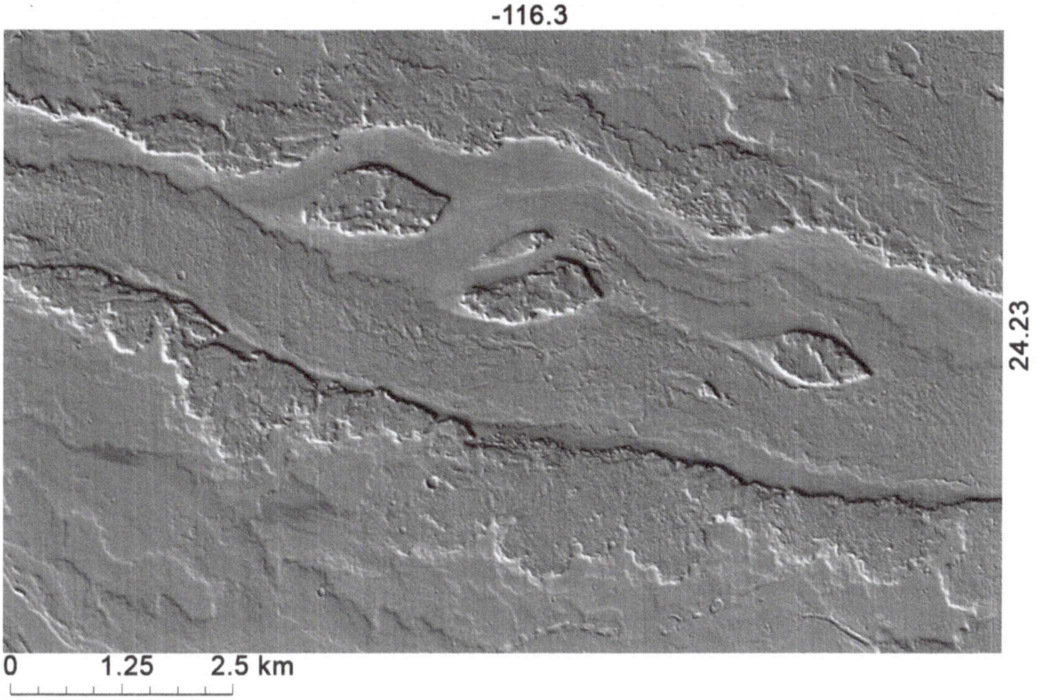

Fig. 5 Lobate-margin channelized flow in Northeast Tharsis with streamlined islands and additional in-channel flow. CTX image

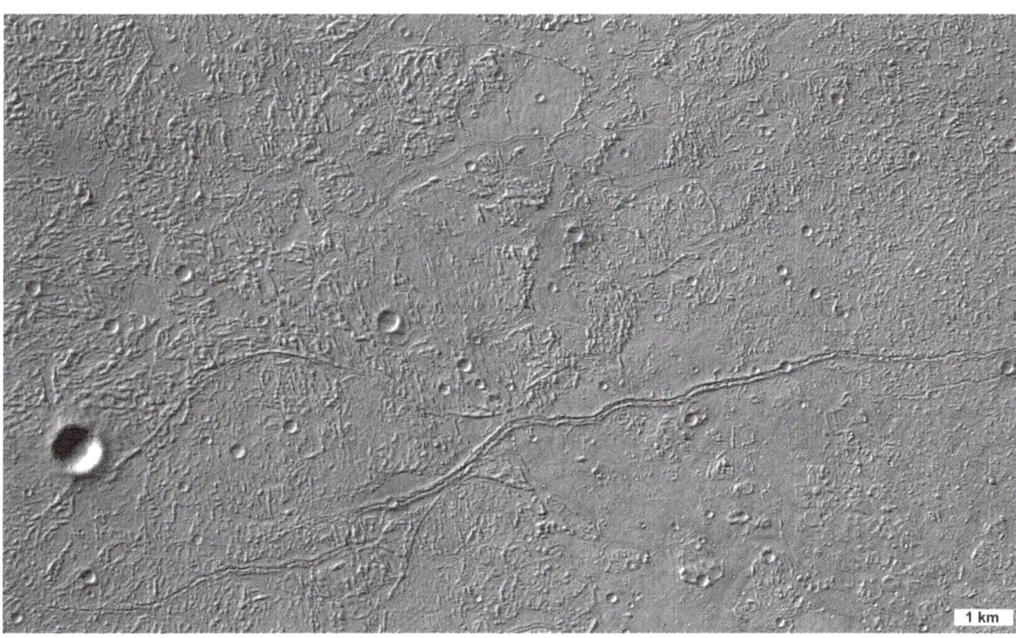

Fig. 6 The double lineament is a lava channel with levees in Western Arsia Mons near Mangala Fossa, Mars near 17° S, 212°E. CTX: B20_017625_1641_XI_15S147W (NASA/JPL/MSSS)

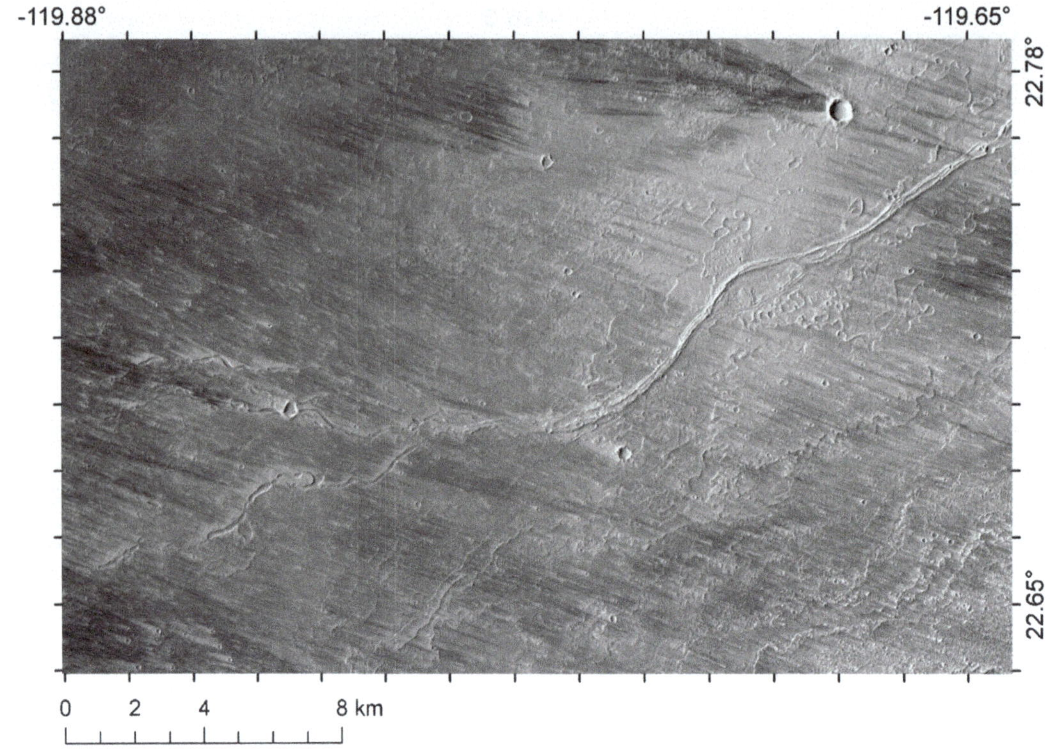

Fig. 7 Serpens feature near Olympica Fossae. CTX image

surface ponding aquifers were subsequently proposed (e.g. [67]). According to such fluvio-volcanic model, flood channels (in Northeast Tharsis, for example) would still form from the volcanic thermal energy that would mobilize confined ground water, or other subsurface volatiles, and then released in extensive hydrothermal systems surrounding the large volcanoes. Depending on the volume and the frequency of additional magmatic pulses, these magmatically-driven hydrothermal systems could have lasted several tens of millions of years [68]. In the same way, several other outflow channels in volcanic terrain are interpreted to have formed and carved by aqueous processes, where the original channel provided a pathway for different liquids, including lava or hyperconcentrated flow, on the surface under putative different atmospheric or subsurface conditions over time. These are Athabasca Valles [69, 70], Grjótá Valles [69] and Hrad Vallis [71]. However, an alternative formation by lava flows was also proposed [31, 72–74]. Similarly multiple-process geologic evolution was proposed for less prominent channels, such as an unnamed channel (Channel D) east of Navua Valles, near Hadriacus Mons volcano [59]. Also, other mechanisms involving alternate fluvial and lava episodes were proposed. It was suggested that for Athabasca Valles volcanic and fluvial floods formed repeatedly for more than two billion years ending with a fluvially modified volcanic resurfacing event at 30 Ma and renewed volcanic activity as recently as 3 Ma [75]. While the same mechanism was also proposed for Mangala Valles [76], an alternative formation mechanism by volcanic lava only also provided a plausible scenario [17, 77].

The direct formation of outflow channels from volcanic eruptions of lava is supported by their general location close to or spreading from volcanic centres (Fig. 8), and by the global lack of alteration of the olivine naturally contained in erupted lavas [15–17, 50, 78]. Several studies identified lava on the bedfloor of many outflow

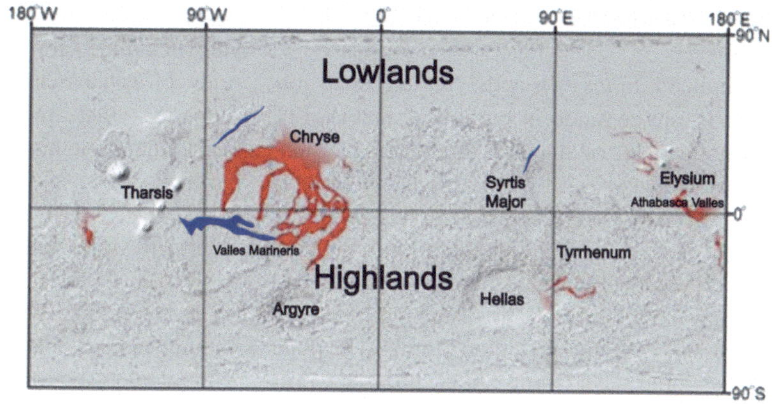

Fig. 8 The outflow channels of Mars. Modified by Leone (blue) after Leverington (red) [15]

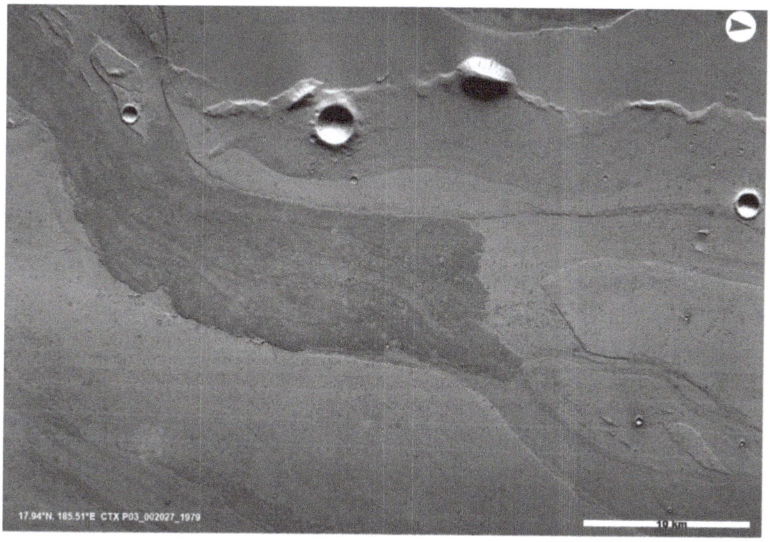

Fig. 9 Lava flowing into an outflow channel with streamlined forms. Marte Vallis. CTX mosaic image

channels [15, 16, 78] such as in Marte Vallis (Fig. 9). Although water in underground hydrothermal system can be stable as soon as the lithostatic pressure allows its stability [17, 50], under current atmospheric conditions, due to low atmospheric pressure, liquid water or even H_2O ice is unstable on the surface of Mars [15, 50, 79]. The chaos regions, though, where the major outflow channels originate, are generally at low elevations (−3 to −4 km) where pressure is 500–700 Pa. However, Olympica Fossae, proposed to be hydrothermal and volcanic [57] is located as high as 3 km height (Pressure 400–500 Pa).

Alternative models interpret outflow channels, including those in Chryse Planitia, as lava carved channels. Lunar, Venusian and Mercurian [80] channelized lava flows show morphologic features that have been typically attributed to fluvial processes on Mars (streamlined islands, anastomosing channels) and many lunar rilles commence abruptly at depressions similar to outflow channels on Mars. Their emergence at volcanic provinces and location of mouth at volcanic plains is consistent with volcanic origin [15–17, 73, 77, 81, 82] with reported evidence for local lava erosion in Kasei Valles [83].

4.6 Streamlined Forms and Kipukas

Streamlined forms—'islands' during their formation—are common in outflow channels [14]. Streamlined forms may be used as a morphological indicator for features produced by currents and they were mapped in various volcanic provinces (Fig. 10). Morphometric investigation may differentiate volcanic kipukas from aeolian yardangs, but is unable to distinguish between fluvial and volcanic streamlined forms. The point is that fluid lava forms branching networks exactly how water would do [62]. While streamlined forms were mostly associated with fluvial [84] or ice erosional processes on Earth [85], the different environmental conditions on Mars suggest that terrestrial drained lava channels, similar to incised fluvial channels, can also produce streamlined forms [62]. The area east–northeast of Olympus Mons contains especially numerous streamlined forms occurring predominantly in terrains exhibiting lava flow morphology [59].

Based on terrestrial analogs in Hawaii, Garry et al. [86] suggested that in-channel islands in a channel near Ascraeus Mons are formed from material rafted and deposited by surges of lava flowing down the channel, and eventually eroded into streamlined forms by the flowing lava. In a study of Athabasca Valles, Cataldo et al. [74] suggested that thermal erosion played an insignificant role in eroding the channels because mechanical erosion is predominant in a steep slope.

In terrestrial volcanic settings, any residual surface surrounded by lava flows is called a kipuka. We may distinguish kipukas located within channels, and anastomosing channel belts with kipukas between lava channels. Kipukas may be streamlined or have irregular edges. They have been individuated on the Moon [87] and on Mars [26, 88]. The Columbia Hills inside Gusev crater are the classic example of a kipuka [89] but other examples were individuated on the eastern flank of Tharsis, just west of Kasei Valles [90], and in any volcanic province of Mars where lava flowed over pre-existing terrains [88].

4.7 Platy Terrains in Volcanic Channels

Some outflow channel associated features' morphology alone may be misleading: polygonal

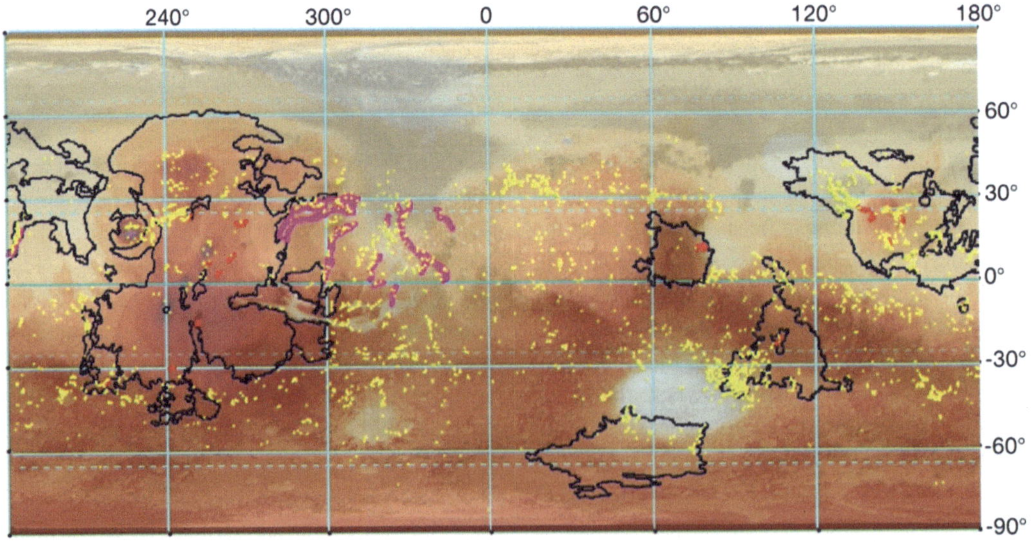

Fig. 10 Distribution of the largest volcanic provinces (black outlines), rilles (red lines), and major outflow channels (pink lines), after Tanaka et al. [91]. Background: topographic map (MOLA) with albedo markings. Yellow dots show the centerpoints of streamlined forms

plates were interpreted as ice slabs [92–94] although they were previously interpreted as lava rafts [95, 96]. Long and narrow ridges of up-buckled blocks of the solidified crust of lava [97] is similar in morphology to ice pressure ridges [98]. Finger rafting pattern of pressure ridges occurs in both materials [97]. Similarly, round platy features may resemble pancake ice (e.g., [99]) but the coils and polygonal crust found in Athabasca Valles were unequivocally attributed to lavas [100]. Platy-ridged flat terrain is another evidence for ice-lava similarity. Platy flows were found in Echus Chasma and Kasei Valles [90, 101], southwest of Arsia Mons [102], and south Amazonis Planitia [103] but, again, they were attributed to lava flows.

4.8 Valleys and Valley Networks (VNs)

Valley and channels are distinct geomorphologic features. Valleys and valley networks formed by channel activity and in many cases it is difficult to distinguish channels from valleys especially where they are filled with dust. The Martian valley networks (VNs), originally termed runoff channels, were already studied since the time of the Mariner and Viking missions and the volcanic interpretation [10, 106] prevailed over the fluvial [8] interpretation at that time. After then the hydrological interpretation prevailed [108, 110], despite evidence against water [107].

Valleys on Mars consists of several morphologies and patterns proposed to be formed by a variety of endogenic and exogenic processes [108]. Some valleys, mostly in subparallel systems, are associated with volcanic forms, or steep slopes [59], others, many in dendritic, branching systems (Valley Networks), are located on the densely cratered highlands and yet others, mostly single valleys with amphitheater terminations, are suggested to originate from sapping [108]. There is a clear relationship between valleys and volcanic terrain (Figs. 10 and 11). Many VNs appear mantled by material interpreted as eolian or volcanic at the resolution of the Viking images [105].

VNs that are characterized by anastomosing branches can be formed by fluid lava flowing on the surface as it also occurs in volcanic fields on Earth [62]. The morphology of dendritic VNs on Mars is similar to that of fluvial networks on Earth [109] and other channel morphologies are similar to channels on Venus [110]. Although such a similar geomorphology was certainly more misleading at the time of the Viking Orbiter images [105] than it could be with images at higher resolution received from the Mars Reconnaissance Orbiter (MRO) mission or the Magellan radar images, there is no doubt that lava was involved in the formation of VNs on Venus [110] or sinuous rilles on the Moon [10, 38]. Today, it is possible to see very clearly how some VNs on Mars are filled by lava that flowed inside them [88].

The valley networks have narrower bedfloors compared to the wider outflow channels. In a volcanic-origin framework, such a difference is due to the smaller amount of lava, and thus of both thermal and mechanical erosion, involved in their formation with respect to the outflow channels [16]. VNs have a typical latitudinal distribution confined to highland regions. Either substantial channeled terrain is buried under subsequently emplaced material (which is yet to be identified) or rainfall was unlikely involved in their formation [104] otherwise their presence must have been more ubiquitous and uniformly distributed all over the Martian surface [50]. Some valleys are located at high altitudes like those on Ascraeus Mons [47], basically at stratospheric heights [79] where liquid water is even less likely to exist than at the surface, unequivocally demonstrating that some valleys were formed by lava rather than by water [15, 50].

The highest-elevation dendritic valley network is also the most prototypical (densest) one, Warrego Valles, which is located as high as 8 km, on the Thaumasia Highlands, where the atmospheric pressure is only 330–380 Pa. This region was proposed as formed by rifting and intrusive magmatism and Warrego Valles was proposed to be a magmatic-related hydrothermal system [111].

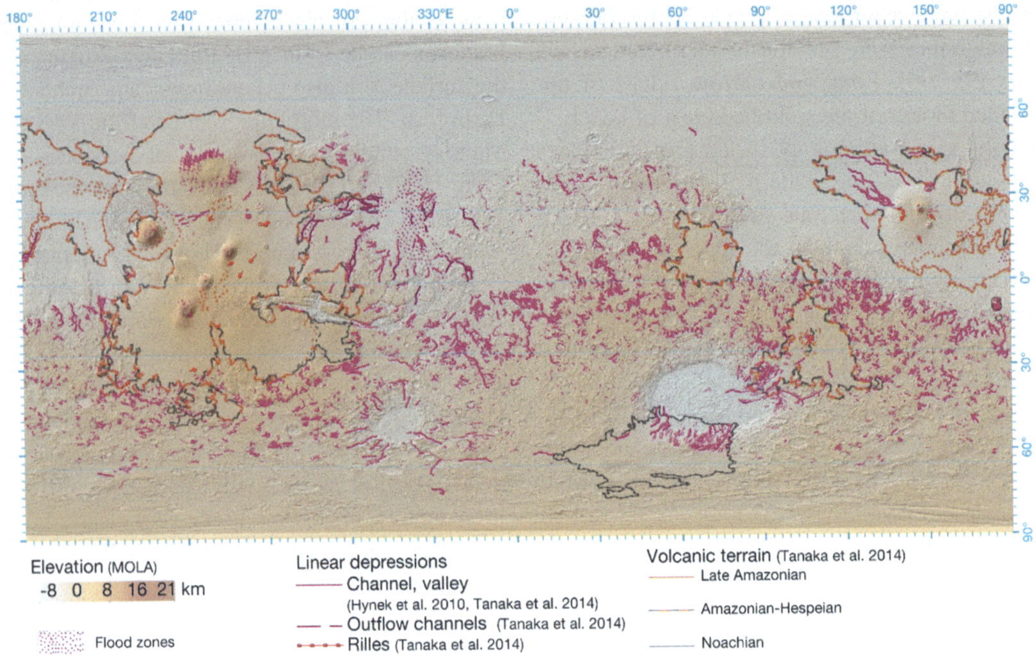

Fig. 11 Valley networks, channels and volcanic terrain on Mars

On Meridiani Planum a lack of alteration of jarosite was observed that otherwise would produce ferric oxyhydroxides with the humidity present on the surface and in the atmosphere, suggesting dry conditions [112]. Furthermore, the interior of Meridiani Planum lacks valley networks, many of which appear along its borders.

Despite problems in finding definitive fluvial source for the valley networks, for more than two decades rainfall or groundwater were generally suggested for their formation instead of lava [48, 113–116]. Most VNs were mapped in regions of Mars where no prominent volcanic centres were apparently observed [117]. However, volcanic centres were subsequently observed in many regions thus providing a possible source for those VNs [88]. Furthermore, the evidence that VNs were not found on terrains younger than Lunae Planum suggested that they must be older than those terrains and of the intercrater plains [104, 105, 107]. Indeed, although geologically recent volcanism was put forward [57, 118], it is confined to a few locations and volcanic processes peaked during the first 500 Ma of the Martian history with a declining phase during the Noachian and complete end by the Hesperian in many regions of Mars except maybe in e.g., Tharsis and Elysium Rises [17, 50, 88, 118].

5 Summary

The channels of Mars over the eons formed under similar dry and cold climatic conditions that we see today [119–121] and in in a volcanic geological setting. The morphology of the currently exposed surface reflects the cumulative traces of the Martian history. Volcanism is consistent with a large, if not the total, part of the observed channel morphologies along with morphologies observed within and around the channels. The volcanic interpretation is also confirmed by the observations of similar morphologies on planetary bodies without an atmosphere such as the Moon and Mercury, and on Venus, where surface temperature rules out the role of water as formative agent. However, due to the nature of morphology-based comparative planetology, on Mars only careful consideration of the geologic setting, the time of formation and modification and, perhaps most importantly, spectral analysis or (future) in situ sampling of the channel materials can provide proof for the formative process(es) in each case.

References

1. Schiaparelli GV (1879) Osservazioni astronomiche e fisiche sull'asse di rotazione e sulla topografia del pianeta Marte fatte nella Reale Specola di Brera in Milano coll'equatoriale di Merz. Memoria seconda. Brera
2. Lowell P (1907) Mars and its canals. Science 25 (639):499–500
3. Öpik EJ (1950) Mars and the asteroids. Irish Astron J 1:22
4. McLaughlin DB (1955) Changes on Mars, as evidence of wind deposition and volcanism. Astron J. 60:261
5. Schmidt JFJ (1866) Über Rillen auf dem Monde. Verlag von JA Barth, Leipzig
6. Urey HC (1967) Water on the Moon. Nature 216:1094–1095
7. Schubert G, Lingenfelter RE, Peale SJ (1970) The morphology, distribution, and origin of lunar sinuous rilles. Rev Geophys. 8(1):199–224
8. Masursky H (1973) An overview of geological results from Mariner 9. J Geophys Res 78:4009–4030
9. Carr MH (1973) Volcanism on Mars. 78:4049–4062
10. Carr MH (1974) The role of lava erosion in the formation of lunar rilles and Martian channels. Icarus 22:1–23
11. Hodges CA, Moore HJ (1994) Atlas of volcanic landforms on Mars. Prof Pap 1534. https://doi.org/10.3133/PP1534
12. Thomas RJ (2013) Identification of possible recent water/lava source vents in the Cerberus plains: stratigraphic and crater count age constraints. J Geophys Res E Planets. 118:789–802
13. Bleacher JE, de Wet AP, Garry WB, Zimbelman JR, Trumble ME (2010) Volcanic or fluvial; comparison of an Ascraeus Mons, Mars, braided and sinuous channel with features of the 1859 Mauna Loa flow and Mare Imbrium flows [abs.]. Lunar Planet. Sci. Conf. 41st, Abstr. v. 41
14. Leverington DW (2004) Volcanic rilles, streamlined islands, and the origin of outflow channels on Mars. J Geophys Res 109:E10011
15. Leverington DW (2011) A volcanic origin for the outflow channels of Mars: key evidence and major implications. Geomorphology 132:51–75
16. Leone G (2014) A network of lava tubes as the origin of Labyrinthus Noctis and Valles Marineris on Mars. J Volcanol Geotherm Res. 277:1–8
17. Leone G (2017) Mangala Valles, Mars: a reassessment of formation processes based on a new geomorphological and stratigraphic analysis of the geological units. J Volcanol Geotherm Res 337:62–80
18. Masursky H, El-Baz F, Doyle FJ, Kosofsky LJ (1978) Apollo over the Moon: a view from orbit. NASA SP 362:251
19. Larson CV (1991) Nomenclature of lava tube features. In: 6th international symposium on vulcanospeleology
20. Hargitai H (2015) Lava channel. Encycl Planet Landforms. https://doi.org/10.1007/978-1-4614-3134-3_221
21. Gioia G, Chakraborty P, Kieffer S (2006) Lava channel formation via the viscoplastic indentation of hot substrates. Geophys Res Lett. 33:L19305
22. Greeley R, Fagents SA, Harris RS, Kadel SD, Williams DA, Guest JE (1998) Erosion by flowing lava: field evidence. J Geophys Res Solid Earth 103:27325–27345
23. Head JW, Wilson L (2020) Rethinking lunar mare basalt regolith formation: new concepts of lava flow protolith and evolution of regolith thickness and internal structure. Geophys Res Lett. 47: e2020GL088334
24. Hurwitz DM, Fassett CI, Head JW, Wilson L (2010) Formation of an eroded lava channel within an Elysium Planitia impact crater: distinguishing between a mechanical and thermal origin. Icarus 210:626–634
25. Lang NP, Hansen VL (2006) Venusian channel formation as a subsurface process. J Geophys Res. 111:E04001
26. Hauber E, Bleacher J, Gwinner K, Williams D, Greeley R (2009) The topography and morphology of low shields and associated landforms of plains volcanism in the Tharsis region of Mars. J Volcanol Geotherm Res. 185(1–2):69–95
27. Gasparri D, Leone G, Cataldo V, Punjabi V, Nandakumar S (2020) Lava filling of Gale crater from Tyrrhenus Mons on Mars. J Volcanol Geotherm Res 389:106743
28. Harris A, Rowland S (2001) FLOWGO: a kinematic thermo-rheological model for lava flowing in a channel. Bull Volcanol. 63:20–44
29. Roberts CE, Gregg TKP (2019) Rima Marius, the Moon: formation of lunar sinuous rilles by constructional and erosional processes. Icarus 317:682-688
30. ALGI Team (1972) Geologic setting of the Apollo 15 samples. Science 175(4020):407–415
31. Keszthelyi L, Jaeger W, McEwen A, Tornabene L, Beyer RA, Dundas C, Milazzo M (2008) High resolution imaging science experiment (HiRISE)

images of volcanic terrains from the first 6 months of the Mars reconnaissance orbiter primary science phase. J Geophys Res E Planets. 113:E04005
32. Komatsu G, Hargitai H (2015) Sinuous Rille. Encycl Planet Landforms. https://doi.org/10.1007/978-1-4614-3134-3_351
33. Hurwitz DM, Head JW, Hiesinger H (2013) Lunar sinuous rilles: distribution, characteristics, and implications for their origin. Planet Space Sci. 79–80:1–38
34. Hurwitz DM, Head JW, Wilson L, Hiesinger H (2012) Origin of lunar sinuous rilles: modeling effects of gravity, surface slope, and lava composition on erosion rates during the formation of Rima Prinz. J Geophys Res Planets 117:E00H14
35. Cameron WS (1964) An interpretation of Schröter's Valley and other lunar sinuous rills. J Geophys Res. 69(12):2423–2430
36. Greeley R (1971) Lunar Hadley Rille: considerations of its origin. Science 172(3984):722–725
37. Gornitz V (1973) The origin of sinuous rilles. The Moon 6:337–356
38. Hulme G (1982) A review of lava flow processes related to the formation of lunar sinuous rilles. Geophys Surv. 5:245–279
39. Williams DA, Fagents SA, Greeley R (2000) A reassessment of the emplacement and erosional potential of turbulent, low-viscosity lavas on the Moon. J Geophys Res Planets 105:20189–20205
40. Baker VR, Komatsu G, Parker TJ, Gulick VC, Kargel JS, Lewis JS (1992) Channels and valleys on Venus: preliminary analysis of Magellan data. J Geophys Res 97:13421–13444
41. Komatsu G, Baker VR, Gulick VC, Parker TJ (1993) Venusian channels and valleys: distribution and volcanological implications. Icarus 102:1–25
42. Komatsu G, Baker VR (1994) Meander properties of Venusian channels. Geology 22(1):67–70
43. Oshigami S, Namiki N, Komatsu G (2009) Depth profiles of venusian sinuous rilles and valley networks. Icarus 199(2):250–263
44. Peale SJ, Schubert G, Lingenfelter RE (1968) Distribution of sinuous rilles and water on the Moon. Nature 220:1222–1225
45. Scott ED, Wilson L (1999) Evidence for a sill emplacement event on the upper flanks of the Ascraeus Mons shield volcano, Mars. J Geophys Res Planets 104:27079–27089
46. Mouginis-Mark PJ, Christensen PR (2005) New observations of volcanic features on Mars from the THEMIS instrument. J Geophys Res Planets 110: E08007
47. Murray JB, van Wyk de Vries B, Marquez A, Williams DA, Byrne P, Muller J-P, Kim J-R (2010) Late-stage water eruptions from Ascraeus Mons volcano, Mars: implications for its structure and history. Earth Planet Sci Lett 294:479–491
48. Gulick VC, Baker VR (1989) Fluvial valleys and martian palaeoclimates. Nature 341:514–516
49. Byrne PK, van Wyk de Vries B, Murray JB, Troll VR (2012) A volcanotectonic survey of Ascraeus Mons, Mars. J Geophys Res Planets. 117:E01004
50. Leone G (2020) The absence of an ocean and the fate of water all over the Martian history. Earth Sp Sci 7:1–16
51. Cushing GE, Okubo CH, Titus TN (2015) Atypical pit craters on Mars: new insights from THEMIS, CTX, and HiRISE observations. J Geophys Res Planets. 120:1023– 1043
52. Wyrick D, Ferrill DA, Morris AP, Colton SL, Sims DW (2004) Distribution, morphology, and origins of Martian pit crater chains. J Geophys Res E Planets. 109:E06005
53. Ferrill DA, Wyrick DY, Morris AP, Sims DW, Franklin NM (2004) Dilational fault slip and pit chain formation on Mars. GSA Today 14(10):4–12
54. Schultz RA (1998) Multiple-process origin of Valles Marineris basins and troughs, Mars. Planet Space Sci. 46(6-7):827–829, 831–834
55. Barta G, De Hon R, Hargitai H (2014) Sapping valley. Encycl Planet Landforms. https://doi.org/10.1007/978-1-4614-9213-9_642-1
56. Leverington DW (2006) Volcanic processes as alternative mechanisms of landform development at a candidate crater-lake site near Tyrrhena Patera, Mars. J Geophys Res E Planets 111:E11002
57. Hargitai HI, Gulick VC (2018) Late Amazonian–aged channel and island systems located East of Olympus Mons, Mars. In: Dynamic mars. Elsevier, pp 121–154
58. Crown DA, Bleamaster LF, Mest SC (2005) Styles and timing of volatile-driven activity in the eastern Hellas region of Mars. J Geophys Res 110:E12S22
59. Hargitai HI, Gulick VC, Glines NH (2017) Discontinuous drainage systems formed by highland precipitation and ground-water outflow in the Navua Valles and southwest Hadriacus Mons regions, Mars. Icarus 294:172–200
60. Riker JM, Cashman KV, Kauahikaua JP, Montierth CM (2009) The length of channelized lava flows: insight from the 1859 eruption of Mauna Loa Volcano, Hawai'i. J Volcanol Geotherm Res. 183 (3–4):139–156
61. Glaze LS, Baloga SM, Garry WB, Fagents SA, Parcheta C (2009) A hybrid model for leveed lava flows: implications for eruption styles on Mars. J Geophys Res 114:E07001
62. Dietterich HR, Cashman KV (2014) Channel networks within lava flows: formation, evolution, and implications for flow behavior. J Geophys Res Earth Surf 119:1704–1724
63. Williams R (2014) Serpens, [Serpentes]. Encycl Planet Landforms. https://doi.org/10.1007/978-1-4614-9213-9_378-1
64. Williams RME, Irwin RP, Burr DM, Harrison T, McClelland P (2013) Variability in martian sinuous ridge form: case study of Aeolis Serpens in the

Aeolis Dorsa, Mars, and insight from the Mirackina paleoriver, South Australia. Icarus. 225(1):308–324
65. Williams RME, Moersch JE, Fergason RL (2018) Thermophysical properties of Martian fluvial sinuous ridges: inferences on "Inverted Channel" induration agent. Earth Sp Sci. 5:516–528
66. Bleacher JE, Orr TR, de Wet AP, Zimbelman JR, Hamilton CW, Brent Garry W, Crumpler LS, Williams DA (2017) Plateaus and sinuous ridges as the fingerprints of lava flow inflation in the Eastern Tharsis Plains of Mars. J Volcanol Geotherm Res 342:29–46
67. Carr MH (1979) Formation of Martian flood features by release of water from confined aquifers. J Geophys Res. 84(B6):2995–3007
68. Gulick VC (1998) Magmatic intrusions and a hydrothermal origin for fluvial valleys on Mars. J Geophys Res E Planets. 103(E8):19365–19387
69. Burr DM, McEwen AS, Sakimoto SEH (2002) Recent aqueous floods from the Cerberus Fossae. Mars. Geophys Res Lett 29:1013
70. Burr DM, Enga MT, Williams RME, Zimbelman JR, Howard AD, Brennand TA (2009) Pervasive aqueous paleoflow features in the Aeolis/Zephyria Plana region, Mars. Icarus. 200(1):52–76
71. Hamilton CW, Mouginis-Mark PJ, Sori MM, Scheidt SP, Bramson AM (2018) Episodes of aqueous flooding and effusive volcanism associated with Hrad Vallis, Mars. J Geophys Res Planets. 123:1484–1510
72. Jaeger WL, Keszthelyi LP, McEwen AS, Dundas CM, Russell PS (2007) Athabasca Valles, Mars: a lava-draped channel system. Science 317:1709–1711
73. Hopper JP, Leverington DW (2014) Formation of Hrad Vallis (Mars) by low viscosity lava flows. Geomorphology 207:96–113
74. Cataldo V, Williams DA, Dundas CM, Keszthelyi LP (2015) Limited role for thermal erosion by turbulent lava in proximal Athabasca Valles, Mars. J Geophys Res Planets 120:1800–1819
75. Werner SC, van Gasselt S, Neukum G (2003) Continual geological activity in Athabasca Valles, Mars. J Geophys Res E Planets. 108:8081
76. Keske AL, Hamilton CW, McEwen AS, Daubar IJ (2015) Episodes of fluvial and volcanic activity in Mangala Valles, Mars. Icarus 245:333–347
77. Leverington DW (2007) Was the Mangala Valles system incised by volcanic flows? J Geophys Res 112:E11005
78. Leverington DW (2020) Incision of Ma'adim Vallis (Mars) by dry volcanic megafloods effused from multiple highland sources. Planet Space Sci 191:105021
79. Moyer EJ, Irion FW, Yung YL, Gunson MR (1996) ATMOS stratospheric deuterated water and implications for troposphere-stratosphere transport. Geophys Res Lett 23:2385–2388
80. Byrne PK, Klimczak C, Williams DA, Hurwitz DM, Solomon SC, Head JW, Preusker F, Oberst J (2013) An assemblage of lava flow features on Mercury. J Geophys Res E Planets. 118:1303–1322
81. Leverington DW, Maxwell TA (2004) An igneous origin for features of a candidate crater-lake system in western Memnonia. Mars. J Geophys Res 109: E06006
82. Leverington DW (2009) Reconciling channel formation processes with the nature of elevated outflow systems at Ophir and Aurorae Plana, Mars. J Geophys Res Planets 114:E10005
83. Dundas CM, Keszthelyi LP (2014) Emplacement and erosive effects of lava in south Kasei Valles, Mars. J Volcanol Geotherm Res 282:92–102
84. McClenagan JD (2013) Streamlined erosional residuals and drumlins in central British Columbia, Canada. Geomorphology 189:41–54
85. Krabbendam M, Eyles N, Putkinen N, Bradwell T, Arbelaez-Moreno L (2016) Streamlined hard beds formed by palaeo-ice streams: a review. Sediment Geol. 338:24–50
86. Garry WB, Zimbelman JR, Gregg TKP (2007) Morphology and emplacement of a long channeled lava flow near Ascraeus Mons Volcano, Mars. J Geophys Res Planets. 112:E08007
87. Head JW (1976) Lunar volcanism in space and time. Rev Geophys. 14(2):265–300
88. Leone G (2016) Alignments of volcanic features in the southern hemisphere of Mars produced by migrating mantle plumes. J Volcanol Geotherm Res 309:78–95
89. Ruff SW, Niles PB, Alfano F, Clarke AB (2014) Evidence for a Noachian-aged ephemeral lake in Gusev crater, Mars. Geology. 42(4):359–362
90. Chapman MG, Neukum G, Dumke A et al (2010) Noachian-Hesperian geologic history of the Echus Chasma and Kasei Valles system on Mars: New data and interpretations. Earth Planet Sci Lett. 294 (3–4):256–271
91. Tanaka KL, Robbins SJ, Fortezzo CM, Skinner JA, Hare TM (2014) The digital global geologic map of Mars: chronostratigraphic ages, topographic and crater morphologic characteristics, and updated resurfacing history. Planet Space Sci 95:11–24
92. Page DP (2007) Recent low-latitude freeze–thaw on Mars. Icarus. 189(1):83–117
93. Dundas CM, Keszthelyi LP, Bray VJ, McEwen AS (2010) Role of material properties in the cratering record of young platy-ridged lava on Mars. Geophys Res Lett. 37:L12203
94. Woodworth-Lynas C, Guigné JY (2003) Ice keel scour marks on Mars: evidence for floating and grounding ice floes in Kasei Valles. Oceanography. https://doi.org/10.5670/oceanog.2003.15
95. Keszthelyi L, Mcewen AS, Thordarson T (2000) Terrestrial analogs and thermal models for Martian flood lavas. J Geophys Res Planets 105:27–42
96. Hartmann WK, Berman DC (2000) Elysium Planitia lava flows: crater count chronology and geological implications. J Geophys Res Planets 105:15011–15025

97. Hargitai H, Korteniemi J, Diniega S (2015) Pressure ridge. Encycl Planet Landforms. https://doi.org/10.1007/978-1-4614-3134-3_285
98. Timco GW, Johnston M (2004) Ice loads on the caisson structures in the Canadian Beaufort Sea. Cold Reg Sci Technol. 38(2–3):185–209
99. De La Rosa S, Maus S (2012) Laboratory study of frazil ice accumulation under wave conditions. Cryosphere. 6:173–191
100. Ryan AJ, Christensen PR (2012) Coils and polygonal crust in the Athabasca Valles region, Mars, as evidence for a volcanic history. Science 336 (6080):449–452
101. Dundas CM, Cushing GE, Keszthelyi LP (2019) The flood lavas of Kasei Valles, Mars. Icarus. 321:346–357
102. Crown DA, Ramsey MS (2017) Morphologic and thermophysical characteristics of lava flows southwest of Arsia Mons, Mars. J Volcanol Geotherm Res 342:13–28
103. Harmon JK, Nolan MC, Husmann DI, Campbell BA (2012) Arecibo radar imagery of Mars: the major volcanic provinces. Icarus 220:990–1030
104. Pieri D (1976) Distribution of small channels on the Martian surface. Icarus 27:25–50
105. Pieri DC (1980) Martian valleys: morphology, distribution, age, and origin. Science (80-) 210:895–897
106. Gulick VC (2001) Origin of the valley networks on Mars: a hydrological perspective. Geomorphology 37:241–268
107. Carr MH, Clow GD (1981) Martian channels and valleys: their characteristics, distribution, and age. Icarus 48:91–117
108. Sharp RP, Malin MC (1975) Channels on Mars. Geol Soc Am Bull 86:593
109. Baker VR, Hamilton CW, Burr DM, Gulick VC, Komatsu G, Luo W, Rice JW, Rodriguez JAP (2015) Fluvial geomorphology on earth-like planetary surfaces: a review. Geomorphology 245:149–182
110. Komatsu G, Gulick VC, Baker VR (2001) Valley networks on Venus. Geomorphology 37:225–240
111. Dohm JM, Tanaka KL (1999) Geology of the Thaumasia region, Mars: plateau development, valley origins, and magmatic evolution. Planet Space Sci 47:411–431
112. Madden MEE, Bodnar RJ, Rimstidt JD (2004) Jarosite as an indicator of water-limited chemical weathering on Mars. Nature 431:821–823
113. Gulick VC, Baker VR (1990) Origin and evolution of valleys on Martian volcanoes. J Geophys Res 95:14325–14344
114. Ivanov MA, Head JW (2006) Alba Patera, Mars: topography, structure, and evolution of a unique late Hesperian-early Amazonian shield volcano. J Geophys Res E Planets 111:E09003
115. Mouginis-Mark PJ, Wilson L, Zimbelman JR (1988) Polygenic eruptions on Alba Patera, Mars. Bull Volcanol 50:361–379
116. Kereszturi A, Petrik A (2020) Age determination for valley networks on Mars using tectonic-fluvial interaction. Planet Space Sci 180:104754
117. Hynek BM, Beach M, Hoke MRT (2010) Updated global map of Martian valley networks and implications for climate and hydrologic processes. J Geophys Res. 115:E09008
118. Leone G, Tackley PJ, Gerya TV, May DA, Zhu G (2014) Three-dimensional simulations of the southern polar giant impact hypothesis for the origin of the martian dichotomy. Geophys Res Lett 41:8736–8743
119. Wordsworth R, Forget F, Millour E, Head JW, Madeleine J-B, Charnay B (2013) Global modelling of the early martian climate under a denser CO2 atmosphere: water cycle and ice evolution. Icarus 222:1–19
120. Wordsworth RD, Kerber L, Pierrehumbert RT, Forget F, Head JW (2015) Comparison of "warm and wet" and "cold and icy" scenarios for early Mars in a 3-D climate model. J Geophys Res Planets 120:1201–1219
121. Wordsworth RD (2016) The climate of early Mars. Annu Rev Earth Planet Sci 44:381–408

General Concepts on Porous-Media Flow on Mars—Thermal H_2O-CO_2-fers–

Andrea Borgia, Curtis M. Oldenburg, Massimo Calcara, Arturo Catalano Gonzaga, and Giovanni Leone

Abstract

Fluids in the pores of the shallow crust of Mars are commonly two-phase mixtures: a H_2O-rich CO_2 phase usually occupying the upper (shallower) parts of the crust and a CO_2-rich H_2O phase occupying the lower parts of the upper crust. By analogy to terrestrial aquifers, these H_2O- and CO_2-filled formations are referred to as Martian CO_2-fers or H_2O-fers (or in general fluidfers). Here we present numerical experiments of heat and fluid flow in several large-scale fluidfer systems relevant to the Martian shallow crust. Given the lack of data on the Martian subsurface, the model systems and simulation results are presented in the spirit of conceptual models answering questions about what is possible rather than about precisely what is happening in any given Martian system. The simulations are carried out using the TOUGH2 and TOUGH3 numerical codes, developed at LBNL, with the Aquaveo-GMS graphical user interface to model the ground-fluid flow. We use the equation of state module ECO2N, which models phase equilibria between the components CO_2-H_2O-NaCl from 3 °C to 240 °C and pressures from 0.1 MPa to 60 MPa. These simulators were developed for terrestrial systems and we chose for experimentation here specific Martian fluidfer systems where P–T conditions conform to the limits of the simulators. For simplicity in demonstrating Martian fluid-flow phenomena, we use homogeneous porosity and permeability but we include the effects of topography as a key control on large-scale thermo-fluid flow. We also include the coupled thermo-fluid flow effects produced by an igneous intrusion at depth. Because the flows are two-phase porous media flows, we model the effects of capillary pressure, relative permeability, and the Klinkenberg effect. As boundary conditions we use lithostatic pressure at the base of the Martian cryosphere all along the top boundary of the models, while the remaining boundaries are impermeable (closed), apart from above the intrusion where we set a constant temperature of 240 °C and a pressure 1 bar above ambient. In total, we present 11 different simulations for various broad classes of scenarios. For each scenario, we show the time evolution of pressure,

A. Borgia (✉)
EDRA, Rome, Italy
e-mail: aborgia@lbl.gov

A. Borgia · C. M. Oldenburg
Lawrence Berkeley National Laboratory, Berkeley, CA, USA

M. Calcara · A. C. Gonzaga
Istituto Nazionale di Geofisica e Vulcanologia, Roma, Italia

G. Leone
Instituto de Investigaciòn en Astronomía y Ciencias Planetarias, Universidad de Atacama, Copiapó, Chile

G. Leone (ed.), *Mars: A Volcanic World*,
https://doi.org/10.1007/978-3-030-84103-4_6

temperature, gas saturation and gas density. We find that, (a) as expected, the time scale for convective cooling is much faster than conductive cooling. Therefore, we speculate that the current upper Martian crust has water-freezing temperatures down to the depth of pore closure (10–15 km), unless magmatic intrusions are occurring in the upper crust. (b) The CO_2-fers tend to overlie the H_2O-fers apart from the deep upper-crustal layers where the CO_2 can reach densities comparable to those of the water. Because of this, CO_2 eruptions at the surface are possible as both gaseous or liquid (unstable) phases. H_2O may also have erupted at the surface if the fluidfers were water-saturated, particularly at the base of relevant topographic relief (e.g., base of a mesa) and above deep intrusions that would have triggered strong hydrothermal convection. This phenomenon, though, would only last for times on the order of 100 ka. (c) In addition to water, also high-density CO_2 could perhaps mobilize and precipitate metals potentially leading to ore deposits and carbonate rocks.

1 Introduction

Given the extremely rarefied (a few mbars), dry, and cold atmosphere (mean annual surface temperatures well below the freezing point of water), the existence of surficial liquid water (or brines) on Mars even for short ephemeral periods has been disputed for the past over 50 years in the scientific literature (e.g., recent opposite stands are presented in [1–3] on the "wet" side, and Leone [4, 5] this issue for the "dry" side). Despite this controversy about the relevance of hydrogeological processes at the surface of Mars, there is a more general agreement on the possible existence of a variably thick and continuous cryosphere extending from 1 km of depth downward into the crust for many kilometers [6–13]. Below the cryosphere there is again disagreement on the existence or non-existence of liquid water—perhaps as saturated brines—in variably deep aquifers, even though, in contrast with the surface, pressures and temperatures could allow existence of liquid water (cf. [13–19]).

The H_2O and CO_2 on Mars are thought to originate mainly from planetary accretion, followed by the differentiation of the core, mantle, and crust, and subsequent magmatism during the Pre-Naochian times [20, 21]. During Noachian, Hesperian (and perhaps Amazonian) times, mantle degassing and magma ascent and crystallization with release of volatiles are considered the main source of these species in the underground below the cryosphere (cf. [13, 21]). Primary contributions of liquid H_2O are not expected to be derived from the cryosphere itself because from the Pre-Noachian Mars has been cooling; thus, the cryosphere has been and should still be growing downward if there is any liquid-phase water available at depth. In the absence of a liquid water supply to form additional ice, temperatures below the freezing point of water can extend well below the cryosphere bottom [13].

At this point in the evolution of knowledge of the Martian subsurface, there are not enough data to constrain unambiguously the actual subsurface composition of the fluid present in the pores and fractures of the rocks below the cryosphere. Certainly, below the cryosphere, there is CO_2 and it will coexist with H_2O, if present. The CO_2 may be in solid, gaseous, liquid or supercritical state across the possible ranges of pressures and temperatures of the Martian crust [13, 21]. H_2O may be found in the gaseous and or liquid state in the crust below the cryosphere; temperatures are probably not high enough (>374 °C) for the existence of supercritical H_2O above the depth at which fracture and pores will be closed (approximately 10–15 km), that limit the deeper circulation of underground fluids [13]. Among the possible solid phases if H_2O is present, clathrates (CO_2-hydrates) may form and they have been hypothesized as being present in the cryosphere [22]. A CO_2-H_2O phase diagram showing these relations is presented in Fig. 1 (from [21] and references therein). Finally, if there is liquid water, salts in variable concentrations should be present in solution in the liquid

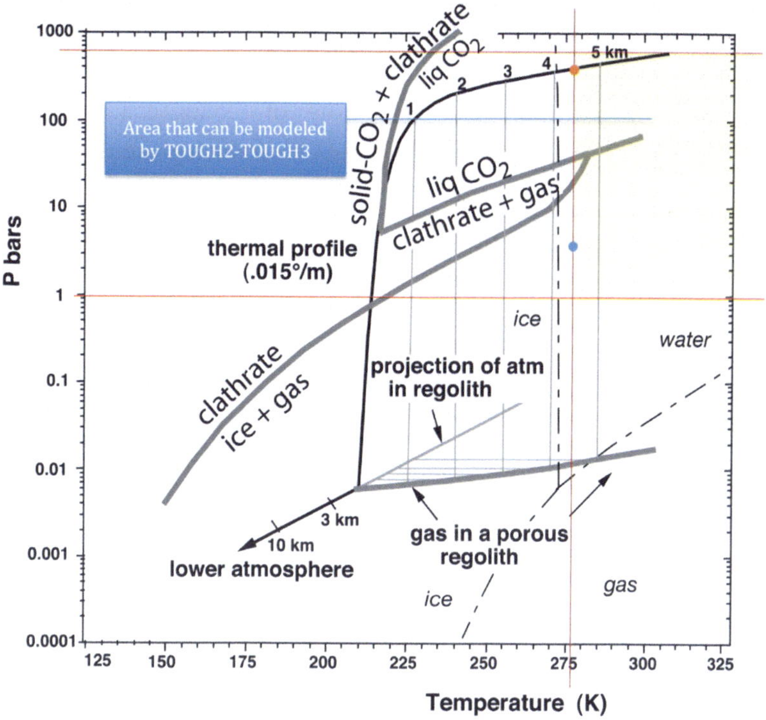

Fig. 1 Phase diagram of the system CO_2-H_2O ([20, 24, 25], adapted from [21]). The black line is a probable pressure/temperature gradient within the Martian crust; thick gray and dashed lines are the CO_2 and H_2O phases stability fields, respectively. The light yellow box shows the range of applicability of TOUGH2-TOUGH3 (range of temperatures of about 277–516 K and pressures of about 1–676 MPa). The light blue and orange dots are, respectively, the gaseous and liquid CO_2 top boundary conditions for the TOUGH simulations. See text for explanation

H_2O-rich phase or as solid coatings, covering both rock grains and fracture surfaces [23]. To date, no comprehensive numerical modeling has been done of the porous/fractured media coupled thermo-fluid dynamics of the CO_2-H_2O-salt system on Mars, even though many attempts have been made to model various aspects of Martian hydrogeology (see for instance [16] and references therein).

In this chapter, we address a variety of Martian subsurface thermo-fluid flow scenarios in two-dimensional cross sections. Our approach is to assume a simple homogeneous porosity–permeability-compressibility system with and without topography from the base of the cryosphere down to the pore-closure depth. We focus on the first order coupled thermo-fluid flow behavior of Martian CO_2- and H_2O-filled porous media (CO_2-H_2O-fers). We use the TOUGH 2 v.2 [26] and TOUGH3 [27] codes as a first attempt at simulating the behavior of Martian CO_2-H_2O-fers. Because the constitutive equations of TOUGH2 are made for terrestrial temperature and pressure fields, we have to restrict the range of our simulations to Earth-like values, imposing appropriate boundary conditions in order avoid those rock volumes with out-of-range P–T conditions. In addition, of the many salts present on Mars [23], we use only Sodium Chloride (NaCl) to simulate the water density changes due to variable salt concentration and salt precipitation from oversaturated water (Salt precipitation is important because of its effect on reducing absolute permeability and potentially fully clogging the rock matrix).

In our two-dimensional cross-sectional models, we consider a "no-topography" and a "mesa-like-topography" relief models to show the potential effect of topography on fluid-flow. For each of these, we compare the results of models that have a high and low volume fraction of CO_2, at low and high confining pressure, where the CO_2 gas-phase and liquid phase, respectively, are stable for a total of six base models. In addition, we add to each of these models the thermal effects of magmatic intrusions. Although this set of models cannot be considered a complete description of all the possible CO_2-H_2O-fers on Mars, it comprises a useful preliminary set for demonstrating variability and complexity of coupled thermo-fluid flow in the Martian crust.

2 Governing Equations

TOUGH2 is a multiphase, multicomponent, non-isothermal numerical simulator for flow in porous and fractured media (see [26] for a detailed description). It solves the integral mass and energy balance equations that can be written as.

$$\frac{d}{dt}\int_{V_n} M^K dV_n = \int_{\Gamma_n} F^K \cdot n d\Gamma_n + \int_{V_n} q^K dV_n \quad (1)$$

where V_n is an arbitrary subdomain of the flow system, bounded by the closed surface Γ_n; M is mass or energy per unit volume with $K = 1$, …, NK are the mass components in the system and $K = NK + 1$ is the heat "component". $\boldsymbol{F}$ is either mass or heat flux, and q represents sink/source terms; $\boldsymbol{n}$ is the normal vector on the surface element $d\Gamma_n$ positive into the volume V_n. In turn, M is written as:

$$M^K = \phi \sum_\beta S_\beta \rho_\beta X_\beta^K \quad (2)$$

that is, the total mass of component K is the sum of the mass in each phase β (liquid, gas and solid). In Eq. (2) ϕ is porosity, S is the volumetric pore fraction occupied by the phase β; ρ is the phase density and X is the mass fraction of component k in the phase β. Likewise, for heat accumulation M is given by:

$$M^{NK+1} = (1-\phi)\rho_R C_R T_R + \sum_\beta S_\beta \rho_\beta u_\beta \quad (3)$$

where ρ_R and C_R are, respectively, the rock grain density and the specific heat of the rock R, T is temperature, and u is the specific energy of each phase.

The advective mass flux of each component, $\mathbf{F}^K$, is given by the sum over all phases:

$$F^K\big|_{adv} = \sum_\beta X_\beta^K F_\beta \quad (4)$$

where the flux of each phase is given by the multiphase Darcy's law:

$$F_\beta = \rho_\beta u_\beta = -k\frac{k_{r\beta}\rho_\beta}{\mu_\beta}(\nabla P_\beta - \rho_\beta g) \quad (5)$$

In Eq. (5) u is the volume flux of each phase, k is the absolute permeability, which varies following [28], $k_{r\beta}$ is the relative permeability of phase β, and μ is the viscosity of the fluid phase. In addition, P is the sum of the pressure of a reference phase (usually the gas phase) and the negative capillary pressure, and, finally, $\mathbf{g}$ is the gravity acceleration vector. Finally,

$$F^{K+1} = -\lambda\nabla T + \sum_\beta h_\beta F_\beta \quad (6)$$

where λ is the thermal conductivity, and h is the specific enthalpy of each phase.

In TOUGH2 space is discretized using the "integral finite difference" method (IFDM [29, 30]) with a Voronoi grid. The IFDM makes it possible to implement double- and multiple-porosity methods for fractured media [31–33]. Time is discretized fully implicitly as a first-order backward finite difference.

TOUGH3 is the multi-processor, millions-of-grid-blocks version of TOUGH2; thus, it is much faster for large problems. TOUGH3 has also a number of improvements relative to TOUGH2, such us a more flexible labeling and coordinates format of cell, and a user-defined output. Depending on the complexity and size of the problem, we have used both codes for the various simulations presented here. As a graphical-user interface for geologic modeling and data/results visualization we use GMS from Aquaveo™ [34].

3 Constitutive Equation

As stated in the introductory chapter, within the Martian crust, we are concerned with a fluid system that can be adequately described by the three mass components: water (H_2O), carbon dioxide (CO_2) and sodium chloride (NaCl). In this system, TOUGH2 uses an equation of state module (named ECO2N V2.0; see [35] for detailed description) applicable to this system over the range of temperatures of about 277–516 K and pressures of about 1–676 MPa.

This equation of state describes the fluid as composed of a maximum of four phases:

(1) a water-rich phase with dissolved variable amounts of CO_2 and NaCl (called “aqueous” or “a”),
(2) a carbon dioxide-rich phase with dissolved variable amount of H_2O (called “liquid” or “l”),
(3) a gas phase with variable amount of H_2O and CO_2 (called “gas” or “g”),
(4) a solid sodium chloride phase (halite) that is assumed immobile (e.g., as a coating on rock grains) and that may precipitate from or dissolve into the aqueous phase.

Thus, there can be seven possible combinations of fluid phases (Fig. 2) that can coexist with the solid halite phase. While ECO2N can calculate the thermo-physical properties of gaseous and liquid CO_2 phases, it does not model the gas–liquid phase change (across the saturation curve). Thus, it can represent only the combinations in Fig. 2 numbered 1 (single-phase aqueous with or without dissolved CO_2 and salt), 2 and 3 (a single CO_2-rich phase that may be either liquid or gaseous that may include dissolved water), and 4 and 5 (two-phase conditions consisting of an H_2O-rich and a single CO_2-rich phases, with no distinction being made as to whether the CO_2-rich phase is liquid or gas). It will not compute the conditions 6 (two-phase mixture of liquid and gaseous CO_2) and 7 (three-phase conditions).

A different equation of state module is available, ECO2M [36], which can be used to calculate gas–liquid phase change and two phase CO_2 conditions. However, ECO2M cannot simulate temperatures higher than about 383 K. In this work we use ECO2N V.2.0 and avoid conditions in the Martian crust that necessitate simulating the CO_2 liquid–gas phase transition.

ECO2N V2.0 models, for each grid cell, the equilibrium phase partitioning H_2O and CO_2 between the “liquid” and the “gas” as a function of temperature, pressure, and salinity following [37, 38]. In addition, it models the dissolution and precipitation of NaCl using local equilibrium solubility. Liquid and gaseous CO_2 may coexist along the saturated vapor pressure curve of CO_2, which ends at the critical point (T_{crit}, P_{crit} = 31.04 °C, 73.82 bar, [39]). Above critical temperatures or pressures there is just a single CO_2-rich “gas” phase. Technically this is not correct, because the critical curve of this system is at much higher temperature and pressure [40]. However, because the three-phase line and its critical end point for the CO_2-H_2O system follows closely the vapor saturation curve for pure CO_2 [41], it is practical to refer to “subcritical” and “supercritical” conditions in reference to the pure CO_2. ECO2N does not consider hydrates phases such as clathrates.

4 Simulating the H_2O-CO_2-fers of the Martian Crust

The set of simulation cases presented here was chosen to demonstrate some fundamental aspects of thermo-dynamical behavior of fluid flow in the first few kilometers of the Martian crust. We consider idealized systems that capture representative conditions of some of the interesting Martian conditions that have not terrestrial analogue. The time period for these flows is pre-Noachian, Noachian and perhaps early-Hesperian times on Mars. The conceptual models chosen and their simulation are presented in the spirit of exploration of potentially important fluid and heat flow phenomena rather than explanations of particular observations.

Because there are no definite data on the underground fluid conditions on Mars in relation to temperature, pressure, CO_2-H_2O phase saturation, and salinity we approach the problem in a deterministic way using “reasonable” assumptions about the unknown conditions and their variability. The full set of models are specified in Table 1 and Fig. 3.

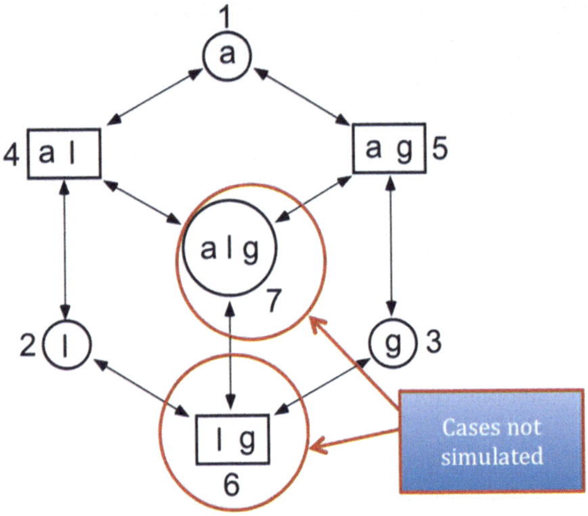

Fig. 2 The seven potential phase combinations in the system H_2O-CO_2 which can be associated with halite (from [35]). a = water-rich liquid (aqueous) phase; l = carbon dioxide-rich liquid phase; g = gaseous or supercritical H_2O-CO_2 phase

Table 1 Set of models developed

Models			
	Low Top Pressure (LTP)		High Top Pressure (HTP)
No Topography (NT)	Low gas saturation (98%)	High gas saturation (50%)	Low gas saturation (98%)
Mesa Relief (MR)	Low gas saturation (98%)	High gas saturation (50%)	Low gas saturation (98%)

We assume the presence of a cryosphere and a lithostatic pressure in the fluid right below this cryosphere [14, 16, 21]. Farther downward the fluid will maintain a hydrostatic pressure gradient. Accordingly (see also Fig. 1), we will have two cases:

(1) the first with low top pressure (LTP) of 0.1 MPa with CO_2 in its gaseous phase corresponding to a thin cryosphere, and
(2) the second with high top pressure (HTP) of 10 MPa with CO_2 in its liquid phase corresponding to a thick cryosphere.

In both cases, the CO_2 will become supercritical downward at higher pressure and temperature. These models extend to a depth where the pressure reaches about 50–60 MPa.

Therefore, the first case with LTP may extend to a depth of 16 km, while the HTP can extend downward only to 10 km below the cryosphere. In the thicker LTP model the cryosphere is obviously thinner, in the order of a many meters thick, while in the thinner HTP model the cryosphere is accordingly thicker and in the order a many hundred of meters thick. In practice both models extend to about 11–16 km from the surface of Mars, which is thought to be the depth at which the permeability becomes too low to be significant [16].

The temperature at the top of the model is always set to 277 K, while temperature at the bottom of both model types is set to 437 K. Therefore, in the LTP models the thermal gradient is about 10 K/km, while the gradient is 16 K/km in the HTP type of models.

The salinity of the water in all models is set to a high value (0.17 salt mass fraction) below the NaCl-saturation level at the beginning of the simulations. During the evolution of the simulations, if the CO_2 volume fraction increases, the salinity may reach saturation leading to the precipitation of halite that, in turn, reduces permeability.

In the LTP models, we use either a high (98%) or a low (50%) gas saturation at the top, while in the HTP models we use only the high

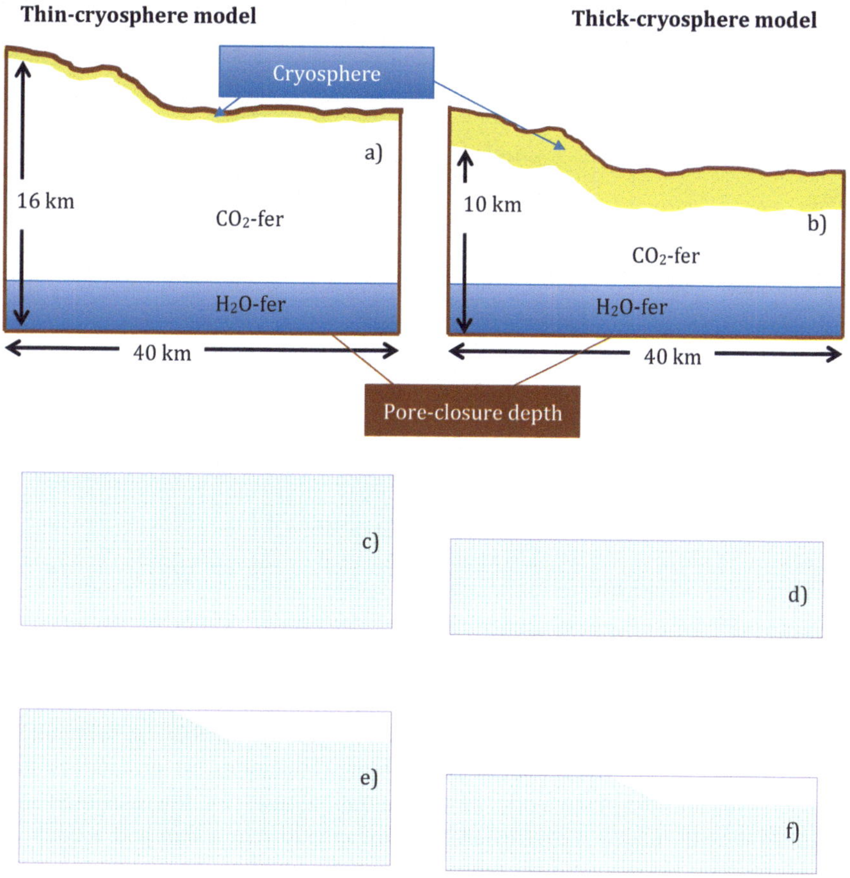

Fig. 3 Conceptual models of the Martian crust developed in the simulations. **a** Thin-cryosphere model; **b** Thick-cryosphere model. The four grids used in the simulations. **c** and **e** are 16 km high by 40 km long, and are used for thin-cryosphere "low top-pressure" simulations; **d** and **f** are 10 km high and 40 km long, and are used for thick-cryosphere "high top-pressure" simulations. **c** and **d** are for simulations with no topography and **e** and **f** are for simulations with "mesa-relief" topography; the topographic gradient is the common gradients on the scarps of Mons Olympus and Valles Marineris

gas saturation. This is because, in the case of a low gas saturation, the pressure at 16 km depth increases above the maximum pressure allowed by the TOUGH equation of state.

In addition, we present a set of models with no topography (NT) and a set with a mesa-like topographic relief (MR) of 3 km above the lowland. Despite the fact that Mars has much larger topographic relief, e.g., the Olympus Mons scarps or the scarps of Valles Marineris that both may reach heights of 10 km, our model is limited, in the height of relief simulated by the fact that we need to simulate ground-fluid flow within a thickness of the crust between the pressure limits of the simulator.

We impose a Dirichlet top boundary condition with pressure, temperature, salinity and CO_2-phase gas saturation that do not change in time. This top boundary condition models the presence of a cryosphere somewhere above the boundary, which can absorb or release water, respectively by freezing or melting. We realize that this may be considered a gross approximation. However, it is the base for future improvements. The other three boundaries are non-conductive and impermeable. The initial conditions are fluido-static pressure and constant geothermal gradient.

Finally, we simulate the ground-fluid flow induced by an intrusion occurring at some depth below the bottom of the model by increasing and

Table 2 Values of the parameters used in TOUGH simulations [26]

Rock grain density (kg m^{-3})	2300				
Porosity	0.1				
Permeability (m^2)	$x = 10^{-13}$	$y = 10^{-13}$	$z = 10^{-13}$		
Formation heat conductivity (W m^{-1} ° C^{-1})	2.1				
Rock grain specific heat (J kg^{-1} $°C^{-1}$)	1000				
Compressibility (Pa^{-1})	0.0				
Pore expansivity ($°C^{-1}$)	0.0				
Tortuosity	1.0				
Klinkenberg parameter b	$1.0 * 10^5$				
Relative permeability	Corey	$S_{lr} = 0.3$	$S_{gr} = 0.05$		
Capillary pressure	van Genuchten	1 = 0.4438	$S_{lr} = 0.0$	$S_{ls} = 2.402 * 10^{-4}$	$S_{gr} = -1.0 * 10^8$

maintaining, in a few cells of the bottom the temperature from 437 to 413 K and the pressure by 0.1 MPa above ambient. In the NT models, the "intrusion" is set at the left side of the model for a length of 2 km. In the MR models the intrusion is set either at the far right below the lower topography, in the center in correspondence of the scarp, or at the far left side below the relief.

In the following presentation of modeling data, for simplicity, we refer to the CO_2-fluid phase (usually H_2O saturated) as the "gas phase" or "gas"; conversely, we call the H_2O-fluid phase (usually CO_2 saturated) as "liquid water phase". Only by considering the local P–T conditions one may distinguish true liquid, gas or supercritical conditions in the CO_2-phase and liquid water or gas for the H_2O-phase. The full set of parameters used in the simulations is indicated in Table 2.

5 Effects of an Intrusion Below a Flat Topography

The following set of three numerical experiments shows the potential underground CO_2 and H_2O flow during an igneous intrusion event occurring at some depth below the left-hand side of the model without any influence of other factors such as topography. In this, as with all of the following simulations, we present first the high gas-saturation cases for low and high top pressure, corresponding to a thin and a thick cryosphere, respectively. Then we present the high aqueous-phase-saturation case only for high top pressure corresponding to a thick cryosphere.

5.1 High Gas-Saturation, Low Pressure

In this first experiment the top boundary condition mimics the pressure of a thin cryosphere. The fluids and the heat rising from the hot intrusion below the bottom left-hand side of the model cause pressure to increase in the rising plume (Fig. 4). Also the pressure increases at the bottom of the model where the liquid–water phase is present. On the other hand, the pressure decreases in most part of the model where a 50–70% gas saturation is also present.

The temperature increases significantly above the intrusion and at the bottom of the system (Fig. 5) where the hot liquid water expands on top of colder water (Fig. 6); in the gas phase

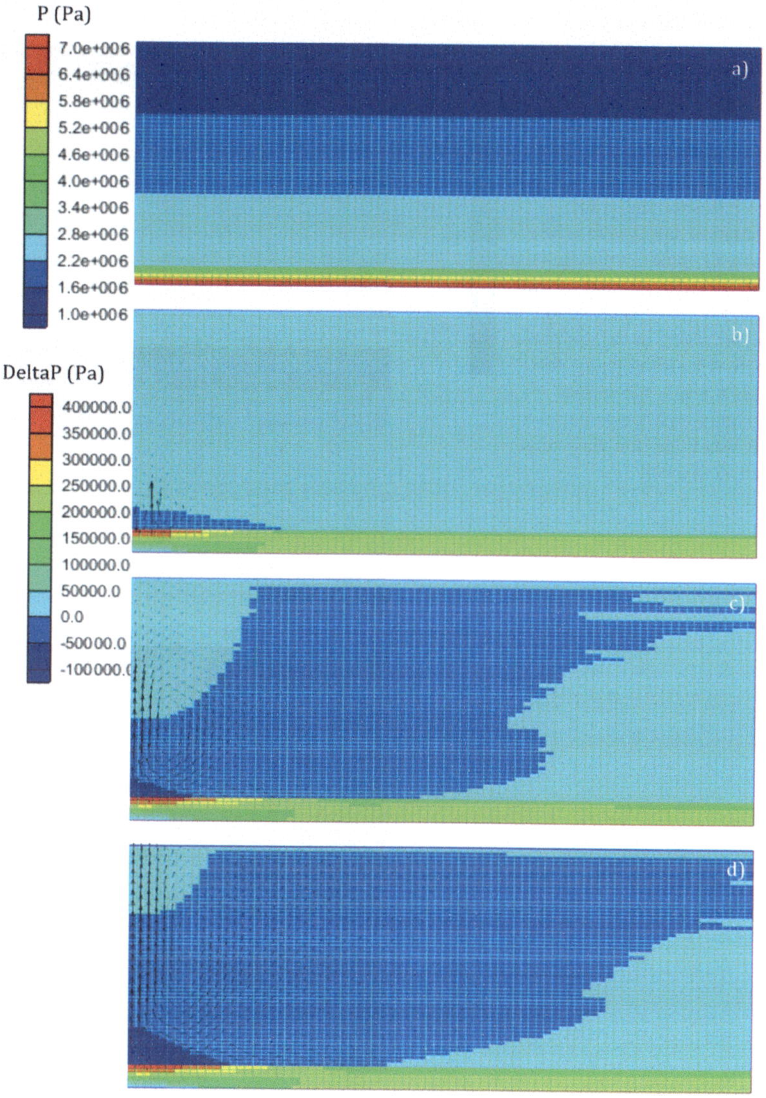

Fig. 4 Variations in pressure due to the effects of an intrusion somewhere below the left-hand side of the model for the high gas-saturation low-pressure case with flat topography. The model is 40 km long and 16 km high; the vectors are the gas velocity. **a** Normal distribution pressure at initial conditions, gas static in the upper part and liquid–water-static at the very bottom; note how the pressure gradient is much higher towards the very bottom below the water table. Variations in pressure after **b** 10 ka, **c** 50 ka, **d** 100 ka. In time, relative to the normal pressure gradient, the pressure increases significantly above the intrusion and at the bottom of the model where the liquid water phase is present (see following Fig. 6), but pressure decreases where gas phase is predominant

above the intrusion there is a strong plume that rises towards the surface bringing high temperatures all the way to the top via a single convective cell. Even though the cryosphere is expected to be very thin, due to adiabatic expansion of the gas phase, the temperatures at the top of the convective cells do not appear to be sufficient to melt the cryosphere and erupt at the surface for any relevant amount of time.

In time, the gas saturation increases significantly above the water-saturated layer, and much water is "lost" to the upper boundary accreting the cryosphere bottom. The gas density decreases all along the rising hot plume, but gas density increases significantly below the water-saturated layer (Fig. 7). Given the insignificant interaction with the upper layer for this particular model, little effect is expected to be observed at the surface.

5.2 High Gas Saturation, High Pressure

In this second experiment, the top boundary condition mimics the pressure at the bottom of a thick cryosphere. The fluids and the heat rising

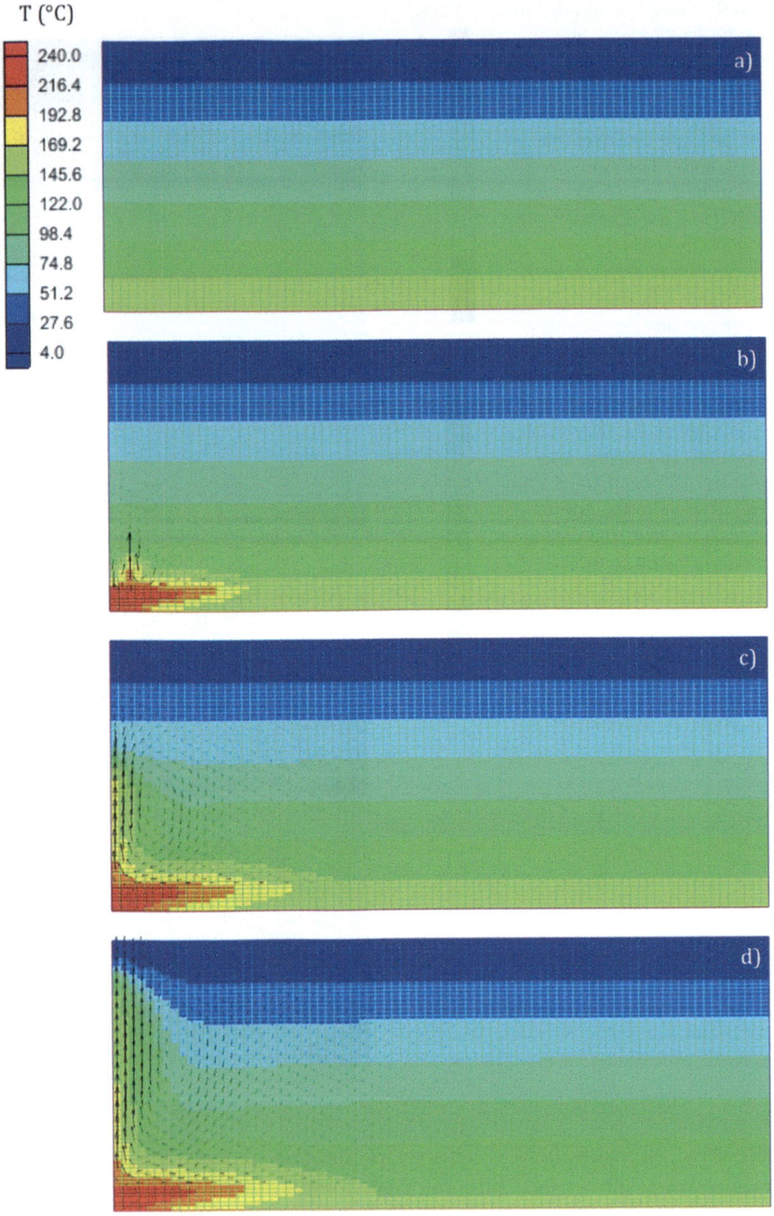

Fig. 5 Variations in temperature. **a** Initial temperature distribution. Temperature distribution after **b** 10 ka, **c** 50 ka, **d** 100 ka. In time, the temperature increases significantly above the intrusion and at the bottom of the system where the water phase is present (see following Fig. 6) and the hot liquid water expands on top of colder water; in the gas phase there is a strong plume that rises towards the surface bringing high temperatures all the way to the top via a single convective cell

from the intrusion below the bottom left-hand side of the model make the pressure decrease all along the rising hot plume, but pressure increase significantly only below the water saturated layer. After 5 ka some gas is absorbed by the bottom boundary at the point of intrusion (Fig. 8).

The temperature first increases rapidly bringing a hot plume to the base of the cryosphere, then it decreases in the whole model due to the convecting-downward cold-gas phase that quenches the intrusion (Fig. 9). Due to convection, the cold gas tends to fill in the whole model with 50–60% gas saturation (Fig. 10). Only patches of water-saturated (aqueous) regions remain at the center of convection cells. The gas density increases significantly at the bottom becoming comparable to that of the liquid water (Fig. 11).

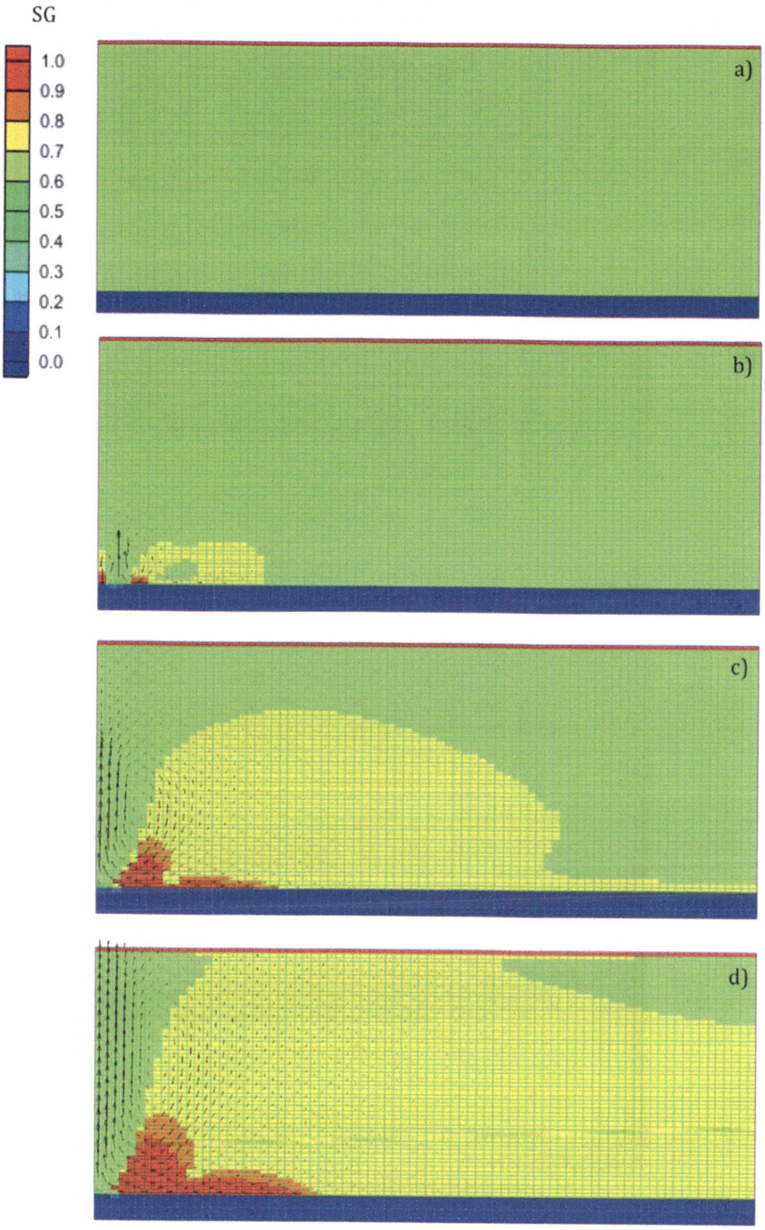

Fig. 6 Variations in gas saturation. **a** Equilibrium gas saturation distribution for the P/T conditions imposed at the start of the simulation (see Figs. 4 and 5); note that the upper part of the model has about 50% of gas saturation, while at the bottom the system saturated with liquid water. Gas saturation distribution after **b** 10 ka, **c** 50 ka, **d** 100 ka. In time, the gas saturation increases significantly above the water saturated layer, and much water is "lost" to the upper boundary

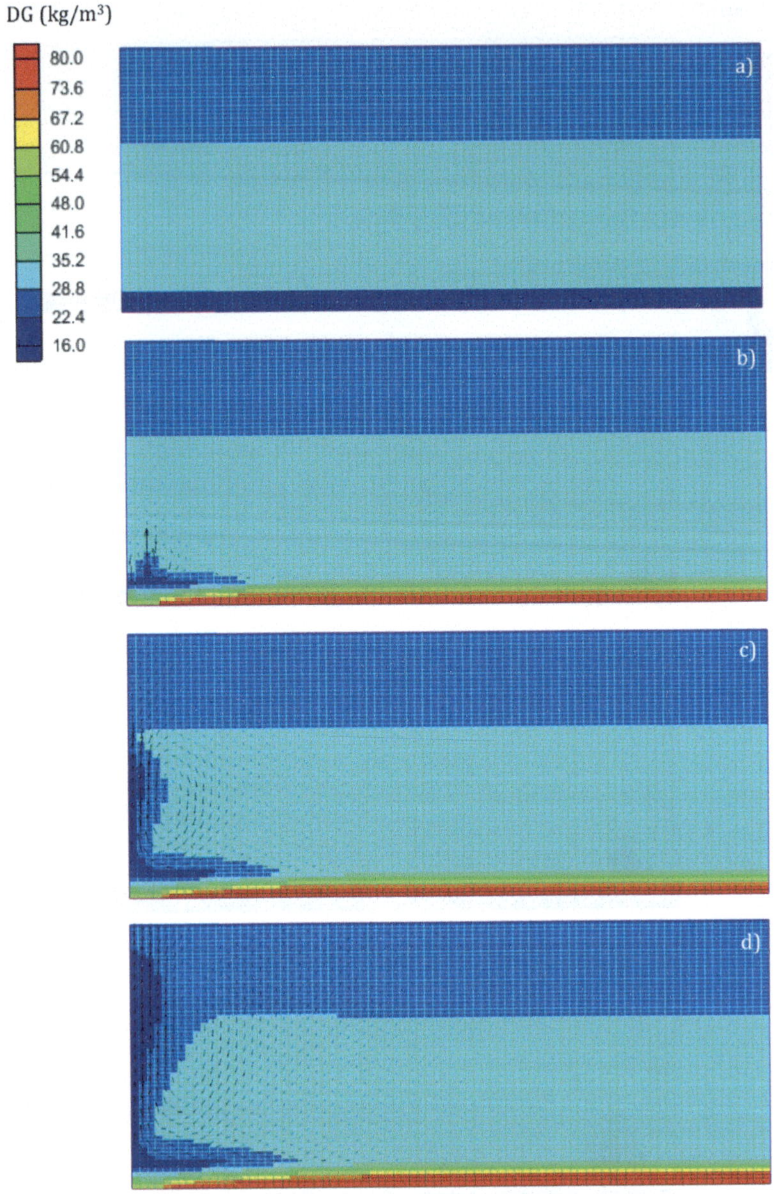

Fig. 7 Variations in gas density. **a** Equilibrium gas density distribution for the P–T conditions imposed at the start of the simulation (see Figs. 4 and 5); note that the upper part of the model that has about 50% gas saturation has low densities, while the bottom of the system saturated by liquid water has high densities. Gas density distribution after **b** 10 ka, **c** 50 ka, **d** 100 ka. In time, the gas density decreases all along the rising hot plume, but increases significantly only below the liquid–water-saturated layer

For this particular simulation there appears to be not enough buoyancy in the rising plume to affect the cryosphere and to produce any significant evidence of the igneous intrusion at the surface. The fluid convection mechanism is very effective at transferring heat from the bottom to the top boundary, rapidly quenching the whole system.

5.3 Low Gas Saturation, High Pressure

In this third experiment, the top boundary condition mimics the pressure of a thick cryosphere, but the whole system is mostly saturated with aqueous phase (gas saturation is 5–10%). In time, the pressure first increases above the intrusion, than it decreases due to heating of the rock formation as a result of convection. Water is lost at the top boundary (Fig. 12).

A strong hot plume develops above the intrusion interfering along a significant part of the bottom of the cryosphere (Fig. 13). Farther away from the intrusion, convection cools down the system. Gas saturation increases in the hot rising plume (Fig. 14) and more significantly at the top of the model far away from the plume forming a thick gas-saturated layer (98%). The gas density decreases all along the rising hot plume, but increases significantly far away from it (Fig. 15).

Such a pronounced and long-lasting plume of hot liquid water could interact with the cryosphere and locally melt it. Given these conditions, it is conceivable that the rising plume of hot water could erupt at the surface for a significant amount of time.

6 Mesa Topography

The following set of experiments show the CO_2-H_2O ground-fluid flow under the influence of a 3-km-high relief with real Martian topographic gradient. First we discuss the effect of topography only. Then we introduce the effect of an intrusion (a) below the lowland far away from the scarp (right-hand-side of model), (b) below the scarp (center of model), and (c) below the relief far away from the scarp (left-hand-side of model).

6.1 High Gas Saturation, Low Pressure, no Intrusion

In the high gas-saturation low pressure experiment, convection is imposed upon the system by the fact that at the same depth in the model, there are higher pressures below the topographic relief compared with the lowland, because the pressures at the bottom of the cryosphere (the top of the model) are equal. This phenomenon is well known in hydrology on Earth where hot water springs tend to be close to the bottom of hillslopes [42]. While on Earth this is because of recharge, on Mars (where there is no recharge) it is because similar thicknesses of cryosphere are supposed to exist at similar latitudes.

With time, the pressure decreases below the relief and increases below the lowland and all along the bottom where an aqueous phase (water) layer is formed (Fig. 16). A plume of hot gas develops in the lowland interfering for a significant distance with the bottom of the cryosphere (Fig. 17). Such a pronounced and long-lasting plume of hot gas could locally melt the cryosphere and given its thinness reach the surface as liquid or gaseous CO_2 even in the absence of an intrusion.

The gas saturation increases toward the bottom below the relief (Fig. 18). In addition, liquid water is condensed at the bottom of the model. The gas density increases toward the bottom below the relief and decreases within the rising plume (Fig. 19). Obviously, gas density is also high within the liquid water-dominated volume at the bottom of the model.

6.2 High Gas Saturation, Low Pressure, Deep Intrusion Below Slope

The following simulation is identical to the former one apart from having added the effects of an intrusion below the scarp at some depth below the bottom of the model. In this case, as in the

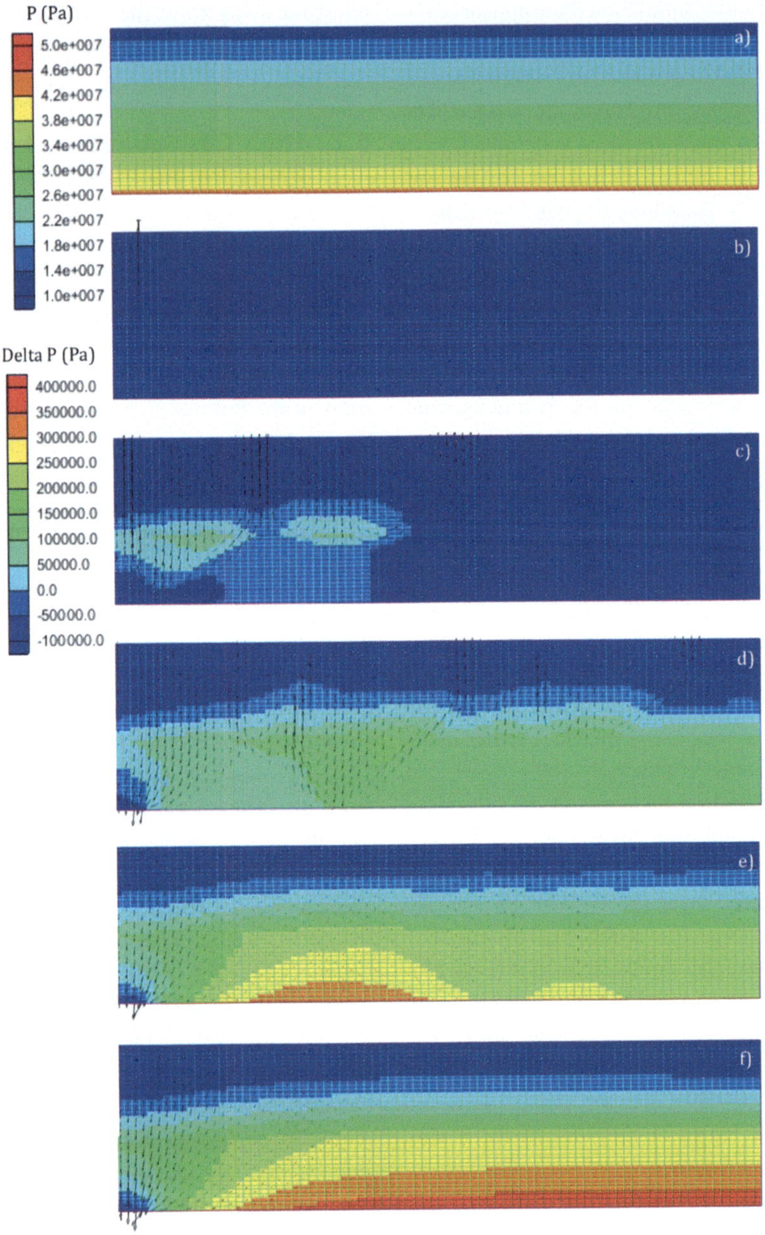

Fig. 8 Variations in pressure from initial conditions due to an intrusion on the left-hand side below the bottom of the model for the high gas-saturation high-pressure case with flat topography. The model is 40 km long and 10 km high; the vectors are gas velocities. **a** Pressure distribution imposed at the start of the simulation the gradient is gas-static (50% gas saturation) in the top half of the model and hydrostatic for the lower half (see Fig. 10). Variations in pressure after **b** 1 ka, **c** 5 ka, **d** 10 ka, **e** 60 ka, **f** 200 ka. In time, the pressure decreases all along the rising hot plume, but increases significantly only below the water saturated layer. After 5 ka some gas is absorbed by the bottom boundary at the point of intrusion

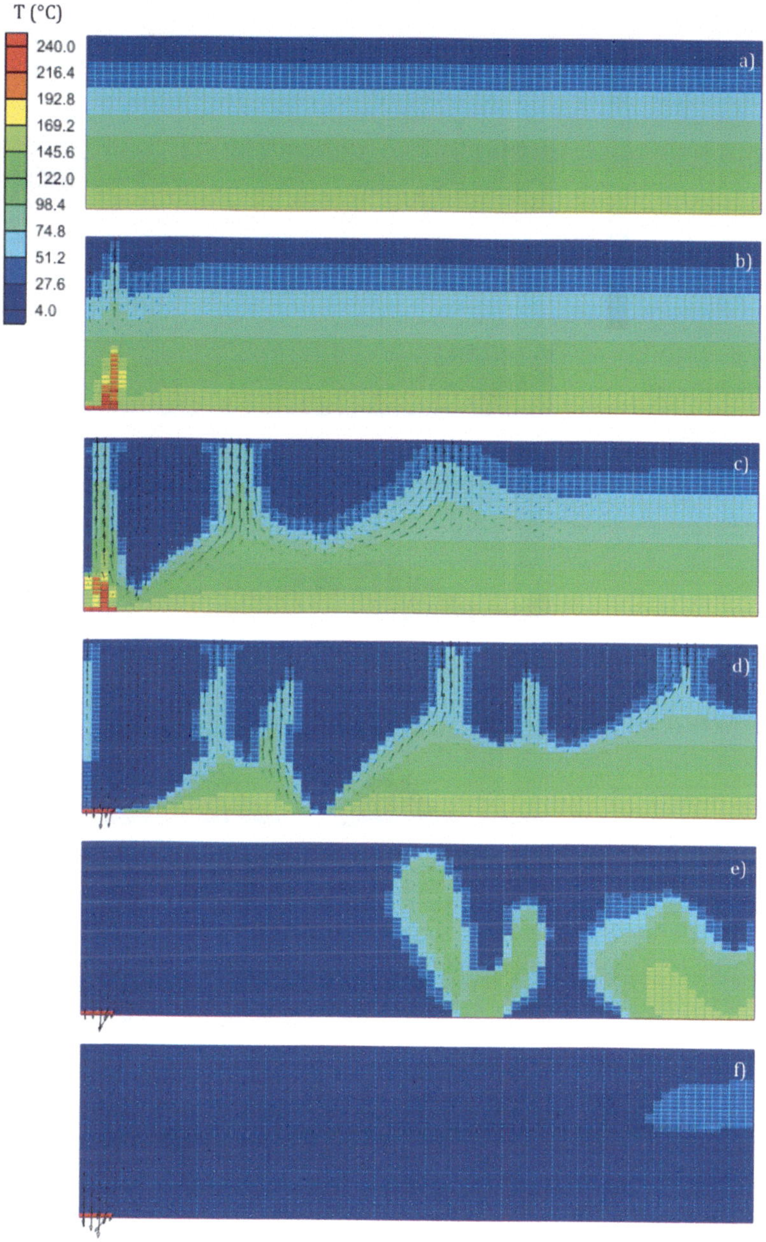

Fig. 9 Variations in temperature. **a** Simple temperature gradient distribution imposed at the start of the simulation. Temperatures after **b** 1 ka, **c** 5 ka, **d** 10 ka, **e** 60 ka, **f** 200 ka. In time, the temperature first increases rapidly bringing a hot plume to the top, the base of the cryosphere, than it decreases in the whole model due to the convecting cold-gas phase quenching the whole system

former one, the pressure decreases below the relief and increases below the lowland all the way to the bottom and downstream of the intrusion (Fig. 16).

In time, a plume of hot gas develops in the lowland far away from the bottom of the scarp interfering for a significant area with the bottom of the cryosphere. This interaction may lead to

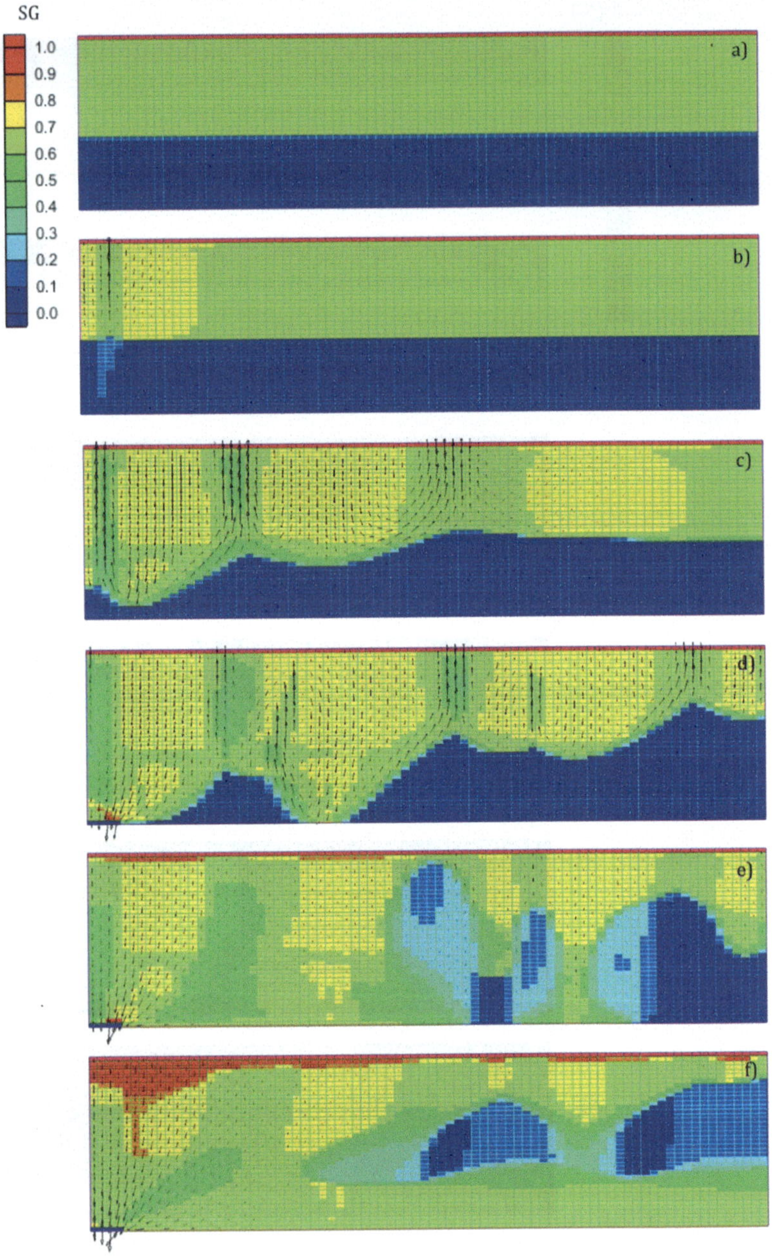

Fig. 10 Variations in gas saturation. **a** Equilibrium gas saturation distribution for the P/T conditions imposed at the start of the simulation (see Figs. 9 and 10); note that the upper part of the model has about 50% gas saturation, while the bottom of the is liquid–water saturated. Gas saturation after **b** 1 ka, **c** 5 ka, **d** 10 ka, **e** 60 ka, **f** 200 ka. In time, due to convection the cold gas tends to fill in the whole model. Only patches of water-saturated regions remain at the center of the convection cells. For this particular case the CO_2-fer underlies the H_2O-fers

the melting of the cryosphere and the eruption of liquid or gaseous CO_2 at the surface.

The temperature of the intrusion, though, is rapidly quenched by the convecting downward cold gas phase (Fig. 17). The gas saturation increases below the high topography moving toward the lowland (Fig. 18). The gas density increases below the high topography, later increasing also towards the lowland (Fig. 19).

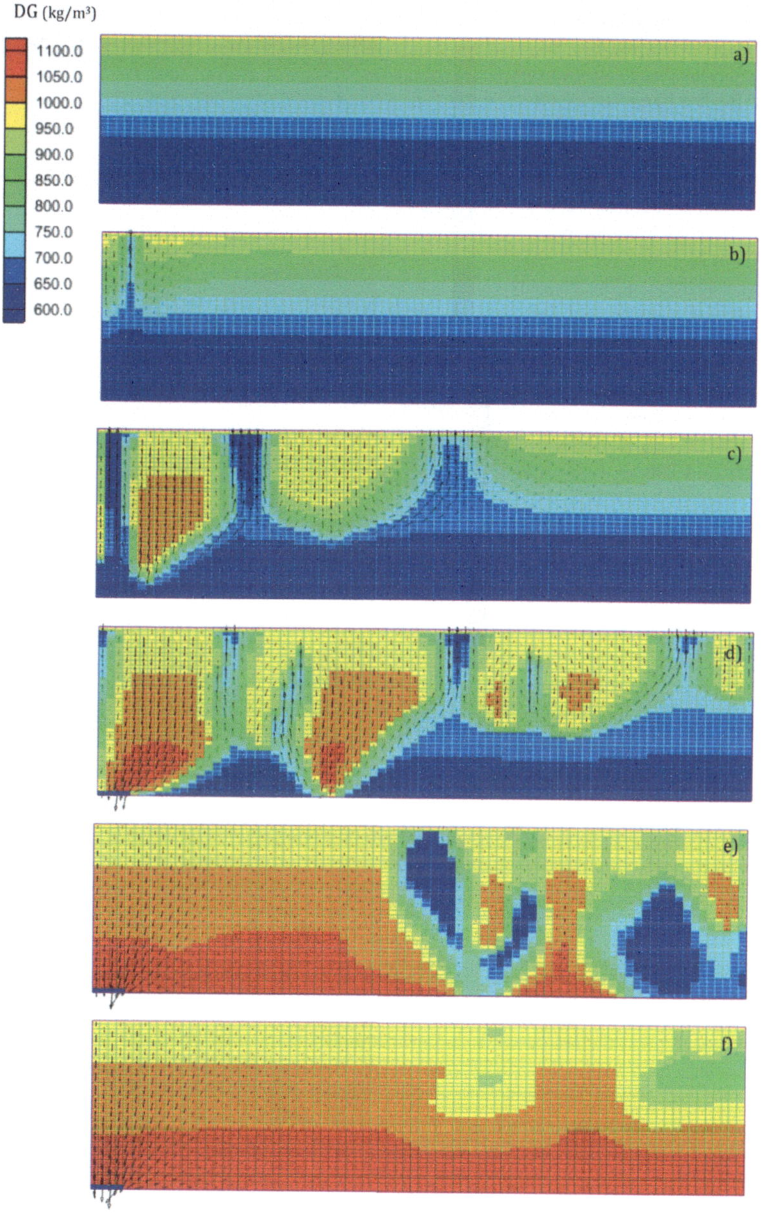

Fig. 11 Variations in gas density. **a** Equilibrium gas density distribution for the P–T conditions imposed at the start of the simulation (see Figs. 9 and 10); note that the upper part of the model has about 50% gas saturation, while the bottom of the model is liquid–water-saturated. Gas density after **b** 1 ka, **c** 5 ka, **d** 10 ka, **e** 60 ka, **f** 200 ka. In time, the gas density increases significantly at the bottom becoming comparable and larger to that of the liquid water, and forming a thick CO_2-fer at the bottom of the model

6.3 High Gas Saturation, High Pressure, Deep Intrusion

In the following simulations with high top pressures we show the influence of three different locations for the intrusion in relation to the scarp: (a) below the lowland at the far right-hand side of the model, (b) beneath the scarp, (c) below the high relief (mesa) at the far left-hand side of the model.

6.3.1 Intrusion Below Lowland

Compared to the former simulation, in this simulation the effect of the intrusion disturbs the effect of the topographic relief. In practice, the pressure decreases at the beginning everywhere

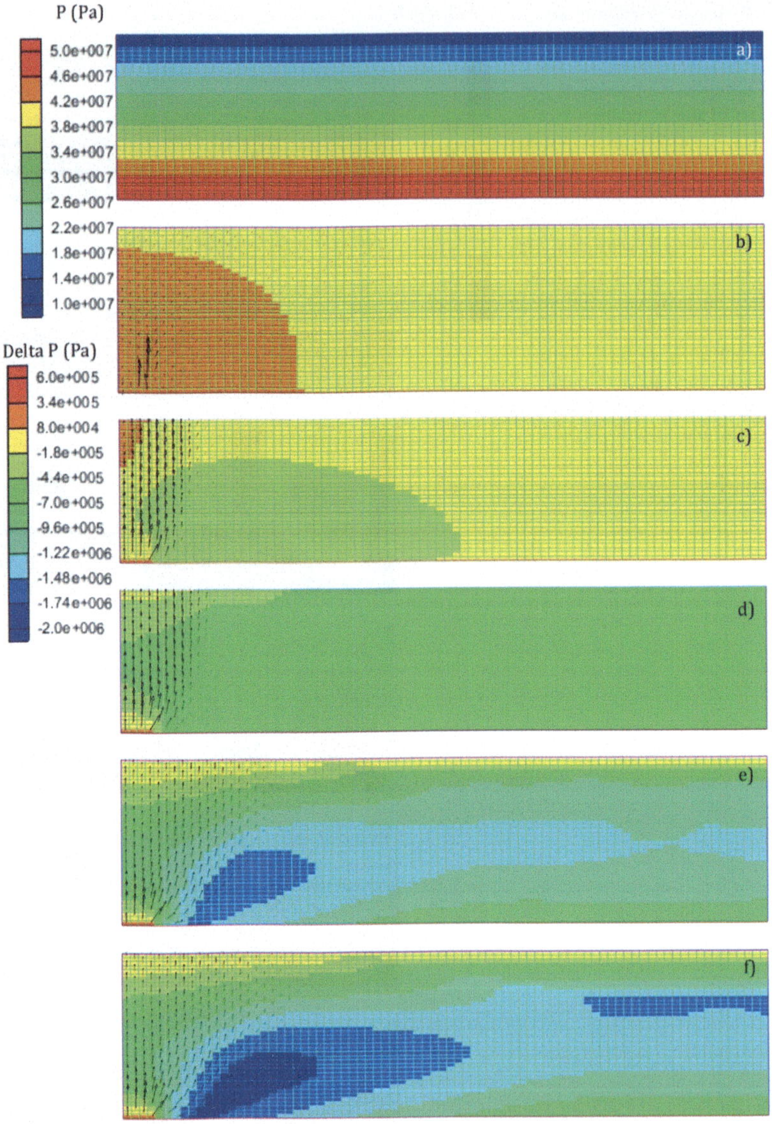

Fig. 12 Variations in pressure from initial conditions due to an intrusion on the left hand side of the model at some depth below the bottom for the low gas-saturation high-pressure case with flat topography. The model is 40 km long and 10 km high; the vectors are gas velocities. **a** The pressure distribution imposed at the start of the simulation is liquid water-static all across the model (see Fig. 14). Variations in pressure after **b** 1 ka, **c** 5 ka, **d** 10 ka, **e** 60 ka, **f** 200 ka. In time, the pressure first increases above the intrusion, then it decreases due to heating of the rock formation as a result of convection. Water is lost at the top boundary

apart from below the bottom of the scarp, where a gas plume rises from the bottom to the cryosphere. Later the pressure increases again below the high relief and the lowland particularly close to the scarp (Fig. 24).

In time, two hot gas and water (gas saturation 50%) plumes develop at the bottom of the scarp and at the far right-hand side in the lowland, interfering for many 10's of ka with the bottom of the cryosphere (Fig. 25). Such plumes of hot gas and water could locally reach the Martian surface and erupt. Later in time, the temperature of the whole system, including the intrusion, is quenched by the downward-convecting cold gas–water phases.

The gas saturation increases in the central area at the bottom of the model and close to the scarp where a very dry zone forms closer to the top (Fig. 26). An ephemeral aqueous-phase-saturated zone forms at the bottom of the model below the

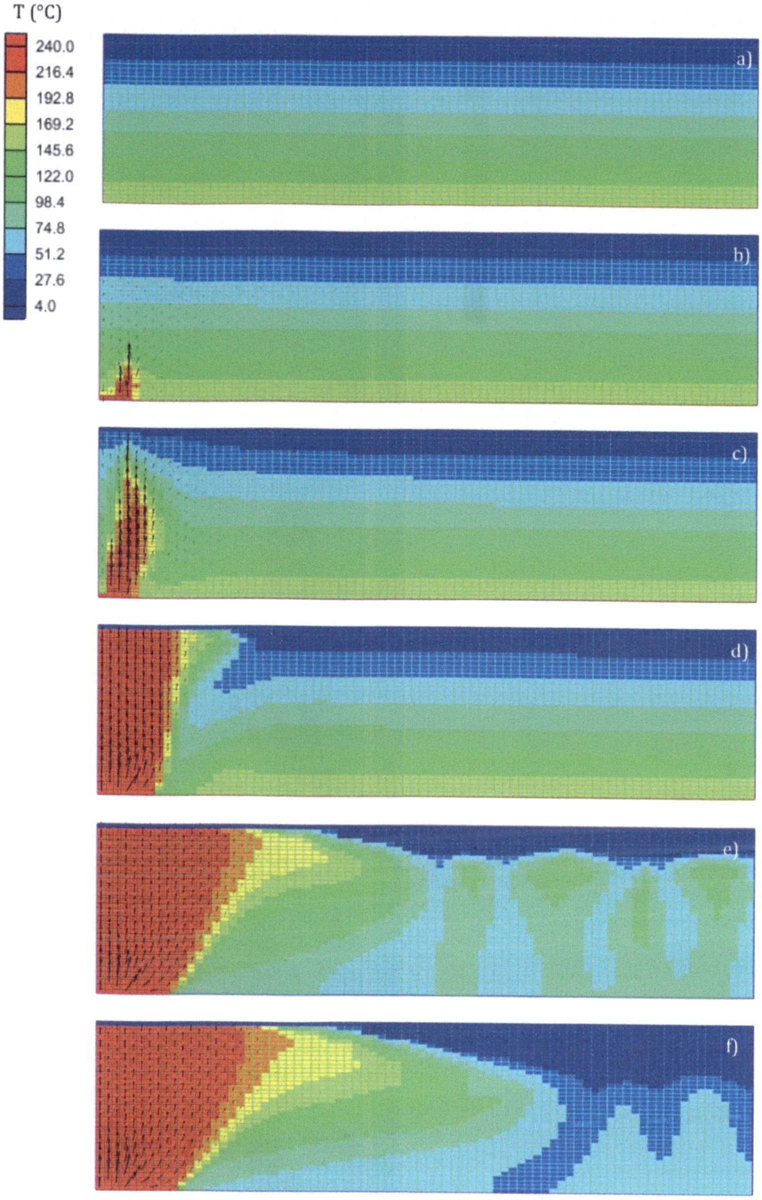

Fig. 13 Variations in Temperature. **a** Constant temperature gradient imposed at the start of the simulation. Variations in temperature after **b** 1 ka, **c** 5 ka, **d** 10 ka, **e** 60 ka, **f** 200 ka. In time, a strong hot plume develops above the intrusion interfering for a significant area with the bottom of the cryosphere. Further away from the intrusion, convection cools down the system. Such a pronounced and long-lasting plume of hot water below the cryosphere could locally melt it and perhaps reach the surface as an outpouring of liquid water

lowland and a more permanent one on the far left-hand side. The gas density increases significantly at first below the scarp but later in the whole model (Fig. 27).

6.3.2 Intrusion Below Slope

This experiment is similar to the preceding one apart from the position of the intrusion that is located below the scarp. During the simulation, at first the pressure increases below the lowland and particularly at the base of the scarp (Fig. 28); later the pressure increases also below the high-relief region (mesa). In practice, the effect of the topographic relief on the underground fluid flow produced by the scarp adds to the convective effect induced by the intrusion.

A hot-gas plume starting at the bottom of the model rises up to the base of the scarp reaching and, for a few 10's of ka, interfering with the bottom of the cryosphere. Such plumes of hot gas

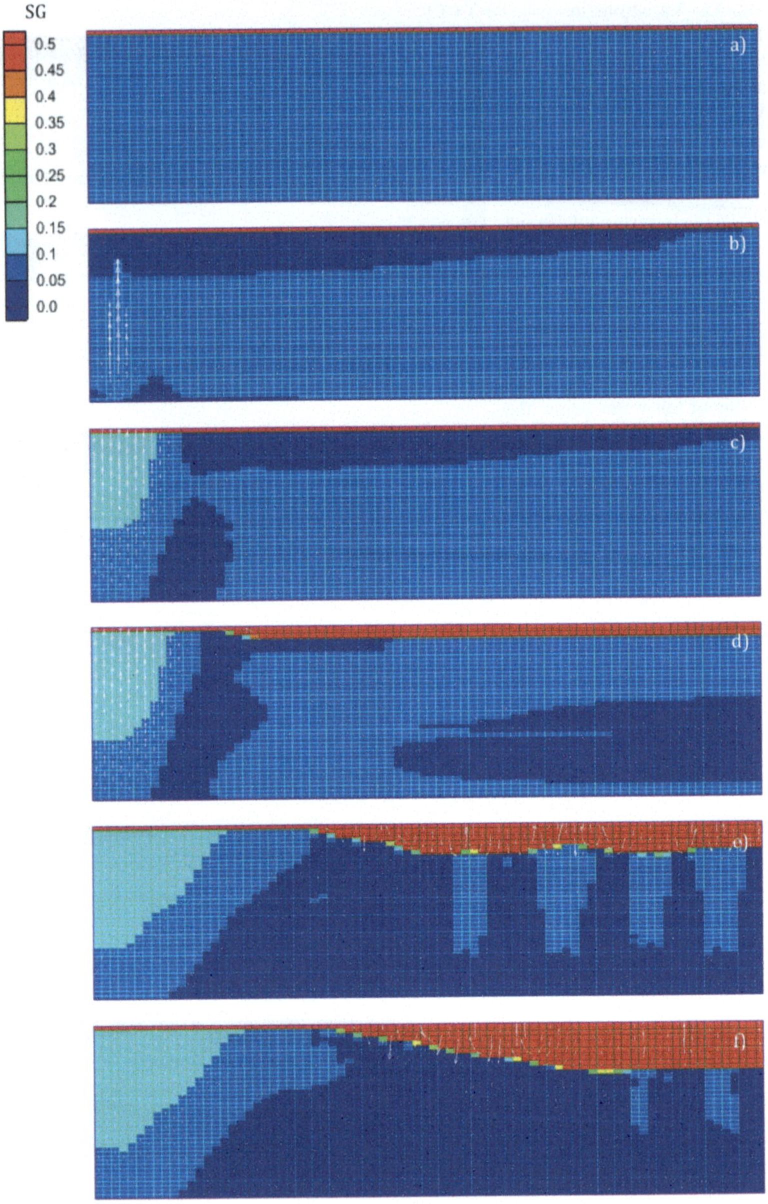

Fig. 14 Variations in gas saturation. **a** Equilibrium gas saturation calculated at the beginning of the simulation according to the P–T conditions given in Figs. 12 and 13; the whole system is aqueous-phase (water)-saturated up to 92%. Gas saturation after **b** 1 ka, **c** 5 ka, **d** 10 ka, **e** 60 ka, **f** 200 ka. In time, the gas saturation increases in the hot plume and more significantly at the top of the model far away from the plume

could perhaps locally reach and erupt at the surface (Fig. 29). The temperature of the whole system, including the intrusion, though, is then rapidly quenched by the downward-convecting cold gas phase.

In time, aqueous-phase-saturated regions form below the highlands and lowlands (Fig. 30); these areas later reduce in volume but persist through to the end of the simulation. Volumes with high gas-phase concentrations develop on top of the intrusion and below the top of the scarp. During the simulation, the gas density increases significantly in the whole model (Fig. 31).

6.3.3 Intrusion Below Top of Mesa

In this experiment the intrusion is below the left-hand margin of the model. In time, the pressure

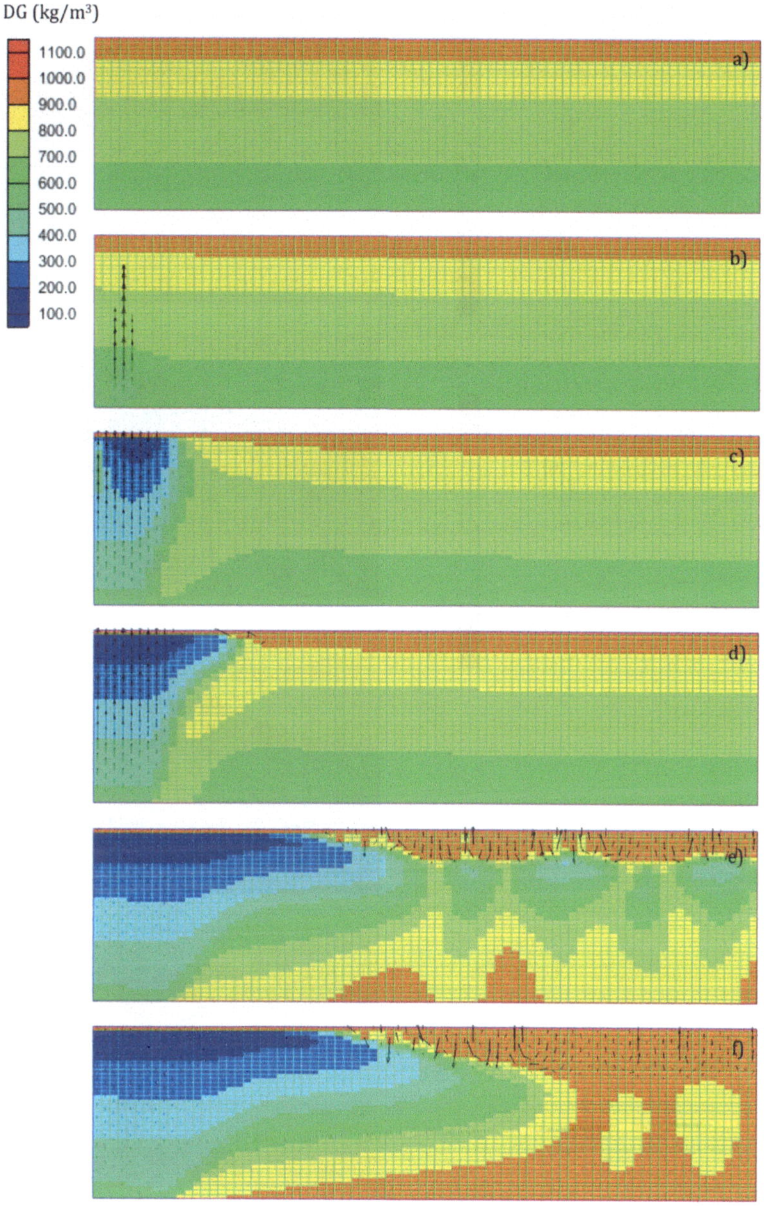

Fig. 15 Variations in gas density from initial conditions. **a** Equilibrium gas density calculated at the beginning of the simulation according to the P–T conditions given in Figs. 12 and 13; the equilibrium gas density distribution imposed at the start of the simulation follows the distribution of temperature. Variations in gas density after **b** 1 ka, **c** 5 ka, **d** 10 ka, **e** 60 ka, **f** 200 ka. In time, the gas density decreases all along the rising hot plume, but increases significantly far away from it

decreases below the relief (mesa) and increases below the lowland, particularly at the bottom of the scarp (Fig. 32).

At first a hot-gas plume develops at the bottom of the left-hand-side boundary for a few 10's of ka and interacts with the bottom of the cryosphere (Fig. 33); such plumes of hot gas could locally, for limited time, reach and erupt at the surface. The temperature of the whole system, including the intrusion, is rapidly quenched by the downward-convecting cold gas phase.

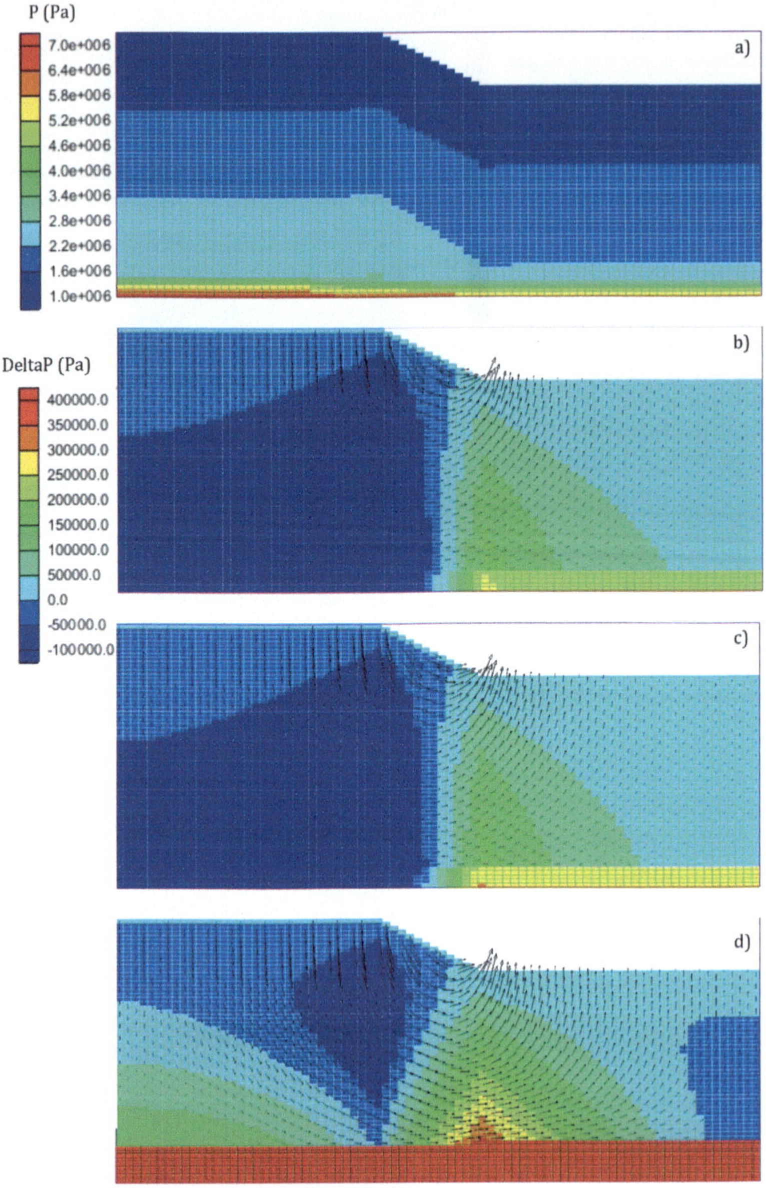

Fig. 16 Variations in pressure from initial conditions due to a mesa-like topographic relief. The model is 40 km long and 16 km high; the vectors are gas velocities. **a** The pressure distribution imposed at the start of the simulation is gas-static apart from the very bottom where the pressure gradient becomes liquid water-static. (see Fig. 18). Variations in pressure after **b** 10 ka, **c** 50 ka, **d** 100 ka. In time, the pressure decreases below the relief and increases below the lowland and all along the bottom where a water layer is formed (see Fig. 18)

Also, at first water condenses and saturates most of the volume below the lowland (Fig. 34); then the water is evaporated again and a relatively dry layer is formed at the bottom below the relief and at the top of the scarp. The density increases significantly below the top of the scarp forming a very dense layer all throughout by the end of the simulation (Fig. 35).

6.4 Low Gas Saturation, High Pressure, Deep Intrusion

In the following simulations we show the influence of three different locations for the intrusion in relation to the scarp for the case of low gas saturation (high aqueous phase saturation): (a) below the lowland at the far right-hand side of

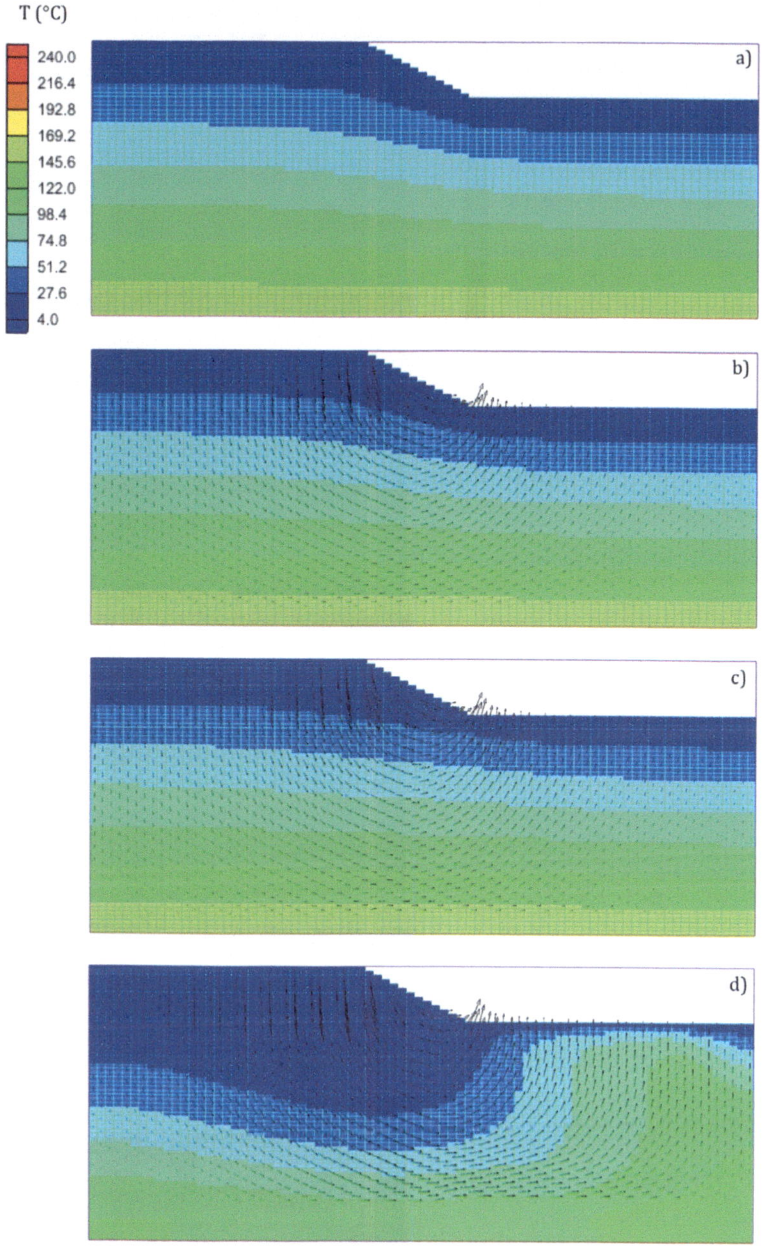

Fig. 17 Variations in temperature. **a** Constant temperature gradient imposed at the start of the simulation. Variations in temperature after **b** 10 ka, **c** 50 ka, **d** 100 ka. In time, a plume of hot gas develops in the lowland away from the bottom of the scarp interfering for a significant area with the bottom of the cryosphere. Such a pronounced and long lasting plume of hot gas below the cryosphere could locally melt it and reach the surface even in the absence of an intrusion

the model, (b) below the scarp, (c) below the high relief (mesa) at the far left-hand side of the model.

6.4.1 Intrusion Below Lowland

In the first of these experiment the intrusion is set at some depth below the right-hand margin of the model. The pressure decreases below the relief and increases only for a relative short period of time below the lowland, particularly at the bottom of the scarp (Fig. 36). At first a hot liquid–water plume develops at the bottom of the scarp for about 10 ka (Fig. 37); later on, a very-hot liquid–water plume rises at the left margin above the intrusion and interacts with the bottom of the cryosphere; such plumes of hot water persisting

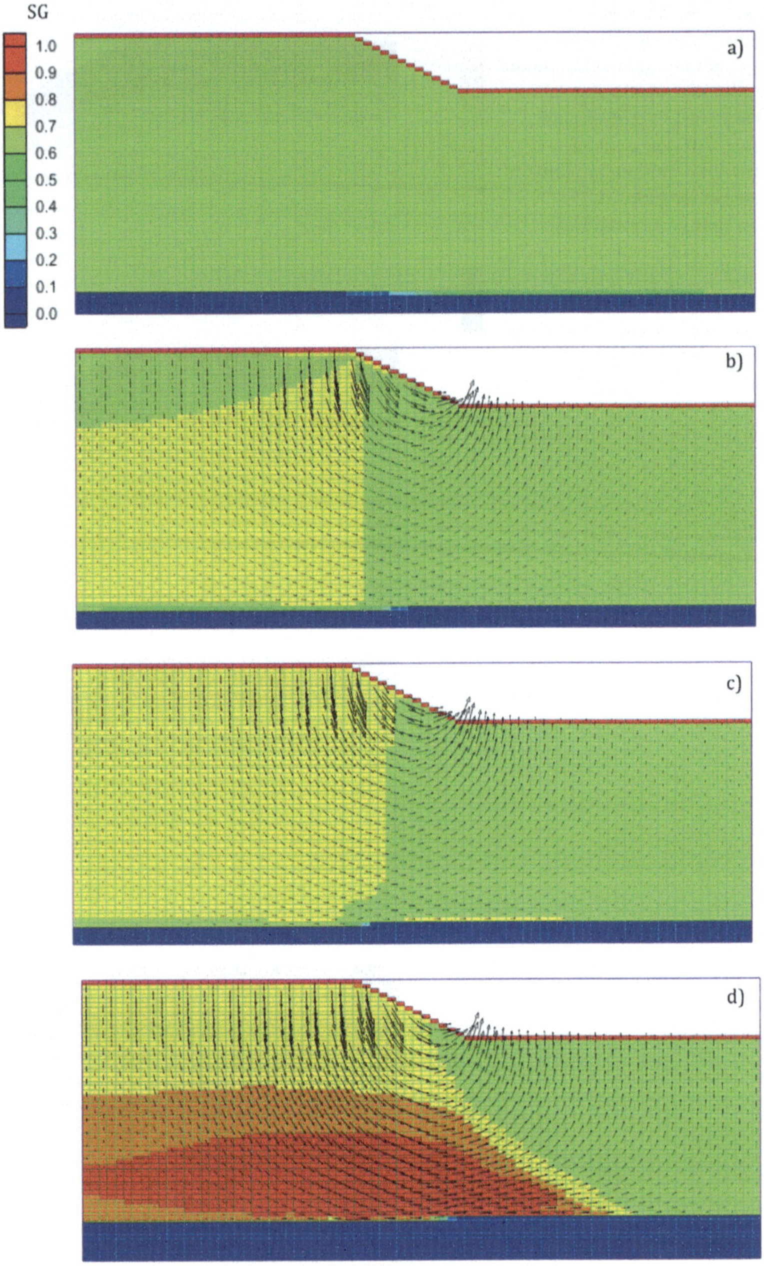

Fig. 18 Variations in gas saturation. **a** Equilibrium gas saturation obtained at the start of the simulation given the P–T conditions in Figs. 16 and 17. Variations in gas saturation after **b** 10 ka, **c** 50 ka, **d** 100 ka. In time, the gas saturation increases toward the bottom below the relief. In addition, the water is condensed at the bottom of the model

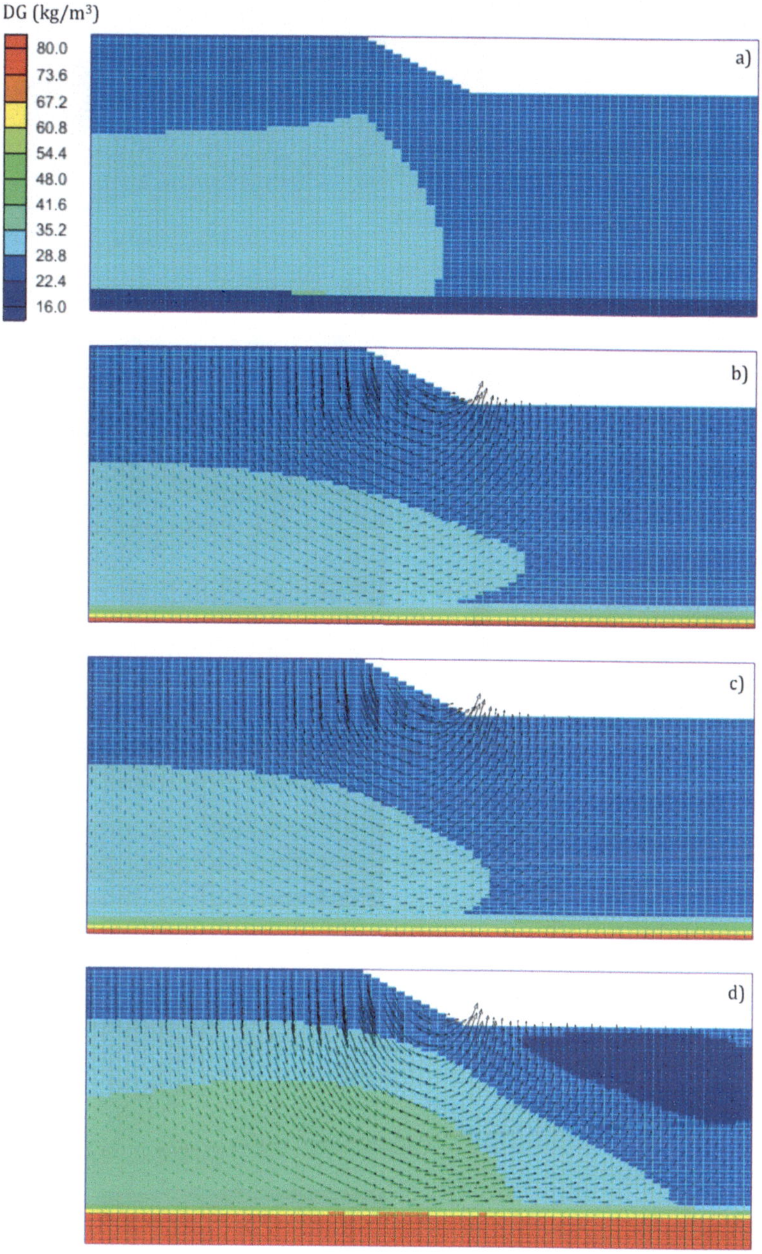

Fig. 19 Variations in gas density. **a** Equilibrium gas density obtained at the start of the simulation given the P–T conditions in Figs. 16 and 17. Variations in gas saturation after **b** 10 ka, **c** 50 ka, **d** 100 ka. In time, the gas density increases toward the bottom below the relief and decreases within the rising plume. Gas density is also high within the liquid water dominated volume at the bottom of the model

for 200 ka could easily reach and erupt at the surface. The temperature below the highland, though, is rapidly quenched by the downward-convecting cold gas phase.

In time, a large gas-saturated area forms below the relief and extends to the bottom of the model forming an upper-crust thick CO_2-fer (Fig. 38). The right-hand side remains water-saturated. The gas density increases significantly in the gas-saturated area (Fig. 39).

6.4.2 Intrusion Below Slope

In this experiment with the intrusion located below the scarp, the pressure increases

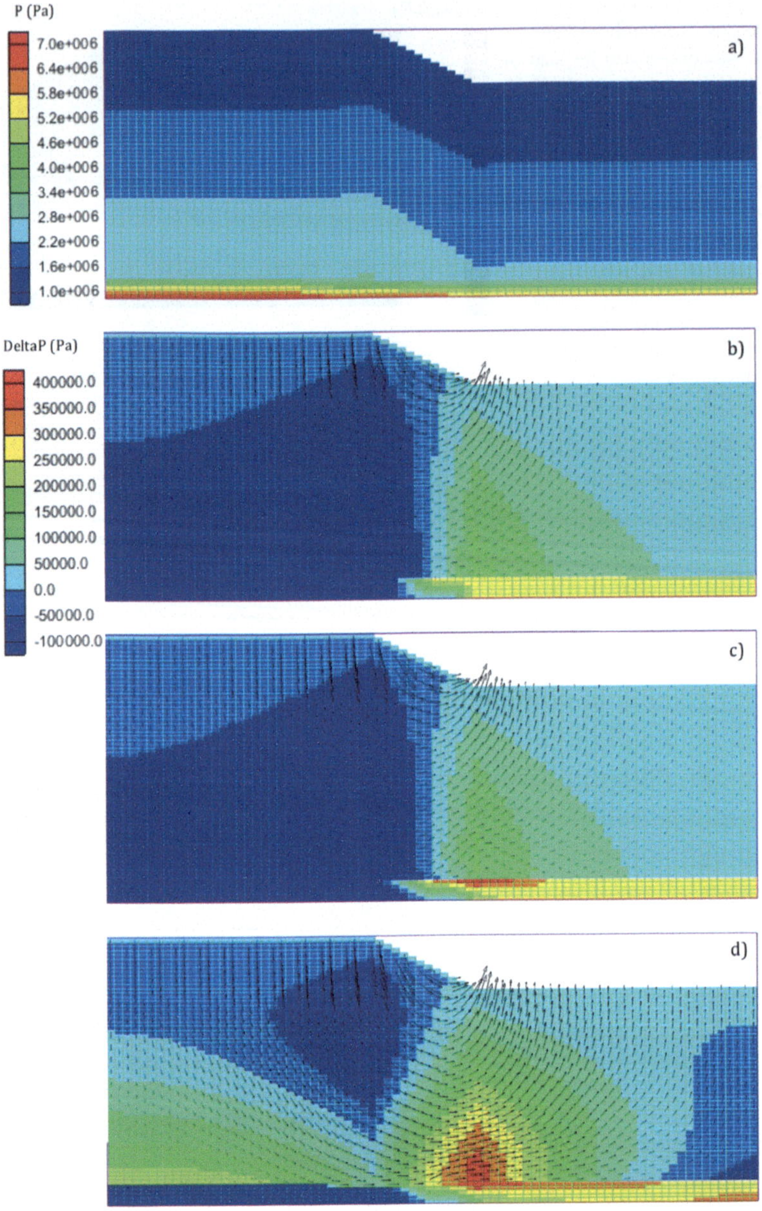

Fig. 20 Variations in pressure from initial conditions due to a mesa-like topographic relief and an intrusion located in correspondence of the scarp at some depth below the bottom of the model. The model is 40 km long and 16 km high; the vectors are gas velocities. **a** The pressure distribution imposed at the start of the simulation is gas-static apart from the very bottom where the pressure gradient becomes liquid water-static. (see Fig. 22). Variations in pressure after **b** 10 ka, **c** 50 ka, **d** 100 ka. In time, the pressure decreases below the relief and increases below the lowland all the way to the bottom downstream of the intrusion

significantly below the base of the scarp, subsequently decreasing slowly in time (Fig. 40). A strong hot plume of liquid water and gas (see Fig. 41) develops at the bottom reaching the base of the scarp and extending, after many 10's of ka to the lowland. The persistence of this plume is so firm, that it can melt the cryosphere with aqueous phase in addition to a CO_2-phase reaching and erupting at the surface for significant time.

A large 50% gas-saturated region develops below the topographic relief, while a 15% saturated area develops within the rising hot plume (Fig. 42). The gas density increases below the topographic relief (where the 50% gas-saturated area is developing—see Fig. 43) all the way to the bottom of the model giving rise to a thick CO_2-fer, which is side-by-side with the H_2O-fer remaining below the lowland.

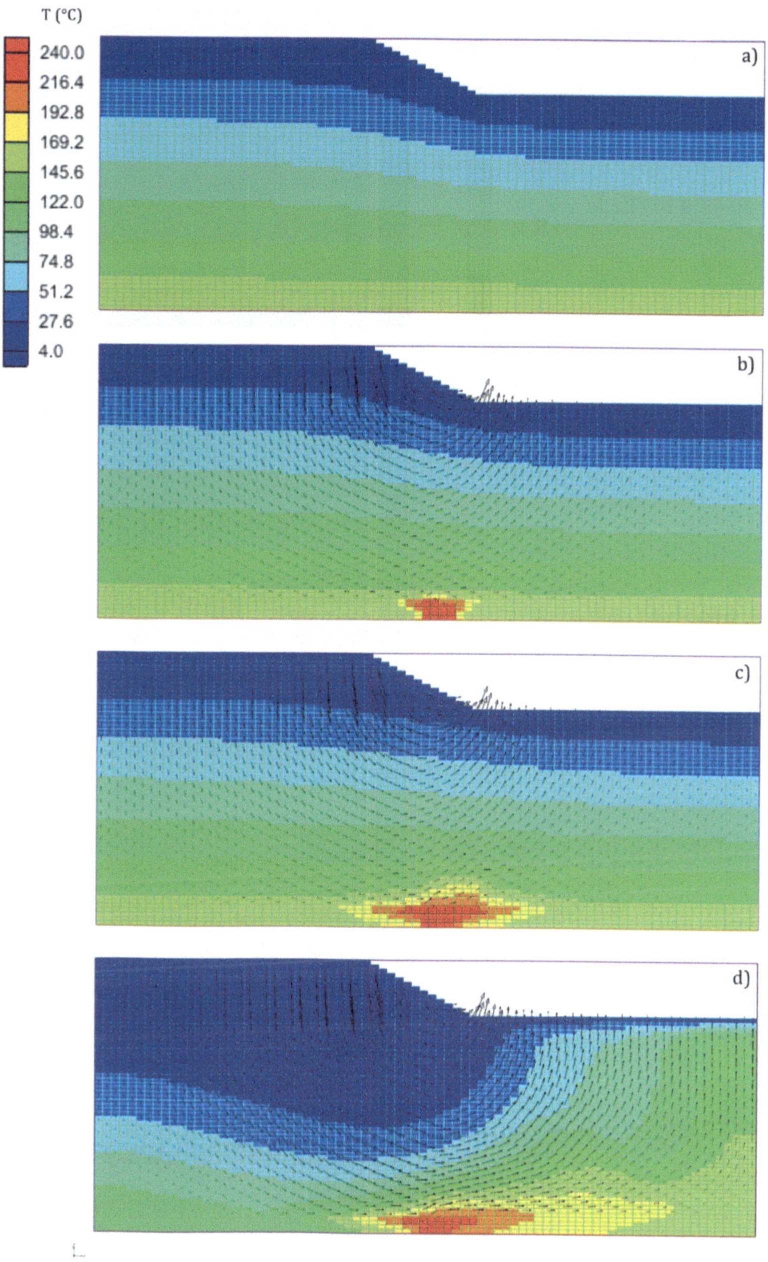

Fig. 21 Variations in temperature. **a** Constant vertical temperature gradient imposed at the start of the simulation. Variations in temperature after **b** 10 ka, **c** 50 ka, **d** 100 ka. In time, a plume of hot gas develops in the lowland away from the bottom of the scarp interfering for a significant area with the bottom of the cryosphere. Such a pronounced and long-lasting plume of hot gas could locally reach the surface. The temperature of the intrusion, though, is quenched by the downward-convecting cold gas phase

6.4.3 Intrusion Below Top of Mesa

The third and last case is with the intrusion located at the far left-hand side at some depth below the bottom of the model. In time, the pressure decreases above the intrusion and below the top of the scarp (Fig. 44). On the contrary, the pressure increases significantly below the base of the scarp at the beginning of the simulation, decreasing slowly in time later on.

A strong hot plume of liquid water and some gas develops above the intrusion, reaching the cryosphere (Fig. 45). After many 10's of ka, a second hot plume from the intrusion also joins the plume that forms at the base of the scarp interacting with the cryosphere. The persistence

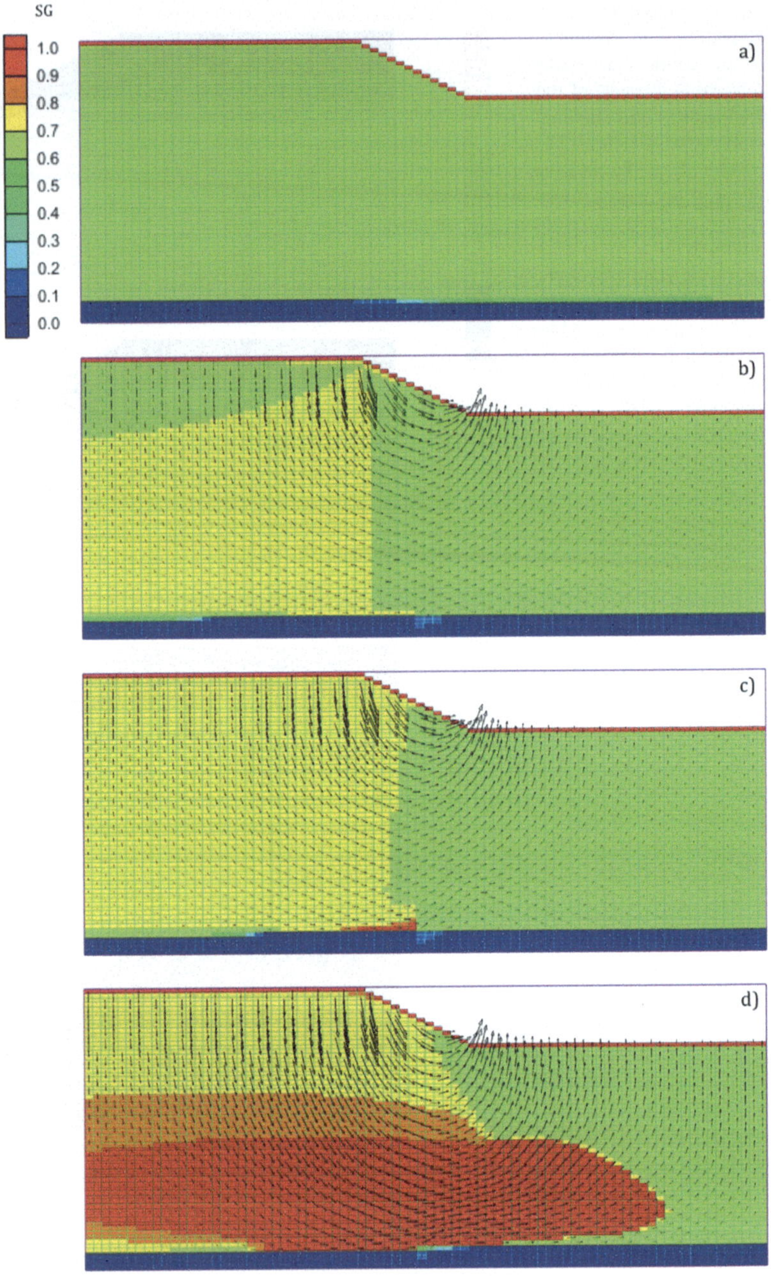

Fig. 22 Variations in gas saturation. **a** Equilibrium gas saturation calculated at the start of the simulation from the P–T conditions shown in Figs. 20 and 21. Variations in gas saturation after **b** 10 ka, **c** 50 ka, **d** 100 ka. In time, the gas saturation increases below the high topography moving toward the lowland

of these plumes may melt the cryosphere allowing aqueous phase to reach and erupt for significant time at the surface.

A 50% saturated gas zone develops below the high topography and the scarp (Fig. 46). The gas density increases in the 50% gas saturated zone below the high relief and the scarp (Fig. 47). It decreases on the left-hand side of the model above the intrusion and in the lowland.

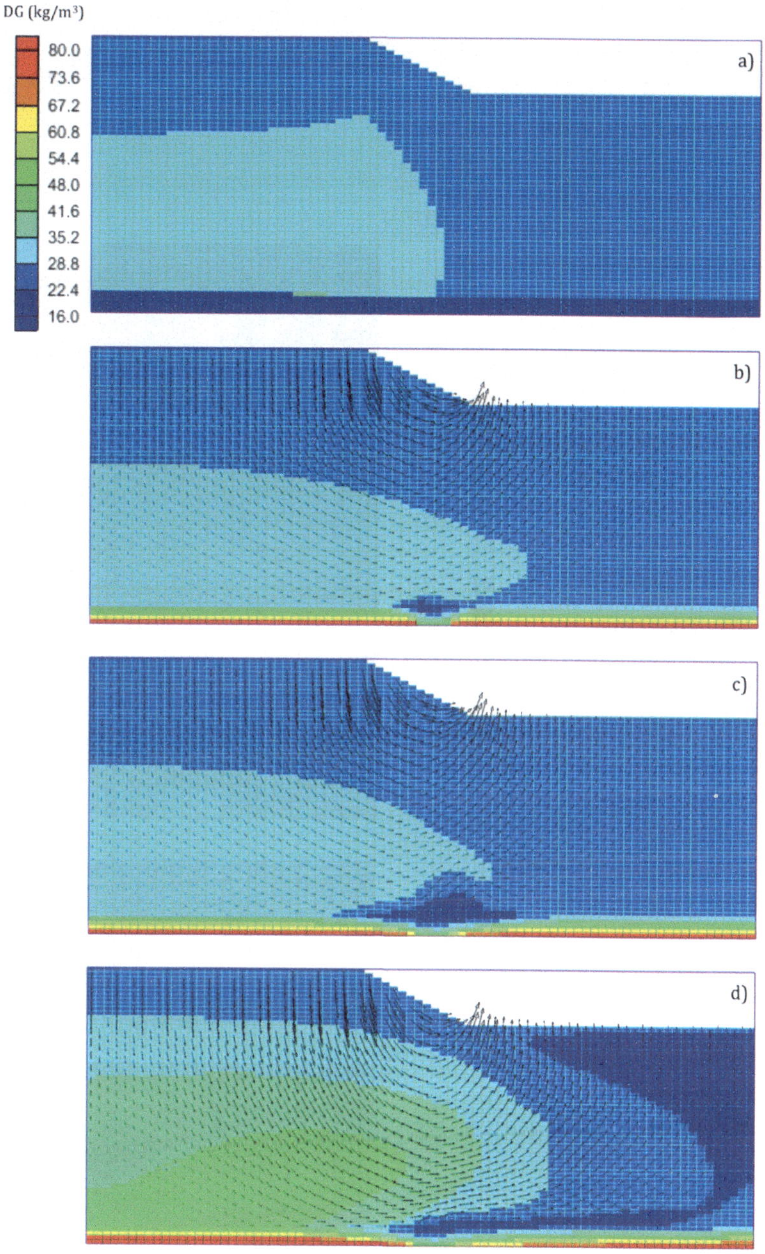

Fig. 23 Variations in gas density. **a** Equilibrium gas density calculated at the start of the simulation from the P/T conditions shown in Figs. 20 and 21. Variations in gas density after **b** 10 ka, **c** 50 ka, **d** 100 ka. In time, the gas density increases below the high topography moving towards the lowland

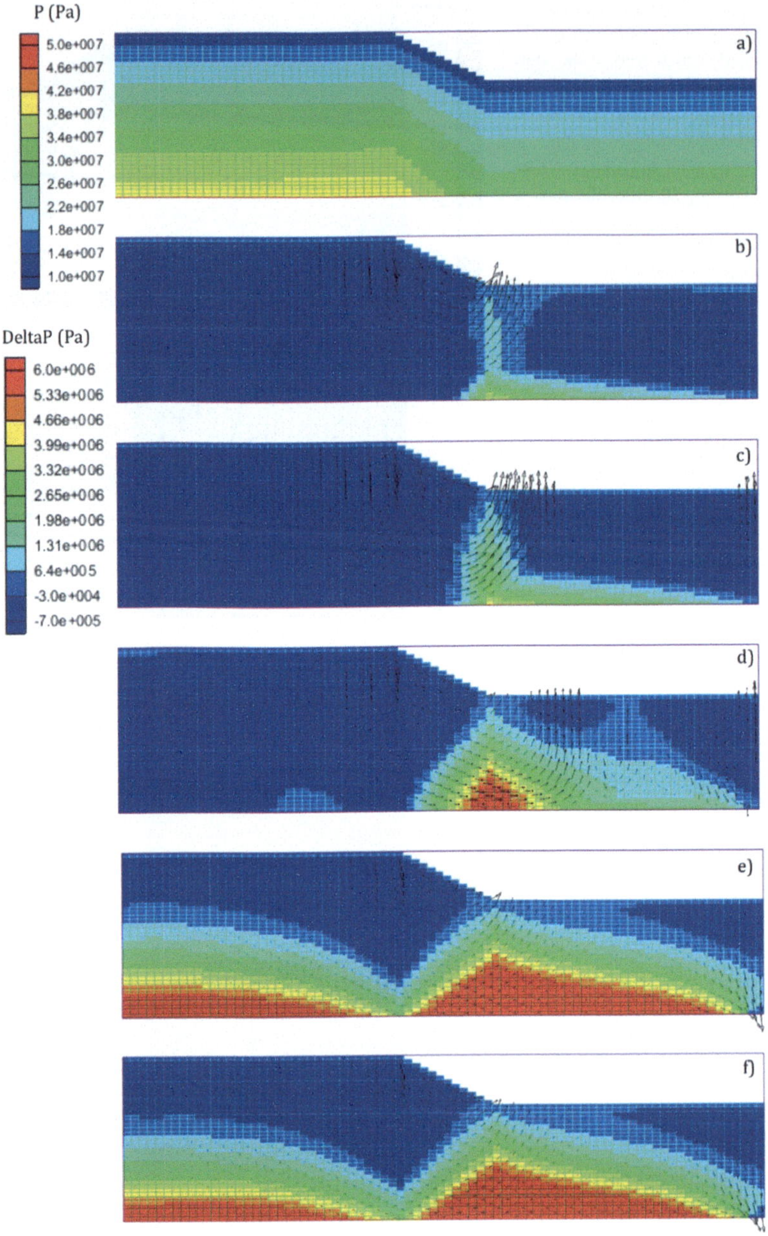

Fig. 24 Variations in pressure from initial conditions due to a mesa-like topographic relief and an intrusion located in the far right at some depth below the bottom of the model. The model is 40 km long and 10 km high; the vectors are gas velocities. **a** The pressure distribution imposed at the start of the simulation is gas-static. (see Fig. 26). Variations in pressure after **b** 1 ka, **c** 5 ka, **d** 10 ka, **e** 60 ka, and **f** 200 ka. In time, at first the pressure decreases everywhere apart from below the bottom of the scarp, where a gas plume rises to the cryosphere, but later the pressure increases again below the high relief and the lowland particularly close to the scarp

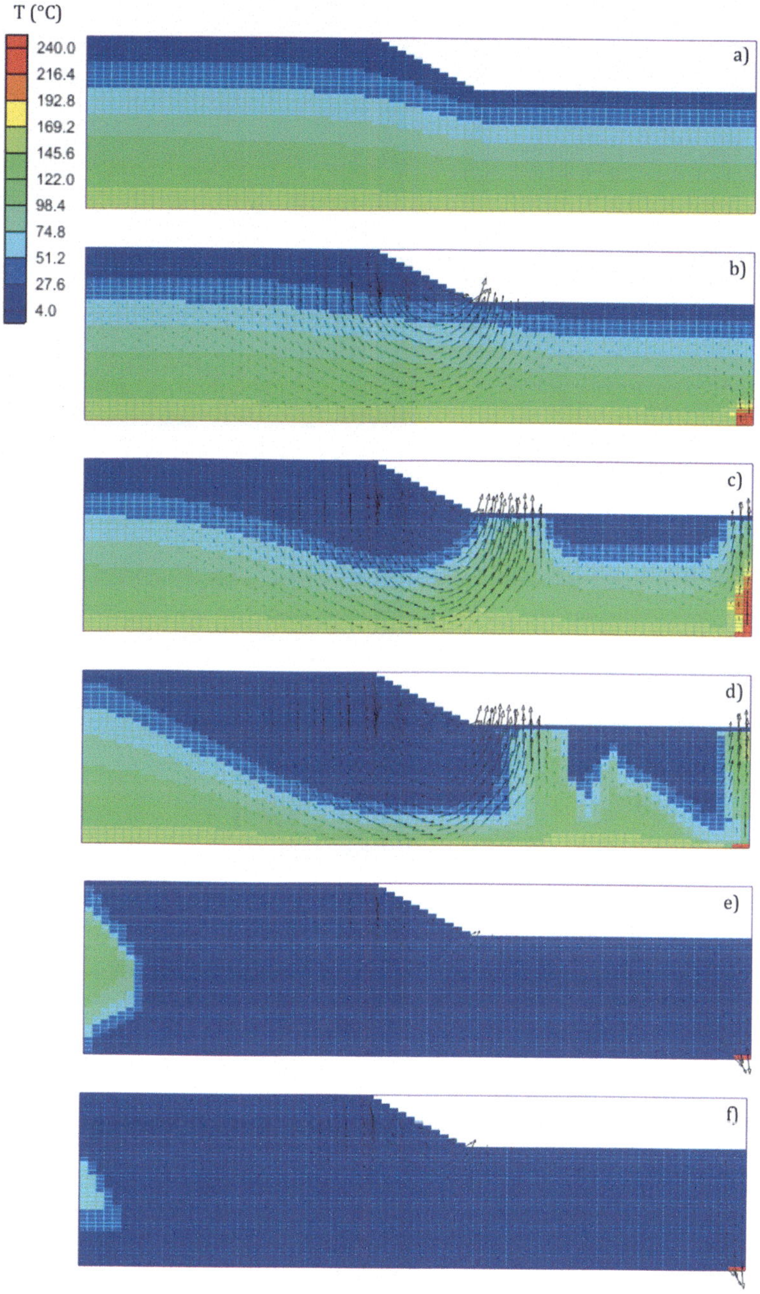

Fig. 25 Variations in temperature. **a** Constant vertical temperature gradient imposed at the start of the simulation. Variations in temperature after **b** 1 ka, **c** 5 ka, **d** 10 ka, **e** 60 ka, and **f** 200 ka. In time, two hot-gas plumes develop at the bottom of the scarp and at the far right in the lowland interfering for many 10's of ka with the bottom of the cryosphere. Such plumes of hot gas could locally reach and erupt at the surface. The temperature of the whole system, including the intrusion, is quenched by the convecting downward cold gas phase

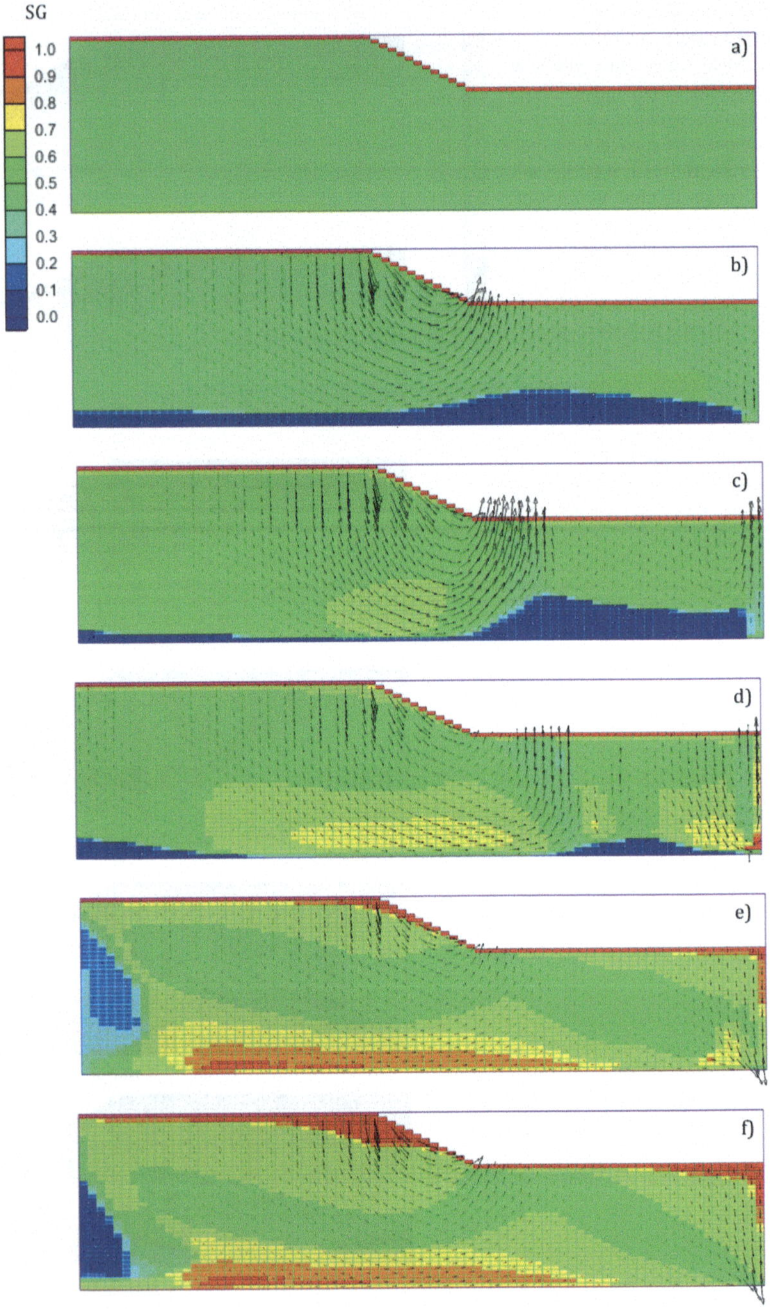

Fig. 26 Variations in gas saturation. **a** Equilibrium gas saturation imposed at the start of the simulation according to the P–T conditions indicated in Figs. 24 and 25. Variations in gas density after **b** 1 ka, **c** 5 ka, **d** 10 ka, **e** 60 ka, and **f** 200 ka. In time, the gas saturation increases in the central area at the bottom of the model and close to the scarp where a very dry zone forms closer to the top. An ephemeral water saturated zone forms at the bottom of the model below the lowland and a more permanent one on the far left

Fig. 27 Variations in gas density. **a** Equilibrium gas density imposed at the start of the simulation according to the P–T conditions indicated in Figs. 24 and 25. Variations in temperature after **b** 1 ka, **c** 5 ka, **d** 10 ka, **e** 60 ka, and **f** 200 ka. In time, the gas density increases at first below the scarp but later in the whole model

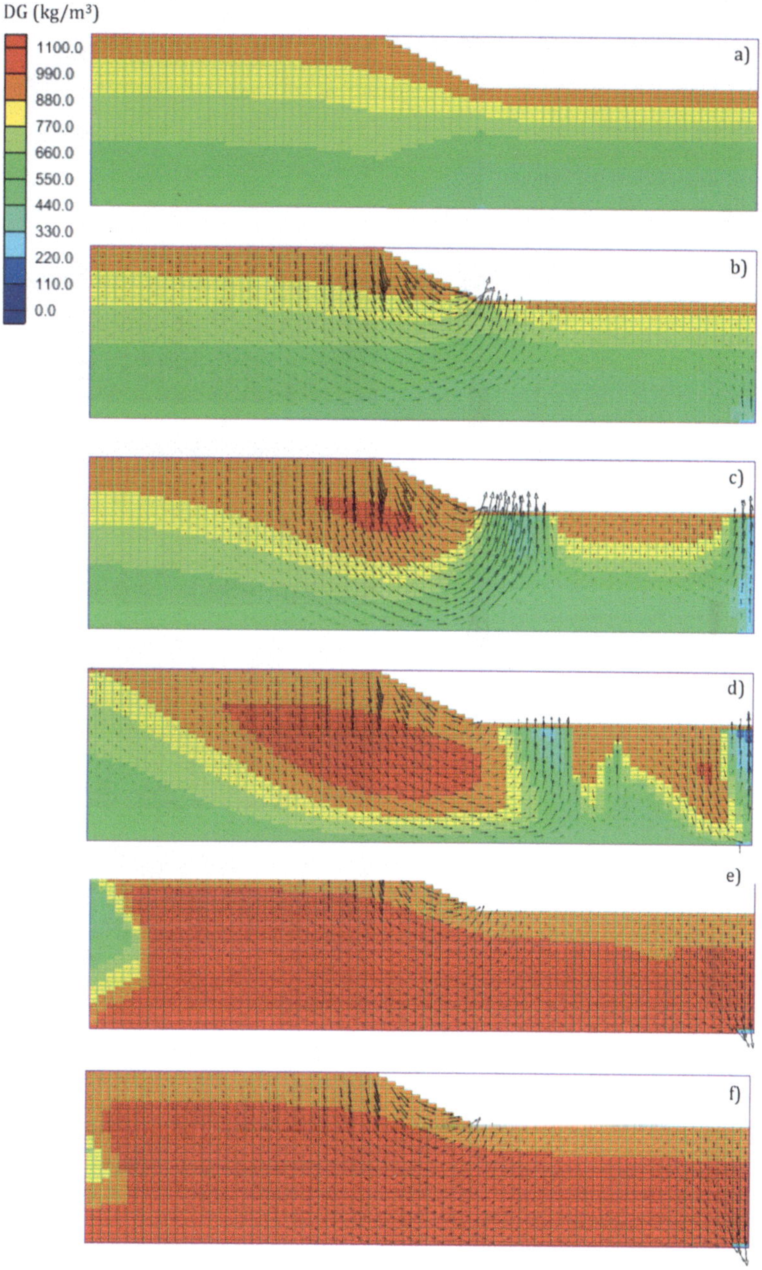

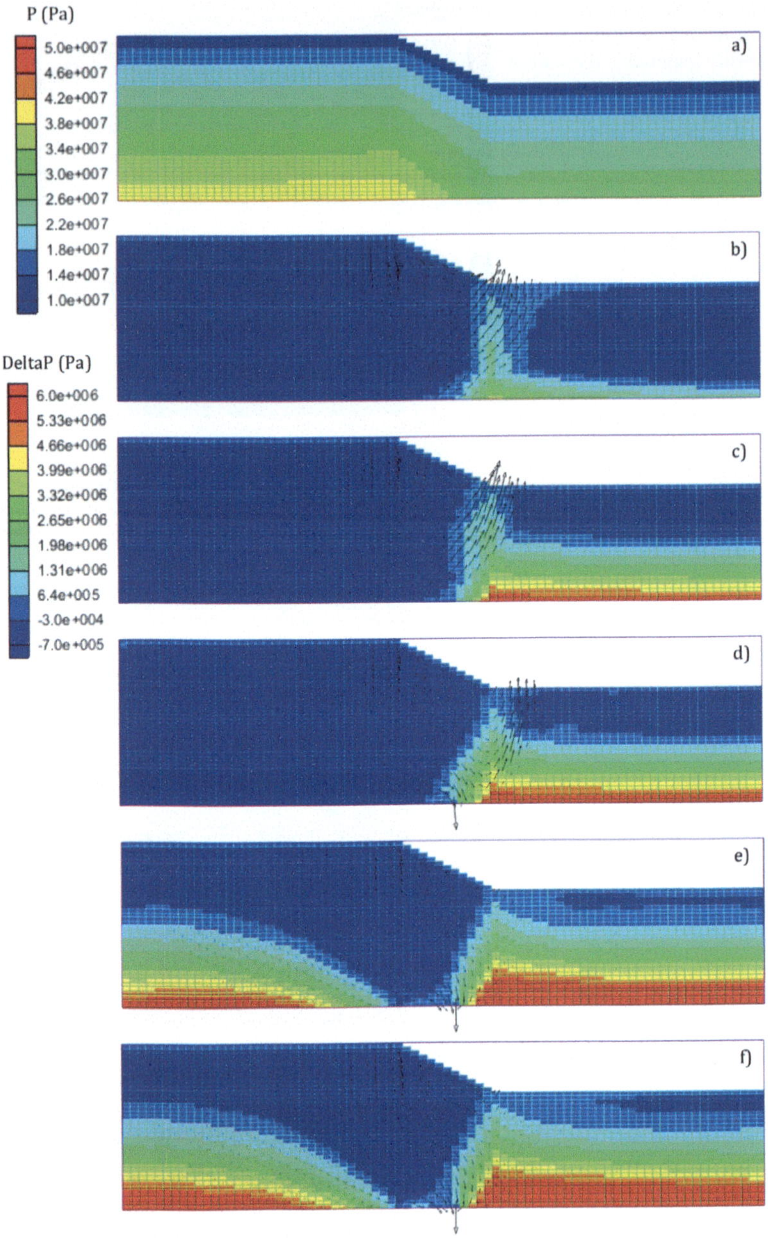

Fig. 28 Variations in pressure from initial conditions due to a mesa-like topographic relief and an intrusion located in correspondence of the scarp at some depth below the bottom of the model. The model is 40 km long and 10 km high; the vectors are gas velocities. **a** The pressure distribution imposed at the start of the simulation is gas-static. (see Fig. 30). Variations in pressure after **b** 1 ka, **c** 5 ka, **d** 10 ka, **e** 60 ka, and **f** 200 ka. In time, at first the pressure increases below the lowland and particularly at the bottom of the scarp; later the pressure increases also below the high relief

7 Discussion and Conclusions

Martian CO_2-H_2O-fers (fluidfers) are distinct from terrestrial aquifers because of the ubiquitous presence of high CO_2 concentrations in the fluid phases that permeate the underground rock porosity. In the Martian underground *P–T* conditions, gaseous and liquid CO_2, in addition to supercritical CO_2, may coexists with gaseous and liquid water. Solid phases that constitute the cryosphere are H_2O and CO_2 ice in addition to clathrates ([21] and references therein).

In most underground conditions both a H_2O-rich CO_2 phase and a CO_2-rich H_2O phase may coexist with variable degrees of reciprocal saturations. According to their densities, the CO_2-phase will tend to occupy with a greater

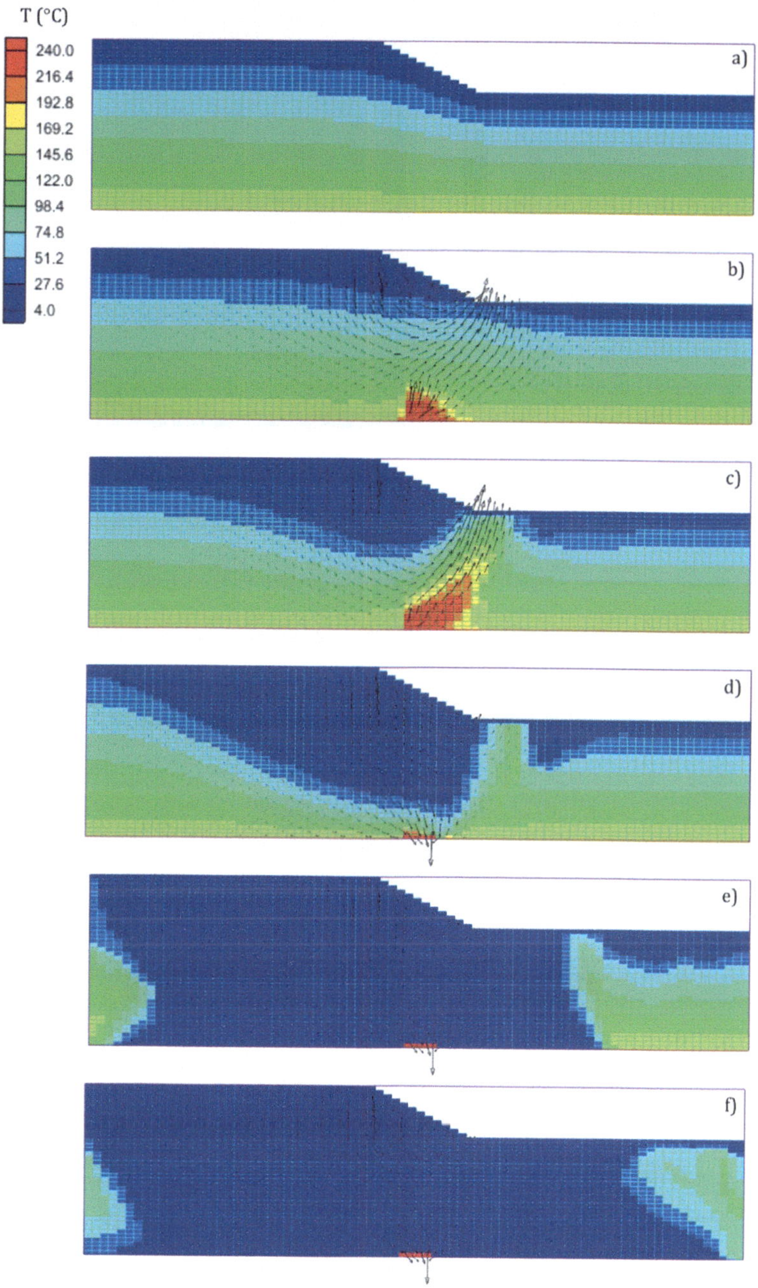

Fig. 29 Variations in temperature. **a** Constant vertical temperature gradient imposed at the start of the simulation. Variations in temperature after **b** 1 ka, **c** 5 ka, **d** 10 ka, **e** 60 ka, and **f** 200 ka. In time, a hot-gas plume develops at the bottom of the scarp interfering for few 10's of ka with the bottom of the cryosphere. Such plumes of hot gas could locally reach and erupt at the surface. The temperature of the whole system, including the intrusion, is rapidly quenched, though, by the convecting downward cold gas phase

saturation the upper parts of the fluidfers, while the H_2O-phase will tend to saturate the lower parts of the fluidfers. Therefore, by gravitational flow, CO_2 will tend to flow upward at the same time that H_2O flows downward. Only at relatively high pressure and low temperature will the CO_2 reach densities that are comparable to those of liquid water. For these cases one may find the CO_2-phase deeper than the H_2O-phase. In the presence of a topographic relief feature (e.g., a mesa), the CO_2-fer will tend to occupy the volume below the highland, while almost side-by

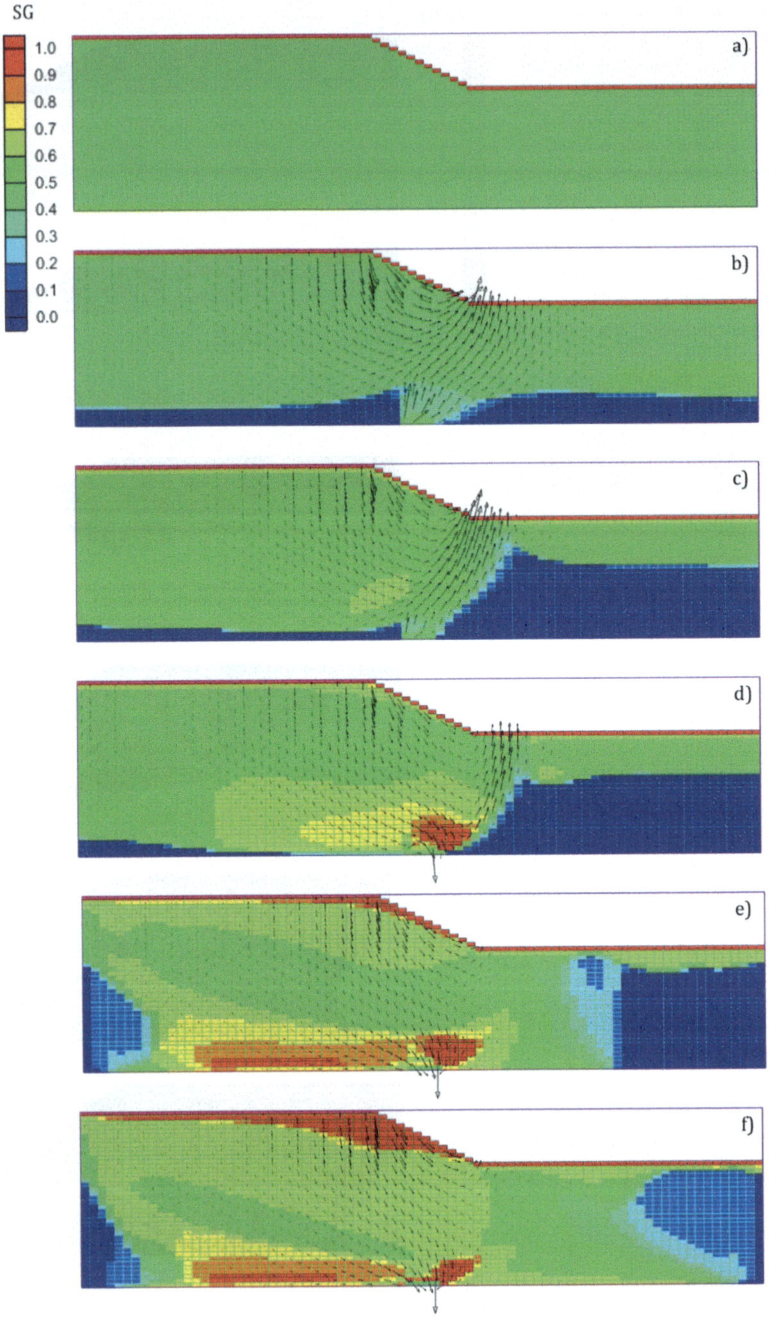

Fig. 30 Variations in gas saturation. **a** Equilibrium gas saturation imposed at the start of the simulation according to the P–T conditions given in Figs. 28 and 29. Variations in gas saturation after **b** 1 ka, **c** 5 ka, **d** 10 ka, **e** 60 ka, and **f** 200 ka. In time, water saturated regions form below the highlands and lowlands; these areas later reduce in volume still remaining at the end of the simulation. High concentrations of the gas phase develop on top of the intrusion and below the top of the scarp

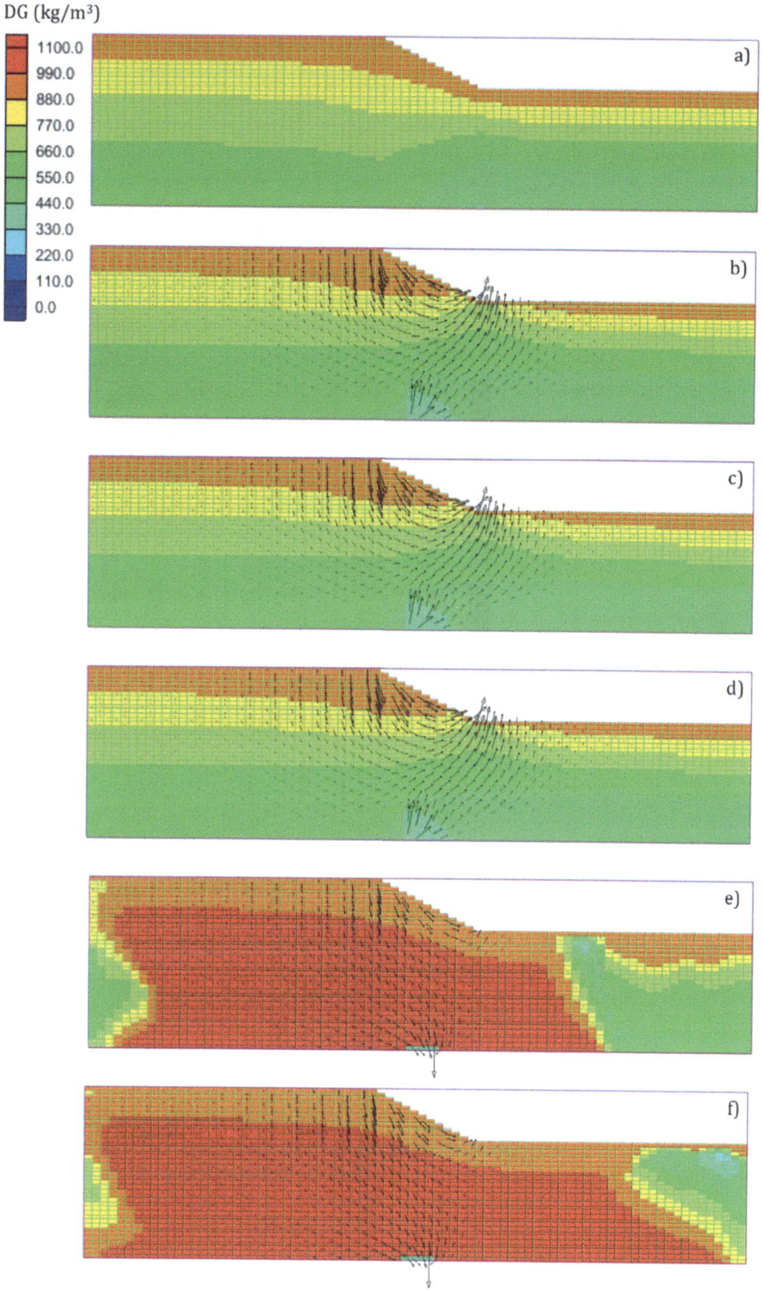

Fig. 31 Variations in gas density. **a** Equilibrium gas density imposed at the start of the simulation according to the P–T conditions given in Figs. 28 and 29. Variations in gas density after **b** 1 ka, **c** 5 ka, **d** 10 ka, **e** 60 ka, and **f** 200 ka. In time, gas density increases in the whole model, apart from the water saturated volumes

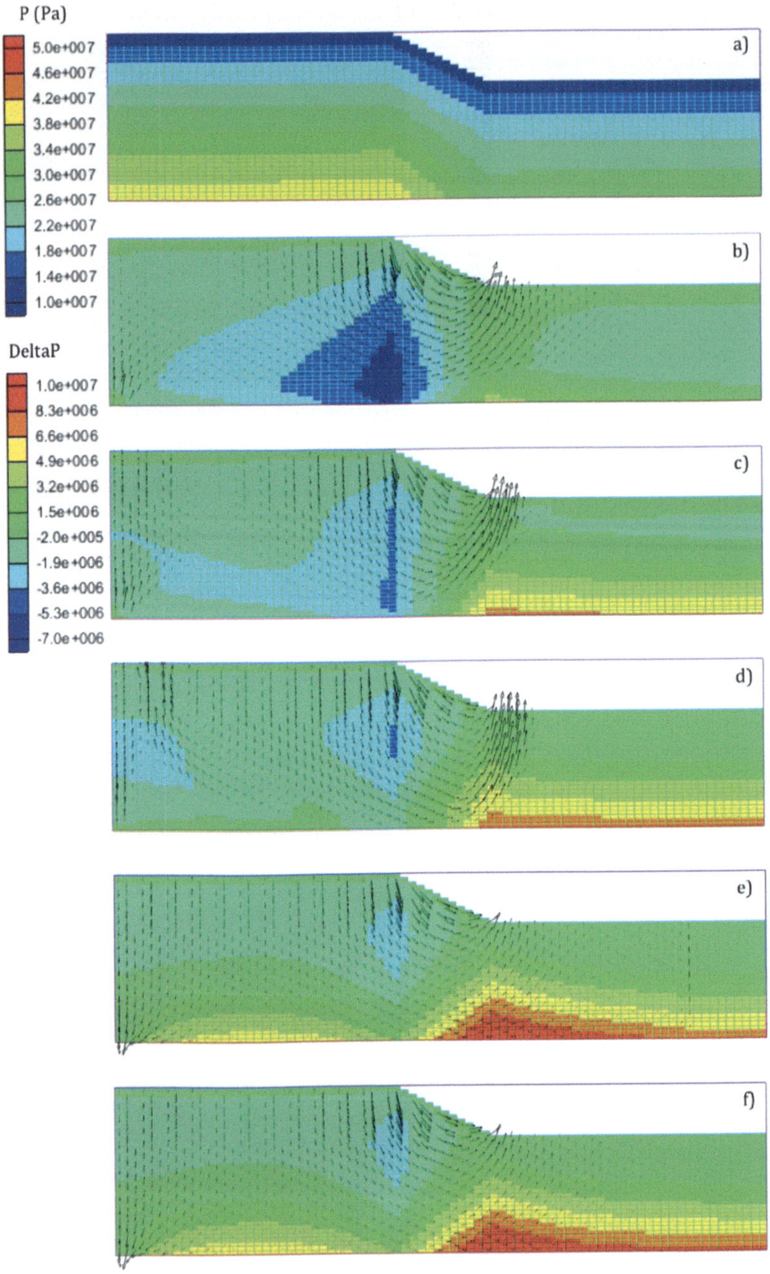

Fig. 32 Variations in pressure from initial conditions due to a mesa-like topographic relief and an intrusion located at the far left at some depth below the bottom of the model. The model is 40 km long and 10 km high; the vectors are gas velocities. **a** The pressure distribution imposed at the start of the simulation is gas-static. (see Fig. 34). Variations in pressure after **b** 1 ka, **c** 5 ka, **d** 10 ka, **e** 60 ka, and **f** 200 ka. In time, the pressure decreases below the relief and increases below the lowland, particularly at the bottom of the scarp

side the H_2O-fer will be found below the lowland.

The high-density CO_2-phase (liquid or supercritical) is particularly relevant in the context of Martian fluidfers, because in this thermophysical condition CO_2 becomes a polar molecule—the $O = C = O$ bond-angle deviates from 180° reaching values down to about 160° [43, 44]. Therefore, as a polar molecule, it can solvate, transport and eventually precipitate metal cations (particularly alkaline-earth and alkaline metals) perhaps forming actual geochemical selective metal enrichments. This possibility appears to be relevant to the high-gas saturation models because it is considered possible that rising CO_2-plumes, in addition to H_2O

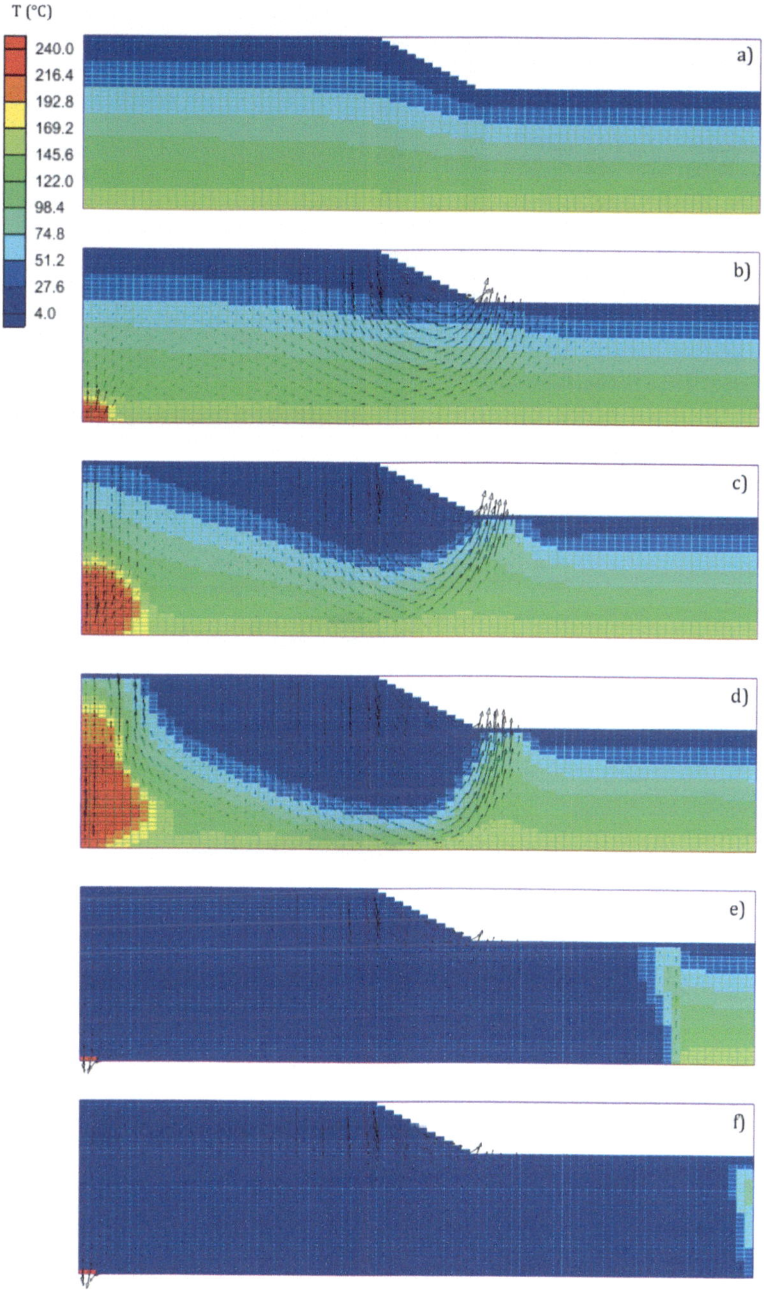

Fig. 33 Variations in temperature. **a** Constant vertical temperature gradient imposed at the start of the simulation. Variations in temperature after **b** 1 ka, **c** 5 ka, **d** 10 ka, **e** 60 ka, and **f** 200 ka. In time, at first a hot-gas plume develops at the bottom of the left boundary for few 10's of ka and interacts with the bottom of the cryosphere; such plumes of hot gas could momentarily locally reach and erupt at the surface. The temperature of the whole system, including the intrusion, though is rapidly quenched, by the convecting downward cold gas phase

plumes, could have created, during Martian geologic history, metal-rich ore deposits in specific areas of Mars (Figs. 43 and 47).

The high-density CO_2-phase can also react with hydroxyl-bearing minerals to free water molecules and precipitate carbonate minerals according to the general unbalanced "type" reaction:

$$CO_2 + Me(OH)_n \rightarrow MeCO_3 + H_2O \quad (7)$$

In fact, rare carbonates are found on Mars in small veins at cm scale [21, 45]. This type of reaction, though, may occur even in the absence of a liquid water-phase and supports the inferences for dry carbonated formation [46].

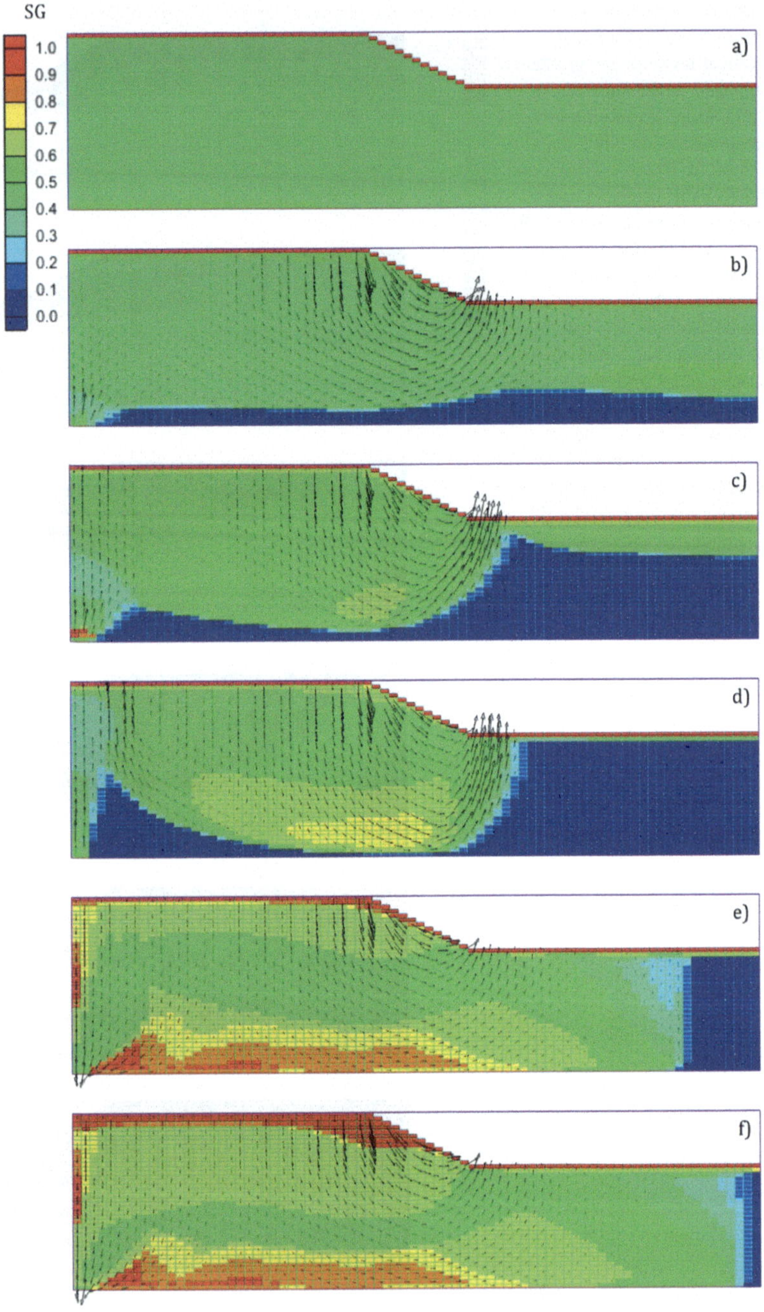

Fig. 34 Variations in gas saturation. **a** Equilibrium gas saturation calculated at the start of the simulation from the initial P–T conditions indicated in Figs. 32 and 33. Variations in gas saturation after **b** 1 ka, **c** 5 ka, **d**, 10 ka, **e** 60 ka, and **f** 200 ka. In time, at first water condenses and saturates most the volume below the lowland; than the water is evaporated again and a relatively dry layer is formed at the bottom below the relief and at the top of the scarp

The first general observation that can be derived from the simulation is that underground fluid convection both of H_2O or CO_2 will cool the Martian crust, down to the depth of pore closure (about 10–15 km [16]), much faster than what conduction-only models may predict. In fact, the characteristic time-scale of fluido-thermal convective cooling is only on the order of a hundred thousand years (100 ka), a very small fraction of the billion-year time scale of Martian eons. In practice, we are forced to deduce that today the Martian upper-crustal

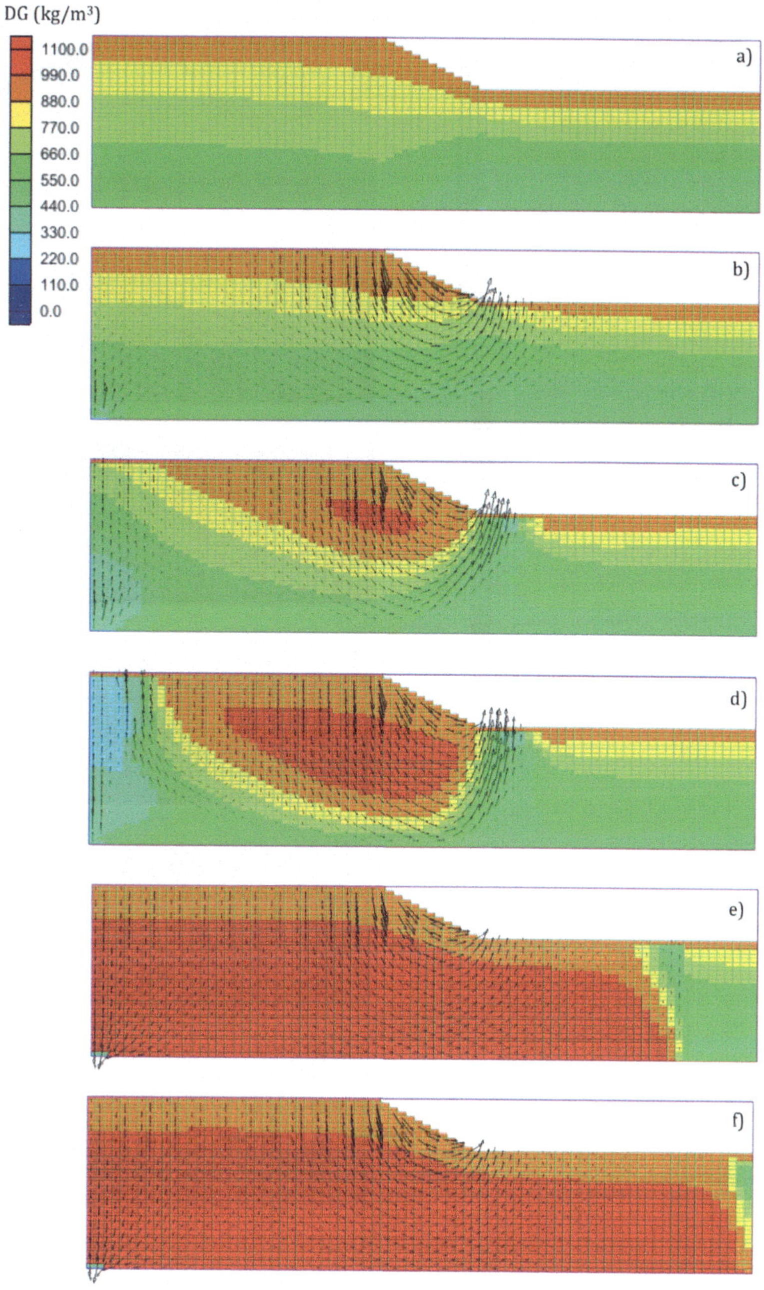

Fig. 35 Variations in gas density. **a** Equilibrium gas density calculated at the start of the simulation from the initial P–T conditions indicated in Figs. 32 and 33. Variations in gas density after **b** 1 ka, **c** 5 ka, **d**, 10 ka, **e** 60 ka, and **f** 200 ka. In time, the density increases significantly below the top of the scarp forming a very dense layer all throughout by the end of the simulation

temperatures are probably very cold and that water-freezing temperatures have probably already reached the bottom of the permeable thickness. This is true unless magmatic intrusive activity is still able to bring heat into the upper crust. In fact, there appear to be evidences of volcanic activity on Mars as recent as a few million years [47, 48] as some active hydrothermal activity seems to show [49]. There could even be the possibility that this magmatic activity is far more ancient going back to Noachian [5].

Both location of intrusion and topographic reliefs may have played a significant role in the

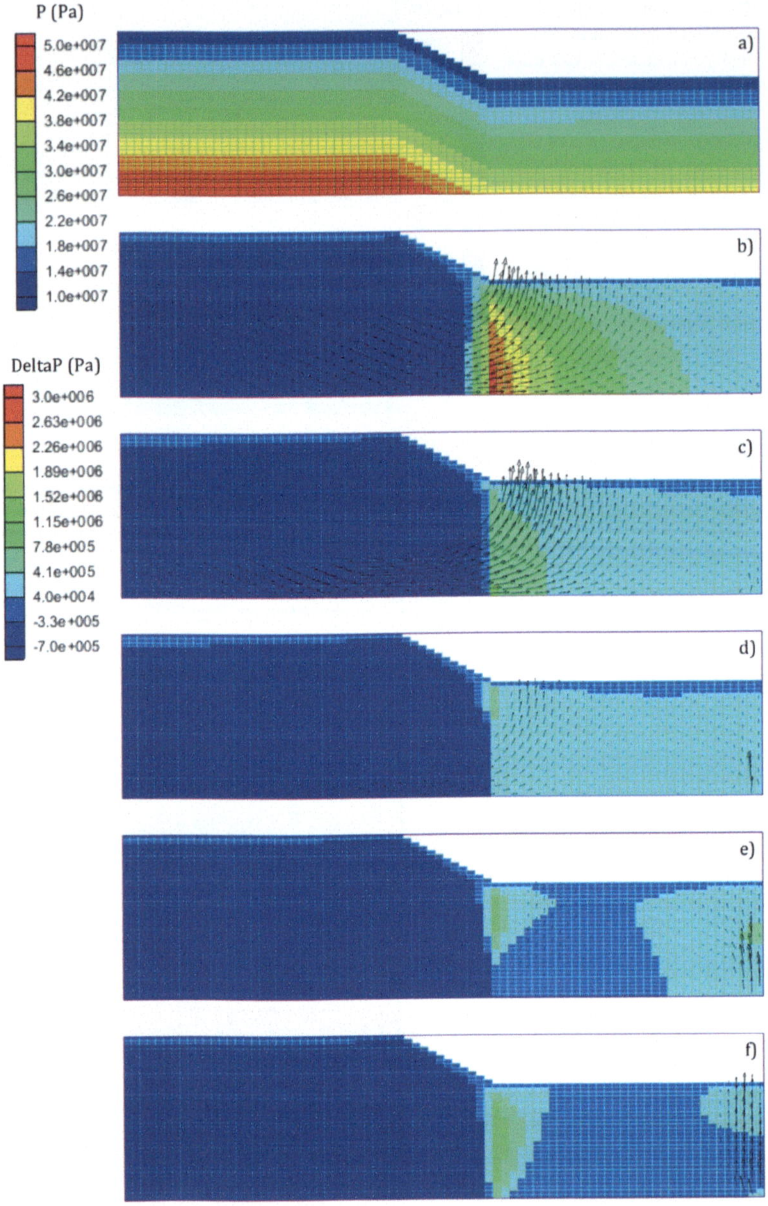

Fig. 36 Variations in pressure from initial conditions due to a mesa-like topographic relief and an intrusion located at the far right-hand side at some depth below the bottom of the model. The model is 40 km long and 10 km high; the vectors are gas velocities. **a** The pressure distribution imposed at the start of the simulation is liquid water-static. (see Fig. 38). Variations in pressure after **b** 1 ka, **c** 5 ka, **d**, 10 ka, **e** 60 ka, and **f** 200 ka. In time, the pressure decreases below the relief and increases only momentarily below the lowland, particularly at the bottom of the scarp

development of fluidothermal circulation on Mars when it was still active. Because of the presence of the cryosphere, which at any given latitude will tend to have the same thickness independent of topography, the upper pore pressure of any CO_2- and H_2O-fers will be equal to the lithostatic pressure at the bottom of the cryosphere [14, 16, 21]. Therefore, the underground fluid flow will tend to be from highlands to lowlands. This flow pattern is topologically similar, although induced by a different cause, to what is happening on Earth, where rainfall and recharge create aquifers that have higher water tables below higher-elevation areas [42].

The location of intrusions may affect this natural underground fluid circulation. An intrusion located in the low topography will tend to reduce the effect the topographic relief. An intrusion located below a topographic scarp, and even more an intrusion located below a highland,

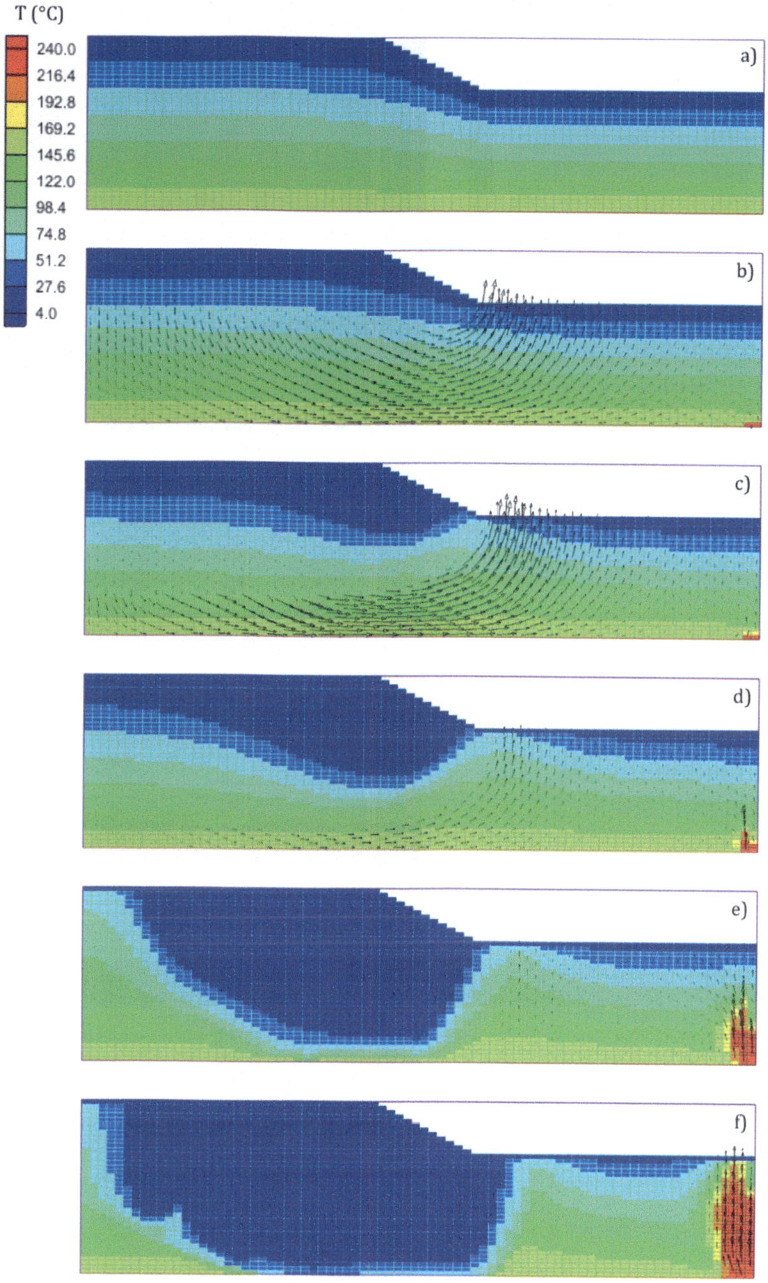

Fig. 37 Variations in temperature. **a** Constant vertical temperature gradient imposed at the start of the simulation. Variations in temperature after **b** 1 ka, **c** 5 ka, **d** 10 ka, **e** 60 ka and **f** 200 ka. In time, at first a hot liquid–water plume develops at the bottom of the scarp for about 10 ka; later on, a very-hot liquid–water plume rises at the left margin above the intrusion and interacts with the bottom of the cryosphere; such plumes of hot water persisting for 200 ka could easily reach and erupt at the surface. The temperature below the high land, though, is rapidly quenched, by the convecting-downward cold gas phase

will tend to enhance the fluid circulation, induced by the relief, from the high- to the low-elevation area.

Because the CO_2-fers tend to be stratified on top of the H_2O-fers particularly in the presence of topographic relief closer to the scarp, it appears to be relatively easy (practically all models show this feature) to have CO_2 emissions at the surface of Mars and it appears reasonable that some of the long run-out avalanches on Mars may have benefitted from having a liquid-CO_2 lubricating layer permeating the avalanche material [50].

For liquid water to be able to erupt at the surface, the upper crust below the cryosphere

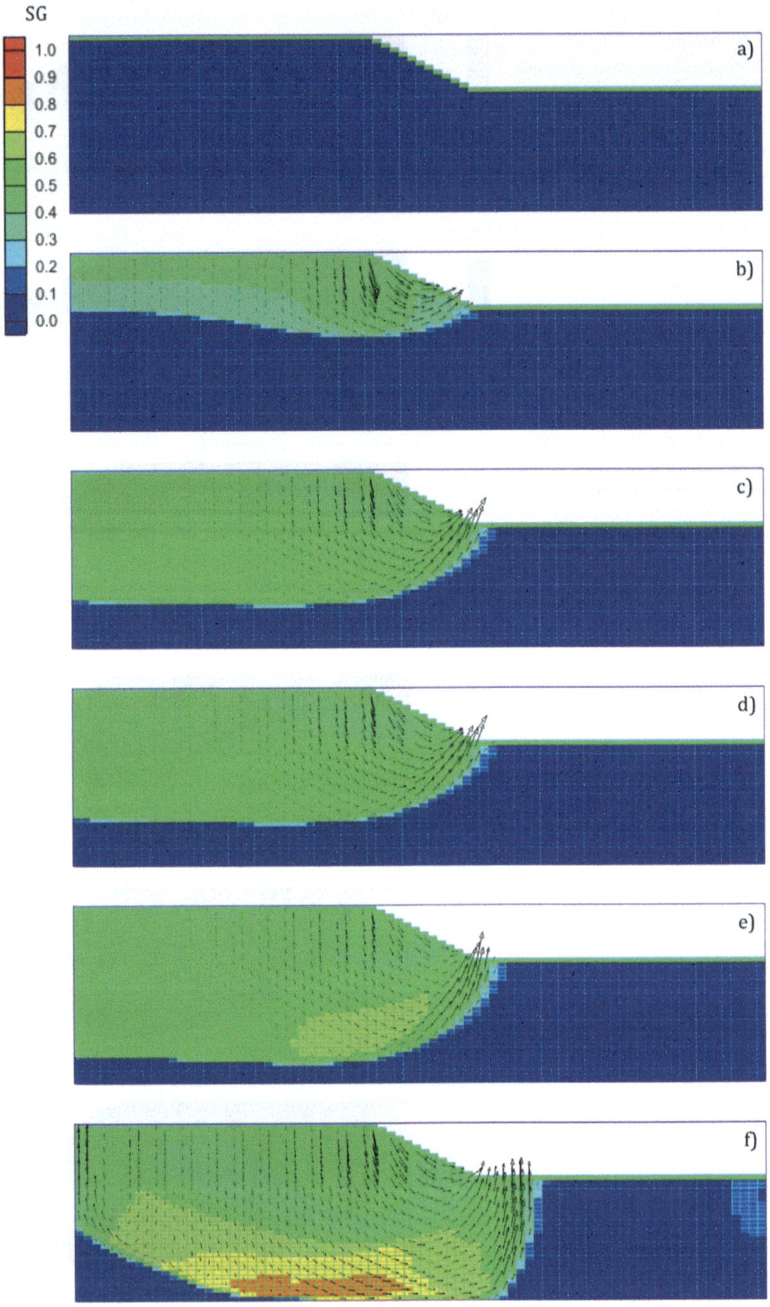

Fig. 38 Variations in gas saturation. **a** Equilibrium gas saturation calculated at the start of the simulation from the P–T conditions given in Figs. 36 and 37. Variations in gas saturation after **b** 1 ka, **c** 5 ka, **d**, 10 ka, **e** 60 ka, and **f** 200 ka. In time, a large gas-saturated area forms below the relief and extends to the bottom of the model forming a thick CO_2-fer. The right-hand side H_2O-fer remains water saturated

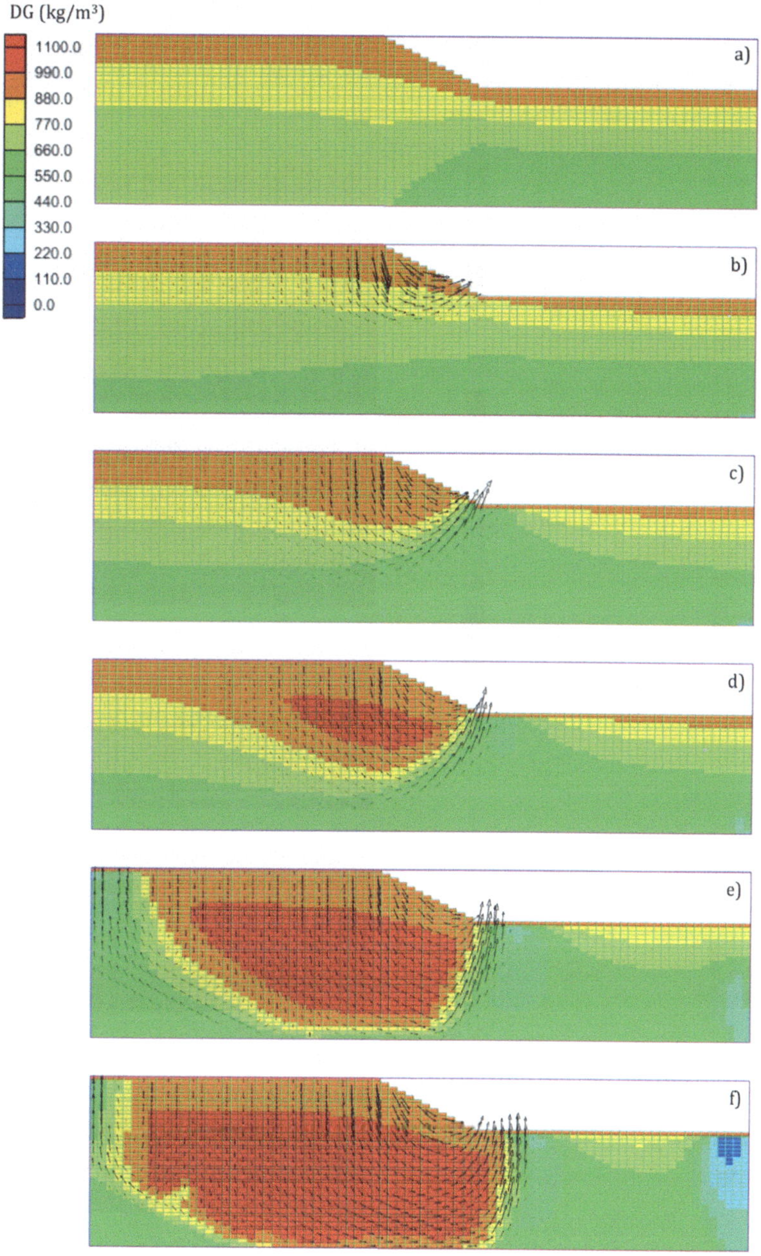

Fig. 39 Variations in gas density. **a** Equilibrium gas density calculated at the start of the simulation from the P–T conditions given in Figs. 36 and 37. Variations in gas density after **b** 1 ka, **c** 5 ka, **d**, 10 ka, **e** 60 ka, and **f** 200 ka. In time, the gas density increases significantly in the gas saturated area

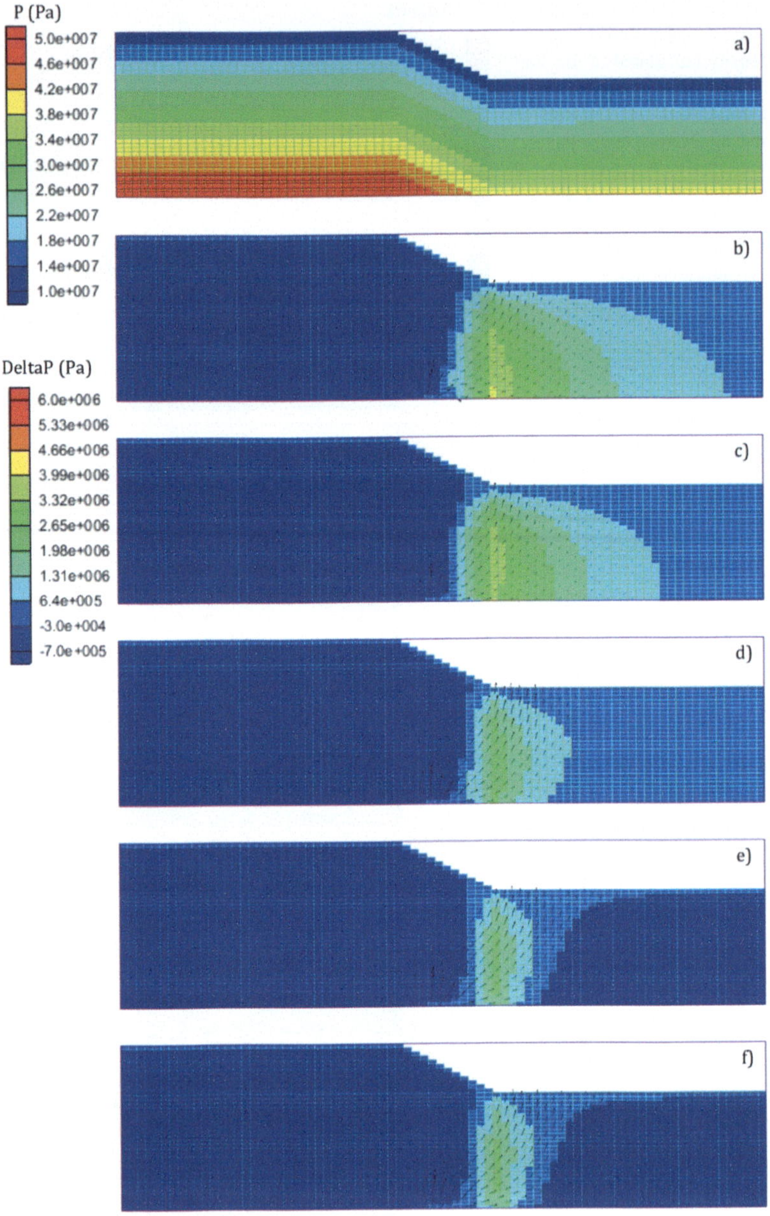

Fig. 40 Variations in pressure from initial conditions due to a mesa-like topographic relief and an intrusion located in correspondence of the scarp at some depth below the bottom of the model. The model is 40 km long and 10 km high; the vectors are gas velocities. **a** The pressure distribution imposed at the start of the simulation is liquid water-static. (see Fig. 42). Variations in pressure after **b** 1 ka, **c** 5 ka, **d**, 10 ka, **e** 60 ka, and **f** 200 ka. At the start, the pressure increases significantly below the base of the scarp, decreasing then slowly in time

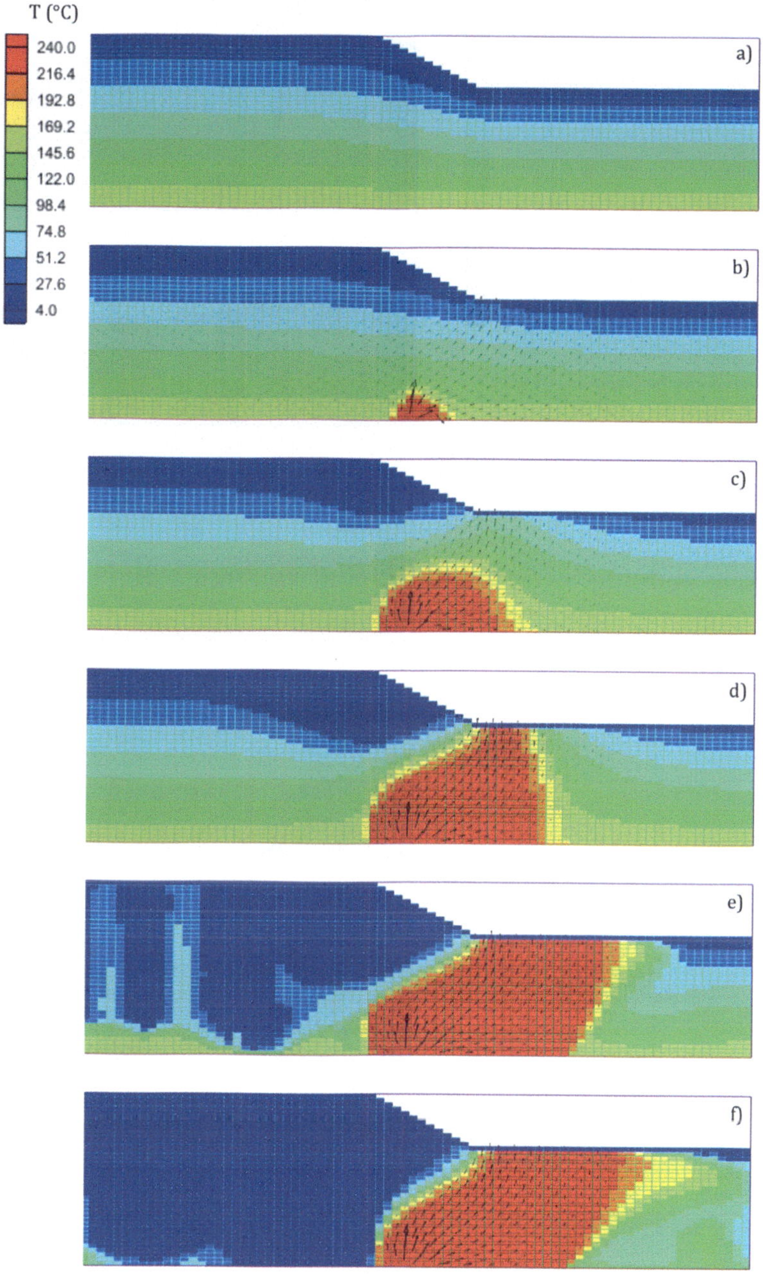

Fig. 41 Variations in temperature. **a** Constant vertical temperature gradient imposed at the start of the simulation. Variations in temperature after **b** 1 ka, **c** 5 ka, **d**, 10 ka, **e** 60 ka, and **f** 200 ka. In time, a strong hot plume of liquid water and gas (see Fig. 42) develops at the base of the scarp extending after many 10's of ka to the lowland. The persistence of this plume may melt the cryosphere reaching and erupting water and gas for significant time at the surface

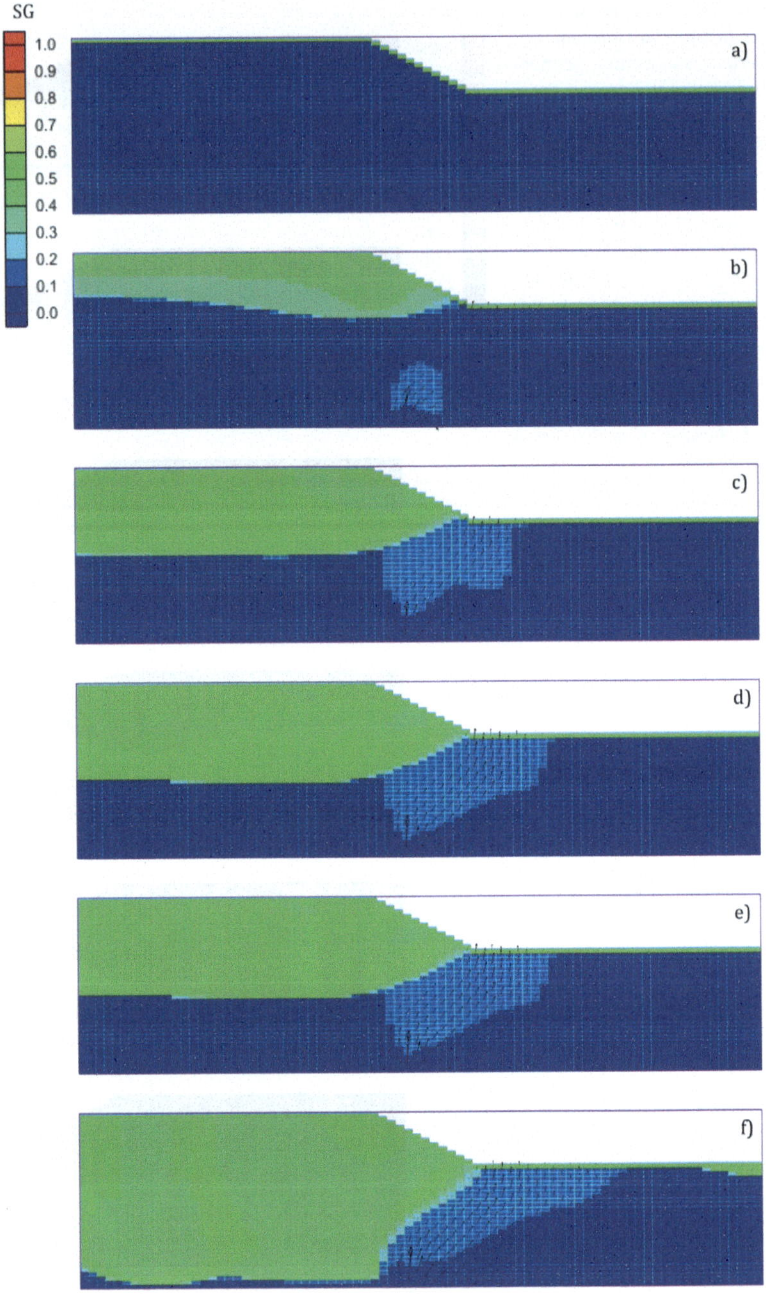

Fig. 42 Variations in gas saturation. **a** Gas saturation distribution calculated for the start of the simulation from the P–T conditions given in Figs. 40 and 41. Variations in gas saturation after **b** 1 ka, **c** 5 ka, **d**, 10 ka, **e** 60 ka, and **f** 200 ka. In time, a large 50% gas-saturated region develops below the topographic relief. A 15% saturated area develops within the rising hot plume. It is worth noting the dichotomy between the CO_2-fer below the relief and the H_2O-fer below the lowland

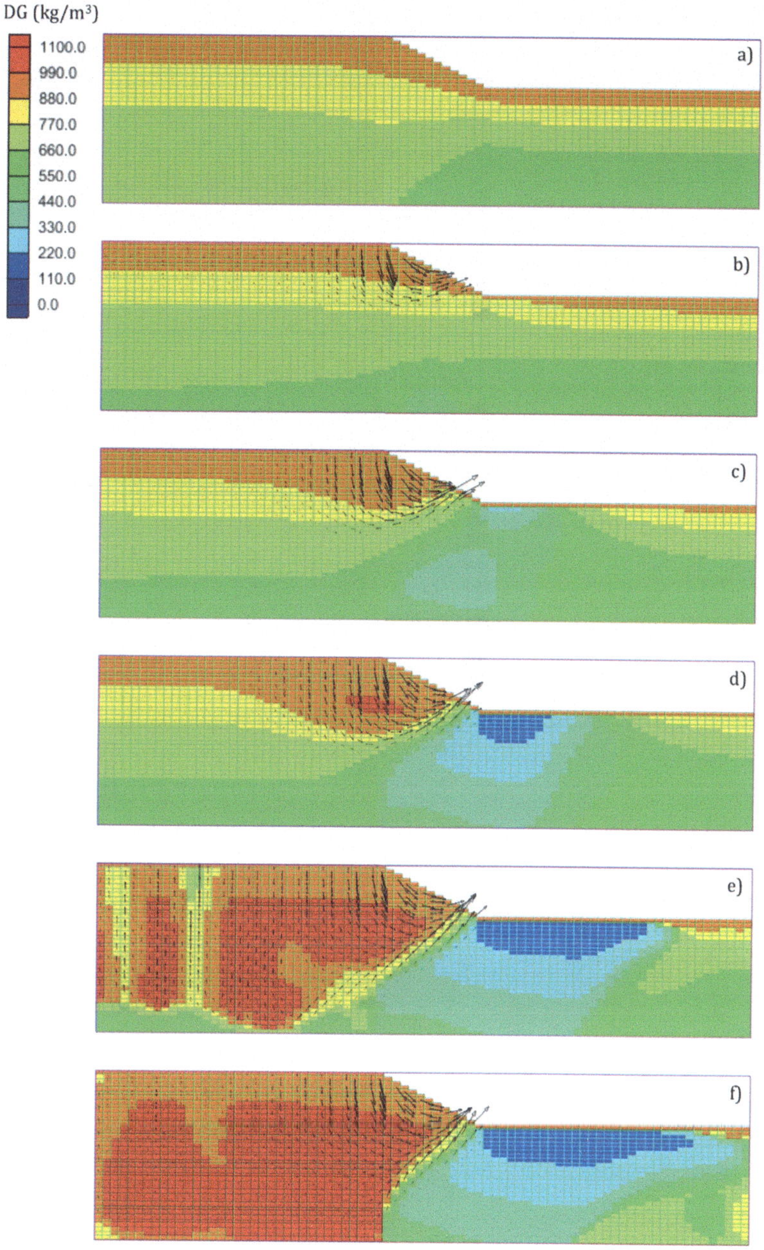

Fig. 43 Variations in gas density. **a** Gas density distribution calculated for the start of the simulation from the P–T conditions given in Figs. 40 and 41. Variations in gas density after **b** 1 ka, **c** 5 ka, **d**, 10 ka, **e** 60 ka, and **f** 200 ka. In time, the gas density increases below the topographic relief (where the 50% gas-saturated area is developing—see Fig. 39) all the way to the bottom of the model

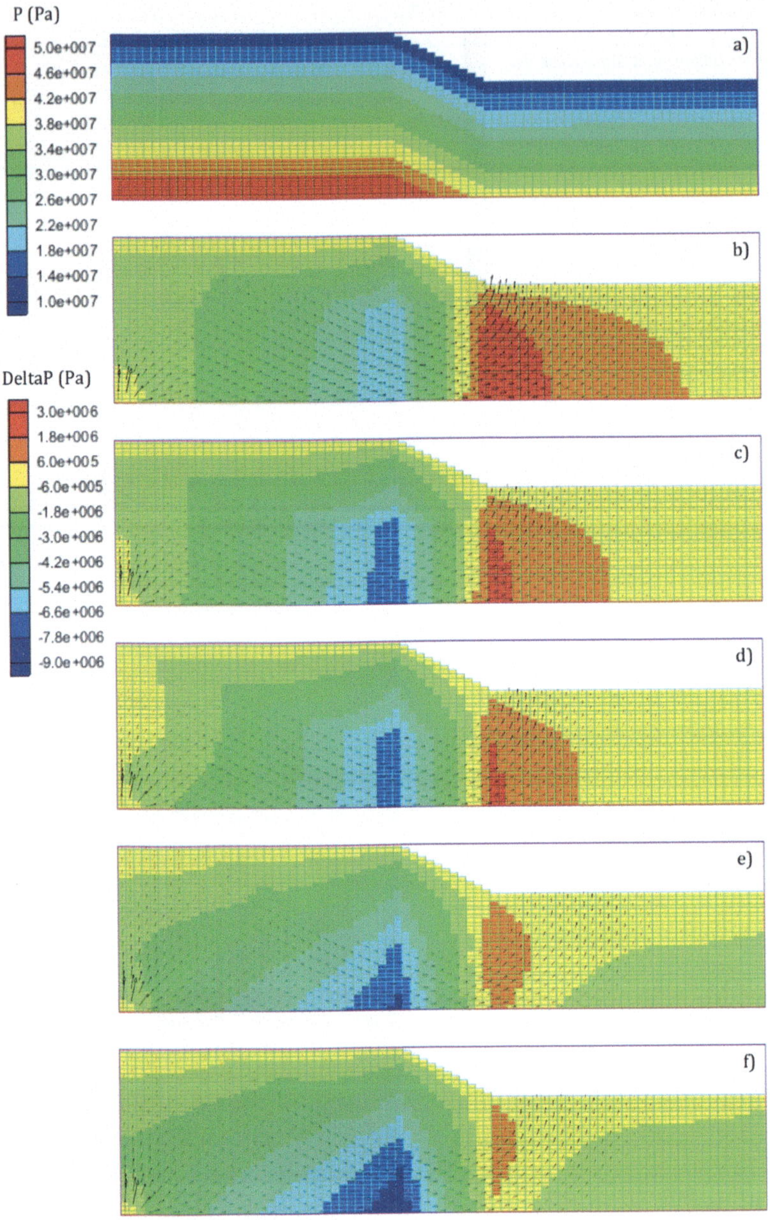

Fig. 44 Variations in pressure from initial conditions due to a mesa-like topographic relief and an intrusion located at the far left-hand side at some depth below the bottom of the model. The model is 40 km long and 10 km high; the vectors are gas velocities. **a** The pressure distribution imposed at the start of the simulation is liquid water-static. (see Fig. 46). Variations in pressure after **b** 1 ka, **c** 5 ka, **d** 10 ka, **e** 60 ka and **f** 200 ka. In time, the pressure decreases above the intrusion and below the top of the scarp. At the start it increases significantly below the base of the scarp, decreasing slowly in time

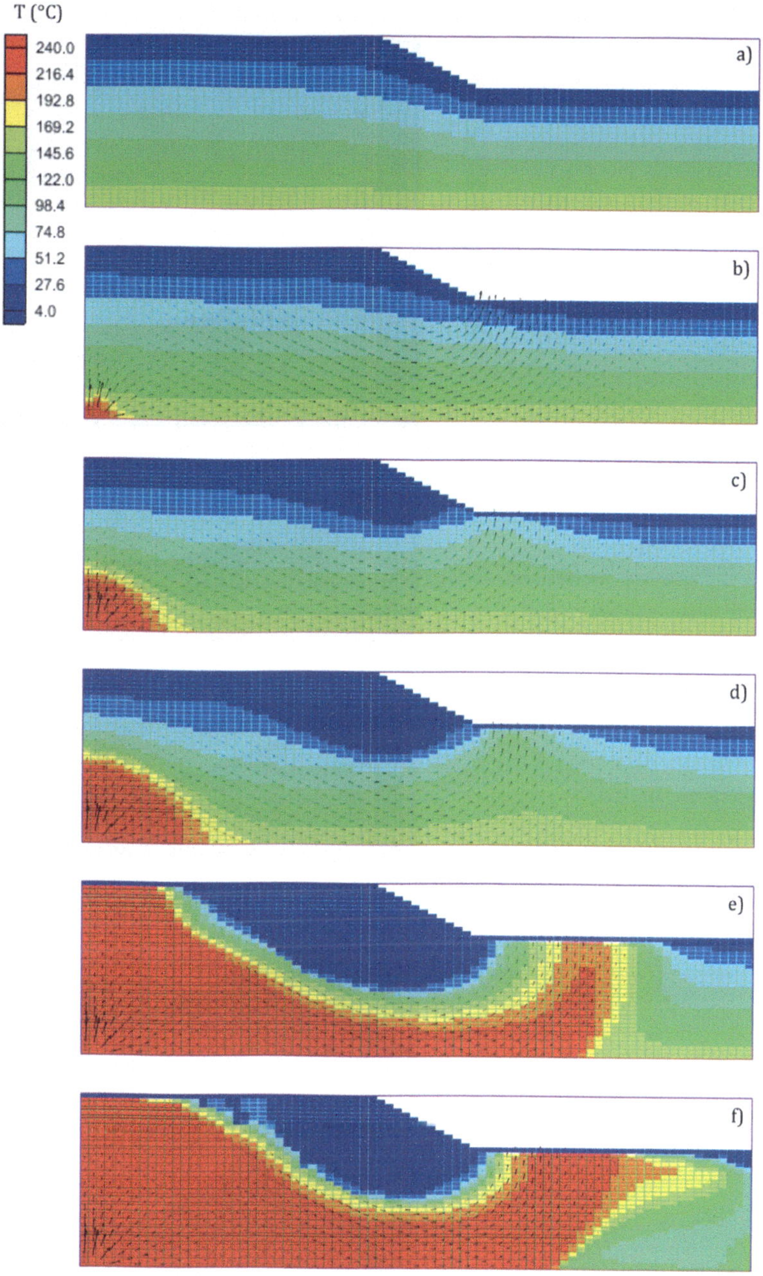

Fig. 45 Variations in temperature. **a** Constant vertical temperature gradient imposed at the start of the simulation. Variations in temperature after **b** 1 ka, **c** 5 ka, **d** 10 ka, **e** 60 ka and **f** 200 ka. In time, a strong hot plume of liquid water and some gas (see Fig. 46) develops above the intrusion, reaching the cryosphere. After many 10's of ka the hot plume also joins the plume that forms at the base of the scarp interacting with the cryosphere. The persistence of these plumes may melt the cryosphere allowing the aqueous phase to reach and erupt for significant time at the surface

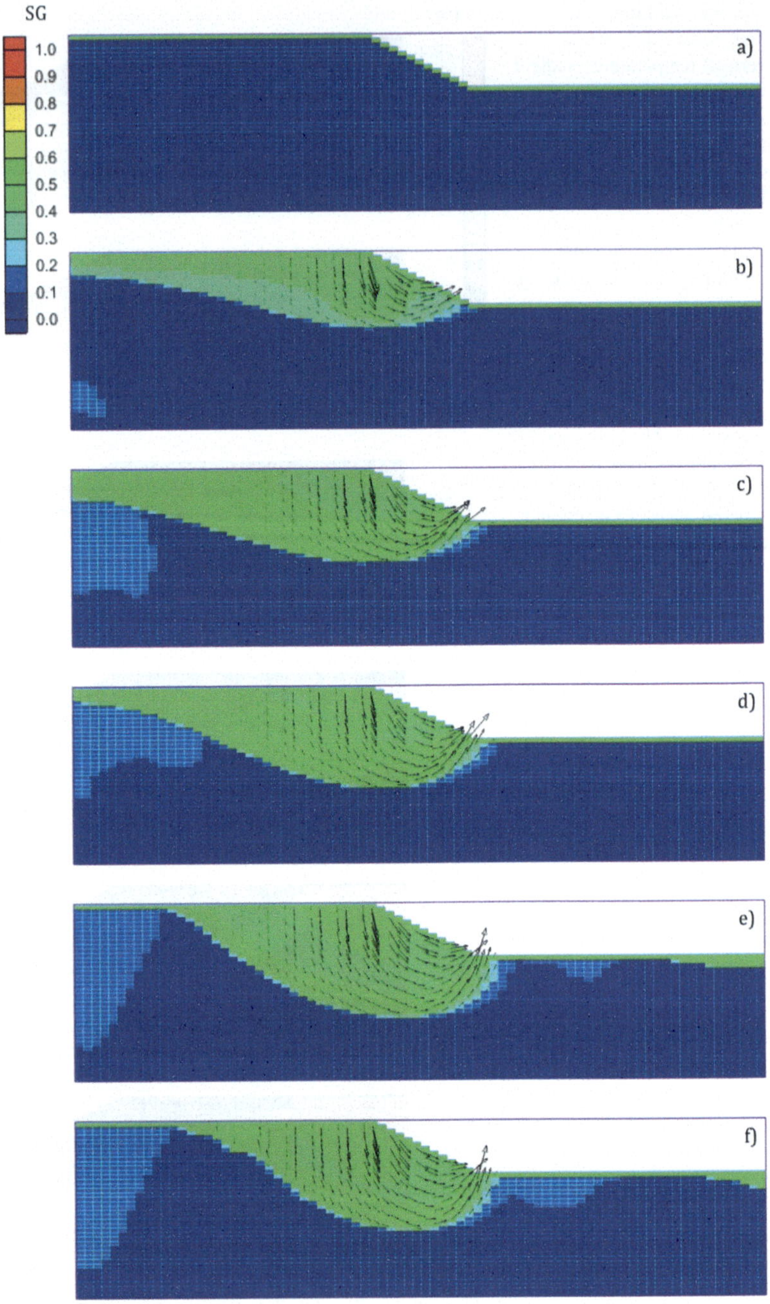

Fig. 46 Variations in gas saturation. **a** Equilibrium gas saturation at the start of the simulation calculated from the P–T conditions described in Figs. 44 and 45. Variations in gas saturation after **b** 1 ka, **c** 5 ka, **d**, 10 ka, **e** 60 ka, and **f** 200 ka. In time, a 50% saturated gas zone develops below the high topography and the scarp

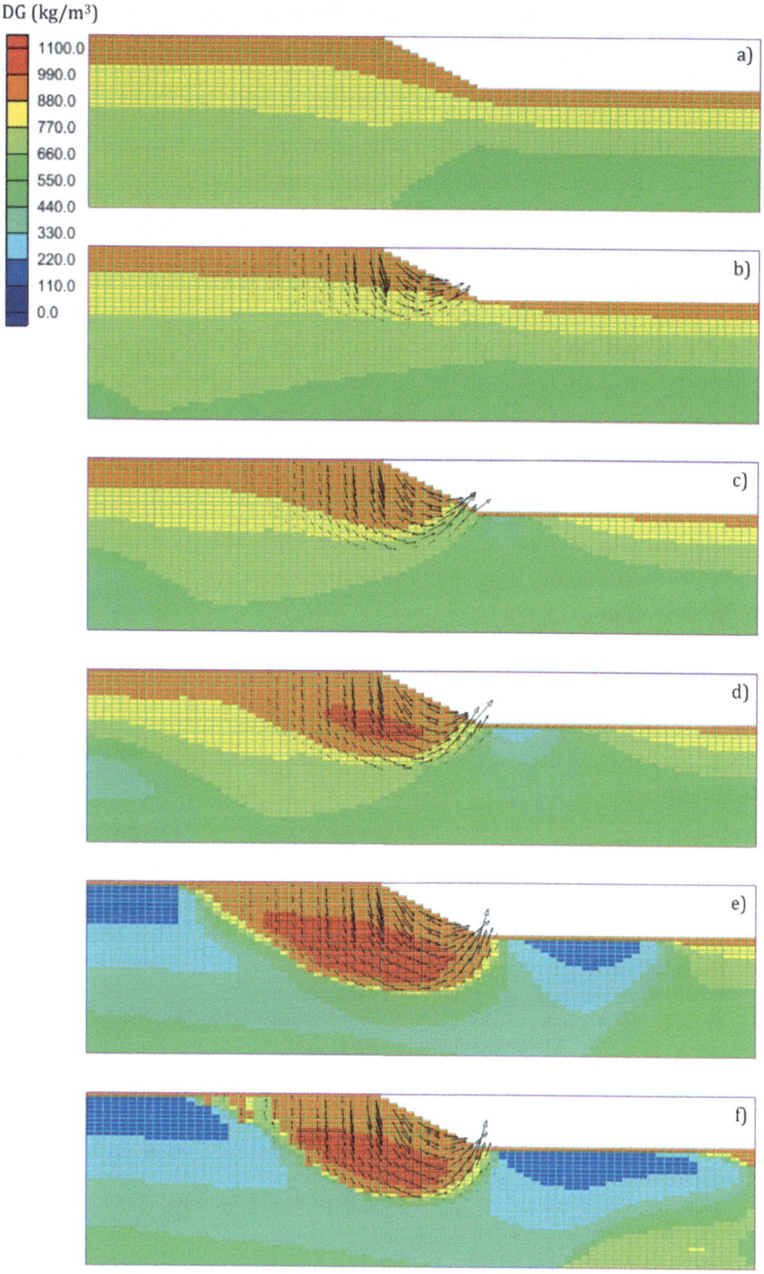

Fig. 47 Variations in gas density. **a** Equilibrium gas density at the start of the simulation calculated from the P–T conditions described in Figs. 44 and 45. Variations in gas density after **b** 1 ka, **c** 5 ka, **d**, 10 ka, **e** 60 ka, and **f** 200 ka. In time, the gas density increases in the 50% gas saturated zone below the high relief and the scarp. It decreases on the left-hand side of the model above the intrusion and in the lowland

should be closer to water saturation, so that liquid water can actually pierce through the cryosphere all the way to the surface.

There is a need to extend the range of applicability of numerical reservoir simulators to conditions relevant to Mars. Specifically, temperature ranges of the TOUGH codes need to be extended to lower temperatures, down to about −56.4 °C (216.8 K), and include the effects of solid phases like water ice and clathrates. ECO2N should also extend to Martian atmospheric pressure of a few millibars and higher underground pressure of up to at least 50 MPa. In addition, the equation of state must include explicitly the CO_2 liquid–gas phase transition. Finally, numerical modeling of specific Martian areas also including the effect of primary and secondary permeability may allow prediction of sources of volatiles and useful minerals on Mars.

Acknowledgements A. Borgia acknowledges financial support by EDRA.

References

1. Grau Galofre A, Jellinek AM, Osinski GR (2020) Valley formation on early Mars by subglacial and fluvial erosion. Nat Geosci. https://doi.org/10.1038/s41561-020-0618-x
2. Lauro SE, Pettinelli E, Caprarelli G, Guallini L, Rossi AP, Mattei E, Cosciotti B, Cicchetti A, Soldovieri F, Cartacci M, Di Paolo F, Noschese R, Orosei R (2020) Multiple subglacial water bodies below the south pole of Mars unveiled by new MARSIS data. Nat Astron. https://doi.org/10.1038/s41550-020-1200-6
3. Carr MH, Head JW (2015) Martian surface/near-surface water inventory: Sources, sinks, and changes with time. Geophys Res Lett 42:726–732. https://doi.org/10.1002/2014GL062464
4. Leone G, Tackley PJ, Gerya TV, May DA, Zhu G (2014) Three-dimensional simulations of the southern polar giant impact hypothesis for the origin of the Martian dichotomy. Geophys Res Lett 41:8736–8743. https://doi.org/10.1002/2014GL062261
5. Leone G (2020) The absence of an ocean and the fate of water all over the Martian history. Earth Space Sci 7:e2019EA001031. https://doi.org/10.1029/2019EA001031
6. Fanale FP (1976) Martian volatiles: their degassing history and geochemical fate. Icaurs 28:179–202
7. Clifford SM (1993) A model for the hydrologic and climatic behavior of water on Mars. J Geophys Res 98(E6):10973–11016. https://doi.org/10.1029/93JE0022
8. Mellon MT, Jakosky B, Postawko S (1997) The persistence of equatorial ground ice on Mars. J Geophys Res 102(E8):19357–19370
9. Kuzmin RO (2005) Ground ice in the martian regolith. In: Tokano T (ed) Water on Mars and life. Berlin, Springer, pp 155–189
10. Grimm RE, Painter SL (2009) On the secular evolution of groundwater on Mars. Geophys Res Lett 36(24):L24803. https://doi.org/10.1029/2009GL041018
11. Clifford SM, Lasue J, Heggy E, Boisson J, McGovern P, Max MD (2010) Depth of the Martian cryosphere: revised estimates and implications for the existence and detection of subpermafrost groundwater. J Geophys Res 115(E7):E07001. https://doi.org/10.1029/2009JE003462
12. Lasue J, Mangold N, Hauber E, Clifford S, Feldman W, Gasnault O, Grima C, Maurice S, Mousis O (2013) Quantitative assessments of the Martian hydro-sphere. Space Sci Rev 174(1–4):155–212. https://doi.org/10.1007/s11214-012-9946-5
13. Weiss DK, Head JW (2017) Evidences for stabilization of the ice-cemented cryosphere in earlier Martian history: implication for the current abundaces of growndwater at depth on Mars. Icarus 288:120–147. https://doi.org/10.1016/j.icarus.2017.01.018
14. Hoffman N (2001) Modern geothermal gradients on Mars and implications for subsurface liquids. In: Conference on the geophysical detection of subsurface water on Mars. Earth sciences, La Trobe University, Melbourne, vol 3086, p 7044
15. Mellon M, Phillips R (2001) Recent gullies on Mars and the source of liquid water. J Geophys Res 106:23165–23179
16. Hanna JC, Phillips RJ (2005) Hydrological modeling of the Martian crust with application to the pressurization of aquifers. J Geophys Res 110(E1):E01004. https://doi.org/10.1029/2004JE002330
17. Bargery AS, Wilson L (2010) Dynamic of the ascent and eruption of water containing dissolved CO_2 on Mars. J Geophys Res 115:E05008. https://doi.org/10.1029/2009JE003403
18. Goldspiel JM, Squyres SW (2011) Groundwater discharge and gully formation on Martian slopes. Icarus 211(1):238–258. https://doi.org/10.1016/j.icarus.2010.10.008
19. Grott M, Morschhauser A, Breuer D, Hauber E (2011) Volcanic outgassing of CO_2 and H_2O on Mars. Earth Planet Sci Lett 308:391–400. https://doi.org/10.1016/j.epsl.2011.06.014
20. Kieffer H, Jakosky BM, Snyder CM (1992) The planet Mars: from antiquity to present. In: Mars, Kieffer HH et al (ed). Univ. of Ariz. Press, Tucson, p 1–33

21. Longhi J (2006) Phase equilibrium in the system Co2-H2O: application to Mars. J Geophys Res 111: E06011. https://doi.org/10.1029/2005JE002552
22. Carr MH (1996) Water on Mars. Oxford University Press, New York, p 229
23. Martínez GM, Renno NO (2013) Water and Brines on Mars: current evidence and implications for MSL. Space Sci Rev 175:29–51
24. Weast RC, Selby SM, Hodgman CD (1964) Handbook of chemistry and physics, 45th edn. Chem. Rubber Co., Cleveland, Ohio
25. Angus S, Armstrong B, de Reuck KM (1976) International thermodynamic tables of the fluid state carbon dioxide. Elsevier, New York, p 385
26. Pruess K, Oldenburg C, Moridis G (2012) TOUGH2 user's guide, version 2. Lawrence Berkeley Laboratory Report, Berkeley, CA, LBNL-43134. p 197
27. Jung Y, Pau GSH, Finsterle S, Doughty C (2018) TOUGH3 user's guide, version 1.0. Lawrence Berkeley Laboratory Report, Bekeley, CA, LBNL-2001093. p 163
28. Klinkenberg LJ (1941) The permeability of porous media to liquids and gases. In: API Drilling and Production Practice. p 200–213
29. Edwards AL (1972) TRUMP: a computer program for transient and steady state temperature distributions in multidimensional systems. National Technical Information Service, National Bureau of Standards, Springfield, VA
30. Narasimhan TN, Witherspoon PA (1976) An integrated finite difference method for analyzing fluid flow in porous media. Water Resour Res 12(1):57–64
31. Pruess K, Narasimhan TN (1982) On fluid reserves and the production of superheated steam from fractured vapor-dominated geothermal reservoirs. J Geophys Res 87(B11):9329–9339
32. Pruess K, Narasimhan TN (1985) A practical method for modeling fluid and heat flow in fractured porous media. Soc Pet Eng J 25(1):14–26
33. Pruess K (1983) GMINC—a mesh generator for flow simulations in fractured reservoirs. Lawrence Berkeley Laboratory Report, Berkeley, CA, LBL-15227
34. https://www.aquaveo.com/software/gms-groundwater-modeling-system-introduction
35. Pan L, Spycher N, Daughty C, Pruess K (2015) ECO2N V.2.0: a New TOUGH2 fluid property module for mixtures of water, NaCl, and CO_2. Lawrence Berkeley Laboratory Report, Berkeley, CA, LBNL-6930E, p 118
36. Pruess K (2011) ECO2M: a TOUGH2 fluid property module for mixture of water, NaCl, and CO_2, including super- and sub-critical conditions, and phase change between liquid and gaseous CO_2. Laurence Berkeley National Laboratory Report, Berkeley, CA, LBNL-4590E, p 84
37. Spycher N, Pruess K (2005) CO_2-H_2O mixtures in the geological sequestration of CO_2. II. partitioning in chloride brines at 12–100 ℃ and up to 600 bar. Geochim Cosmochim Acta 69(13):3309–3320. https://doi.org/10.1016/j.gca.2005.01.015
38. Spycher N, Pruess K (2010) A phase partitioning model for CO_2-brine mixtures at elevated temperatures and pressures: application to CO_2-enhanced geothermal systems. Transp Porous Med 82:173–196. https://doi.org/10.1007/s11242-009-9425-y
39. Vargaftik NB (1975) Tables on the thermophysical properties of liquids and gases, 2nd edn. Wiley, New York, NY
40. Takenouchi S, Kennedy GC (1964) The binary system H_2O–CO_2 at high temperatures and pressures. Am J Sci 262:1055–1074
41. Wendland M, Hasse H, Maurer G (1999) Experimental pressure temperature data on three- and four-phase equilibria of fluid, hydrate, and ice phases in the system carbon dioxide-water. J Chem Eng Data 44(5):901–906
42. Hurwitz S, Kipp KL, Ingebristen SE, Reid ME (2003) Groundwater flow, heat transport, and water table position within volcanic edifices: Implication for volcanic processes in the Cascade Range. J Geophys Res 108(B12):2557. https://doi.org/10.1029/2003JB002565
43. Saharay M, Balasubramanian S (2007) Evolution of intermolecular structure and dynamics in supercritical carbon dioxide with pressure: an ab Initio molecular dynamics study. J Phys Chem B 111:387–392. https://doi.org/10.1021/jp065679t
44. Calcara M, Borgia A, Caricaterra M (2021) CO2 dipole moment: a simple model—implications for CO2-rock interactions. Submitted to Geofluids
45. Valley JW, Eiler JM, Graham CM, Gibson EK, Romaneck CS, Stolper EM (1997) Low-temperature concretions in the Martian meteorite ALH84001: evidence from stable isotopes and mineralogy. Science 275:1633–1638
46. Stephens SK, Stevenson DJ (1990) Dry carbonate formation on Mars: a plausible sink for an early dense CO2 atmosphere? Lunar and Planetary Science Conference. Houston, TX, vol 21, p 56
47. Màrquez A, Fernandez C, Anguita F, Farelo A, Anguita J, de la Casa MA (2004) New evidences for a volcanically, tectonically, and climatically active Mars. Icaurs 172(2):573581. https://doi.org/10.1016/j.icarus.2004.07.015
48. Neukum G, Jaumann R, Hoffmann H, Hauber E, Head JW, Basilevsky AT, Ivanov BA, Werner SC, van Gasselt S, Murray JB, McCord T (2004) The HRSC Co-investigator team. recent and episodic volcanic and glacial activity on mars revealed by the high resolution stereo camera. Nature 432:971–979
49. Malin MC, Edgett KS, Posiolova LV, McColley SM, Noe Dobrea EZ (2006) Present-day impact cratering rate and contemporary gully activity on Mars. Science 314:1573. https://doi.org/10.1126/science.1135156
50. Pilorget C, Forget F (2015) Formation of gullies on Mars by debris flows triggered by CO_2 sublimation. Nat Geosci 9:65–69

Buried Volcano-Hydrothermal Systems and Minerals on Mars

Dmitri Rouwet, Claudio Inguaggiato, and Giovanni Leone

Abstract

The pressure and temperature conditions on planet Mars have probably never permitted the presence of liquid water at its surface, as witnessed by unaltered olivines of any age since the Naochian. Nevertheless, iron sulphates at the surface suggest hydrothermal activity at some point in time and space, arguably at depth, from where minerals were expulsed by volcanic activity and/or "sweating out" during phases of heating upon meteorite impacts. Beneath the cryosphere, at depths between 4 and 10 km, the reigning P–T conditions meet the stability conditions of water and vapour, originating from a degassing magma. The sulphate dominance in Mars' mineralogy mimics the terrestrial analogues of hyperacid and hypersaline fluids and minerals hosted in Earth's most extreme crater lakes (e.g. Kawah Ijen, Poás, Copahue, Ruapehu, White Island) and ore depositing systems (e.g. high-sulphidation and porphyry systems). The observed similarity enables a bidirectional learning process for future research: on one hand, the more extreme systems on Mars can teach us on how highly active magmatic-hydrothermal and ore depositing systems on Earth might work; on the other hand, metal transport and deposition at terrestrial systems opens perspectives for "space mining" on Mars.

1 Volcano-Hydrothermal Systems and Minerals on Mars

The discovery on Mars of iron sulphates (e.g. szomolnokite and jarosite) that on Earth are typically found in a hydrothermal environment has raised the question whether hydrothermal systems may have existed on the surface of Mars [1–4]. The association of iron sulphates with other minerals like alunite, anhydrite, hematite, and kaolinite, has been claimed as an indicator of acidic (low-pH) sulphurous fluids at or near the Martian surface [2, 5]. However, the lack of alteration observed in olivine [6] and jarosite [7] found in surface deposits of any age at global level since the Noachian [8] indicates that likely water was never present in its liquid state on the surface of Mars [9]. Only a few detections of serpentine, indicating alteration of the olivine,

D. Rouwet (✉) · C. Inguaggiato
Istituto Nazionale di Geofisica e Vulcanologia, Sezione di Bologna, Bologna, Italy

C. Inguaggiato
Departamento de Geología, Centro de Investigación Científica y de Educación Superior de Ensenada, Baja California (CICESE), Carretera Ensenada-Tijuana 3918, Ensenada, Baja California, Mexico

G. Leone
Instituto de Investigacion en Astronomía y Ciencias Planetarias, Universidad de Atacama, Copiapó, Chile

G. Leone (ed.), *Mars: A Volcanic World*,
https://doi.org/10.1007/978-3-030-84103-4_7

were found in Noachian terrains [8], all characterized by volcanic activity [10]. Another striking discovery was that phyllosilicates showed scarce or no correlation to all the locations where water was claimed to exist [11]. The obvious conclusion would be that hydrothermal environments might have developed in a few locations of the Martian subsurface [12], where the lithostatic pressure still allowed the stability and the circulation of liquid water in fractures and pores [13, 14], well repaired from the cold and harsh environment for water on the surface. Then, the hydrothermal minerals formed in such environments would have been exhumed or expelled from the subsurface [15] by erosional processes during volcanic eruptions [9, 16].

Martian mineralogy has been derived from rover data [6, 17–25] and from SNC meteorites [26, 27] (see also chapter "Petrologic Evolution of Martian Volcanism and Clues from Meteorites") to infer the source region [28]. The crust of Mars now shows a compositional complexity [23] that was unexpected in the previous analyses showing a basaltic composition with a variable content of pyroxenes, plagioclases, and olivine [29, 30]. Compositions of unaltered soils dated from Noachian to Hesperian ages range from basalts to andesite and even rhyolite [17, 21, 31–36]. Mineralogy at Meridiani Planum was found consistent with shergottites [20]. Orthopyroxene was more commonly observed in Noachian lavas while clinopyroxene was more abundant in Hesperian lavas indicating a possible magmatic evolution over time [37]. Minerals of potential hydrothermal origin were reported in various places of Mars: Home Plate [36, 38, 39] and Paso Robles [40] at Gusev crater; the Eridania basin [41], which is the source of the Ma'adim Vallis—Gusev crater system [10]; Meridiani Planum [42, 43]; Mawrth Vallis [44–46]; Nili Fossae [47]; Gale crater [48, 49]. However, the local distribution and the small amounts found in all these places do not suggest surface aqueous processes at global level but rather localized underground hydrothermal environments [9, 12, 15].

Some information about the presence of rare earth elements (REE) on Mars comes from Martian meteorites [50, 51]. Interestingly, although REE were found in KREEP (potassium—rare earths-phosphorus) terrains on the surface of the Moon [52–54], which is a well-known dry surface body, not many surface findings were reported on Mars with the exception of an unusually high P_2O_5 abundance (>5 wt%), suggesting mechanical admixture of merrillite [55]. The analysis of this outcrop even suggested that the high-P_2O_5 tephrite signature is not due to secondary aqueous alteration, but to a primitive igneous rock composition [55]. Merrillite is often preferred to apatite in REE studies because it represents 90% of a rock's REE budget, and it is quite common in many extra-terrestrial basalts [51]. The presence of merrillite does not necessarily mean a depletion in volatiles at the moment of the formation of the rock, but the low contents in whitlockite and bobdownsite suggest the lack of alteration by hydrothermal fluids [56]. However, various studies postulated a Martian mantle depleted in volatiles due to the lack of crustal recycling [57–65]. The W isotopic data are consistent with the Martian dichotomy in ages and composition and inconsistent with putative plate tectonics on Mars [66, 67]. Yet, REE are the classic elements that form in a hydrothermal environment (see below), transported and deposited by fluids [68], and yet REE are the lithophile elements that would remain into the cooled crust after the solidification of the magma ocean. So, how is it possible that so many claimed hydrothermal environments on Mars do not show the presence of REE? Probably an easy answer would be that no water was present even in the shallow underground and that the possibility of existence of hydrothermal environments might just be another speculation like it was for water on the surface for the past two decades. Or maybe we were extremely unlucky in all the missions to Mars, probably we never landed on an outcrop containing REE.

2 From Basaltic Magma to Ore Minerals

A review of the physical principles ruling the rise of magma within the Martian crust was made by Wilson and Head [69]. All the various processes for the generation of mantle plumes feeding magma rise in the crust of Mars were already discussed in the chapter "The Formation of Mars and the Origin of Its Volcanic Provinces". As already discussed in the chapter "Volcanic Eruptions on Mars, Lava Flow Morphology, and Thermodynamics", various factors influence the rise of magma in the crust and these are essentially composition, density, and juvenile content of the magma compared to those of the host rock [70–72]. We will focus here on the specific mechanisms that form magma reservoirs and the associated volcano-hydrothermal systems (VHS hereafter) in the Martian crust. On Earth, water-rich magma reservoirs forming ore-generating fluids are generally found between 2 and 8 km depth [73], or even at the surface in most active VHS.

On Mars, this depth is generally double than on Earth, with magma reservoirs between 4 and 16 km depth; dikes spreading from these reservoirs will be 1.35 times wider and 1.8 times longer in outcrop than on Earth, thus also being able to reach the surface, if density and volatile content allow it [69]. Upon magma stalling, ore deposits will be injected in the fractures surrounding the magma reservoirs and dikes will pervasively find their way into the crust [68]. The VHS generated by the fluids leaving the magma reservoirs will be responsible for the formation of various classes of ore deposits: large Cu-Mo-Au deposits, generally known as porphyries, are associated to shallow intrusions; Ag-Zn-Pb-W and other elements form at greater distances from the magma reservoir [68]. Fluids play a fundamental role in the formation of ore deposits. Mineralization of porphyries is enhanced by magmatic fluids during the early stages of ore formation and the deposits are generally localized in fractures or veinlets, which commonly form at depths of 2–5 km on Earth [68]. By comparison, these fractures and veinlets rich in ore deposits might be found at double depth, 4–10 km on Mars where the lithostatic pressure allows the stability of magmatic fluids.

Magma composition also plays an important role in ore forming processes. Komatiitic and picritic magmas, for example, are ultramafic types undersaturated in sulphur that prevent the depletion of chalcophile elements, whereas mafic magmas produced by less than 25% of partial melting are generally saturated in sulphur thus depleting ore deposits with chalcophile elements [74]. This process happens because original contents of Cu are lost by the extraction into magmatic sulphides [75]. Nevertheless, most of the siderophile elements can be found in Fe and Ni sulphides present in the upper mantle that are incorporated in basaltic magmas during partial melting, and then ascend to the crust [76].

3 The Role of Volcanic Fluids in the Martian Crust

On Earth, the source of ore deposits was suggested since the sixteenth century by Agricola, as rainwater heated underground would leach metals from rocks to deposit them in ore (i.e. an economically extractable material with a utility). A century later, this view was challenged by René Descartes, who proposed a mechanism of vapour release upon cooling of the Earth's interior responsible for filling the fractures with ore deposits. This debate between Neptunists and Plutonists is still ongoing today [68].

Regarding Mars, this debate will not happen because rainwater never existed during its whole history [9] so that Neptunists would eventually have no argument to point this out. The possibility that water in the subsurface was frozen during the volcanic eruptions is unlikely, although a possible cryosphere has been postulated for a long time [77–83]. The latest revised estimates for the depth of occurrence of such a cryosphere range from 0.5 to 4.5 km at equatorial latitude up to 10.5 km at subpolar latitude [83]. Depending on magma composition and juvenile content of H_2O and CO_2, volatile degassing on Mars is expected to occur already at depths of ~1.8–0.06 km, the higher the initial content of volatile the deeper the degassing starts [69, 84]. Thus, the effective depth of the cryosphere may shrink further to a depth between 1.8 and 4.5 km in case of high gas content (i.e. 5 wt % of water and or carbon dioxide), or arrive straight to the surface. The higher depths have been beyond detection of the MARSIS radar sounding in several experiments [85–87], despite the nominal ~3–5 km achievable in optimal conditions (i.e. presence of pure ice) [83, 88]. However, even the shallowest depths have not returned the expected results in the search for water [89], except for a possible detection in the southern polar area [90] after previous attempts [91]. The juvenile (or entrained) content of fluids also determines the rise of Martian magmas in the crust with variable depth of magma reservoirs in magma buoyancy zones [69].

Magmas can have different compositions, densities, and volatile contents depending on the degree of partial melting. The solubility of volatiles in magmas is also dependent on composition, temperature, and pressure. For example, H_2O has a greater solubility in magmas of rhyolitic composition than basaltic composition, but much of this difference is more due to the effect of temperature rather than composition [92]. Due to its high solubility, H_2O is the volatile phase that last leaves evolved magmas, so that it represents only a small fraction of the circulating fluids in the deep crust that will thus be more dominated by CO_2 [68]. The presence of SO_2 in the magma may decrease the solubility of H_2O and CO_2, hence increasing their activity as possible circulating fluids at higher depths [68].

From high-temperature fumarolic degassing on Earth it is known that magmatic gas mixtures (H_2O-CO_2-SO_2) are able to transport metals, to be deposited as minerals when the temperature drops near the fumarole outlet (e.g. Colima and Kydryavy volcanoes, [93, 94]). Alkali and alkali earth sulphates are common products, as well as barite, Cu, Pb and Al sulphates, Cl-F-bearing Cu, Ca, Fe, Zn, Pb, and Fe, Ni, W, Co, Ti-oxides besides native Au [94]. Highly volatile trace elements (Cl, S, I, F, Bi, Cd, B, Br) are enriched in mineral depositions when the temperature drops; less volatile elements (Re, Mo, W, Cu, Co), however, are also transported in high-T magmatic gases [93], as well as REE.

As such, high temperature magmatic gases are ideal transport media of metals, although an efficient deposition mechanism, besides temperature drop, is needed to explain high ("ore-grade") metal deposition as sulphides. On the one hand, a kinetically and thermodynamically favorable reaction in the H_2O-SO_2-CO_2 system is the chemisorption of SO_2 in the presence of plagioclase to create anhydrite, andalusite, quartz and H_2S, hence dictating the needed reducing conditions to drop sulphide minerals [95]. On the other hand, the most abundant gas species besides H_2O for terrestrial volcanism and Mars's main gas species, CO_2, has the capacity to drastically lower the solubility of metals (e.g. for Cu by a factor 10,000), in a more efficient way than a pressure and temperature drop has. Flushing of CO_2 in a vapor-dominated system will create conditions to deposit from, or resolve metals into the gas phase, depending on the dP_{H2O}, caused by varying proportions of H_2O/CO_2 [96]. The P–T conditions, the CO_2-SO_2-dominance, "sweating out" of H_2O, and the absence of O_2 in the subsurface of Mars, beneath the cryosphere could even more accentuate metal transport and deposition than it does on Earth. In fact, sulphides are 2–3 times more abundant in Martian shergottites than in terrestrial igneous rocks [97, 98], possibly providing evidence for the above hypotheses.

4 Liquid Analogues for Mars on Earth

Physical–chemical processes occurring on the surface on Earth are not adaptable to the situation for Mars. Hence, the most active VHS on Earth cannot simply be considered surface analogs for Mars, but only so if "buried" to depths of 4–10 km, given the P–T stability field of water. Moreover, according to our knowledge on Earth's porphyries, their generation processes and intrinsic connection to active VHS proof that fluids do not necessarily need to have flushed at or near the surface to be observed at the surface anyway (see previous section). The strongest evidence of the absence of water on Mars's surface is the lack of alteration of the olivines found at the surface into serpentine, despite the 4.5 Ga geological history of Mars [9]. However, volcanism and meteorite impacts can violently expel the VHS and convert them into outcrops, witnessed by various minerals that enable to track the history and formation processes of VHS.

The S-abundance in Martian lavas suggests that prior to cooling and crystallization, lavas should have released considerably high amounts of sulphur-rich volatiles. In the geological record, the mineral assemblages encountered on Mars (Fe-rich sulfates, alunite, anhydrite, goethite, hematite, jarosite and kaolinite) strongly resemble those on Earth manifested as the liquid-mineral assemblages in its most extreme natural form: high-acidity VHS, and peak-activity crater lakes and their surrounding drainage systems [99–104]. Let us now imagine burying peak-activity crater lake fluids to depths of 4–10 km to meet the P–T stability conditions for water and vapour beneath the cryosphere on Mars. Figure 1 shows four of the most studied and renowned hyperacid crater lakes on Earth: Kawah Ijen (Java, Indonesia, Fig. 1a), Ruapehu (Fig. 1b), White Island (both New Zealand, Fig. 1c), Poás (Costa Rica, Fig. 1d).

Besides their spectacular colours, varying from grey to turquoise to yellow-green, that unveil the high contents of solutes and solids [105–107], their chemistry proofs them to be (1) efficient absorbers of magmatic volatiles

Fig. 1 Arguably the four most acidic and saline crater lakes on Earth: **a** Kawah Ijen, Java, Indonesia; **b** Ruapehu, New Zealand/Aotearoa; **c** White Island/Whakaari, New Zealand/Aotearoa; **d** Poás, Costa Rica. (**a**)–(**c**) pictures by D.R., (**d**) picture by R. Mora-Amador

when rising gases hit the cooler surface water body, and (2) aggressive rock leachers. The former process, also called scrubbing, massively increases the anion content, and drastically drops the solution's pH. The latter process adds rock forming elements to the hyper-acid solution, in an isochemical manner. Besides an increase in the most mobile major elements Na, K, Ca, and Mg, Al, Fe, trace metals and REE are easily leached [108, 109]. The ranges in pH and SO_4/Cl ratios for the most studied hyper-acidic peak-activity crater lakes are illustrated in Fig. 2a. The crater lake at White Island/Whakaari, New Zealand/Aotearoa, shows the widest range in pH and the most Cl-rich composition (Fig. 2b); Copahue, Argentina-Chile, instead, is the most SO_4-rich and has a pH > 0; the wide ranges in pH and Cl and SO_4 concentrations for Poás's crater lake, Laguna Caliente, Costa Rica, strongly depend on the state of activity, with highest concentrations and lowest pH values observed during phreatic eruption cycles; Yugama crater lake at Kusatsu-Shirane volcano, Japan, is less extreme and the most constant in its chemistry; the crater lake chemistry of Kawah Ijen, Indonesia, and Ruapehu, New Zealand/Aotearoa, volcanoes is more stable and less extreme, because they are the two most voluminous crater lakes on Earth, hence less sensitive to variations in volcanic degassing activity (Fig. 2a, b).

As long as magmatic degassing continues, the crater lake will maintain its high acidity, will never neutralize, nor reach steady-state conditions in terms of water–mineral equilibrium. Nevertheless, evaporation from the lake surface will ulteriorly concentrate the brine solution, which could lead to mineral precipitation, besides a pH drop down to negative values (Fig. 2a). Anhydrite, gypsum, jarosite, (Na)alunite are most common minerals [99, 102, 108, 109, 115].

Elemental sulfur is a notable "byproduct", especially in the temperature range 119–159 °C, as sulfur turns liquid at T > 119 °C and increases its viscosity by four orders of magnitude at T > 159 °C (e.g. yellow sulfur mats in Fig. 1c, d) [116, 117]. This sulfur dynamics interacts with the already chaotic nature of the crater lake, and will affect fluid migration, pressure regimes, changes in porosity and permeability in fractures beneath the lakes, inside the underlying magmatic-hydrothermal system. Sulfur sealing also triggers phreatic eruptions from VHS and peak-activity crater lakes [102, 113, 116, 118, 119], a process that is not excluded for "Martian buried lakes", and furthermore a potential candidate to transport hydrothermal minerals towards the surface of the Red Planet.

5 Sulphate Minerals on Mars and Their Potential Association with VHS on Earth

Hyperacid/acid waters in VHS are extreme environments well known on the Earth to precipitate sulphate minerals as gypsum, anhydrite, alunite-jarosite minerals. Acidic sulphate solutions formed by the interaction between the SO_2-rich gas and H_2S under oxidizing conditions, as well as the contribution of other acidic volcanic gases as HCl and HF, are responsible for the strong acidity of the waters in VHS, with pH up to ~0 [101, 120]. The interaction of these acidic/hyperacidic waters with igneous rocks mobilize the elements from the primary rocks to the fluids. The precipitation of gypsum and several Al-sulphate minerals were identified in Kawah Ijen hyperacid crater lake (Indonesia), while gypsum precipitation was recognized at Poás crater lake, Costa Rica [101, 103, 108, 109]. Hematite, alunite, jarosite, anhydrite/gypsum minerals precipitated from acidic waters in Copahue volcano, Argentina [111, 121]. Alunite minerals were found at the Kusatsu-Shirane volcano area [122]. Natroalunite, jarosite, pyrite, marcasite, gypsum, anhydrite, hematite, goethite are found in eruptive products thrown out by phreatic eruptions from Ruapehu's Crater Lake Te Wai a Moe [102].

Several sulphate minerals were identified on the Mars surface, among which gypsum, anhydrite, Mg-sulfates and jarosite-alunite minerals [19, 123, 124]. As reiterated from earlier, liquid water does not exist on the surface of Mars, due

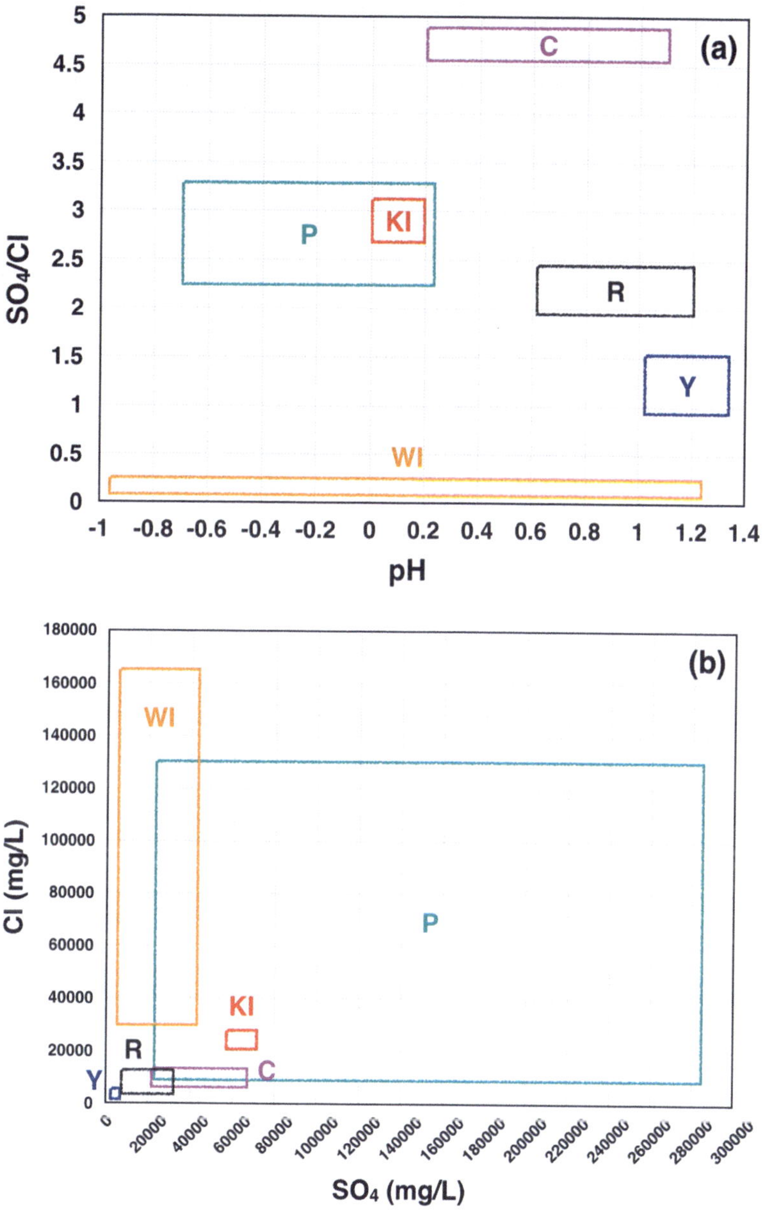

Fig. 2 Showing the chemical composition in terms of **a** pH versus SO_4/Cl (weight ratios) and **b** SO_4 versus Cl, of well studied peak-activity crater lakes and related fluids (Kawah Ijen-KI, White Island-WI, Copahue-C, Poás-P, Yugama-Y, and Ruapehu-R). *Data sources* KI-[100, 103]; WI-[110]; C-[111]; P-[112, 113]; Y-[114]; R-[102, 115]

to the low atmospheric pressure and temperature. However, the presence of sulphate minerals on Mars, more generally evaporitic minerals usually precipitating from a liquid phase, suggests the chemical tracers of ancient subterranean fluids from which the minerals precipitated [123–125] and then brought to the surface by volcanism. The discovery of alunite-jarosite minerals on the surface of Mars implicates the existence in the past of acidic SO_4 fluids with $pH < 4$ and oxidizing conditions [98, 124]. Acidic VHS on Earth are hence good candidates as analogue of the environments from which the precipitation of sulphate minerals on Mars occurred [98, 121, 126].

6 Rare Earth Elements in Acid/hyperacid Waters Associated to VHS

The high demand of REE increased in the last years due to the wide application field of REE, boosting the quest for unconventional potential REE reservoir [127–130]. The REE are used in several application fields: renewable energy, electronics, high-technology, medical science etc. To the best of our knowledge, REE ore deposits on Mars were not (yet) identified [131]. Nevertheless, higher REE concentrations with respect to other Martian meteorites with igneous origins were identified in the NWA 7034 meteorite (La = 13.7 ppm; Lu = 4.3 ppm), a regolith breccia that could represent the Martian surface [131, 132]. As mentioned above, merrillite is the mineral that hosts most of the REE in the extraterrestrial sample rocks. Lunar merrillite is characterized by higher REE concentrations with respect to the Martian Merrillite [131, 133].

Extrapolating the analogy introduced in section "Liquid Analogues for Mars on Earth", several studies were carried out regarding the REE mobility in active, hyperacid VHS on Earth, showing that the $\sum$REE concentration in waters is a function of the pH (Fig. 3), with REE concentrations increasing with decreasing pH [109, 134–137].

The high concentration of REE in the hyperacidic waters is mainly due to the intense rock leaching during water–rock interaction [103, 136, 141]. On Earth, higher REE concentrations were found in hyperacid lakes associated with active volcanoes (e.g. Kawah Ijen crater lake water up to ~6.5 mg kg^{-1}; Poás crater lake water up to ~2.2 mg kg^{-1}; [103, 109, 111, 136, 141–143]). Consequently, fluids enriched in REE could precipitate secondary minerals with high REE concentrations. Each element belonging to the REE group, can have a different concentration as a function of its affinity with the secondary minerals precipitated. Some studies on

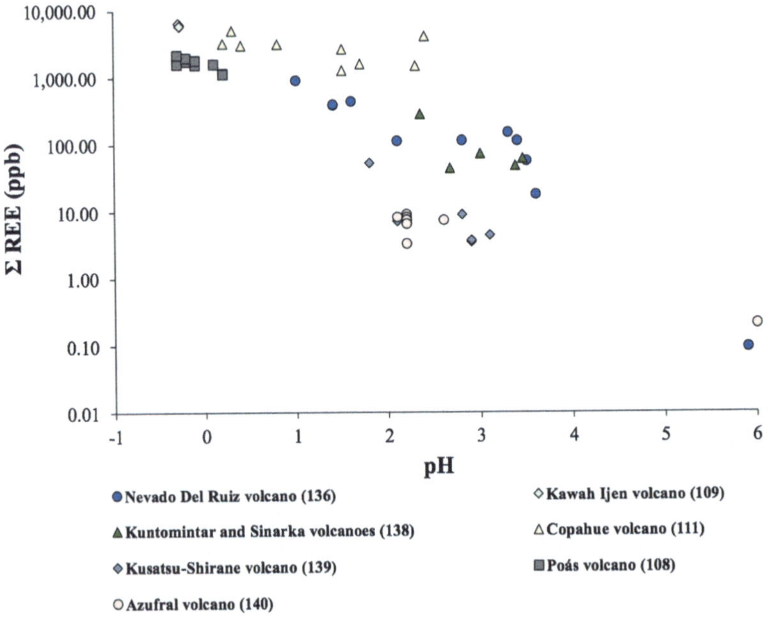

Fig. 3 Total REE concentrations dissolved in waters versus pH values in the following VHS: (1) Nevado del Ruíz volcano, Colombia [136]; (2) Kawah Ijen volcano, Java, Indonesia [109]; (3) Kuntomintar and Sinarka volcanoes, Shiashkotan Island, Kuril island arc [138]; (4) hot springs of Copahue volcano, Argentina-Chile [111]; (5) Kusatsu-Shirane volcano, Japan [139]; (6) Poás volcano, Costa Rica [108]; and (7) Azufral volcano, Colombia [140]

natural systems as well as experiments in the laboratory have identified that alunite-jarosite family minerals and gypsum prefer to remove LREE (Light Rare Earth Elements) with respect to HREE (Heavy Rare Earth Elements) from the waters [104, 108, 109, 122, 137, 144, 145]. The hyperacid waters belonging to VHS show REE patterns normalized to the primary rocks (e.g. andesite rocks) that are characterized by quite flat patterns showing a near-congruent dissolution of the primary rock, or patterns depleted in LREE [108, 136, 141]. Patterns depleted in LREE normalized to the local rock (andesite) are the consequence of the precipitation of sulphate secondary minerals such as gypsum and/or alunites [108, 122, 136, 137, 140, 141].

The identification of the VHS on Earth as potential analogue to justify the formation of sulphate rich minerals on Mars [98, 121, 126], allows to propose that the REE fractionation processes already identified in VHS on Earth during the interaction between rocks and acidic fluids, could also be true for Mars. The concentration of REE in the hypothetical aqueous phase of Mars can be a function of the pH of the fluids and the REE concentration of the rocks interacting with the fluids. Moreover, REE-enriched secondary minerals could precipitate from hypothetical REE-enriched fluids. Acidic leaching experiments in the laboratory on a basaltic shergottite Zagami shows the mobility of REE, Th, U and K in a range of pH from 1 to 5 [146]. The REE show a higher mobility at lower pH, especially the LREE [146]. The interaction of the hyperacid fluids with the local rocks on Mars, in the subsurface at depth where an aqueous phase existed or exists, could generate fluids enriched in dissolved elements, including REE. Furthermore, if waters on Mars were/are more acidic than on Earth, a high mobilization of REE from the rock to the hyperacid waters could be hypothesized. Within this assumption, and given the importance of REE in technological advancements, considering Mars as a potential mining site of REE, as well as porphyry-type deposits, could be put forward as an idea for future exploration and research.

7 Conclusions

This chapter summarizes the theoretical possibilities that VHS type minerals found on Mars may have formed in volcano-hydrothermal system subsurface environments. We have compared Martian analogs with several well documented terrestrial volcano-hydrothermal systems, with a special focus on hyper-acid magmatic-hydrothermal systems, well represented by peak activity crater lakes, such as at Kawah Ijen, Poás, Ruapehu, Whakaari/White Island, Copahue and Kusatsu-Shirane volcanoes and acid spring systems at Nevado del Ruíz, Kuntomintar and Sinarka, Copahue, Azufral, and Kusatsu-Shirane volcanoes. The distribution of major, minor and trace metals, including REE, between mineral and fluid phases in such extreme systems opens promising perspectives for the potential of future "space mining" of the Mars subsurface. On the contrary, the CO_2- and S-dominance of the Martian fluids with respect to terrestrial fluids at active volcanoes can provide insights on how these two gas species determine metal transport and eventual mineral deposition on Mars and on Earth.

References

1. Klingelhöfer G, Morris RV, Bernhardt B, Schröder C, Rodionov DS, de Souza PA et al (2004) Jarosite and hematite at Meridiani Planum from Opportunity's Mössbauer spectrometer. Science (80–)306 (5702)
2. Ehlmann BL, Edwards CS (2014) Mineralogy of the Martian surface. Annu Rev Earth Planet Sci 42 (1):291–315
3. Ming DW, Mittlefehldt DW, Morris RV, Golden DC, Gellert R, Yen A et al (2006) Geochemical and mineralogical indicators for aqueous processes in the Columbia Hills of Gusev crater, Mars. J Geophys Res Planets 111(E2)
4. Gendrin A, Mangold N, Bibring JP, Langevin Y, Gondet B, Poulet F et al (2005) Sulfates in Martian layered terrains: the OMEGA/Mars express view. Science (80–) [Internet] 307(5715):1587–1591 [cited 2020 Nov 28]. http://science.sciencemag.org/
5. Ehlmann BL, Swayze GA, Milliken RE, Mustard JF, Clark RN, Murchie SL et al (2016) Discovery of alunite in Cross crater, Terra Sirenum, Mars:

evidence for acidic, sulfurous waters. Am Mineral 101(7):1527–1542
6. McSween HY, Wyatt MB, Gellert R, Bell JF, Morris RV, Herkenhoff KE et al (2006) Characterization and petrologic interpretation of olivine-rich basalts at Gusev Crater, Mars. J Geophys Res Planets 111(E2):E02S10
7. Madden MEE, Bodnar RJ, Rimstidt JD (2004) Jarosite as an indicator of water-limited chemical weathering on Mars. Nature 431:821–823
8. Ehlmann BL, Mustard JF, Murchie SL (2010) Geologic setting of serpentine deposits on Mars. Geophys Res Lett 37:L06201
9. Leone G (2020) The absence of an ocean and the fate of water all over the martian history. Earth Space Sci 7(4):e2019EA001031
10. Leone G (2016) Alignments of volcanic features in the southern hemisphere of Mars produced by migrating mantle plumes. J Volcanol Geotherm Res 309:78–95
11. Michalski JR, Cuadros J, Bishop JL, Darby Dyar M, Dekov V, Fiore S (2015) Constraints on the crystal-chemistry of Fe/Mg-rich smectitic clays on Mars and links to global alteration trends. Earth Planet Sci Lett [Internet] 427:215–225 [cited 2019 Apr 8]. https://www.sciencedirect.com/science/article/pii/S0012821X15003775
12. Ehlmann BL, Mustard JF, Murchie SL, Bibring J-P, Meunier A, Fraeman AA et al (2011) Subsurface water and clay mineral formation during the early history of Mars. Nature [Internet] 479(7371):53–60 [cited 2019 Oct 2]. http://www.nature.com/articles/nature10582
13. Evans BW (2004) The serpentinite multisystem revisited: chrysotile is metastable. Int Geol Rev [Internet] 46(6):479–506 [cited 2019 Apr 6]. https://www.tandfonline.com/doi/full/https://doi.org/10.2747/0020-6814.46.6.479
14. Evans BW (2010) Lizardite versus antigorite serpentinite: magnetite, hydrogen, and life(?). Geology [Internet] 38(10):879–882 [cited 2019 Apr 6]. http://pubs.geoscienceworld.org/geology/article/38/10/879/130064/Lizardite-versus-antigorite-serpentinite-Magnetite
15. Michalski JR, Onstott TC, Mojzsis SJ, Mustard J, Chan QHS, Niles PB et al (2018) The Martian subsurface as a potential window into the origin of life. Nat Geosci 11(1):21–26
16. Leone G (2017) Mangala Valles, Mars: a reassessment of formation processes based on a new geomorphological and stratigraphic analysis of the geological units. J Volcanol Geotherm Res 337:62–80
17. Bell JF, McSween HY, Crisp JA, Morris RV, Murchie SL, Bridges NT et al (2000) Mineralogic and compositional properties of Martian soil and dust: results from Mars Pathfinder. J Geophys Res Planets [Internet] 105(E1):1721–1755 [cited 2018 Nov 10]. https://doi.org/10.1029/1999JE001060
18. McSween HY, Ruff SW, Morris RV, Bell JF, Herkenhoff K, Gellert R et al (2006) Alkaline volcanic rocks from the Columbia Hills, Gusev crater, Mars. J Geophys Res [Internet] 111(E9):E09S91 [cited 2018 Oct 18]. https://doi.org/10.1029/2006JE002698
19. Ming DW, Gellert R, Morris RV, Arvidson RE, Brückner J, Clark BC et al (2008) Geochemical properties of rocks and soils in Gusev Crater, Mars: results of the alpha particle X-ray spectrometer from cumberland ridge to home plate. J Geophys Res [Internet] 113(E12):E12S39 [cited 2018 Nov 8]. https://doi.org/10.1029/2008JE003195
20. Zipfel J, Schröder C, Jolliff BL, Gellert R, Herkenhoff KE, Rieder R et al (2011) Bounce Rock—a shergottite-like basalt encountered at Meridiani Planum, Mars. Meteorit Planet Sci 46(1):1–20
21. Christensen PR, Wyatt MB, Glotch TD, Rogers AD, Anwar S, Arvidson RE et al (2004) Mineralogy at Meridiani Planum from the Mini-TES experiment on the Opportunity Rover. Science [Internet] 306 (5702):1733–1739 [cited 2018 Nov 10]. http://www.ncbi.nlm.nih.gov/pubmed/15576609
22. Cousin A, Sautter V, Payré V, Forni O, Mangold N, Gasnault O et al (2017) Classification of igneous rocks analyzed by ChemCam at Gale crater, Mars. Icarus 288:265–283
23. Sautter V, Toplis MJ, Beck P, Mangold N, Wiens R, Pinet P et al (2016) Magmatic complexity on early Mars as seen through a combination of orbital, in-situ and meteorite data. Lithos 254–255:36–52
24. Payré V, Fabre C, Cousin A, Sautter V, Wiens RC, Forni O et al (2017) Alkali trace elements in Gale crater, Mars, with ChemCam: calibration update and geological implications. J Geophys Res Planets [Internet] 122(3):650–679 [cited 2019 Mar 27]. https://doi.org/10.1002/2016JE005201
25. Yen AS, Gellert R, Schröder C, Morris RV, Bell JF, Knudson AT et al (2005) An integrated view of the chemistry and mineralogy of martian soils. Nature 436:49–54
26. McSween HY, Murchie SL, Crisp JA, Bridges NT, Anderson RC, Bell JF et al (1999) Chemical, multispectral, and textural constraints on the composition and origin of rocks at the Mars Pathfinder landing site. J Geophys Res Planets [Internet] 104 (E4):8679–8715 [cited 2018 Oct 26]. https://doi.org/10.1029/98JE02551
27. Papike JJ, Karner JM, Shearer CK, Burger PV (2009) Silicate mineralogy of martian meteorites. Geochim Cosmochim Acta 73(24):7443–7485
28. Hamilton VE, Christensen PR, McSween HY, Bandfield JL (2003) Searching for the source regions of martian meteorites using MGS TES: integrating martian meteorites into the global distribution of igneous materials on Mars. Meteorit Planet Sci 38(6):871–885
29. Hamilton VE, Wyatt MB, McSween HY, Christensen PR (2001) Analysis of terrestrial and Martian volcanic compositions using thermal emission spectroscopy: 2. Application to Martian surface

spectra from the Mars Global Surveyor Thermal Emission Spectrometer. J Geophys Res Planets [Internet] 106(E7):14733–14746 [cited 2019 Jan 19]. https://doi.org/10.1029/2000JE001353
30. McSween HY, Taylor GJ, Wyatt MB (2009) Elemental composition of the Martian crust. Science (80–) 324(5928):736–739
31. Sautter V, Fabre C, Forni O, Toplis MJ, Cousin A, Ollila AM et al (2014) Igneous mineralogy at Bradbury Rise: the first ChemCam campaign at Gale crater. J Geophys Res Planets [Internet] 119 (1):30–46 [cited 2019 Mar 27]. https://doi.org/10.1002/2013JE004472
32. Cousin A, Sautter V, Mangold N, Fabre C, Forni O, Rapin W et al (2015) Igneous rock classification at Gale (Sols 13–800). In: Lunar and planetary science conference, p 2452
33. Ollila AM, Newsom HE, Clark B, Wiens RC, Cousin A, Blank JG et al (2014) Trace element geochemistry (Li, Ba, Sr, and Rb) using Curiosity's ChemCam: early results for Gale crater from Bradbury Landing Site to Rocknest. J Geophys Res E Planets 119:255–285
34. Anderson R, Bridges JC, Williams A, Edgar L, Ollila A, Williams J et al (2015) ChemCam results from the Shaler outcrop in Gale crater, Mars. Icarus 249:2–21
35. Meslin PY, Gasnault O, Forni O, Schroder S, Cousin A, Berger G et al (2013) Soil diversity and hydration as observed by ChemCam at Gale Crater, Mars. Science (80–) 341(6153):1238670–1238670
36. Morris RV, Klingelhöfer G, Schröder C, Fleischer I, Ming DW, Yen AS et al (2008) Iron mineralogy and aqueous alteration from Husband Hill through Home Plate at Gusev Crater, Mars: results from the Mössbauer instrument on the Spirit Mars Exploration Rover. J Geophys Res 113(E12):E12S42
37. Chevrier V, Mathé PE (2007) Mineralogy and evolution of the surface of Mars: a review. Planet Space Sci 55(3):289–314
38. Schmidt ME, Ruff SW, McCoy TJ, Farrand WH, Johnson JR, Gellert R et al (2008) Hydrothermal origin of halogens at Home Plate, Gusev Crater. J Geophys Res 113(E6):E06S12
39. Ruff SW, Farmer JD, Calvin WM, Herkenhoff KE, Johnson JR, Morris RV et al (2011) Characteristics, distribution, origin, and significance of opaline silica observed by the Spirit rover in Gusev crater, Mars. J Geophys Res 22:116
40. Yen AS, Morris RV, Clark BC, Gellert R, Knudson AT, Squyres S et al (2008) Hydrothermal processes at Gusev Crater: an evaluation of Paso Robles class soils. J Geophys Res [Internet] 113 (E6):E06S10 [cited 2018 Nov 7]. https://doi.org/10.1029/2007JE002978
41. Michalski JR, Dobrea EZN, Niles PB, Cuadros J (2017) Ancient hydrothermal seafloor deposits in Eridania basin on Mars. Nat Commun [Internet] 8 (1):1–10 [cited 2020 Aug 24]. www.nature.com/naturecommunications
42. Golden DC, Ming DW, Morris RV, Graff TG (2008) Hydrothermal synthesis of hematite spherules and jarosite: implications for diagenesis and hematite spherule formation in sulfate outcrops at Meridiani Planum, Mars. Am Mineral [Internet] 93 (8–9):1201–1214 [cited 2019 Jun 1]. https://pubs.geoscienceworld.org/ammin/article/93/8-9/1201-1214/44769
43. Hynek BM, Arvidson RE, Phillips RJ (2002) Geologic setting and origin of Terra Meridiani hematite deposit on Mars. J Geophys Res [Internet] 107(E10):5088 [cited 2019 May 26]. https://doi.org/10.1029/2002JE001891
44. McKeown NK, Bishop JL, Noe Dobrea EZ, Ehlmann BL, Parente M, Mustard JF et al (2009) Characterization of phyllosilicates observed in the central Mawrth Vallis region, Mars, their potential formational processes, and implications for past climate. J Geophys Res [Internet] 114:E00D10 [cited 2019 Dec 9]. https://doi.org/10.1029/2008JE003301
45. Bishop JL, Dobrea EZN, McKeown NK, Parente M, Ehlmann BL, Michalski JR et al (2008) Phyllosilicate diversity and past aqueous activity revealed at Mawrth Vallis, Mars. Science [Internet] 321 (5890):830–833 [cited 2019 Apr 8]. http://www.ncbi.nlm.nih.gov/pubmed/18687963
46. McKeown N, Bishop JL, Cuadros J, Hillier S, Amador E, Makarewicz HD et al (2011) Interpretation of reflectance spectra of clay mineral-silica mixtures: implications for Martian clay mineralogy at Mawrth Vallis. Clays Clay Miner 59(4):400–415(16)
47. Brown AJ, Hook SJ, Baldridge AM, Crowley JK, Bridges NT, Thomson BJ et al (2010) Hydrothermal formation of clay-carbonate alteration assemblages in the Nili Fossae region of Mars. Earth Planet Sci Lett 297(1–2):174–182
48. Berger JA, Schmidt ME, Gellert R, Boyd NI, Desouza ED, Flemming RL et al (2017) Zinc and germanium in the sedimentary rocks of Gale Crater on Mars indicate hydrothermal enrichment followed by diagenetic fractionation. J Geophys Res Planets 122(8):1747–1772
49. Morris RV, Rampe EB, Vaniman DT, Christoffersen R, Yen AS, Morrison SM et al (2020) Hydrothermal precipitation of sanidine (adularia) having full Al,Si structural disorder and specular hematite at Maunakea Volcano (Hawai'i) and at Gale Crater (Mars). J Geophys Res Planets 125(9): e2019JE006324
50. Liu Y, Ma C, Beckett JR, Chen Y, Guan Y (2016) Rare-earth-element minerals in martian breccia meteorites NWA 7034 and 7533: implications for fluid–rock interaction in the martian crust. Earth Planet Sci Lett 451:251–262
51. Shearer CK, Papike JJ, Burger PV, Sutton SR, McCubbin FM, Newville M (2011) Direct determination of europium valence state by XANES in extraterrestrial merrillite: implications for REE crystal chemistry and martian magmatism. Am Mineral 96(8–9):1418–1421

52. Warren PH, Wasson JT (1979) The origin of KREEP. Rev Geophys [Internet] 17(1):73 [cited 2020 Oct 29]. https://doi.org/10.1029/RG017i001p00073
53. Tartèse R, Anand M, McCubbin FM, Elardo SM, Shearer CK, Franchi IA (2014) Apatites in lunar KREEP basalts: the missing link to understanding the H isotope systematics of the Moon. Geology 42 (4):363–366
54. Snyder GA, Taylor LA, Halliday AN (1995) Chronology and petrogenesis of the lunar highlands alkali suite: cumulates from KREEP basalt crystallization. Geochim Cosmochim Acta 59(6):1185–1203
55. Usui T, McSween HY, Clark BC (2008) Petrogenesis of high-phosphorous Wishstone Class rocks in Gusev Crater, Mars. J Geophys Res 113(E12): E12S44
56. Shearer CK, Burger PV, Papike JJ, McCubbin FM, Bell AS (2015) Crystal chemistry of merrillite from Martian meteorites: mineralogical recorders of magmatic processes and planetary differentiation. Meteorit Planet Sci 50(4):649–673
57. Morgan JW, Anders E (1979) Chemical composition of Mars. Geochim Cosmochim Acta 43 (10):1601–1610
58. McCubbin FM, Nekvasil H, Harrington AD, Elardo SM, Lindsley DH (2008) Compositional diversity and stratification of the Martian crust: inferences from crystallization experiments on the picrobasalt Humphrey from Gusev Crater, Mars. J Geophys Res 113(E11):E11013
59. Fraeman AA, Korenaga J (2010) The influence of mantle melting on the evolution of Mars. Icarus 210 (1):43–57
60. Grott M, Morschhauser A, Breuer D, Hauber E (2011) Volcanic outgassing of CO_2 and H_2O on Mars. Earth Planet Sci Lett 308(3–4):391–400
61. Morschhauser A, Grott M, Breuer D (2011) Crustal recycling, mantle dehydration, and the thermal evolution of Mars. Icarus 212(2):541–558
62. Balta JB, McSween HY (2013) Water and the composition of Martian magmas. Geology [Internet] 41(10):1115–1118 [cited 2019 Oct 1]. https://pubs.geoscienceworld.org/geology/article/41/10/1115-1118/131070
63. Jones JH (2015) Various aspects of the petrogenesis of the Martian shergottite meteorites. Meteorit Planet Sci 50(4):674–690
64. Wang Z, Becker H (2017) Chalcophile elements in Martian meteorites indicate low sulfur content in the Martian interior and a volatile element-depleted late veneer. Earth Planet Sci Lett 463:56–68
65. Hu S, Lin Y, Zhang J, Hao J, Yamaguchi A, Zhang T et al (2020) Volatiles in the martian crust and mantle: clues from the NWA 6162 shergottite. Earth Planet Sci Lett 530:115902
66. Lee D-C, Halliday AN (1997) Core formation on Mars and differentiated asteroids. Nature [Internet] 388(6645):854–857 [cited 2019 Oct 12]. http://www.nature.com/articles/42206
67. Leone G, Tackley PJ, Gerya TV, May DA, Zhu G (2014) Three-dimensional simulations of the southern polar giant impact hypothesis for the origin of the Martian dichotomy. Geophys Res Lett 41 (24):8736–8743
68. Hedenquist JW, Lowenstern JB (1994) The role of magmas in the formation of hydrothermal ore deposits. Nature 370(6490):519–527
69. Wilson L, Head JW (1994) Mars: review and analysis of volcanic eruption theory and relationships to observed landforms. Rev Geophys [Internet] 32(3):221 [cited 2018 Sep 22]. https://doi.org/10.1029/94RG01113
70. Wilson L (1980) Relationships between pressure, volatile content and ejecta velocity in three types of volcanic explosion. J Volcanol Geotherm Res 8(2–4):297–313
71. Wilson L, Head JW (1981) Ascent and eruption of basaltic magma on the Earth and Moon. J Geophys Res Solid Earth [Internet] 86(B4):2971–3001 [cited 2019 Aug 30]. https://doi.org/10.1029/JB086iB04p02971
72. Head JW, Wilson L (1992) Magma reservoirs and neutral buoyancy zones on Venus: implications for the formation and evolution of volcanic landforms. J Geophys Res [Internet] 97(E3):3877 [cited 2019 Apr 5]. https://doi.org/10.1029/92JE00053
73. von Quadt A, Erni M, Martinek K, Moll M, Peytcheva I, Heinrich CA (2011) Zircon crystallization and the lifetimes of ore-forming magmatic-hydrothermal systems. Geology 39(8):731–734
74. Keays RR (1995) The role of komatiitic and picritic magmatism and S-saturation in the formation of ore deposits. Lithos 34(1–3):1–18
75. Czamanske GK, Moore JG (1977) Composition and phase chemistry of sulfide globules in basalt from the Mid-Atlantic Ridge rift valley near 37°N lat. Bull Geol Soc Am 88(4):587–599
76. Lorand JP (1990) Are spinel Iherzolite xenoliths representative of the abundance of sulfur in the upper mantle? Geochim Cosmochim Acta 54 (5):1487–1492
77. Rossbacher LA, Judson S (1981) Ground ice on Mars: inventory, distribution, and resulting landforms. Icarus [Internet] 45(1):39–59 [cited 2019 Aug 8]. https://www.sciencedirect.com/science/article/pii/0019103581900051
78. Clifford SM (1993) A model for the hydrologic and climatic behavior of water on Mars. J Geophys Res [Internet] 98(E6):10973 [cited 2019 Apr 5]. https://doi.org/10.1029/93JE00225
79. Clifford SM, Parker TJ (2001) The evolution of the Martian hydrosphere: implications for the fate of a primordial ocean and the current state of the northern plains. Icarus [Internet] 154(1):40–79 [cited 2019 Apr 5]. https://www.sciencedirect.com/science/article/abs/pii/S0019103501966710

80. Wilson L, Mouginis-Mark PJ (2003) Phreatomagmatic dike-cryosphere interactions as the origin of small ridges north of Olympus Mons, Mars. Icarus 165(2):242–252
81. Hanna JC, Phillips RJ (2005) Hydrological modeling of the Martian crust with application to the pressurization of aquifers. J Geophys Res [Internet] 110(E1):E01004 [cited 2019 Apr 5]. https://doi.org/10.1029/2004JE002330
82. Head JW, Wilson L (2007) Heat transfer in volcano-ice interactions on Mars: synthesis of environments and implications for processes and landforms. Ann Glaciol 45:1–13
83. Clifford SM, Lasue J, Heggy E, Boisson J, McGovern P, Max MD (2010) Depth of the Martian cryosphere: revised estimates and implications for the existence and detection of subpermafrost groundwater. J Geophys Res [Internet] 115(E7): E07001 [cited 2019 Apr 5]. https://doi.org/10.1029/2009JE003462
84. Bargery AS, Wilson L (2010) Dynamics of the ascent and eruption of water containing dissolved CO_2 on Mars. J Geophys Res [Internet] 115:5008 [cited 2019 Apr 22]. https://doi.org/10.1029/2009JE003403
85. Heggy E, Clifford SM, Grimm RE, Dinwiddie CL, Wyrick DY, Hill BE (2006) Ground-penetrating radar sounding in mafic lava flows: assessing attenuation and scattering losses in Mars-analog volcanic terrains. J Geophys Res 111(E6):E06S04
86. Grimm RE, Heggy E, Clifford S, Dinwiddie C, McGinnis R, Farrell D (2006) Absorption and scattering in ground-penetrating radar: analysis of the Bishop Tuff. J Geophys Res 111(E6):E06S02
87. Boisson J, Heggy E, Clifford SM, Frigeri A, Plaut JJ, Farrell WM et al (2009) Sounding the subsurface of Athabasca Valles using MARSIS radar data: exploring the volcanic and fluvial hypotheses for the origin of the rafted plate terrain. J Geophys Res Planets 114(E8):E08003
88. Picardi G, Plaut JJ, Biccari D, Bombaci O, Calabrese D, Cartacci M et al (2005) Planetary science: radar soundings of the subsurface of Mars. Science (80–) [Internet] 310(5756):1925–1928 [cited 2020 Aug 24]. http://science.sciencemag.org/
89. Watters TR, Campbell B, Carter L, Leuschen CJ, Plaut JJ, Picardi G et al (2007) Radar sounding of the Medusae Fossae Formation Mars: equatorial ice or dry, low-density deposits? Science [Internet] 318 (5853):1125–1128 [cited 2019 Apr 6]. http://www.ncbi.nlm.nih.gov/pubmed/17975034
90. Orosei R, Lauro SE, Pettinelli E, Cicchetti A, Coradini M, Cosciotti B et al (2018) Radar evidence of subglacial liquid water on Mars. Science [Internet] 361(6401):490–493 [cited 2019 Apr 13]. http://www.ncbi.nlm.nih.gov/pubmed/30045881
91. Plaut JJ, Picardi G, Safaeinili A, Ivanov AB, Milkovich SM, Cicchetti A et al (2007) Subsurface radar sounding of the south polar layered deposits of Mars. Science (80–) [Internet] 316(5821):92–95 [cited 2020 Aug 31]. http://science.sciencemag.org/
92. Wallace P, Plank T, Edmonds M, Hauri E (2018) Volatiles in magmas. In: Carroll MR, Holloway JR (eds) Volatiles in Magmas, vol 30. De Gruyter Mouton, 1–517 pp
93. Taran YA, Hedenquist JW, Korzhinsky MA, Tkachenko SI, Shmulovich KI (1995) Geochemistry of magmatic gases from Kudryavy volcano, Iturup, Kuril Islands. Geochim Cosmochim Acta 59 (9):1749–1761
94. Taran YA, Bernard A, Gavilanes J-C, Lunezheva E, Cortés A, Armienta MA (2001) Chemistry and mineralogy of high-temperature gas discharges from Colima volcano, Mexico. Implications for magmatic gas-atmosphere interaction. J Volcanol Geotherm Res 108:245–264
95. Henley RW, King PL, Wykes JL, Renggli CJ, Brink FJ, Clark DA et al (2015) Porphyry copper deposit formation by sub-volcanic sulphur dioxide flux and chemosorption. Nat Geosci 8:210–215. https://doi.org/10.1038/NGEO2367
96. van Hinsberg V, Berlo K, Migdisov AA, Williams-Jones AE (2016) CO_2-fluxing collapses metal solubility in magmatic vapor. Geochem Persp Lett 2:169–177. https://doi.org/10.7185/geochemlet.1617
97. Lorand J-P, Barat J-A, Chevrier V, Sautter V, Pont S (2012) Metal-saturated sulfide assemblages in chaasignite NWA2737; evidence for impact-related sulfur devolatilisation. Meteorit Planet Sci 47:1830–1841
98. Rodríguez A, van Bergen MJ (2016) Volcanic hydrothermal systems as potential analogues of Martian sulphate-rich terrains. Netherlands J Geosci Geologie en Mijnbouw 95(2):153–169
99. Christenson BW, Wood CP (1993) Evolution of a vent-hosted hydrothermal system beneath Ruapehu Crater Lake, New Zealand. Bull Volcanol 55:547–565
100. Delmelle P, Bernard A (1994) Geochemistry, mineralogy, and chemical modeling of the acid crater lake of Kawah Ijen Volcano, Indonesia. Geochim Cosmochim Acta 58(11).2445–2460
101. Delmelle P, Bernard A (2015) The remarkable chemistry of sulfur in hyper-acid crater lakes: a scientific tribute to Bokuichiro Takano and Minoru Kusakabe. In: Rouwet D, Christenson B, Tassi F, Vandemeulebrouck J (eds) Volcanic lakes, Advances in volcanology. Springer. https://doi.org/10.1007/978-3-642-36833-2_10
102. Christenson BW, Reyes AG, Young R, Moebis A, Sherburn S, Cole-Baker J et al (2010) Cyclic processes and factors leading to phreatic eruption events: insights from the 25 September 2007 eruption through Ruapehu Crater Lake, New Zealand. J Volcanol Geotherm Res 191:15–32. https://doi.org/10.1016/j.jvolgeores.2010.01.008
103. van Hinsberg V, Vigouroux N, Palmer S, Berlo K, Mauri G, Williams-Jones A et al (2017) Element

flux to the environment of the passively degassing Kawah Ijen volcano, Indonesia, and implications for estimates of the global volcanic flux. From. In: Ohba T, Capaccioni B, Caudron C (eds) Geochemistry and geophysics of active volcanic lakes. Geological Society, London, Special Publications, p 437. https://doi.org/10.1144/SP437.2
104. van Hinsberg V, Berlo K, Lowenstern J (2020) An experimental investigation of interaction between andesite and hyperacidic volcanic lake water. Minerals 10:1–21
105. Ohsawa S, Saito T, Yoshikawa S, Mawatari H, Yamada M, Amita K et al (2010) Color change of lake water at the active crater lake Aso volcano, Yudamari, Japan: is it in response to change in water quality induced by volcanic activity? Limnology 11:207–215. https://doi.org/10.1007/s10201-009-0304-6
106. Christenson B, Németh K, Rouwet D, Tassi F, Vandemeulebrouck J, Varekamp JC (2015) Volcanic lakes. In: Rouwet D, Christenson B, Tassi F, Vandemeulebrouck J (eds) Volcanic lakes. Springer-Heidelberg. https://doi.org/10.1007/978-3-642-36833-2_1
107. Murphy S, Wright R, Rouwet D (2018) Color and temperature of the crater lakes at Kelimutu volcano through time. Bull Volcanol 80:2. https://doi.org/10.1007/s00445-017-1172-2
108. Inguaggiato C, Iñiguez E, Peiffer L, Kretzschmar T, Brusca L, Mora-Amador R et al (2018) REE fractionation during the gypsum crystallization in hyperacid sulphate-rich brine: the Poás volcano crater lake (Costa Rica) exploited as laboratory. Gondwana Res 59:87–96. https://doi.org/10.1016/j.gr.2018.02.022
109. Inguaggiato C, Pappaterra S, Peiffer L, Apollaro C, Brusca L, De Rosa R et al (2020) Mobility of REE from a hyperacid brine to secondary minerals precipitated in a volcanic hydrothermal system: Kawah Ijen crater lake (Java, Indonesia). Sci Total Environ 740:1–14. https://doi.org/10.1016/j.scitotenv.2020.140133
110. Christenson BW, White S, Britten K, Scott BJ (2017) Hydrological evolution and chemical structure of a hyper-acidic spring-lake system on Whakaari/White Island, NZ. J Volcanol Geotherm Res 346:180–211. https://doi.org/10.1016/j.jvolgeores.2017.06.017
111. Varekamp JC, Ouimette AP, Herman SW, Flynn KS, Bermúdez A, Delpino D (2009) Naturally acid waters from Copahue volcano, Argentina. Appl Geochem 24(2):208–220
112. Rowe GL Jr, Oshawa S, Takano B, Brantley SL, Fernández JF, Barquero J (1992) Using Crater Lake chemistry to predict volcanic activity at Poás Volcano, Costa Rica. Bull Volcanol 54:494–503
113. Rouwet D, Mora-Amador R, Ramírez-Umaña CJ, González G, Inguaggiato S (2017) Dynamic fluid recycling at Laguna Caliente (Poás, Costa Rica) before and during the 2006-ongoing phreatic eruption cycle (2005–10). Geochemistry and Geophysics of Active Volcanic Lakes. In: Ohba T, Capaccioni B, Caudron C (Eds) Geol Soc London, Special Publication, p 437. https://doi.org/10.1144/SP437.11
114. Ohba T, Hirabayashi J, Nogami K (2008) Temporal changes in the chemistry of lake water within Yugama Crater, Kusatsu-Shirane Volcano, Japan: implications for the evolution of the magmatic–hydrothermal system. J Volcanol Geotherm Res 178:131–144
115. Christenson BW (2000) Geochemistry of fluids associated with the 1995–1996 eruption if Mt. Ruapehu, New Zealand: signatures and processes in the magmatic-hydrothermal system. J Volcanol Geotherm Res 97:1–30
116. Takano B, Saitoh H, Takano E (1994) Geochemical implications of subaqueous molten sulfur at Yugama crater lake, Kusatsu-Shirane volcano, Japan. Geochem J 28:199–216
117. Mora-Amador R, Rouwet D, Vargas P, Oppenheimer C (2019) The extraordinary sulfur volcanism of Poás from 1828 to 2018. In: Tassi F, Vaselli O, Mora-Amador R (eds) Poás Volcano, active volcanoes of the world. Springer. https://doi.org/10.1007/978-3-319-02156-0_3
118. Rouwet D, Tassi F, Mora-Amador R, Sandri L, Chiarini V (2014) Past, present and future of volcanic lake monitoring. J Volcanol Geotherm Res 272:78–97. https://doi.org/10.1016/j.jvolgeores.2013.12.009
119. Rouwet D (2021) Volcanic lake dynamics and related hazards. In: Papale P (ed) Forecasting and planning for volcanic hazards, risks and disasters. Elsevier. https://doi.org/10.1016/B978-0-12-818082-2.00011-1
120. Henley RW (2015) Hyperacidic volcanic lakes, metal sinks and magmatic gas expansion in arc volcanoes. In: Rouwet D, Christenson BW, Tassi F, Vandemeulebrouck J (eds) Volcanic Lakes. Springer, Berlin, pp 155–178. https://doi.org/10.1007/978-3-642-36833-2_6
121. Varekamp JC (2004) Copahue volcano: a modern terrestrial analog for the opportunity landing site? Eos 82(41):401–407
122. Kikawada Y, Uruga M, Oi T, Honda T (2004) Mobility of lanthanides accompanying the formation of alunite group minerals. J Radioanal Nucl Chem 261:651–659
123. Klingelhöfer G, Morris RV, Bernhardt B, Schroder S, Rodionov DS, de Souza Jr PA et al (2004) Jarosite and hematite at Meridiani Planum from Opportunity's Mössbauer spectrometer. Science 306:1740–1745
124. Ehlmann BL, Edwards CS (2014) Mineralogy of the martian surface. Annu Rev Earth Planet Sci 42:291–315
125. Tosca NJ, McLennan SM, Clark BC, Grotzinger JP, Hurowitz JA, Knoll AH et al (2005) Geochemical modeling of evaporation processes on Mars: insight

from the sedimentary record at Meridiani Planum. Earth Planet Sci Lett 240:122–148

126. Rodríguez A, Varekamp JC, van Bergen MJ, Kading TJ, Oonk PB, Gammons CH et al. Acid rivers and lakes at Caviahue-Copahue volcano as potential terrestrial analogues for aqueous paleo-environments on Mars. In: Tassi F, Vaselli O, Caselli A (eds) Copahue volcano. Springer, pp 141–172
127. Kato Y, Fujinaga K, Nakamura K, Takaya Y, Kitamura K, Ohta J et al (2011) Deep-sea mud in the Pacific Ocean as a potential resource for rare-earth elements. Nat Geosci 4:535–539
128. Franus W, Wiatros-Motyka MM, Wdowin M (2015) Coal fly ash as a resource for rare earth elements. Environ Sci Pollut R 22:9464–9474
129. Liang H, Zhang P, Jin Z, DePaoli D (2017) Rare earths recovery and gypsum upgrade from Florida phosphogypsum. Miner Metall Process 34:201–206
130. Balaram V (2019) Rare earth elements: a review of applications, occurrence, exploration, analysis, recycling, and environmental impact. Geosci Front 10:1285–1303
131. McLeod CL, Krekeler MPS (2017) Sources of extraterrestrial rare earth elements: to the Moon and beyond. Resources 40:1–28
132. Agee CB, Wilson NV, McCubbin FM, Ziegler K, Polyak VJ et al (2013) Unique meteorite from early Amazonian Mars: water-rich basaltic breccia Northwest Africa 7034. Science 339:780–785
133. Shearer CK, Burger PV, Papike JJ, McCubbin FM, Bell AS (2015) Crystal chemistry of merrillite from Martian meteorites: mineralogical recorders of magmatic processes and planetary differentiation. Meteorit Planet Sci 50:649–673
134. Michard A (1989) Rare earth element systematics in hydrothermal fluids. Geochim Cosmochim Acta 53:745–750
135. Lewis AJ, Palmer MP, Sturchio C, Kemp AJ (1997) The rare earth element geochemistry of acid sulphate and acid sulphate-chloride geothermal system from Yellowstone National Park, Wyoming, USA. Geochim Cosmochim Acta 61:695–706
136. Inguaggiato C, Censi P, Zuddas P, Londoño JM, Chacón Z, Alzate D et al (2015) Geochemistry of REE, Zr and Hf in a wide range of pH and water composition: the Nevado del Ruíz volcano-hydrothermal system (Colombia). Chem Geol 417:125–133. https://doi.org/10.1016/j.chemgeo.2015.09.025
137. Inguaggiato C, Pérez García MA, Meza Maldonado LF, Peiffer L, Pappaterra S, Brusca L (2020) Precipitation of secondary minerals in acid sulphate-chloride waters traced by major, minor and rare earth elements in waters: the case of Puracé volcano (Colombia). J Volcanol Geothermal Res 407:1–14. https://doi.org/10.1016/j.jvolgeores.2020.107106
138. Kalacheva E, Taran Y, Kotenko T (2015) Geochemistry and solute fluxes of volcano-hydrothermal system of Shiashkotan, Kuril Islands. J Volcanol Geotherm Res 296:40–54
139. Kikawada Y, Fukai M, Oi T (2013) Specific REE patterns observed in sulfurous hot springs from a hydrothermal alteration area in Manza, Japan. Procedia Earth Planet Sci 7:428–431
140. Inguaggiato C, Burbano V, Rouwet D, Garzón G (2017) Geochemical processes assessed by Rare Earth Elements fractionation at "Laguna Verde" acidic-sulphate crater lake (Azufral volcano, Colombia). 79:65–74
141. Varekamp JC (2015) The chemical composition and evolution of volcanic lakes. In: Rouwet D, Christenson B, Tassi F, Vandemeulebrouck J (eds) Volcanic Lakes. Springer, pp 93–123. https://doi.org/10.1007/978-3-642-36833-2_4
142. Takano B, Suzuki K, Sugimori K, Ohba T, Fazlullin SM, Bernard A et al (2004) Bathymetric and geochemical investigation of Kawah Ijen crater Lake, East Java, Indonesia. J Volcanol Geotherm Res 135:299–329
143. Inguaggiato C, Censi P, Zuddas P, D'Alessandro W, Brusca L, Pecoraino G et al (2016) Zirconium–hafnium and rare earth element signatures discriminating the effect of atmospheric fallout from hydrothermal input in volcanic lake water. Chem Geol 433:1–11. https://doi.org/10.1016/j.chemgeo.2016.04.002
144. Fulignati P, Gioncada A, Sbrana A (1999) Rare-earth element (REE) behaviour in the alteration facies of the active magmatic–hydrothermal system of Vulcano (Aeolian Islands, Italy). J Volcanol Geotherm Res 88:325–342
145. Dutrizac JE (2017) The behaviour of the rare earth elements during gypsum ($CaSO_4{\cdot}2H_2O$) precipitation. Hydrometallurgy 174:38–46
146. Dreibus G, Haubold R, Huisl W, Spettel B (2008) The loss of K, REE, Th, and U from a Martian and a terrestrial basalt by acidic leaching. Meteorit Planet Sci 43:1895–1908

Explosive Volcanism on Mars

Giovanni Leone

Abstract

Explosive volcanism should be the most common style on a planet characterized by a low atmospheric pressure and by an assumed high volatile content. However, the observations of the surface of Mars show how explosive volcanism is not widespread as it should be in a planet with the above mentioned characteristics. Therefore, one of the two characteristics should be missing. Considering that the atmospheric pressure has been more or less the same all over the Martian history, then the obvious conclusion is that Mars never had such an high volatile content. Yet, low contents of volatile can still trigger explosive volcanism in the low pressure environment of Mars and this is the reason why several pyroclastic deposits have been observed on the surface of the Red Planet.

1 Explosive Volcanism on Mars

The first evidence for volcanism on Mars comes from the images released by the Mariner 4, 6, 7 and 9 missions. These missions, as well as the subsequent Viking missions, revealed how the Martian surface is characterized by volcanoes >10 km-high that have calderas at their summits and lava fields extending over hundreds of kilometers across the surface [1–3]. Geologic mapping of the Martian surface later revealed how these lava fields can extend over thousands of km [4–7]. Initial research was focused on the largest volcanic edifices of Mars (e.g., Olympus, Arsia, Pavonis, Ascraeus, and Elysium Montes), which are associated to extensive volcanic plains forming two main volcanic provinces—Tharsis and Elysium [3, 8]. Morphological, morphometric, and spectroscopic data revealed how the Tharsis, Elysium, as well as the Syrtis Major volcanic provinces were formed by eruption of fluid basaltic lava [2, 9–21]. Besides these large effusive shield volcanoes and associated lava plains, several broad calderas have been discovered in the southern hemisphere of Mars [4, 16, 22–26]. These calderas, indicated as *paterae* in the planetary geologic nomenclature, are different in morphology from Martian large shield volcanoes. Their flanks are low-sloped (generally <1°) and their rims heavily resurfaced by lava overflows. Paterae were also associated to dunes or pyroclastic deposits as an evidence that explosive volcanism may have occurred on Mars [23–29].

After several decades of Mars exploration, the evidence for explosive eruptions still remains significantly smaller than that for effusive ones despite the low atmospheric pressure of the

G. Leone (✉)
Instituto de Investigación en Astronomía y Ciencias Planetarias, Universidad de Atacama, Copiapó, Chile
e-mail: giovanni.leone@uda.cl

G. Leone (ed.), *Mars: A Volcanic World*,
https://doi.org/10.1007/978-3-030-84103-4_8

Martian environment should favour magma fragmentation [30]. On Earth, due to the higher atmospheric pressure, a volatile content higher than 2 wt% would be necessary to trigger an explosive eruption [31, 32]. On Mars, however, only ~0.03 wt% of volatiles are required for such eruptions [30]. This happens because the low atmospheric pressure at the surface, which averages around ~2–6 mbars as measured at Gale crater [33], should cause a rapid expansion of any gases dissolved within the ascending magma [30]. This low pressure would favour explosive volcanism for smaller volumes of dissolved gases than it is common on Earth, and it has been theorized that all steady eruptions under present-day Martian conditions should be explosive [34]. On Earth, volcanic structures formed by explosive volcanism are common and widespread. Examples include scoria cones, also known as cinder cones, kilometer-sized explosive edifices formed by accumulation of pyroclasts resulting from low-energy volcanic eruptions of Strombolian type; together with phreatomagmatic edifices in the form of tuff rings, tuff cones and maars, they represent the overwhelming majority of terrestrial volcanic edifices situated onshore [35, 36]. On Mars, however, such landforms are surprisingly scarce [29, 37].

After the progress in our understanding of the evolution of Mars in the last several decades, the reason for this scarcity might be due to the dry mantle of Mars [38–52]. The aim of this chapter is therefore summarising the available knowledge about explosive volcanism on Mars and providing pathways to resolving this outstanding issue. First, a general background on explosive volcanism is provided in the following subsection followed by an overview of localities where the presence of surface manifestations of explosive volcanism have been proposed based on remote or in-situ investigations. Then, the effect of the Martian environment on volcanic eruptions, pyroclastic transport, and the way this would change the expression of resulting surficial features is discussed. Finally, we address the way explosive volcanism might have affected the evolution of Mars. This overview includes explosive eruptions resulting from silicate melts containing water or CO_2 as the main constituents of the volatiles proposed for Mars [53]. In discussing how volcanism has evolved, I use the geochronologic timescale for Mars [54], which is divided up into four periods of decreasing age, Pre-Noachian (4.567–4.1 Ga), Noachian (4.1–3.7 Ga), Hesperian (3.7–3.1 Ga), and Amazonian (3.1 Ga to present).

2 Effect of Gravity and Atmospheric Pressure on the Volcanic Explosive Activity

The specific environment of Mars plays a crucial role in determining the type of volcanic activity and hence the resulting shapes of volcanic landforms. Mars is a planet with lower gravity and atmospheric pressure than Earth, 3.71 versus 9.81 m/s^2 and ~200–600 versus ~10^5 Pa respectively [33]. Clearly, such a difference in atmospheric pressure affects volcanic eruption styles and the shape of the resulting volcanic features that may significantly vary from their terrestrial counterparts [30, 35, 55]. The lower gravity affects the depth at which magma reservoirs form, the crystal settling processes within them, and the motion of the melt through the surrounding rocks [30]. Larger magmatic bodies can be formed on Mars thus increasing the effusion rates of the volcanic eruptions as well as the amount of material that can be transported to the surface [56]. Finally, lower gravity increases the travel distance of the ejected material [30, 35, 55]. The lower atmospheric pressure increases the depth at which volatiles leave magma and also affects the distance to which the volcanic material can be ejected [53]. The drag on the ejected particles is linearly proportional to the atmospheric density, which is significantly reduced on Mars, hence allowing ejected particles to travel at longer distances [30, 55, 57]. Also, the lower atmospheric pressure favours the initial ejection velocities of pyroclasts [30]. It is estimated that velocities in Martian strombolian eruptions could reach ~2 times the typical velocities on Earth, corresponding to a mean particle velocity of

~92 m/s [55]. Also, the dispersion of ejection angles of pyroclasts should be wider than on Earth [58] hence reducing the height of resulting volcanic features in comparison with their terrestrial counterparts [55]. It was also proposed that dilute pyroclastic density currents should move faster in the Martian atmosphere, thus being able to overcome topographic obstacles up to 5 km high [59] and travelling dozens of kilometers far away from the source area [24, 30]. In addition, the longer distances to which the pyroclastic material can travel on Mars allow more cooling during transport thus producing poorly consolidated, widely dispersed, fine-grained deposits [30]. However, concentrated pyroclastic density currents might remain so hot that they would allow welding at ranges of hundreds of kilometres from their source [24].

Even magmas with very low amounts of gases (~0.1 wt%) can be significantly fragmented on Mars, sufficient to eject pyroclasts at speeds of at least a few tens of meters per second [31]. Consequently, even magmas that do not have sufficient dissolved volcanic gases to trigger explosive eruptions on Earth can cause magma fragmentation on Mars [58]. Numerical modelling shows that ejected particles forming Martian scoria cones should be around ~20 times finer than those typical on Earth, corresponding to a mean particle size of ~2 mm on Mars [55]. Due to the low atmospheric pressure, basaltic Plinian eruptions should be very common on Mars. Even though they are rare for terrestrial basalts with higher columns than their terrestrial counterparts [30, 58]. However, the heights of the Plinian columns on Mars would also have limitations, convecting columns would not be able to rise above 20 km due to the low density of the Martian atmosphere [60]. As a result, the distribution of large volumes of pyroclasts over wide areas should be precluded. On the other hand, the stability of Plinian columns should favour the frequent formation of pyroclastic density currents [24, 60, 61]. The difference of altitude between the lowest and the highest point of Mars is around 30 km, which causes a significant variation in atmospheric pressure, and hence this has a profound effect on the volcanic eruption depending on where this occurs [58]. A Plinian eruption occurring at the bottom of Hellas would be denser and higher than a similar volcanic eruption occurring on top of Olympus Mons. Higher ejected pyroclasts can be distributed by the wind over a wider area thus forming thinner deposits than those known on Earth [28, 62, 63]. Thus, the low pressure of Martian atmosphere may explain the paucity of Plinian eruption deposits recognized so far on Mars [34].

Finally, the low atmospheric pressure and gravity of Mars may also affect the shape of kilometer-sized volcanic features by making their slopes more gentle [57]. On Earth, flanks of young uneroded scoria cones reach an angle of repose of 30°, additional scoria is redistributed by avalanching once this threshold is overcome [64]. On Mars, scoria cones do not reach the angle of repose, regardless of whether this angle is gravity dependent or not, because ejected material spreads over a much larger area than on Earth [65, 66]. As a result, typical erupted volumes on Mars are not large enough for the flank slopes to attain the angle of repose on Mars in contrast to Earth [55, 57].

3 The Effect of the Volcanic Volatiles on the Atmosphere

A better understanding of the nature of past volcanism on Mars and its interaction with the atmosphere is of key importance in assessing the planet's climate evolution. The first question is whether the atmosphere was mainly formed by volcanic contributions or was in major part primordial. Given that the origin occurred in the early past of Mars, we can only make assumptions and see whether these are consistent with the evidence on the ground based on the available data. It has been proposed that the early atmosphere of Mars may have had a pressure greater than the current one [67, 68], perhaps between 140 and even 500 mbar [69], and that the CO_2 lost from the atmosphere to space would account for 800 mbar up to 2000 mbar [70, 71]. The weak gravity together with the demise of magnetic field, which was operating prior to ~4.1 Ga ago [72, 73], contributed to the loss of

CO_2 and other volatiles to space or by sequestration into rocks [71]. Such a loss is currently ongoing at a rate of ~2–3 kg/s as revealed by MAVEN data, although in the past would have been greater [70]. In consideration that the peak of the volcanic eruptions coupled to a strong degassing of volatiles occurred in the first 500 Ma of the Martian history [73], as also confirmed by isotopic studies [74–77], the evidence on the ground shows how such a postulated change of pressure may have somehow inhibited explosive activity. On the other hand, such possibility would rule out any recent volcanism on Mars otherwise the degassing effects in a lower atmospheric pressure would have been more devastating with the result of widespread pyroclastic deposits. Calculations made on the current pressure of CO_2, which is the main constituent of the atmosphere, showed that an original content of CO_2 >800 ppm in the mantle would allow an entirely volcanic origin whereas a content of CO_2 <600 ppm would mean an entirely primordial origin [78]. Low concentrations of CO_2 in the Martian mantle, between 50 and 500 ppm [https://www.sciencedirect.com/science/article/pii/S0012821X08002112, https://www.sciencedirect.com/science/article/pii/S0012821X08002112], indicate a primordial origin.

Obviously, if we consider much higher past pressures as mentioned in the previous section, the content of CO_2 present in the mantle must have been proportionally much higher. But what about other common volcanic volatiles like H_2O and SO_2? Depending on the assumed H_2O content in magma (e.g., 1–3 wt%) and the total volume of erupted volcanic materials on Mars (e.g., ~2×10^7 km^3 by effusive and explosive eruptions), previous estimates for the amount of outgassed water range from few tens of meters to over one hundred metres of global equivalence layer (GEL) [79, 80]. Such an amount is more or less the remaining estimated total inventory of water after all that one degassed during the volcanic activity of Mars [81]. On Earth, pyroclastic deposits formed by explosive volcanism are usually linked with magmas of average volatile content, usually ≤2 wt% mainly consisting of CO_2 and H_2O, but also sulphur-bearing species and H_2 amongst others [82, 83].

On Mars, however, only ~0.03 wt% of volatiles are needed at the current atmospheric pressure [30] that means less water would be available for the global inventory. Nevertheless, approximately 0.5 to 2.8 wt% of volatiles within the magma have been suggested for the explosion of Hecates Tholus in order to form an eruption plume of sufficient height to deposit the mapped 50 × 75 km pyroclastic deposits via airfall [84]. Volatile contents of 1 wt% of CO_2 and H_2O have been proposed for the volcanic deposits located in the southern Elysium region, which would imply ~10^{16} kg (i.e., almost the entire mass of today's atmosphere) of each species as well as other compounds (10^{15} kg S, 10^{13} kg Cl, 10^{13} kg F) [85]. Assuming CO_2 contents of ~0.05 wt% in a model of thermal mantle evolution and dike propagation through a 200 km thick lithosphere, it was suggested that Martian volcanism declined from a magma eruption rate of 0.17 km^3/yr during the Hesperian to around 1×10^{-4} km^3/yr at present implying CO_2 outgassing rate decreased from 8.8×10^7 to 6.6×10^6 kg/yr [86]. Other thermal models end any volcanic activity by the Hesperian [41, 73, 87]. Other estimates, which take into account the mantle oxygen fugacity, limit the CO_2 content in Martian magmas to ~0.1 wt% [45, 88].

However, as indicated by Martian meteorites, the early Martian mantle probably was more volatile-rich, possibly ~1 wt% or more, followed by a depletion over time caused by a lack of recycling via plate tectonics [89–91]. Below, I use the currently estimated inventory of all proposed pyroclastic deposits on Mars (Fig. 1; Table 1) to update the estimated total outgassed volatile volumes on Mars, based on a minimum scenario of 0.03 wt% combined volatile content and a conservative Earth-like value of 2 wt% [82]. Although pyroclastics can be porous, I assume that compaction throughout the majority of the depositional stacks in the extensive and relatively thick deposits listed in Table 1 results in an average, basalt-like density of 3.0 kg/cm^3. This would imply ~2.55 to 2.98×10^{19} kg of pyroclastic material, which corresponds to ~12 to 14% of the entire Martian inventory of extrusive volcanic material (i.e., including the

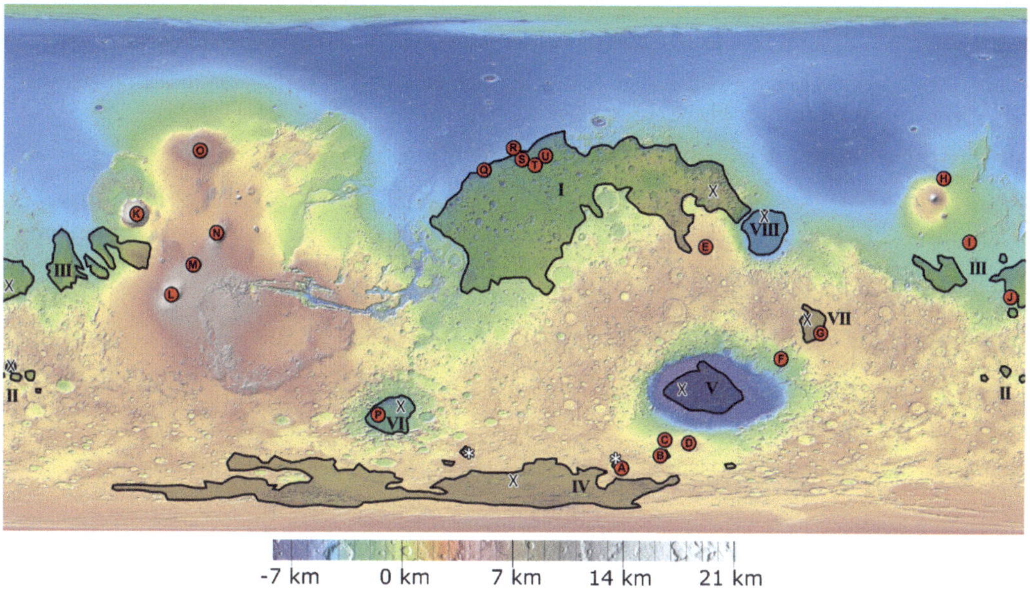

Fig. 1 Global MOLA topographic map showing deposits larger than 10^5 km^2 of potential pyroclastic origin (black outlines; see Table 1 in Sect. 7 for Roman numerals). Black crosses and white asterisks indicate the locations shown in Figs. 8 and 9, respectively. Red dots mark potential centers of explosive volcanism as mentioned in the references listed in Table 1. Letters are: (A) Pityusa Patera; (B) Malea Patera; (C) Peneus Patera; (D) Amphitrites Patera; (E) Nili/Meroe Paterae; (F) Hadriaca Patera; (G) Tyrrhena Patera; (H) Hecates Tholus; (I) Cerberus Fossae; (J) Apollinaris Patera; (K) Olympus Mons; (L) Arsia Mons; (M) Pavonis Mons; (N) Ascraeus Mons; (O) Alba Patera; (P) Argyre Mons; (Q) Eden Patera; (R) Oxus Patera; (S) Ismenia Patera; (T) Siloe Patera; (U) Euphrates Patera. Freely adapted and updated from Tanaka [100] and Kerber et al. [62]

wrinkle-ridged plains), as mapped by remote sensing studies [92]. Multiplied by 2 wt%, this results in a mass of $\sim 10^{15}$ to $\sim 10^{17}$ kg of volatiles that might have been produced by explosive eruptions. According to the upper limit of 0.1 wt% for CO_2 in Martian magmas, this would result in $\sim 10^{16}$ kg or ~6–7 mbar CO_2 [88]. An H_2O content of 2 wt% on the other hand, would result in $\sim 10^{17}$ kg or ~2 m GEL of water. The potentially outgassed CO_2 roughly equals the current mass of Mars' atmosphere. Nevertheless, applying a limit of 0.1 wt% on the currently known potential pyroclastic deposits seems to imply that CO_2 alone from explosive eruptions was an insufficient source for significant climate change as models indicate pressures of several hundreds of millibars of CO_2 are needed for any appreciable temperature increase [93, 94]. Furthermore, the greenhouse properties of sulfur-bearing species such as SO_2 and H_2S that might have been outgassed alongside the CO_2 were probably more than counteracted by the formation of sulfate aerosols causing a cooling effect on the atmosphere [94–98]. Despite this effect, some authors think that the effect by collision-induced absorptions between CO_2, H_2/H_2O and CH_4 outgassed by volcanism might have caused significant warming [93, 99]. Even assuming that H_2 was not outgassed directly (which would require a reducing mantle), but only came from photo-dissociation of outgassed H_2O, the above-mentioned 2 wt% H_2O would result in ~9 mbar of H_2. This would be a significant contribution to the 10s of mbar partial H_2 pressure that models require to raise the average surface temperature above 273 K in a 1.5 bar atmosphere [99]. For an ideal end-member scenario, in which all the potential deposits listed in Table 1 are indeed pyroclastic in nature, I therefore conclude that the CO_2 outgassing from explosive volcanism alone was probably not a significant climate-altering parameter by itself, but that its combination with outgassed H_2/H_2O and CH_4 might have

Table 1 List of possible Martian deposits distributed over territories larger than 10^5 km^2 (black polygons in Fig. 1), for which a pyroclastic origin has been proposed in previous literature (sorted by minimum volume estimate)

Name	Approximate extent (10^6 km^2)	Estimated volume (10^6 km^3)	Indicative observations								Estimated crater count time frame	References
			1	2	3	4	5	6	7	8		
(I) Arabia deposits	5	<5	X	X	X	X			X		Early to mid-Noachian (?)	[26, 100–102]
(II) Electris deposits	>1.8	1	X	X	X	X		X	X		Late Noachian	[100, 103, 104]
(III) MFF	2.5	1–1.4	X	X	X	X	X				Hesperian	[105–111]
(IV) Dorsa Argentea	1.35	0.68–1.35	X	X	X		X		X		Hesperian	[112, 113]
(V) Hellas deposits	0.54–1.4	0.48–1.1	X	X	X	X					Early Hesperian	[114–117]
(VI) Argyre deposits	0.34	0.34–0.68	X	X	X						Early Hesperian to early Amazonian	[118]
(VII) Tyrrhena Patera deposits	0.2–0.32	0.18–0.22	X	X	X						Hesperian (?)	[119, 120]
(VIII) Isidis deposits	0.7	0.03–0.26	X				X			X	Late Hesperian to early Amazonian	[121, 122]

Indicative observations are number-coded as follows: (1) Proximity to potential vent(s); (2) Friability; (3) Layering; (4) Small grain-size; (5) Mantle-like draping over topography; (6) Crater deficiency implying resurfacing; (7) Dark outcrops; (8) Potential degassing structures. Representative CTX images of each deposit are shown in Fig. 8

significantly contributed to atmospheric warming. However, even in this endmember scenario, the 500 Ma time span of peak volcanism over which the potential pyroclastic deposits of Mars were emplaced [73] would result in a very low average production rate of ~0.9 mbar H_2 per 100 million years.

Salts present on the surface of Mars were also proposed as formed by dry deposition from volcanic gases present in the atmosphere via 1-dimensional photochemical reactions, sulphates and nitrates are consistent with estimated emissions whereas perchlorate salts are several orders of magnitude more than any justifiable volcanic emission [123]. Outgassed sulphur-species could also have combined with outgassed H_2O, CO_2, and other compounds to form H_2SO_4, HCl, or HF, which might have significantly contributed to the chemical alteration of the Martian surface [92]. Chemical weathering in a dry or water-limited environment is an alternative explanation [124–127]. At last, volcanic eruptions injected noble gases and their isotopes into the atmosphere, which are then partially lost to space at different rates via sputtering at high altitudes and can therefore be used to assess atmospheric loss over time [128, 129]. The deviation of measured argon and neon isotope volume relations in current Martian atmosphere from the expected values indicates that the atmosphere once contained at least 10 times the volume of these species, hence implying substantial sources that cannot be explained with volcanic outgassing alone [128].

4 Magma Fragmentation

The eruptive style (effusive or explosive) depends on various factors such as magma viscosity, gas loss, conduit geometry [130], and decrease of exsolved volatile pressure [131, 132] that might lead to magma fragmentation, although explosive eruptions are not always a consequence of magma fragmentation [133]. In principle, there are two ways to fragment magma. One is associated with juvenile volcanic gases dissolved within the magma, henceforth referred to as a "dry" or "magmatic" process, and the other one depends on the availability of external volatiles that magma can interact with during its ascent, henceforth referred to as a "wet" or "hydromagmatic" process, being the latter characterized by a much greater fragmentation efficiency [134–141]. The dry process occurs when magma produces bubbles of compressible volcanic gases from its juvenile content in volatiles, on Earth predominantly H_2O and CO_2 but also SO_2 or H_2S, which can rapidly increase in volume as magma is decompressed [82]. The availability of these volcanic gases depends on the various environmental characteristics of each planetary or satellite body. For example, on Io the most common volcanic volatile, at least from what detected so far, today is SO_2 [142, 143]; on Mars probably was a mix of CO_2 and H_2O [53] when it was still active but today the stable volatile on the surface is CO_2 with only ~20–200 ppm of H_2O in the atmosphere [81, 144–148].

Rapidly expanding gases might fragment the magma or may cause a buildup of gas pressure in a high viscosity melt. Gas pressure can viscoelastically deform the magma and finally produce glassy-fractures resulting in violent explosions [149]. The "wet" process is dependent on the amount of external volatiles encountered during its rise to the surface. On Earth the most common volatile is H_2O, for this reason this type of process is often termed as "hydromagmatic". On Earth, magma can encounter groundwater, ground ice, or liquid water at the surface [150]. However, on other bodies within the Solar System different volatiles, mainly CO_2 or SO_2, may contribute to explosive volcanism [53, 84, 151–153]. Volcanic gases, either juvenile or entrained, are able to explode rock fragments (pyroclasts) from the vent to various distances according to their sizes either ballistically [154–156] and/or by turbulent jets [58, 157]. These include Vulcanian or Strombolian style eruptions on other bodies of the Solar System [158, 159]. The pyroclasts fall is influenced by two main factors to form kilometre-sized positive topographic features, like scoria cones or spatter cones: *local temperature* and *accumulation rate*; "features commonly formed at or near the vent include cinder cones, spatter cones, root-less flows, lava ponds, and lava flows" [160]. The build-up within more viscous magma can trigger more energetic eruptions of Plinian, sub-Plinian or Vulcanian types causing the formation of finer pyroclasts, which can be entrained in buoyant convective plumes of hot gas in the atmosphere. Erupted material is then transported far away from the vent by the wind [28, 62, 63]. The rhyolitic composition likely coming from the fallout of these plumes was even found within Gale crater [161–163], which is almost ~2000 km equidistant from Apollinaris Mons, Elysium Mons and Tyrrhenum Mons [164].

Wet processes are generally more energetic than dry processes because there is the conversion of water (or other volatile compound) from liquid/solid to steam. Hydromagmatic fragmentation is also due principally by brittle mechanisms, in which length scale and geometry of fracturing are controlled by the bubble population; this suggests that the fragmentation efficiency "is driven, at least in part, by brittle disintegration of vesicular pyroclasts due to the high thermal stress generated during rapid cooling" [141]. This process causes both magma and surrounding rocks to fragment and subsequently transported from the vent ballistically or by convective plumes out in the atmosphere [165]. Depending on the volume of ejected material and on the dynamic evolution of the explosive eruption, a tuff ring, a tuff cone or a maar could form [150, 166– 168]. Another type of water-lava interaction can lead to the formation of the rootless cones, formerly termed as pseudocraters, these cones are known as rootless due to the lack of a direct feeder from the ground [169–172]. They form when erupted lava flows over surface volatile deposits capable of rapid expansion, this occurs with SO_2 on Io [173] or CO_2 brine that is stable on the surface of Mars [144–146]. It was even suggested that the eruption might be particularly catastrophic if CO_2 is involved [114]. The condensation and sublimation of CO_2 from and to the atmosphere is also the main cause of its variation in pressure [174]. Ground water ice is not likely the cause, due to its instability and inability to form stable deposits on the surface of Mars [81, 147, 175], unless it is located at depths where the lithostatic pressure allows its stability [81, 176]. The resulting near circular edifices commonly form in groups containing hundreds or thousands of edifices between a few tens up to a few hundreds of meters in diameter. They often have a well-developed central crater with partially or fully preserved rim [172]. Both mechanisms can also spread the volcanic material over a large area causing a formation of variously sized pyroclastic deposits instead of positive conical features.

5 The Surface Expression of Explosive Volcanism

This section will focus on different types of explosive eruptions on Mars (Magmatic eruptions: Hawaiian lava fountains, Strombolian, Vulcanian, and Plinian; Phreatomagmatic eruptions: Surtseyan and Subglacial).

Volcanic activity has played a fundamental part in the evolution of Mars, particularly in the first 500 Ma of its history [73, 176], forming twelve alignments of volcanic features as a unique case in the whole Solar System crossing various volcanic provinces [16]. These provinces contain numerous volcanic vents, from which countless lava flows emerged [177], suggesting that effusive volcanism was a dominant style of volcanic activity, particularly early in Mars' history [73, 176]. On the other hand, various sites on Mars, mainly the southern highlands, contain volcanic edifices, also characterized by explosive activity [23, 28, 178]. While the volcanic fields around Tharsis and Elysium seem pristine and hence young [179], a stratigraphic reassessment of the volcanic units coming from Sirenum Mons and Arsia Mons [176] and the analysis of the ages along the main volcanic alignment of Tharsis showed how Arsia, Pavonis, and Ascraeus Mons formed earlier than the cratered Sirenum Mons and Sirenum Tholus located more to the south in the highlands [16]. The older age of the Tharsis volcanoes was also inferred from the study of their tectonic features [180]. Furthermore, the analysis of the isotopic hydrogen showed how Mars lost much of its water during the first 500 Ma of its history [74–77] due to the strong degassing produced by the early intense volcanism [81, 176]. Such a strong degassing unequivocally suggests that explosive volcanism should have been an important process in the early stages of the Martian history. This section contains a summary of putative explosive edifices of various origins (Subsects. 5.1 and 5.2), as well as putative pyroclastic deposits (Sect. 7), observed so far on the surface of Mars. The distribution of the putative pyroclastic fields shows that explosive activity involved a potentially broad family of features with various ages and global occurrence (Fig. 1). This section also describes Martian features for which an explosive origin or activity has been proposed; namely paterae, tholi, scoria cones, spatter cones and hydrovolcanic features, in decreasing order of size and thus amount of ejected material required for their formation.

5.1 Paterae

Paterae can be up to hundreds of kilometres wide and have been interpreted as volcanic edifices due to the presence of shallow, low-relief, central summit calderas and heavily eroded flanks with very gentle slopes (<1°) on which numerous slope-parallel channels can be observed [181]. The first interpretations of pyroclastic activity for Hadriacus and Tyrrhenus Montes (formerly named Hadriaca and Tyrrhena Paterae), were made on the basis of Viking imagery [23, 24, 119]. These first interpretations were later confirmed thanks to Mars Global Surveyor (MGS) and Mars Express imagery [25, 120]. The availability of Mars Reconnaissance Orbiter (MRO) imagery allows a better resolution and the opportunity to refine previous interpretations. The heavy erosion of the flanks may imply that they might consist of friable material, which might not be always the case if effusive activity buried previous pyroclastic deposits. Flow-like features spreading from the central caldera of Hadriacus Mons are visible, which suggests that a transition from dominantly explosive to dominantly effusive eruptions have occurred over time [182]. Some outcrops of friable material located below the latest lava flows (Fig. 2a) suggest that the eruption style changed from explosive to effusive during the latest stages of the evolutive history of Hadriacus and Tyrrhenus Montes. Their flanks and their summits are mainly characterized by lava flow features that can be better seen now at the higher resolution of the MRO imagery. For this reason, we can still confirm the previous interpretations that the flanks and the summit of both Hadriacus Mons and Tyrrhenus Mons might consist of accumulations of pyroclastic material emplaced by airfall and/or gravity-driven pyroclastic flows triggered by water-magma interactions at depth. However, it is harder to say the original extent of the pyroclastic deposits given that subsequent lava flows covered almost everything leaving only some scattered outcrops.

A group of four additional paterae are found aligned to Hadriacus and Tyrrhena Montes on the other side of Hellas [16]: Peneus, Malea, Amphitrites and Pityusa Paterae. They are located on Malea Planum, an ancient part of the southern highlands south of the Hellas basin, and have been interpreted as characterized by pyroclastic activity [115] and some of them (Malea Patera and Pityusa Patera) where proposed to contain ignimbrites [116]. All four of these calderas show signs of heavy modification on their rims, interpreted as periglacial-permafrost, fluvial and aeolian processes [116, 183]. High-resolution MRO images show abundance of lava overflows from them (Fig. 2b) so that the interpretation of modification by lava seems more likely. While Amphitrites Patera (Fig. 2b) is located on a broad topographic rise similar to Hadriacus and Tyrrhenus Montes, the remaining three paterae are characterized by little or no appreciable positive topographic relief (Fig. 2b). Together with Hadriaca Patera and Tyrrhena Patera, these highland paterae form the Circum-Hellas Volcanic Province (CHVP) that covers more than 2.1 million km^2 [116, 184]. The southern part of Hellas Planitia and the northern part of Malea Planum are covered by dissected deposits interpreted as pyroclastic deposits erupted from Amphitrites and Peneus Paterae, from these paterae came also the volcanic filling that formed the wrinkle-ridged plains of the Hellas basin floor [117, 185].

The estimates made through crater-size frequency distribution suggest that Tyrrhenus Mons and Peneus, Malea, and Pityusa Patera were formed around 3.8 Ga, while Hadriacus Mons and Amphitrites Patera at 3.7 and 3.6 Ga respectively, hence from Late Noachian to Early Hesperian [25, 116]. Stratigraphic relationships show that the CHVP paterae formed after the Hellas impact. The Hellas impact was estimated between 4.1 and 4.05 Ga [186]. Several lava overflows modified flanks and rims of the CHVP paterae and flowed inside Hellas after their formation. In the case of Hadriacus Mons its formation began even before the planetary dynamo died out [187], that is 4.09 Ga [72]. However, Hadriacus Mons is superimposed on the rim of Hellas, which does not show magnetic anomalies [72], sign that Hadriacus Mons formed after Hellas somewhere between 4.09 and 4.05 Ga when the magnetic field was already over. Tyrrhenus Mons is even older as its earliest lava flows reached Gale crater and were dated 4.21 Ga through a K–Ar measurement done by the Curiosity rover [16, 164].

An atypical example of a patera is Alba Mons, formerly known as Alba Patera, an enormous shield with a larger diameter than that of Olympus Mons and is located on the northern edge of the Tharsis volcanic province. The volcano is a steeper shield in the upper part reaching an elevation of only 6.8 km at its highest point [181] and is characterized by wide and low-relief flanks with slopes ranging from 1.4° in the upper part to 0.2° at the bottom part of the edifice [188, 189]. The flanks are covered by long lava flows, some extending nearly 1000 km from the summit region [190], but they are also characterized by low thermal inertia material interpreted as friable pyroclastic deposits suggesting at least an explosive phase in its history [191, 192]. Another volcanic centre known as characterized by pyroclastic activity is Apollinaris Mons, which is

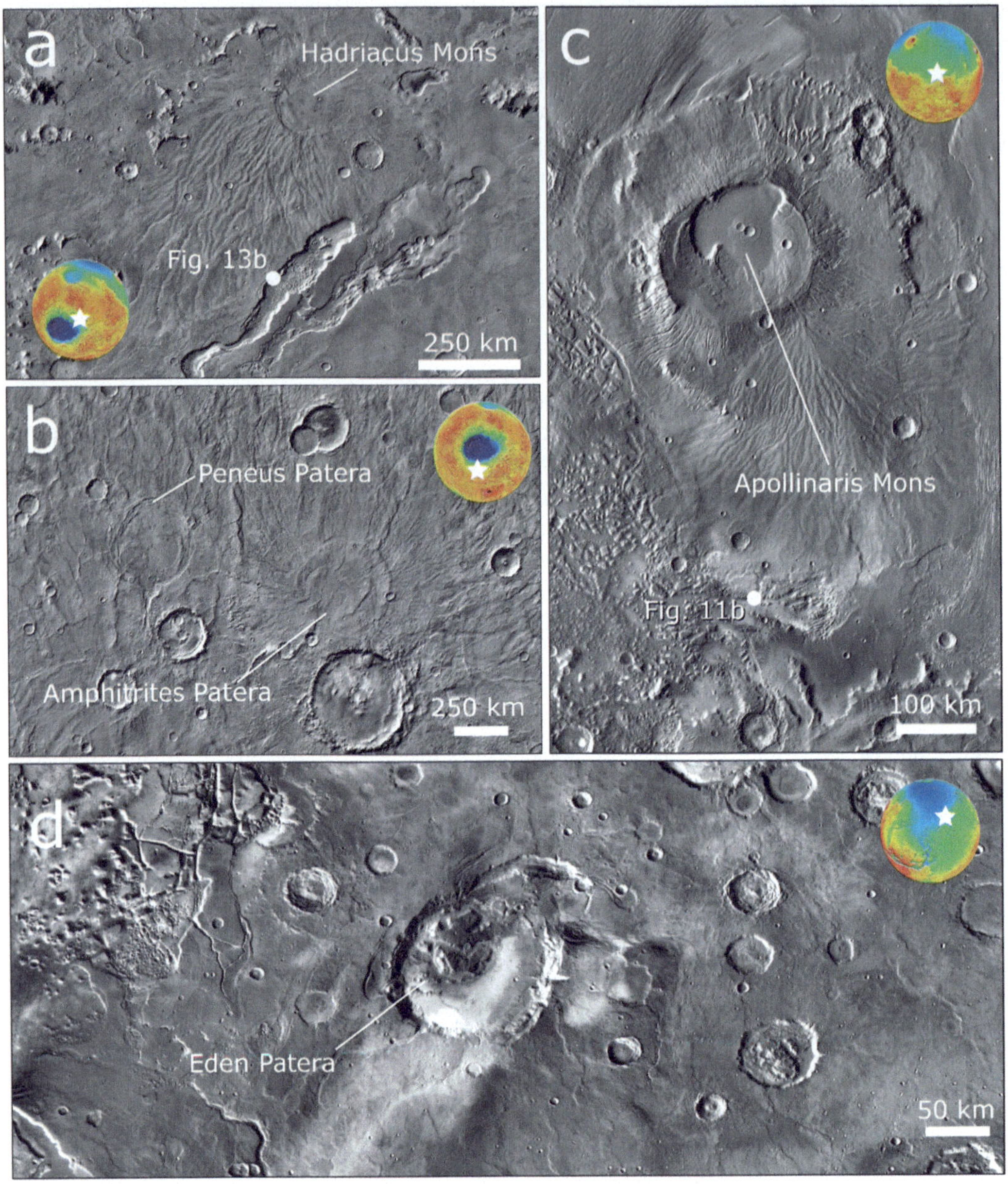

Fig. 2 Image showing examples of paterae on Mars. Panel **a** shows Hadriacus Mons (centred 31.5° S, 91.8° E), **b** Peneus Patera and Amphitrites Patera (centred 58.7° S, 57.4° E), **c** Apollinaris Mons (centred 9.3° S, 174.8° E) and **d** Eden Patera respectively. Note the characteristic heavily eroded flanks and large central calderas on panels **a–c**). All panels are based on THEMIS-day mosaic and north is up in all the images

approximately 190 km-wide and 5 km-high (Fig. 1c), formerly known as Apollinaris Patera and interpreted to be a shield volcano [181]. This volcano is situated near the equator, ~2500 km southeast of Elysium Mons, on the transition topography [193] between northern lowlands and southern highlands. Apollinaris Mons has a ~80 km large, flat-topped summit area interpreted to be a two-stage central caldera, which is breached on the south-eastern side, with an average slope of 1–3°; a large fan of more resistant material appears to originate from this breach and covers the southern flank of the volcano [194]. Based on the resolution of the Viking imagery, the origin of the fan was originally attributed to lavas of unknown composition in contrast with the pyroclastic nature of the edifice [194]. Later, this interpretation and the possibility that Apollinaris Mons is the origin of the pyroclastic deposits of the Medusae Fossae Formation [28] was essentially agreed; it was even proposed that the pyroclastic flows may have formed within the summit area due to hydrothermal activity [195]. This latter hypothesis is more difficult to support given that volatile exsolution starts already at 1 km of depth on Mars [53] and thus cannot occur on its summit area. Age estimates through crater counts suggest that Apollinaris Mons was volcanically active during the Noachian and Hesperian periods [54, 196], involving water-magma interactions and long-lived hydrothermal activity [195]. The volcano may have released more than 10^{15} kg of water vapor into the atmosphere [194] during its activity and large amounts of volcanic ash. Volcanis ashes may be the principal constituent for the formation of some pyroclastic deposits on Mars, such as the Medusae Fossae Formation (MFF) located on the northern and eastern flanks of this volcano [28, 62].

In Arabia Terra several irregularly shaped and low-relief craters, namely Eden (Fig. 2d), Ismenia, Siloe and Euphrates Patera have been proposed as evidence for explosive volcanism [26]. If such an interpretation is correct, these volcanic centres may be the source for the fine-grained, layered deposits that are exposed in various craters throughout Arabia Terra [197, 198], including the characteristic allophane and imogolite found along Mawrth Vallis [199], minerals formed from volcanic ashes [200, 201]. Stratigraphic relationships revealed that they are likely Late Noachian to Early Hesperian in age [26]. Although the volcanic activity of Syrtis Major appears predominantly effusive, a olivine-rich unit situated between Nili Fossae and the north-western edge of Isidis Planitia was interpreted as an ash-fall deposit among other interpretations involving mainly the product of nearby impacts [21, 202, 203]. The ash-fall interpretation was later confirmed by a geomorphological and stratigraphic study, considering the distance ($\sim$1000 km) of this deposit from the closest volcanic centre (the Syrtis Major complex of calderas), the obvious conclusion was that the explosive activity producing the ashes must have come from there [204]. This complex of calderas consists of two separate smaller edifices, Meroe Patera to the south and Nili Patera to the north. The distinctively bright floor of Nili Patera was interpreted either as remnants of welded ignimbrites associated with caldera formation, remnants of lava flows, or part of a felsic pluton [205]. If the ignimbrite interpretation is correct, it might justify the explosive activity associated to the calderas. Syrtis Major might also be the closest source for the locally welded pyroclastic surge deposits that mantled Isidis Basin on which chains of cones formed through the process of devolatilization of these deposits [121]. Despite these observations the evidence for explosive eruptions in association with Syrtis Major remain spatially and temporally limited, with the edifice being dominated by lava flows, which together suggest that explosive volcanism played only a minor role in the evolution of the calderas [206].

5.2 Tholi

The surface of Mars shows other volcanic features of variable sizes interpreted to have had explosive activity. These are the *tholi*, volcanic features characterized by a domical shape [181]. Ceraunius Tholus and Uranius Tholus form the Uranius Group together with Uranius Mons [207], formerly known as Uranius Patera, they are all located within the Tharsis region and were interpreted as mantled by pyroclastic deposits suggesting explosive volcanism [208]. However, the edifice of Uranius Tholus, as well as Ceraunius Tholus, seems more characterized by effusive rather than explosive activity due to the presence of lava flows spreading directly from its summit [207]. No explicit evidence of pyroclastic material was suggested for Tharsis Tholus but rather deposition of wind-blown dust on its flanks [209]. However, explosive activity was included among the possibilities for an episode of magma withdrawal observed within its main caldera [210].

Hecates Tholus (Fig. 3a) was suggested as being characterized by explosive activity because of its lack of impact cratering and because of the presence of valleys that could have been better incised (by water) on volcanic ashes rather than on lava flows [211]. Always on the basis of the low resolution of the Viking imagery a "well-preserved air-fall deposit generated by explosive volcanism on Mars" was proposed "immediately to the west of the summit caldera of Hecates Tholus.." [84]. This conclusion was questioned because "there is no obvious topographic expression associated with the smooth area on the summit of Hecates suggested to be a pyroclastic deposit" [181]. Later, it was suggested that the western flank of Hecates Tholus contains a 10 km wide depression (Fig. 3b), interpreted to be a peripheral caldera formed by an explosive event ~350 million years ago [196, 212, 213]. A geomorphological map of the western flank of Hecates Tholus has included mass wasting among the possible processes forming the depression [214] and thus we investigated on the possibility that such a process might have been facilitated by the friable nature of pyroclastic deposits. In Fig. 3c it is possible to observe how the granulometry of the terrain becomes coarser only close to the rim of the depression with respect to the surrounding slope. This change in granulometry suggests a change in the coherence of the volcanic material. The classic friable material that may be typical of a pyroclastic deposit with respect to the more solid appearance of the slope uphill, which might be composed of effusive lava flows. The HiRISE image PSP_001527_2125/PSP_001527_2125 shows the layer of friable material on the wall of the depression (Fig. 3d). The presence of various small channels along the slope is also indicative of fluid lava flows that may have covered the underlying pyroclastic deposit. The observation of this particular friable material only in the sector where the depression is located strengthens the hypothesis of mass wasting with respect to the hypothesis of a peripheral caldera, which is not a common feature in other volcanic structures of Mars. Actually, calderas are very often well located on the summit of the various volcanic centres.

At last, Zephyria Tholus situated within the Aeolis region. This is a ~36 km wide and ~3 km high circular edifice with a ~8 km wide summit crater. The volcano has steep flanks up to 14° in inclination. Based on its shape, size and stratigraphy it was proposed that this edifice is akin to a terrestrial composite volcano (formerly termed as stratovolcanoes in literature) suggesting alternating explosive and effusive volcanic activity in the early Noachian period [215]. However, this interpretation was based on Viking images accompanied with MOLA altimetry and limited coverage of images obtained by Mars Orbiter Camera of relatively low spatial resolution. A more detailed study using newly available data has yet to be done to confirm or refute this interpretation.

5.3 Shields

A couple of studies linked the origin of the MFF also to the large shields of Tharsis, such as Olympus Mons, Arsia Mons, Pavonis Mons, and Ascraeus Mons [62, 101]. These shields are commonly classified as a typical effusive volcanoes, with postulated episodes of explosive volcanism based more on theoretical rather than observational evidence [58, 216]. Some layering associated to mantling of pyroclastic deposits was observed on Arsia Mons and then extrapolated to Pavonis and Ascraeus Mons [178]. Based on the enormous height of this volcano—Olympus Mons is around 26 km high—it was suggested that the atmospheric pressure would be so low at the summit that most of the eruptions would be explosive even with a small amount of volatiles [30]. In any case, the stratospheric

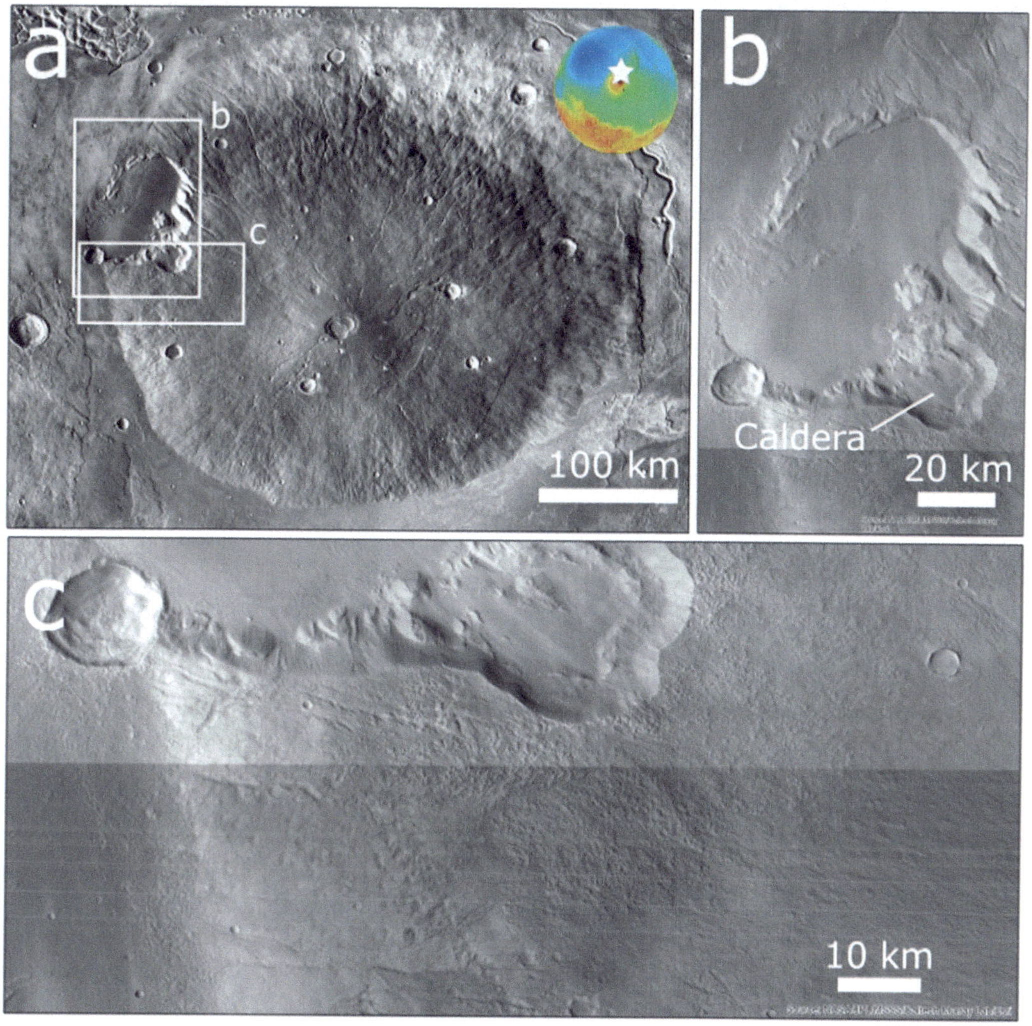

Fig. 3 Image showing an example of a Tholus, Hecates Tholus **a** situated at the northern edge of Elysium Planitia. Panel **b** is a detailed view of the proposed flank caldera is thought to be associated with pyroclastic deposits. Panel **a** is centered at 31.9° N, 150.1° E and uses the THEMIS IR Day 100 m Controlled Partial Mosaic provided by USGS, panel b is based on THEMIS-day mosaic and Murray Beta CTX (Dickson et al. 2018) mosaic and north is up in all panels. Panel **c** CTX mosaic of the western sector of Hecates Tholus where it is located a depression, the limit of the caldera marks the change in granulometry between the terrain surrounding the depression and the remainder of the slope uphill

height of the Tharsis volcanoes would not allow any presence of ice at the summit because it would sublimate [217] before any hypothetical blanket of volcanic ash or dust might isolate it from the very low atmospheric pressure. A detailed study characterizing the morphology and distribution of the dozens of flank vents and grabens on Olympus Mons based on high-resolution images, revealed that effusive eruptions have dominated and that explosive eruptions were rare [218]. This suggests that magmas on Tharsis were already volatile-poor and there were no external volatiles available to trigger the explosive activity at the time of the eruptions.

5.4 Cinder and Spatter Cones

Presence of scoria cones (also known as "cinder cones") was first suggested for several regions of Mars based on the Viking data [2, 8, 35, 219–222] but the interpretations were ambiguous due to the low resolution of available images. Recent high-resolution images allow us to better recognize several sites where Martian scoria cones are present and reveal details on their morphologies and morphometries. Good examples are:

1. On the flanks of the larger shield volcanoes, such as Pavonis Mons [216].
2. As individual cones: East Mareotis Tholus in the Tempe volcanic province [223] along Alignment 3 [16]; and unnamed cones on Syria Planum and Noctis Labyrinthus in the volcanic province of Tharsis [177] along the Alignment 1 and 9 respectively [16].
3. As clusters in large volcanic fields: Ulysses Colles [29, 37, 55] along the Alignment 1 [16].
4. On the floor of Valles Marineris: Coprates Chasma [224, 225], Hydraotes Colles [15, 225, 226] parallel to Alignment 9 [16].
5. On the floor of Dao Vallis [227] along the Alignment 2 [16].
6. At the border of the Martian dichotomy [228] along the Alignment 2 Elysium–Tyrrhenus Mons [16]; at the border between Isidis and Utopia Planitia [229] at the beginning of the Alignment 10 [16] just before Syrtis Major.

Scoria cones were mainly described as subordinate features accompanying other principal processes [177, 226, 229] as mud volcanism [224, 230–233] but the features interpreted as mud volcanoes in reality were just remnant knobs of previous lava fields eroded by subsequent lava flows [15, 16].

MRO high resolution data allows the study and characterization of the shapes of Martian putative scoria cones in detail. The basic characteristics of 40 cones at Hydraotes Colles are that some have central craters and often form small subclusters separated by 5 km. In general, the cones are small, 0.5 to 1.5 km in diameter, with heights between 180 and 230 m, and with a mean slope from 15° to 31° [226]. Additional details came from the study of 29 cones at Ulysses Colles [29, 55]. Detailed morphometrical investigation of these cones revealed they are wider (up to 4080 m), higher (up to 650 m) and more voluminous (up to 4.2×10^9 m^3) than their terrestrial scoria cones, yet have smaller average slopes ranging from 7° to 18° [29, 55]. Despite the evidence of Martian scoria cones, they are not as common as on Earth. Maybe some of them were destroyed or covered by the massive effusive activity on Tharsis [14] but the small number of sites containing these features suggests that pyroclastic volcanism of this type was never dominant on Mars. Furthermore, the issue whether all the cones observed are scoria cones or some might be rootless cones is not resolved yet [15, 55, 226]. "Rootless cone material consists of scoria, spatter and a lesser proportion of lithic material derived from the substrate", which makes still ambiguous distinguishing them from cinder, or spatter, cones on Mars [234] and even on Earth [235]. In order to make a good identification of spatter cones, the spatter particles must have a size above the limit of our current satellite imaging resolution [177]. Despite the difficulty to have such a resolution everywhere on Mars, and even less at the time of the Viking frames, spatter ridges have been reported on the flanks of Alba Patera [236], spatter ramparts on Olympus Mons [218], a spatter cone on the southern flank of Pavonis Mons [223], within Amazonis Planitia [237], and on edifices accompanying the low shield volcanoes within the Tharsis region [177] (Fig. 4).

5.5 Hydrovolcanic Features

Despite the abundant volcanic activity observed and the scarce ground water ice on Mars, particularly at low latitudes [238–244], including its lack of stability on the surface [81], interaction between magma and water was however expected throughout the Martian history [30, 58, 245]. This is because ascending dikes would have to propagate through the putative underground Martian cryosphere and interact with eventual

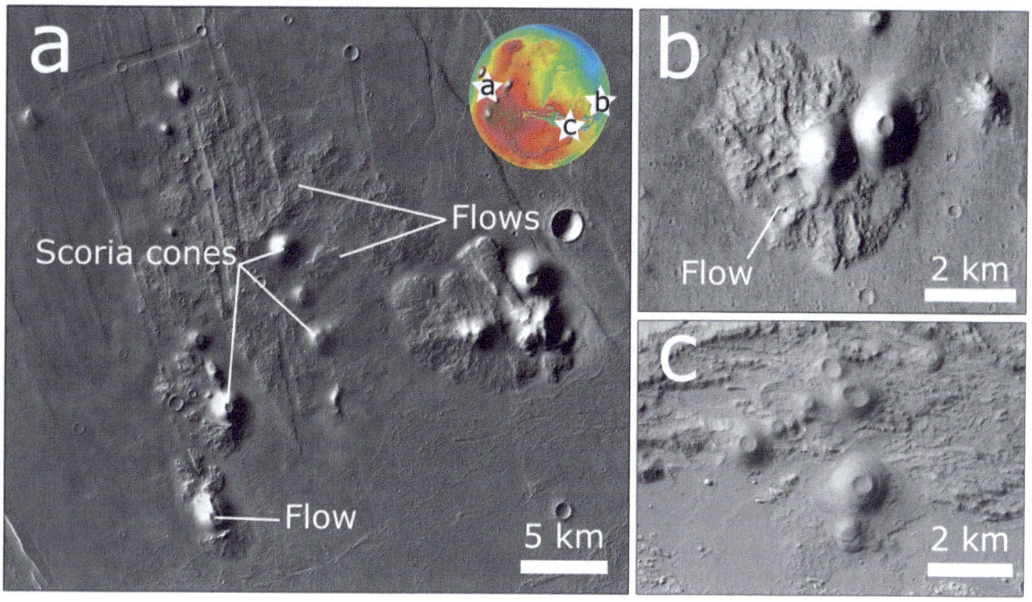

Fig. 4 Image showing three different areas containing promising candidates of martian scoria cones situated within three different areas. **a** A CTX mosaic of a group of cones in Ulysses Colles situated within Tharsis (centered at 5.9° N, 237.1° E), **b** two cones superposed on flow-like unit in Hydraotes Colles situated within Hydraotes Chaos based on CTX image (centered 0.2° N, 326.2° E) and a **c** cluster of small conical features within Coprates Chasma based on HiRISE image ESP_034131_1670 (centered 12.73° S, −297.2° E). North is up in all panels

ground ice [245, 246]. In consideration of the scarce explosive activity observed on the surface of Mars, which shows every geologic process occurred during its whole history, it is logical to conclude that there was less (if any) ground ice than expected [30, 81]. On the water-rich Earth, hydrovolcanism can produce various morphological features depending both on the physical state of the water and the style of the volcanism. The full spectrum of hydrovolcanic edifices have been suggested to be present on Mars [171]: mostly rootless cones and a minority of scoria or spatter cones, tuff rings and tuff cones [37], tuyas [245, 247–249] including lahars on the north-western flank of Elysium Mons [250–252] and in the Galaxias region [253]. On Earth, lahars are triggered by rainfalls [254] or snow present on their summits generally above 4000 m or at latitudes above 35° [255]. The lahar interpretation on Mars is based on the wide misconception that water might exist at its liquid state or as ice on top of the high volcanoes. On Mars, the conditions for the availability of liquid or water ice are not met because it does not rain and the high volcanoes like Elysium Mons are generally at stratospheric height where neither water nor ice can exist [217]. Even at the mid-high latitude of the Phoenix landing site, the ground ice brought to the surface by the excavation of the lander sublimated in 4 sols [242]. Thus, it is more likely that what was observed on the flank of Elysium Mons and in the Galaxias region was a field of lava flows rather than a lahar, lava flows that have carved valleys like Hrad Vallis [18], Valles Marineris and Kasei Valles [15], Mangala Valles [176, 256], Ares Vallis [17] and many others [257–259]. Different situation is for constructional edifices like tuff rings, tuff cones, maars or rootless cones because they rely on underground ice or surface volatiles like CO_2, the latter may form brines that can have limited stability on the surface of Mars [144, 145, 173]. In this case, it is possible albeit not frequent that some lava may have flowed on these brines to form the rootless cones.

Tuyas are a class of hydrovolcanic features reported to be present on the surface of Mars

[245, 247–249]. On Earth, tuyas are flat-topped and steep-sided volcanic edifices formed by subglacial eruptions [249, 260]. When the amount of pillow lava and hyaloclastites completely melts the overburden of ice, explosive activity occurs producing a mound formed mainly by hyalotuff and hydromagmatic tephra [247, 261]. Subsequent slope failures may occur with production of debris avalanches of tephra [247, 262]. Tuyas were widely postulated to occur on Mars in order to explain the abundant mesas commonly found on the Martian surface [169, 171, 249, 263–265], including the lava eroded mounds within Valles Marineris [247], a mound on Arsia Mons [266], the Sysiphi Montes [248], and even to explain the birth of Olympus Mons [267]. However, confirmation of their interpretation as tuyas, such as observations of pillow lavas and columnar jointing, as well the presence of surface water ice [81], is still missing [171]. Hence, this interpretation remains uncertain (Fig. 5).

5.6 Rootless Cones

On Mars, when lava flows over a surface that may contain brines of CO_2 or any other volatile that can be stable on the surface forms rootless cones. The first observations of possible rootless cones on Mars were noted from Viking orbiter images [169, 268] in Acidalia, Utopia, Elysium and Isidis Planitiae, Hephaestus Fossae, the north of Olympus Mons aureole and Arrhenius province [8, 220, 268, 269]. However, the low resolution of the Viking images did not allow for an unambiguous interpretation. Later on, MRO high-resolution imagery provided more detail on the morphology of these cones in other regions [270]. Rootless cones were also identified from mid to northern latitudes (0°–50° N) of Mars [171, 269] justified by the potential presence of volatiles and the evidence of lava flows in this region [6, 271, 272]. Rootless cones in Amazonis Planitia are $\sim$100 m in diameter on average and well preserved compared to those identified in Acidalia and Isidis Planitiae [121, 234].

To understand the surface conditions and to assess the formation mechanism of rootless cones on Mars, terrestrial cones are useful analogues. Icelandic rootless cones groups show comparable morphological appearance and surface characteristics to Mars' cones [237]. The randomness of the spatial distribution of rootless cones was performed by using the nearest neighbour (NN) analysis based on skewness and kurtosis features [273]. According to this analysis, the terrestrial rootless cones are less skewed and have a leptokurtic distribution in comparison to ice mounds. Similarly on Mars, spatial

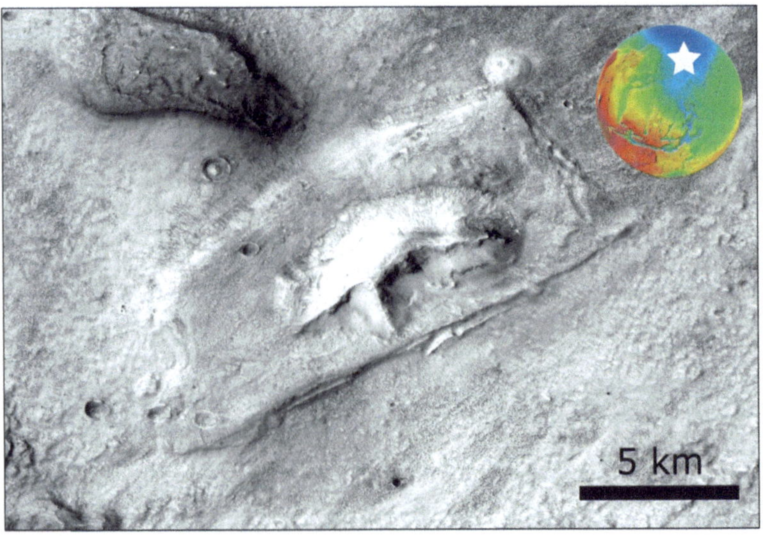

Fig. 5 An example of equivocal Martian tuya from Acidalia Planitia represented by flat-topped and steep-sided mesa with a central knob. CTX image P19_008430_2212_XN_41N036W centred at 41.2° N, 323.57° E, north is up

distribution was used as a tool to distinguish rootless cone groups identified on Cerberus Plains, Acidalia Planitia, Isidis Planitia and Athabasca Valley [274]. Putative rootless cone groups in Cerberus Palus and Central Elysium Planitia align along the direction of the flow in linear chain format and populate densely the center of the Athabasca Valley [275].

For the formation of rootless cones on Mars, a certain amount of CO_2 (or other volatiles) is essential. Thus, the spatial and temporal distribution of rootless cones on Mars can lead to a better understanding of the climatic condition that were present at the time when lava flowed on the surface. Numerical modeling showed how lava flowing above a threshold thickness of the ground, which may contain deeply buried volatile, is not able to produce any phreato-volcanic eruption as the steam stays trapped inside the thick layer of lava [276]. Based on the outcrop and measured thickness of the lava flow in Tartarus Colles region, it was inferred that lava thicknesses between >30 and 60 m with underlying ice layer located at <42 m in depth is optimal to have a rootless cone producing eruption [277]. In this set up, the approximately 4–23 km^3 ground ice should have vaporized and achieved >273 K temperature for $\sim$1300 years. However, the hydrothermal system established by such an event has less endurance in comparison to Earth based analogue [277].

6 Phreatomagmatic Activity

Phreatomagmatic activity, in which magma encounters volatiles during its ascent to the surface is fragmented into fine-grained pyroclasts and then ejected from the vent [278–280], has been proposed to play a role in the formation of various features on Mars [37, 59, 252, 269]. Furthermore, outflow channels on the western flanks of Elysium Mons [251], several linear ridges at the edge of the Olympus Mons aureole materials [246, 281], various martian depressions and pits surrounded by ballistically-emplaced deposits [282, 283], on south-western flank of Ascraeus Mons [284], around the Hellas basin [117] or in the opening of Mangala Valles outflow channel [59]. The evolution of Hrad Vallis, $\sim$800 km long and $\sim$370 m deep lava channel in Elysium Planitia, has been associated with phreatomagmatism [252, 285]. However, the evidence that Hrad Vallis was incised by low-viscosity lavas later challenged this interpretation [18]. The same happened at Mangala Valles [176, 256]. Such previous equivocal interpretation occurred because insufficient image resolution did not allow unambiguous identification of such features before the advent of MRO data. The ambiguity can also be illustrated through an example of cratered cones, which were originally interpreted as tuff cones [8], but high-resolution data revealed no supporting evidence for this interpretation [171]. However, even with the availability of high-resolution data, the interpretation of phreatomagmatic features is still a complicated task. In any case, the observations of constructional phreatomagmatic landforms on Mars are sparse and confirms the paucity of volatiles in the Martian crust.

One of the places where tuff cones and tuff rings were observed is the Nephentes/Amenthes region (Fig. 6), in which a large field of kilometre-sized pitted cones and domes have been described and their morphologies and morphometries have been measured [37], while rootless cones were also observed around Elysium Mons (Fig. 7). These features show some similarities in shape with terrestrial tuff rings and tuff cones but lava flows in the region partially or totally destroyed them so that it is difficult to distinguish them from semidestroyed impact craters. Mud volcanism [230] is also to exclude because of the lack of stability of water at or near the surface [81]. The region is not crossed by any know volcanic alignment but is well flooded by lava flows coming from Syrtis Major through Isidis Planitia, Tyrrhenus Mons through the Amenthes Fossae, and Elysium Mons through Elysium Planitia. This area contains a candidate maar structure on Mars (Fig. 6d). Similar looking conical edifices were also described in Lederberg crater situated on the edge of Chryse Planitia and in the Arena Colles region on the north-western edge of Isidis Planitia [37]. The presence of

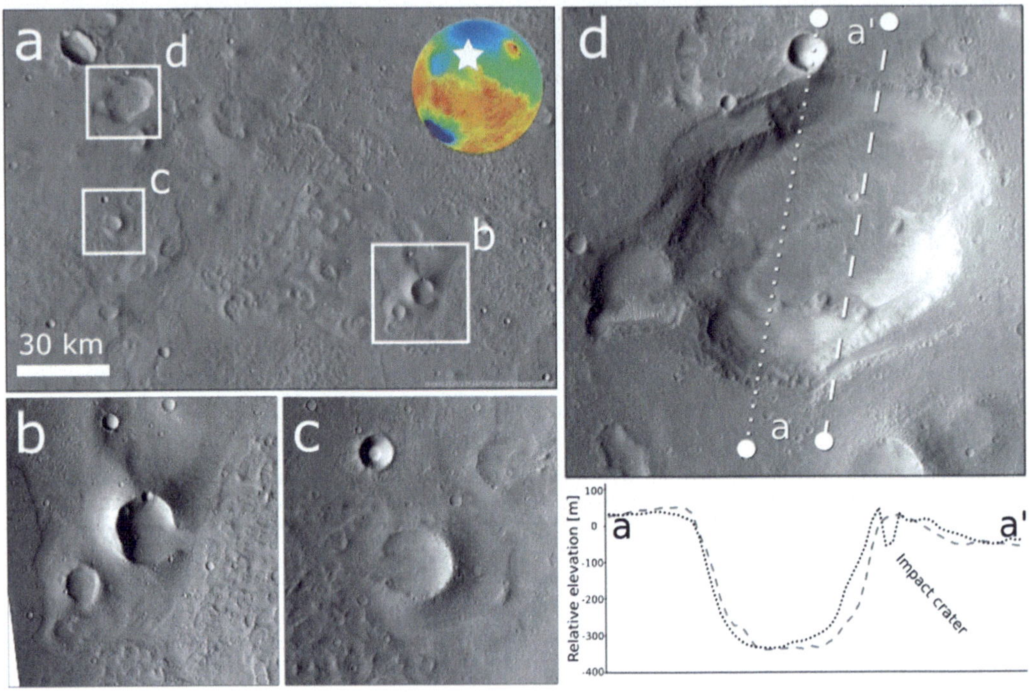

Fig. 6 An example of pitted cones in the Nephentes/Amenthes region (centred 17.4° N, 102.6° E) interpreted as possible martian analogues to terrestrial tuff rings and tuff cones by Brož and Hauber [37] or as sedimentary volcanoes by Skinner and Tanaka [231]. Cones are often breached (**b**, **c**) and they have a wide central crater with floor elevations sometimes below surrounding surface level. Panel **d** shows a non-circular composed depression with slightly elevated rims, which may represent a Martian maar. Topography based on two individual MOLA track profiles. North is up on all panels

maar-diatreme structures and tuff cones has also been proposed in Simud–Tiu Valles [286], Argyre Planitia [118], the floor of Coprates Chasma [225], and Niger and Dao Valles [227]. However, these studies do not contain detailed morphometrical and/or morphological evidence to support their interpretations.

7 Pyroclastic Deposits

Identifying pyroclastic deposits or tephra, the fine-grained products of explosive volcanism, on the surface of Mars even via high-resolution imagery and remote sensing techniques is still a non-trivial problem and alternative formation hypotheses should also be taken in consideration until more ground truth becomes available. So far, the most common observation justifying the interpretation of pyroclastic deposit is the appearance of a friable terrain [105, 194, 195]. This observation must be linked to erosional landforms, yardangs and mass wasting, a small inferred grain-size (based on low thermal inertia), and geographic proximity to potential vent structures [100, 102, 103, 105]. Further characteristics used in order to identify potential pyroclastic deposits were low reflectance layers, potential mafic fines, as well as the dusty mantling of the landscape [84, 100, 102, 106, 107]. Adding up deposits that have previously been interpreted as potentially pyroclastic in origin based on these characteristics (Fig. 1 and Table 1) results in a total area of ~12.4 to −13.4 million km^2, i.e., 8.6 to 9.3% of the martian surface.

Thickness estimates mostly based on stratigraphic and morphometric observations that are occasionally aided by ground-penetrating radar measurements [287,–290] allowed the inference of volumes (Table 1), which potentially total

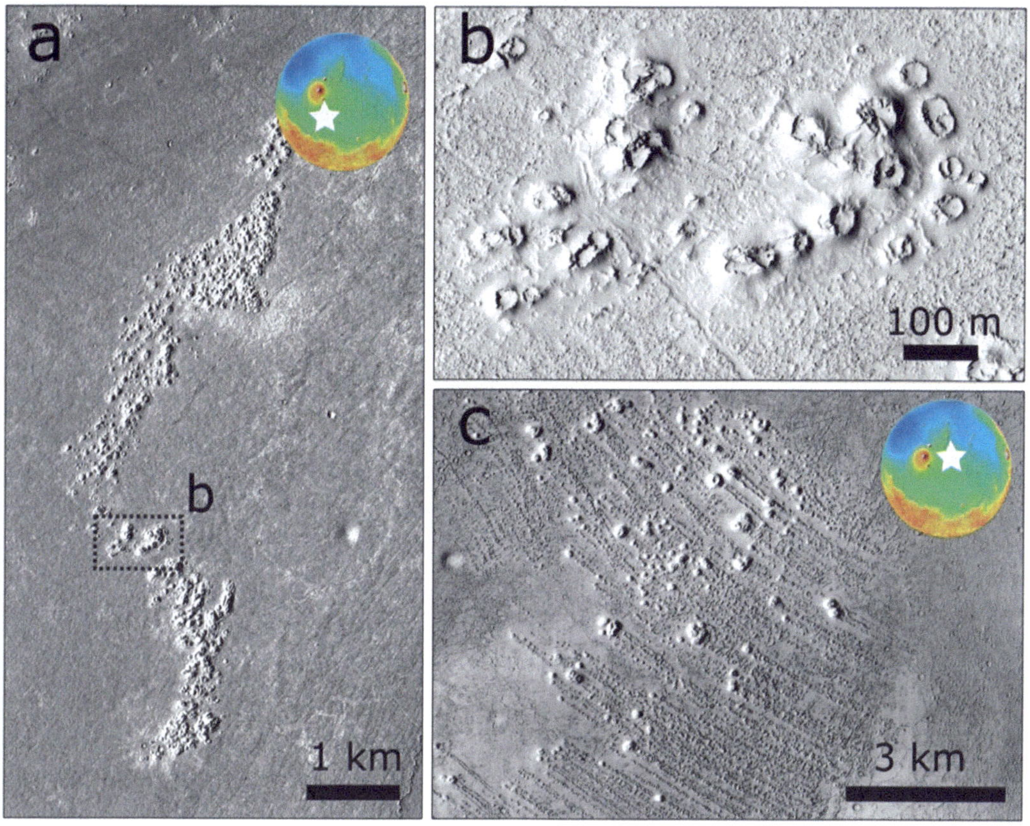

Fig. 7 Examples of putative Martian rootless cones within Athabasca Valles (**a**, **b**) and in the western Tartarus Colles region (**c**). Proposed rootless cones are characterized by small conical edifices with pitted summit craters. Cones are often clustered (**a**, **b**) and associated with chain-like features suggesting the direction of the lava flow movement (**c**). Panel **a** is based on CTX image P03_002147_1865_XI_06N208W, centred at 7.38° N, 152.81° E, **b** is based on HiRISE image PSP_002292_1875, and **c** is based on CTX image G01_018457_2065_XN_26N186W, centred 26.22° N, 173.57° E. North is up on all panels

~8.37–9.93 million km^3 across the entire planet. In general, however, our knowledge of the overall volume of pyroclastic deposits on Mars is limited.

7.1 I—Arabia Deposits

At ~5 million km^2 across a diverse landscape, the proposed Arabia deposits are the most extensive, but also the morphologically most incoherent, of all potential pyroclastic deposits on Mars that have been described so far [100, 101, 272]. Based on their apparent friability, presence of possible bomb sags, and their mantle-like draping over the landscape (Fig. 8), interpretations as remnants of pyroclastic tuff deposits were proposed for the kilometer-sized knobs and irregularly-shaped mesas in north-eastern Arabia Terra [102, 291]. The knobs could also be the result of lava erosion of previous stratifications of lava flows as seen in various regions of Mars [16] so that I preferred to search for deposits of loose terrains. It was even extended the interpretation of the deposits throughout northern Arabia Terra, suggesting a ~1 km thick unit sourced by airfall stretching across an elevation spectrum from ~3 km below the datum to ~1 km above it [100]. It was not possible to reproduce airfall dispersal patterns correlated with the Arabia deposits using Global Circulation models from the far away volcanic edifices of Apollinaris Mons, Elysium Mons, and Tyrrhenus Mons [62]. However, the hypothesis of widespread and thick deposits emplaced by volcanic airfall might find support in the presence of five, up to ~55 km wide, depressions in northern Arabia Terra interpreted as calderas (Fig. 1, letters Q–U) and, thus, potential source vents [26]. Stratigraphic studies of the Arabia deposits suggested airfall dust as possible origin [292]. Regional correlations excluded pyroclastic deposition due to the cyclicity in the layered deposits [293], although failed to explain the lithification of the deposits and why volcanic activity cannot be cyclical [294]. Nevertheless, following the hypothesis that local calderas might be the origin of the Arabia layered deposits [197, 198], I found several craters filled with possible pyroclastic material within a maximum range of 250 km from Euphrates Patera (Fig. 8 a). This range meets the limitations of convective models in pyroclastic ballistic airfall [60]. The craters appear also filled by lava, with hollows in the lava field filled by dark dust of likely pyroclastic origin, and scattered meter-sized blocks (Fig. 8b) that can be hardly moved by the wind [295]. These mantling deposits and scattered blocks superposed on the lava field, with no visible subsequent effusive episode, suggests how explosive activity might likely have been the final stage of Euphrates Patera and of the others Arabia paterae. We will see in Sect. 7.8 ahead, dedicated to the Isidis deposits, how Syrtis Major could also be a possible potential source for the Arabia deposits due to the favour of the wind.

7.2 II—Electris Deposits

The Electris deposits are estimated to cover >1.8 million km^2 ([100, 103]; Table 1) and, as well as the Arabia deposits, are characterised by chaos terrains, i.e., knobs and irregularly-shaped, including flat-topped mesas. The process of deposition of these units among the Ariadnes and Simois Colles, as well as Atlantis, Caralis, and Gorgonum Chaos, includes airfall within the Eridania basin although the origin of the deposited dust remains unexplained [296, 297]. It was suggested that these deposits might be ash fall coming even from Tharsis redistributed through aeolian processes [62, 298]. Relatively small volcanic structures visible closer to the area are Electris Mons, a structure of twin domes belonging to the Alignment 4 like Electris Mons, an unnamed cone just east of the crater Huggins, and a possible caldera from which originated the Drava Valles, both belonging to the Alignment 5 [16]. The volcanic channels of Sirenum Fossae are mostly responsible for the channeling of lava from Arsia Mons [176] so it seems unlikely that these are candidates for explosive activity. Other volcanic features can be possible sources that might justify previous interpretations as volcanic airfall deposits [6, 100, 103], Eridania Mons and Cimmeria Tholi along with some unnamed features located at the end of the Alignment 5. The Eridania basin is located quite far away, ~1800–2000 km from Apollinaris Mons, to consider this volcanic center as possible source. Furthermore, pyroclastic deposits are generally thought to disperse eastward from Apollinaris Mons [62, 63]. Thus the volcanic centres of Electris Mons, or any other potential edifices around the Electris deposits, could easily explain the deposition of the dust. The closest possible candidates, also from the geomorphological point of view, might be the cone near the crater Huggins and the Cimmeria Tholi although we do not exclude Electris Mons. Again, I have thus observed the terrains among these volcanic centres located within a range of 250 km and I have found the deposits shown in Fig. 9a.

The HiRISE image ESP_037367_1340 shows large dunes superposed to fields of boulders likely deposited by pyroclastic activity inside an unnamed crater located at ~181 km east of Electris Mons (Fig. 9a, b). By zooming in the same Fig. 9b, it is also possible to distinguish boulders in dusty deposits partially covered by the field of dunes and embayed by sand ripples (Fig. 9c). The HiRISE image ESP_028690_1455 shows layers of the Electris deposits in one of the knobs located nearly at the centre of the Ariadnes Colles (Fig. 10b). The closest volcanic centre to the Ariadnes Colles is a couple of unnamed twin

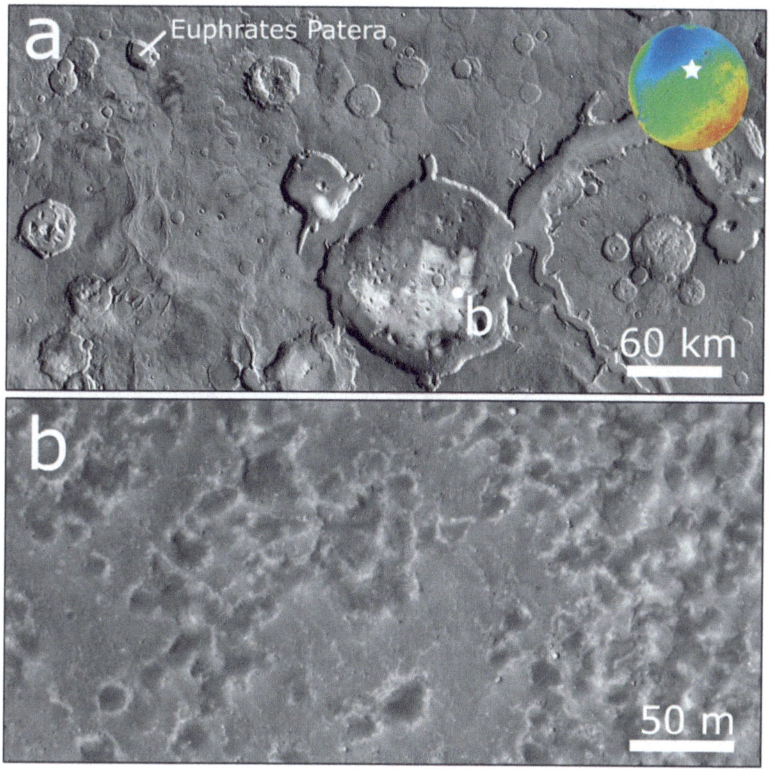

Fig. 8 **a** CTX mosaic showing the context of the Arabia Terra calderas and the erosional basins of Ismenius Lacus. **b** The HiRISE image PSP_009773_2165 shows cavities on the ground filled by dark material, likely pyroclastic sand, and boulders of various sizes, probably bomb sags

domes belonging to the Alignment 4 of Mars [16]. However, the distance of the deposit from the volcanic centre is of ~631 km (Fig. 10a). This means that the distance is above the more restrictive convection models of volcanic explosions [60] but still within the distance of a few hundreds kilometers predicted by less restrictive models [245].

Eventually, any deposits contained within Gorgonum Chaos and Atlantis Chaos would be located at ~333 and ~683 km, respectively, from the volcanic centre of Newton Tholus located in the Alignment 6 of Mars [16]. At last, the phyllosilicate species identified in the Eridania deposits [297, 299] were also found in the Electris deposits close to the Ariadnes Colles [298] and can be explained by differential and/or hydrothermal alteration of volcanic material not necessarily implying a standing body of water on the surface [81, 176, 300–303]. Actually, the formation of colles or chaos terrains is likely due to erosion by lava as already seen for the chaos terrains located inside Valles Marineris [16, 304]. The volcanic sources of the fluid lava flows that eroded and thus formed the chaos terrains containing the Electris deposits and the other chaos terrains in the Eridania basin (Atlantis Chaos, Gorgonum Chaos, Caralis Chaos, and the Ariadnes Colles) are located to the south along the Alignment 3 of Mars [16].

7.3 III—Medusae Fossae Formation

The ~2.5 million km^2 of Medusae Fossae Formation (MFF) extends along the transition topography [193] terrain across ~100° of longitude (nearly 6000 km) between the Tharsis and Elysium bulges and it is one of the most studied pyroclastic deposit on Mars [28, 62, 100, 105, 107– 111, 305–307]. The MFF was recognized already on Viking data as a distinct morphologic and stratigraphic domain characterized by layering, draping over the underlying basement [6, 106]. Yardangs and other erosional features formed by wind and mass wasting, all of which

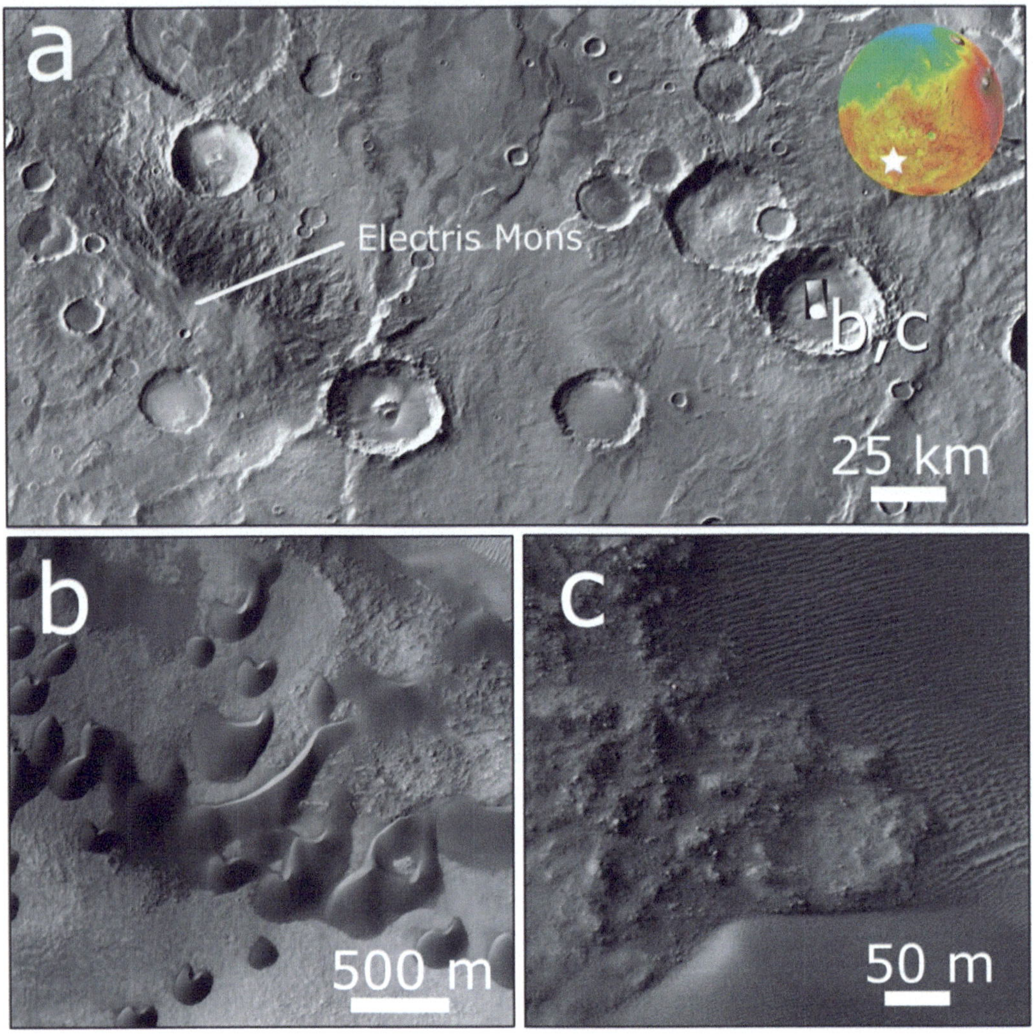

Fig. 9 Context THEMIS mosaic showing the location of Electris Mons and of the pyroclastic deposits visible in the image HiRISE ESP_037367_1340 shown in panels **b** and **c**. **b** Dunes of black sand of possible pyroclastic origin superposed to the candidate Electris deposits. **c** Detail of the Electris deposits, boulders are mixed with friable material

are indicative of its friability, have been studied in the MFF [308–311]. Topography, gravity, and radar data showed that MFF materials range in thickness from tens of meters in western exposures to more than 3 km in eastern exposures, and that it superposes units conformable with both lowland plains and cratered highlands [14, 107, 111, 289]. Based on its stratigraphic relations with neighboring units, the bulk of the MFF was probably emplaced during the Hesperian [110, 306]. Amazonian model ages from crater size-frequency distributions are most likely attributable to resurfacing by intense deflation and eolian redeposition [6, 195, 196]. Various origins have been suggested for the MFF [109], ranging from eolian [272] to paleopolar [312], transcurrent fault [313], lacustrine deposits [197], ash fall [100, 101, 107], ignimbrites [105, 106], and even rafted pumice [314]. The majority of these origins were confuted by new data and a robust body of accumulated evidence, which attributes the origin of the MFF to explosive

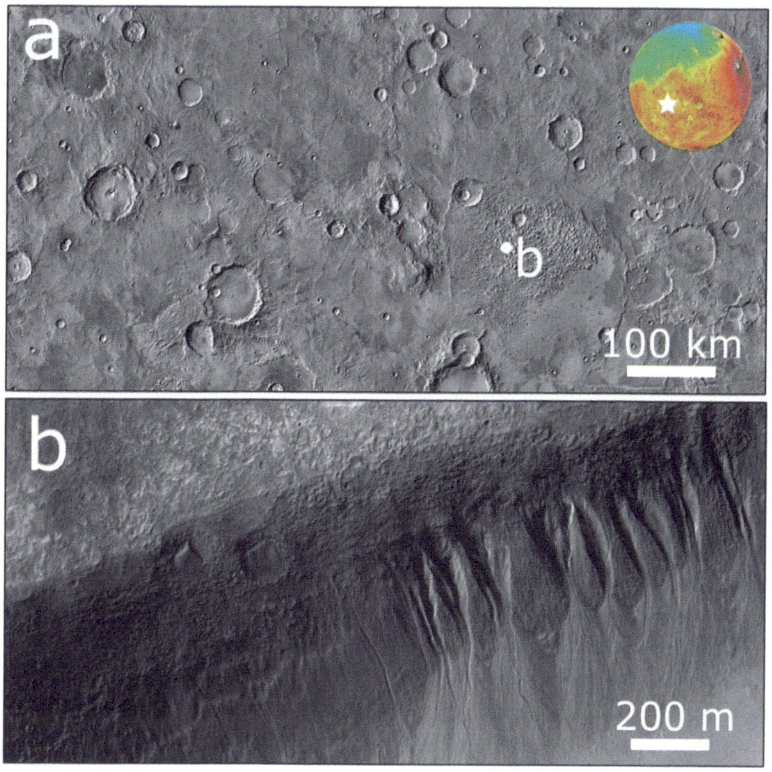

Fig. 10 **a** Themis context image showing the location of the Ariadnes Colles indicated by the letter b, which is also the location of the HiRISE image ESP_028690_1455. **b** Detail of the Electris deposits in Ariadnes Colles, the layers of pyroclastic materials are characterized by mass wasting processes forming the aprons visible in the right lower corner of the image

volcanism [109], being ignimbrite the most viable hypothesis [105]. Spectroscopic observations show the typical iron oxide bands in spectral data along with significant signatures expected in pyroclastic deposits, e.g., feldspars, pyroxenes, or olivines [315–319]. Radar datasets have given information on the composition of the MFF and bolstered its interpretation as volcanic material, as possibly welded tuff [320, 321]. At last, the positive magnetic anomaly that is spatially correlated with Lucus Planum, a part of the western MFF near Apollinaris Patera, is indicative of oxidized, Fe-rich basaltic material consistent with a generally volcanic origin [322].

Given its extension, a unique origin for the MFF is unlikely. Although Apollinaris Mons is the only large volcanic centre in the middle of the MFF, the extreme ends of the MFF are quite far away from it. According to the available convection models of volcanic explosions [58, 60] and to the models of dispersal of the pyroclastic products [28, 62], the activity of Apollinaris Mons cannot be the only one to justify the origin of the MFF. Thus, other volcanic centres must have contributed to the formation of the MFF, at least closer to its extreme ends. Other possible volcanic structures characterized by explosive activity could be Syrtis Major [204], Tyrrhenus Mons [23, 24], Hecates Tholus [84], and the Tharsis volcanoes [101, 178]. However, the question of whether the explosive activity of the Tharsis volcanoes has played a role in the formation of the MFF is linked to the distance of the pyroclastic deposits from the closest volcanic edifices [105]. There were studies that allowed even hundreds of kilometers or more from the deposits to the volcanic centres depending on the atmospheric pressure [58, 63] and another study that placed stricter limitations based on the observed atmospheric pressure [60]. Whilst there is no particular problem for Apollinaris Mons, which is quite close to the MFF, the minimum distance that separates Gordii Dorsum, the easternmost member of the MFF, from the closest volcanic centres of Olympus Mons and Arsia Mons is around 1000 and 1200 km respectively.

For this reason, explosive activity from local fissures was proposed for Gordii and Eumenides Dorsum [100, 105, 106]. Although these volcanic centres are not within the hundreds of kilometers range proposed for the dispersal of pyroclasts [58, 63], the wind may have helped in the case of the Tharsis volcanoes and Elysium at least. The heights of Olympus Mons ($\sim$26 km) and Arsia Mons ($\sim$17 km) are, respectively, beyond and close to the 20 km maximum height of convection of the pyroclastic clouds [62, 323] where the winds are stronger than at the surface an thus more effective at transporting pyroclastic material to greater distances than just ballistic trajectories [62]. Measured wind speeds range from the average $\sim$15 m s^{-1} at Elysium Planitia to the 40–60 m s^{-1} estimated at 60 km of altitude [324] or even the 70–170 m s^{-1} measured at 50 km of altitude [325, 326]. Hecates Tholus is only 4.5 km high but its explosive activity may have raised the pyroclastic cloud sufficiently high to intercept the winds sweeping the surface of Elysium Mons at $\sim$13 km of altitude. Thus, the pyroclastic products dispersed by these volcanic centres may have reached the extreme western end of the MFF. The direction inferred by the streaks found on the surface of Elysium Mons, which is located south of Hecates Tholus, and Olympus Mons indicate winds coming from N to S or from the NE to the SW [327] and from E to W on Arsia Mons [328]. These wind directions are consistent with the eastern end location of the Gordii and Eumenides Dorsum, also associated to the deposition of the radar invisible "Stealth" fine grained dust [329], and the western end of the MFF located to the west of Gale crater that is out of reach of Tharsis and Apollinaris Montes. Moreover, radar-derived bulk dielectric properties of Lucus Planum are more consistent with porous pyroclastic materials, rather than the alternate explanation of ice-rich material [289]. Although the spatial resolution is of $\sim$550 km/px, gamma ray spectroscopy revealed how the MFF has the strongest sulphur and chlorine signatures on Mars (consistent with 2.40–2.88 wt% S and 0.57–0.74 wt% Cl) attributed to volatile phases in pyroclastic materials [28, 111, 330, 331]. Lastly, gravity and topography data analyses using standard loading models found a bulk density of the MFF of $\sim$1.8 g/cm^3, which is consistent with porous rock, e.g., uncompressed pyroclastic deposits [111]. Due to its nature as a sulphur- and chlorine-enriched, friable, and fine-grained deposit, the MFF was even proposed to be a main source of Mars' atmospheric dust inventory and thus a significant factor modulating the current climate [111, 330]. Based on modeled ash plume dispersal, several extant volcanic vents investigated as potential sources for the welded tuff-like deposits of the eastern end of the MFF are: Olympus, Ascraeus, Pavonis, Arsia, and Alba Montes (Fig. 1, letters K–O). As well as Syrtis Major and Apollinaris Patera for the western end of the MFF (Fig. 1, letters E and J, respectively). Other potential additional sources are located in the calderas of north-eastern Arabia Terra (Fig. 1, letters Q–U) and interpreted as possible origin of large-scale explosive eruptions [26].

Although the MFF has been extensively studied, the typical and best gallery of images for the MFF is made of scattered patches between Aeolis Planum and Gordii Dorsum showing knobs and contacts of lava casts partially covering yardangs fields without sufficient context of the source areas [306]. A detailed view of the pyroclastic deposits that are part of the MFF, including a context of the possible source areas, is still missing in the literature. For this reason, I have chosen a representative area of the MFF in which a pyroclastic deposit is visible along with its possible source. Such an area is Apollinaris Sulci, centred south-east from Apollinaris Mons, where the large fan of the volcano is oriented, and north-east from Gusev crater, where there is the debouchment of the lava flows coming from Ma'adim Vallis. At the distal end of the fan there is an outcrop showing the detail of the pyroclastic deposit formed by the explosive activity of Apollinaris Mons (Fig. 11a). Furthermore, Apollinaris Sulci is close enough to Apollinaris Mons in order to study more effectively the result of both effusive and explosive activity through the superposition relationships between lava flows and the MFF. The pyroclastic deposit shown in Fig. 11b is located at the base of Apollinaris Mons at about 186 km of distance from its central caldera. Its location under a cover of lava flows does not give the opportunity to observe its whole extent. However, it is reasonable to think that it might be distributed all over the slope of the volcano in case of surge after the collapse of the initial pyroclastic plume other wise its distribution would be determined by the direction of the wind at the time of the eruption [58]. The HiRISE image ESP_ 017099_1680 shows a detail of the pyroclastic deposit (Fig. 11b). In this image it is possible to observe the friable nature of the deposit including the presence of boulders that can be interpreted as bomb sags. Unfortunately, the lack of additional HiRISE images has not made it possible to observe in detail this pyroclastic deposit along other slopes.

7.4 IV—Dorsa Argentea Formation

The 1.35 million km^2 and on average ~500 m thick Dorsa Argentea Formation (DAF, or "South Polar Pitted Terrain") has been noticed for the presence of friable material in the NPl1, cratered unit of the plateau sequence [332], and in the Parva member of the DAF [333]. A composition that is at least partially pyroclastic in origin, e.g., as thick ash blanket covering glacial deposits, whose deposition models predict that circum-South Pole fallout from ash plumes sourced by the Malea Panum paterae (Fig. 1, letters A–D) would correlate well with today's DAF extent [62]. Another additional potential source might be the numerous isolated buttes of the Sisyphi Montes in the eastern side of the DAF for which a formation as putative tuyas, i.e., subglacial volcanoes, has been proposed [248, 334]. However, radar analyses found that consolidated basalt or dry, porous deposits, i.e.,

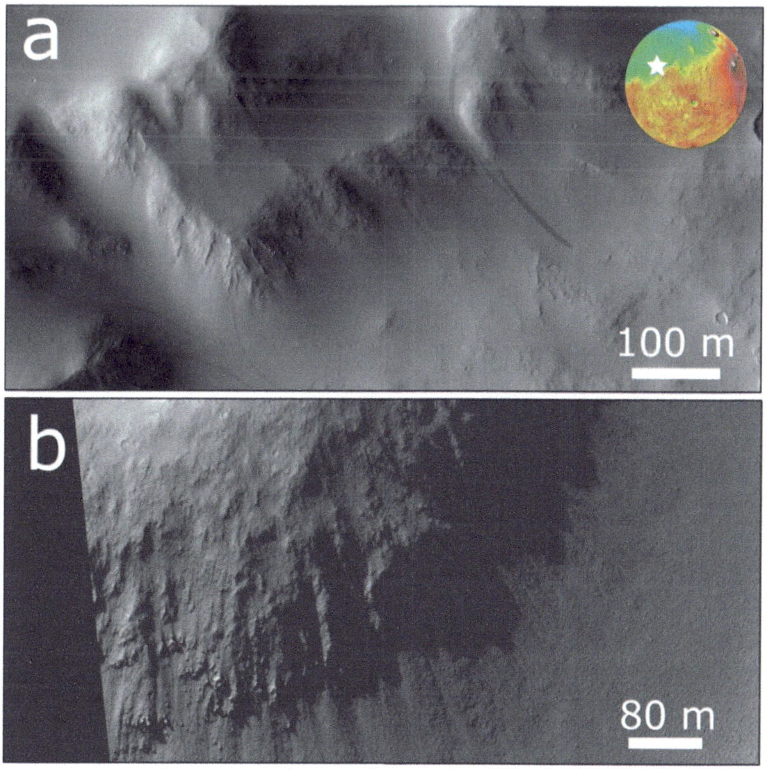

Fig. 11 **a** CTX Mosaic of a possible outcrop of the pyroclastic deposit on the eastern flank of Apollinaris Mons; **b** HiRISE image ESP_017099_1680 showing a pyroclastic deposit on the southern flank of Apollinaris Mons

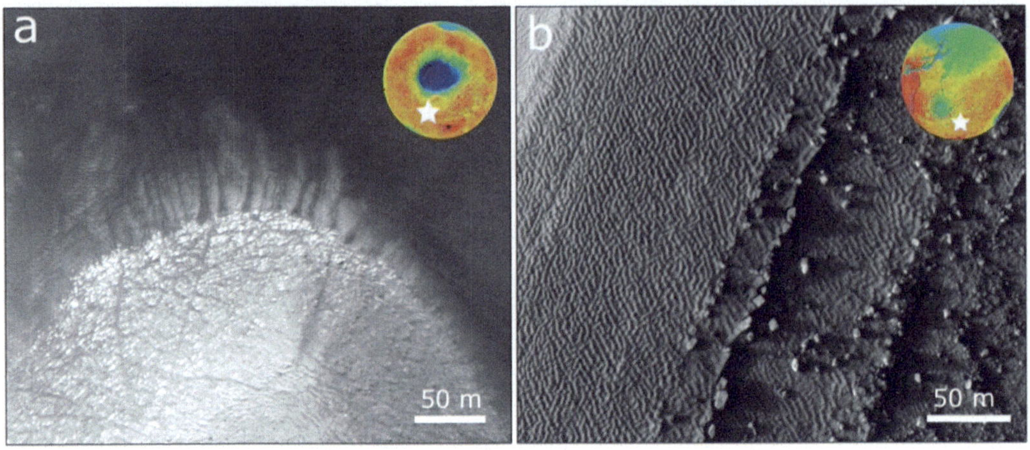

Fig. 12 HiRISE observations of escarpments bounding outlying occurences of the Dorsa Argentea Formation (white asterisks in Fig. 1). **a** shows an escarpment wall, where a layer of fine-grained, dark material crops out below more blocky, bright, and apparently jointed material. **b** shows layers of meter-scale blocks

potentially of pyroclastic origin, would better explain the low dielectric constant and relatively high loss tangent values of the DAF than ice-bearing material [113]. Meter-scale blocks as far north as 62° S and dark material eroding out of the flanks of the DAF (Fig. 12a) indicate that it contains non-volatile components, possibly including mafic fines. In some locations (Fig. 12b), these blocks appear very angular and are elongate, which is indicative of angular jointing, a common structural characteristic of volcanic deposits including pyroclastics [335, 336]. The various outcrops of dark fines (Fig. 12b) indicate that ashfall might have contributed to its depositional stack, which could include significant amounts of pyroclastics from explosive volcanism pre-dating the proposed glacial activity in the region.

7.4.1 V—Hellas Deposits

The up to ~1 km thick hummocky lava filling that covers the central ~0.56 million km^2 of the Hellas basin floor does not allow to see the pyroclastic deposits in its entirety as a whole continuous mantle. However, outcrops of the pyroclastic deposits are visible thanks to the lava channel of Dao Vallis cutting through the slope of Hadriacus Mons (Fig. 13) and thanks to isolated fields of bomb sags close to Hellas Chaos (Fig. 14). Figure 13 shows the detail of friable-layered deposits (FLD) located at ~218 km south from Hadriacus Mons. Figure 14b shows the details of a field of boulders close to Hellas Chaos at ~625 km north from the volcanic centre of Amphitrites Patera. Yardang fields, a high abundance of mass wasting features on shallow to moderate slopes, and a ubiquitous low thermal inertia indicate how the deposits are fine-grained and relatively weakly consolidated [117]. These observations, in addition to spectral signatures of pyroxenes and hydrous minerals [337], the occurrence of layered knobs, the relative vicinity to potential source vents (Peneus, Amphitrites, and Hadriaca Paterae; Fig. 1; letters C–F), favoured the interpretation as volcanic materials emplaced by pyroclastic flows and/or ashfall [114–117].

7.5 VI—Argyre Deposits

The 0.34 million km^2 inner Argyre basin floor, for which early to late Hesperian crater count ages have been derived, is comprised of up to ~2 km thick, at least partially layered deposits [338, 339]. Due to the etched, yardang-like, i.e., possibly friable, appearance of its northern parts (Fig. 15a), and the presence of Argyre Mons (Fig. 1, letter P) on the western basin floor as a potential vent structure resulting from explosive volcanism, an at least partial pyroclastic origin of

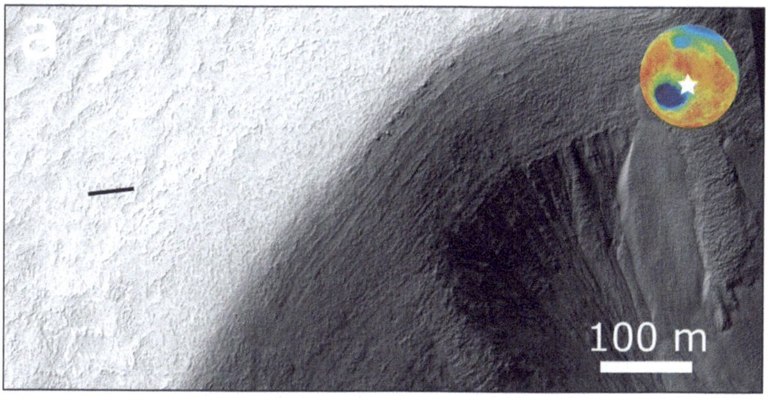

Fig. 13 **a** The HiRISE Image PSP_006659_1460 shows layers of friable material along the slope of Hadriacus Mons and made visible thanks to the incision of Dao Vallis, mass wasting processes form vertical incisions due to the erosion of the material sliding downhill

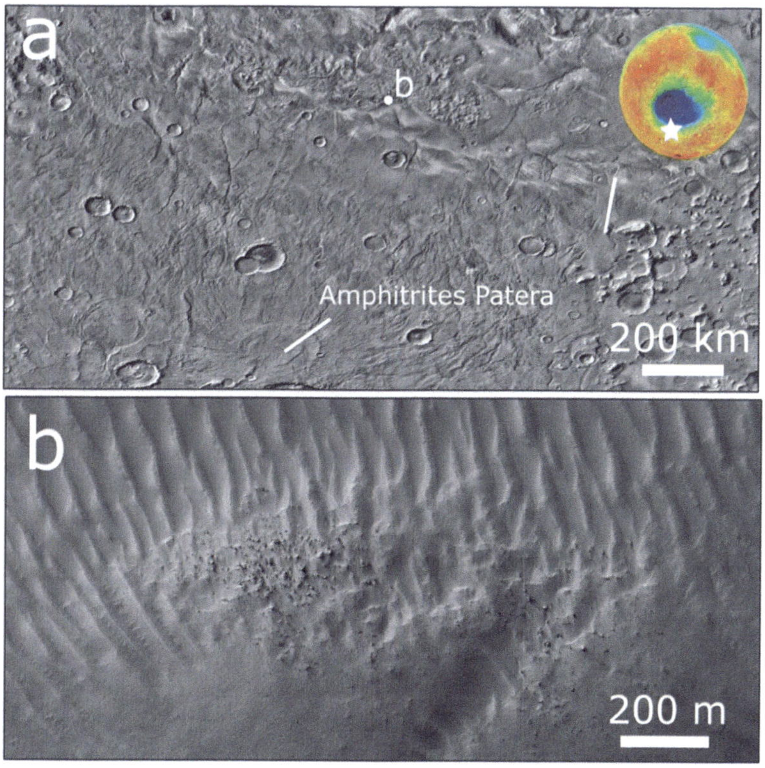

Fig. 14 **a** THEMIS context mosaic showing the location of the Hellas deposits (letter b) observed in detail thanks to the HiRISE image ESP_017024_1310 shown in panel **b**. **b** Detail of the Hellas deposits, a field of boulders mixed with friable material typical of pyroclastic deposits is surrounded by dunes of sand

the basin-fill has been tentatively suggested [6, 54, 118]. Argyre Mons was suggested as a possible source for the pyroclastic deposits present in the etched terrains within the Argyre basin [118]. Based on this suggestion I have explored the terrains around Argyre within a radius of ~120 km in search for pyroclastic deposits under the coverage of HiRISE imagery.

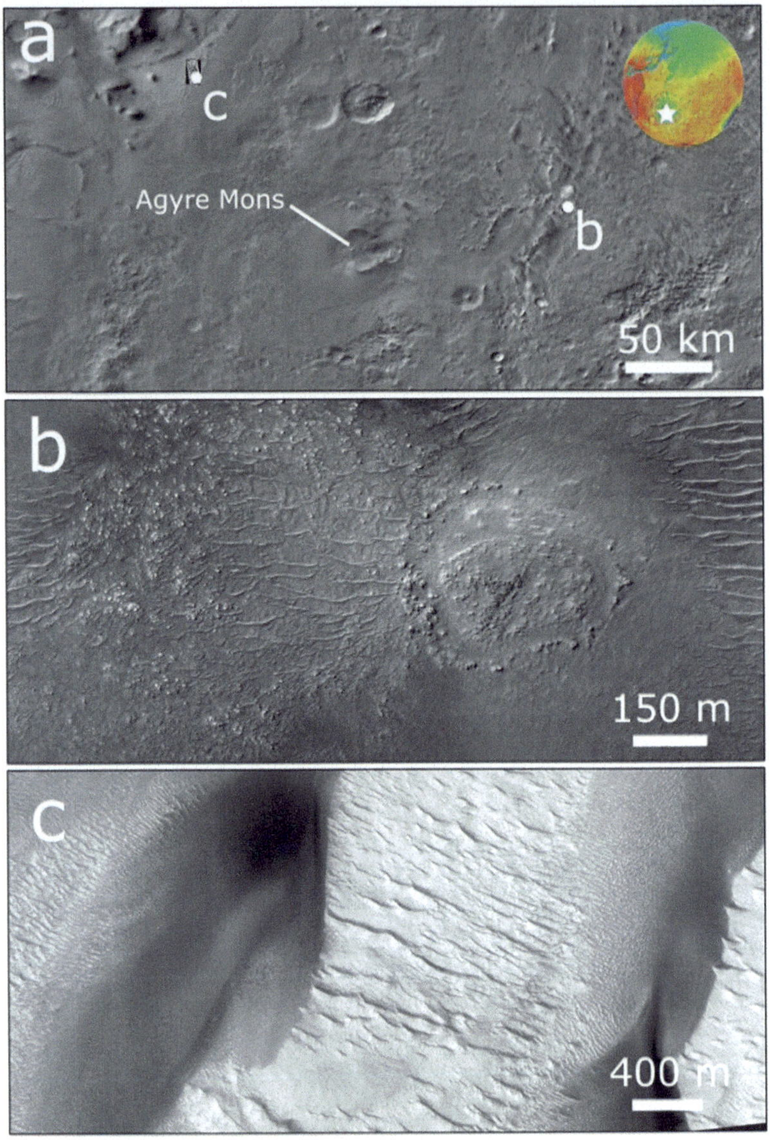

Fig. 15 **a** THEMIS mosaic context image showing the location of the Argyre deposits; **b** Detail of the HiRISE image ESP_023476_1300 showing a putative pyroclastic deposit located east of Argyre Mons, see text for a complete description; **c** Detail of the HiRISE image ESP_019138_1310 showing a putative pyroclastic deposit located northwest of Argyre Mons, see text for a complete description

The HiRISE image ESP_023476_1300 shows a pyroclastic deposit located at ~101 km northeast from Argyre Mons (Fig. 15b). This deposit is also filling an impact crater, boulders mixed to friable material are visible both within the crater and on the field west of the sand ripples located in the middle of the image (Fig. 15b). The Martian wind has formed yardang fields covered with barchan dunes composed of dark sand of likely pyroclastic origin. The wind deposited the dunes mostly in the northwestern and in the northern side of the Argyre basin, sign that the main wind direction is probably coming from the south-southeast. One of the yardangs and barchan dune fields is located in northwestern Argyre at ~121 km from Argyre Mons (Fig. 15c). The HiRISE image ESP_019 138_1310 shows how the dunes postdate the yardangs and are so thick that bury them completely (Fig. 15c).

7.6 VII—Tyrrhena Patera Deposits

The region north of Tyrrhena Patera (Fig. 1, letter G) is characterized by notched mesas having rough surfaces alternated with volcanic plains. The notched mesas were proposed as remnants of friable deposits [119] up to 0.32 million km^2 in Cerberus Dorsa, a region located ~600 km north of Tyrrhenus Mons [120]. However, the available context image of the area shows no pyroclastic deposits but just volcanic plains of indurated lava flows. The many wind streaks present in the area unequivocally indicate how the prevalent direction of the wind is from north to south that is in the opposite direction to which the putative pyroclastic deposits should have travelled from Tyrrhenus Mons (Fig. 16). There is no record of any other wind direction from the visible streaks. On the opposite side of Tyrrhenus Mons there is something that might be interpreted as putative pyroclastic deposits within the craters Trinidad and Kinkora located at ~175 and ~600 km from Tyrrhenus Mons, respectively (Fig. 17a–c). There is a deposit within the crater Trinidad that cannot be made of lava flows because the crater rim is intact and thus not flooded by the surrounding lava fields (Fig. 17c). Actually, the crater postdates the lava fields and it is located downwind in the optimal range for receiving pyroclastic products from Tyrrhenus Mons so that the deposit inside the crater could have been easily formed by airfall. The whole appearance of the deposit suggests friable material accumulated after transport by the wind, some sand ripples are also visible in the right side of the HiRISE image PSP_009427_1565 (Fig. 17b). The crater Kinkora, located farther away with respect to Trinidad, also shows intact crater outer rims and a dusty deposit inside (Fig. 17c). However, given the higher distance from Tyrrhenus Mons, the blocks visible in the HiRISE image PSP_004799_1550 (Fig. 17c) might have fallen there by mass wasting processes of the inner rim rather than to pyroclastic airfall.

7.7 VIII—Isidis Deposits

The 0.7 million km^2 Isidis basin floor hosts thousands of subkilometer-sized pitted cones forming the largest occurrence of "thumbprint terrain" on Mars [121, 122]. Based on buried craters, the underlying unit has been estimated to be up to ~400 m thick and various formation scenarios, usually involving either glacial or volcanic activity, have been proposed [121, 122, 340].

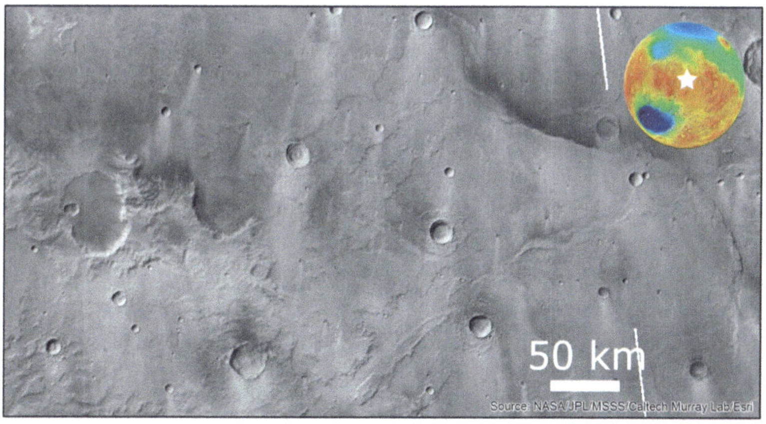

Fig. 16 CTX Mosaic of the Cerberus Dorsa region of Mars showing multiple wind streaks along the north–south direction. The image does not show evident pyroclastic deposits but only lava flows coming from the south, likely from Tyrrhenus Mons. In any case, the wind coming from the north in the image should have pushed eventual pyroclastic clouds back to Tyrrhenus Mons located 600 km to the south

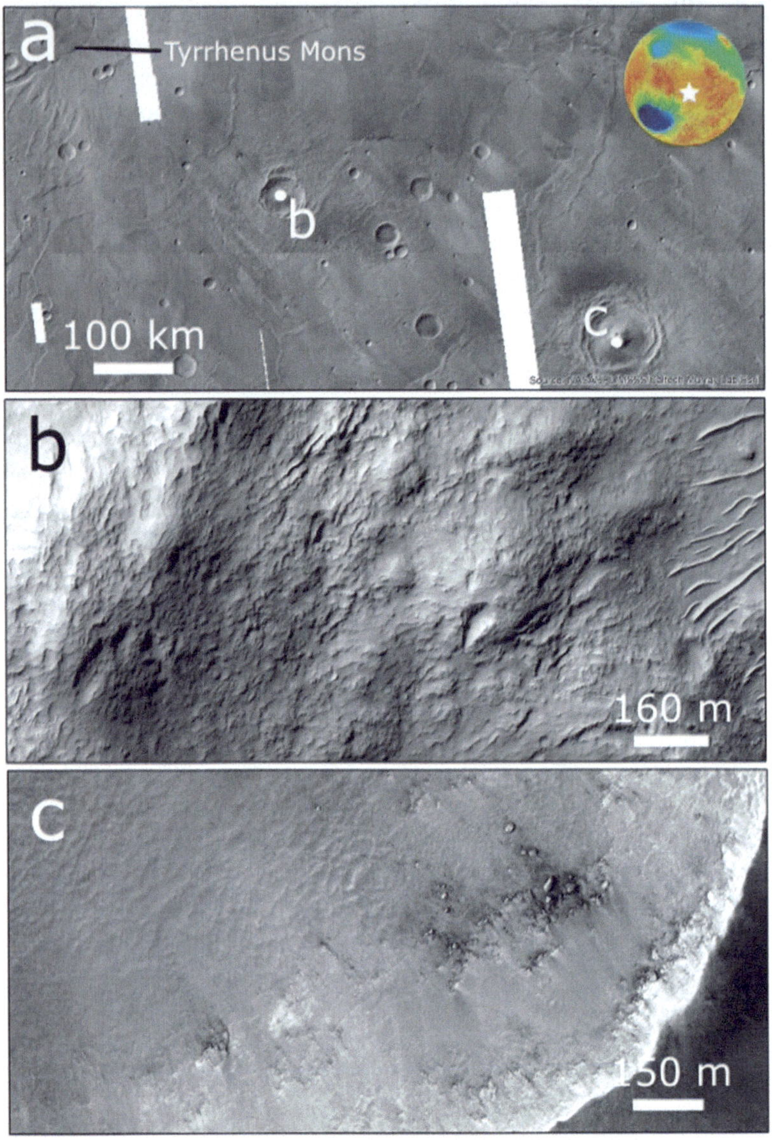

Fig. 17 **a** CTX mosaic context image showing the southeastern lava fields of Tyrrhenus Mons with the craters Trinidad (letter b) and Kinkora (letter c). **b** Detail of the HiRISE image PSP_009427_1565 showing a putative pyroclastic deposit located inside the crater Trinidad. **c** detail of the HiRISE image PSP_004799_1550 showing loose dark material and boulders downhill of the southeastern crater rim, it is not clear whether the presence of such a loose material is due to the friable and sandy (pyroclastic) nature of the rim or to mass wasting processes characterising the rim

Due to its proximity to the potential source of Syrtis Major (Fig. 1, letter e) and to the fine-grained nature of the basin-fill that impact craters indicate to be interrupted only by some shallow, dark, layers it was proposed a scenario in which the pitted cones formed as degassing structures of pyroclastic deposits [121]. Other formation scenarios based on glacial activity [122, 340] do not take into account the lack of stability of ice on the surface of Mars and thus its interpretation remains tenuous at best. I have also explored whether the pyroclastic option might be viable in the climatic conditions of Mars. This means that I have explored whether the direction of the wind may have favoured the transport and subsequent deposition of the pyroclastic products from Syrtis Major to Isidis Planitia. The results are visible in Fig. 18. The wind streaks present in the CTX mosaic of Fig. 18a clearly show how the direction of the wind is actually against the possible transport of the pyroclastic products from Syrtis Major to Isidis Planitia. The closest deposits to Syrtis Major seen on the ground of Isidis Planitia are observable within an unnamed crater located at ~750 km of distance from Syrtis Major (Fig. 18a). The details of the deposit are visible in Fig. 18b. No similar deposits are observed in the middle of Isidis Planitia where they were originally expected [121]. Furthermore, the amount of the Isidis pyroclastic deposits is not minimally comparable to the amount shown by the Medusae Fossae Formation (Table 1). This discrepancy is more striking when we also consider the double size of Syrtis Major (~200 km) with respect to Apollinaris Mons (~100 km). Probably the response is exactly in the direction of the wind. The products of Syrtis Major were probably dispersed in the opposite direction respect to Isidis Planitia, that is Arabia Terra, and only the products of the small pyroclastic cones accumulated at its margin might really have contributed to the formation of the Isidis deposits.

7.8 Other Deposits Interpreted as Produced by Explosive Volcanism

In addition to the largest potential pyroclastic deposits on Mars listed in Table 1 and described in the previous sections, other areas received interpretation as pyroclastic in nature, albeit not confirmed with recent data. The western flank of Hecates Tholus (Fig. 1, letter H) appears very smooth and was interpreted to have been resurfaced by airfall from a ~70 km high plume that emplaced ~100 m, i.e., potentially 60 km^3, of ash [84].

Several constructs, probably originating from explosive volcanic activity, have been reported within the Tharsis bulge [177]. Kilometer-sized, relatively pristine cones on the summit of Pavonis Mons as well as on the plains of western Tharsis are among the most unambiguous landforms indicating Amazonian explosive volcanism in the area, albeit of limited scale [29, 341]. However, large-scale explosive eruptions in the Tharsis province were inferred from theoretical considerations first [178, 101]. Subsequent observations on the ground of ash-like layered deposits (the well-known interior layered deposits, ILD) at heights out of reach from putative water suggest compatibility only with volcanic origin [342] and deposition by airfall [343].

Being the polar caps an evident source of volatiles, the polar areas would had a strong chance to develop explosive volcanism provided there was a volcanic source that could produce it. The possible local presence of pyroclastic deposits has been inferred from the observation of dark material in the north polar layered deposits [344]. Based on Viking thermal inertia data, the lower density of this dark material, compared to the dunefields found at lower latitudes, suggests that it might be composed by dust or unweathered basaltic grains [344–346]. The direction of the wind is both off-pole and on-pole [347], which does not give much help in localizing possible close sources for these deposits. However, volcanic activity close to or at the North Pole was reported, with observations of volcanic cones close to the Scandia region [16, 348] and a volcanic plume that arose directly at the North Pole as a consequence of the southern polar giant impact (SPGI) [16, 73]. Although some volcanic cones are observable in the Scandia region, no large volcanic feature comparable to the large ones found along the Equator is visible at the North Pole. Another important point is that cavi units were mainly associated to volcanic activity at the South Pole [349, 350] whereas glacial or aeolian activity were mainly associated to them at the North Pole [346, 351–354] with the exception of the most ancient Noachian basal units and the Hyperboreus Labyrinthus in which volcanism was invoked [332, 355].

Another interpretation of explosive volcanic activity on Mars came also from the ground observations through rovers. The Mars Exploration Rover (MER) Spirit explored the floor of the ~166 km wide Gusev Crater finding evidence for possible pyroclastic deposits [108, 356–358]. The in-situ evidence for pyroclastic deposits comes from Spirit's investigation of the Columbia Hills, which rise up to ~100 m above the surrounding plains, with a lateral extent of ~6 km (Fig. 19a). Here, bright deposits, which can be seen from orbit as having a roughly

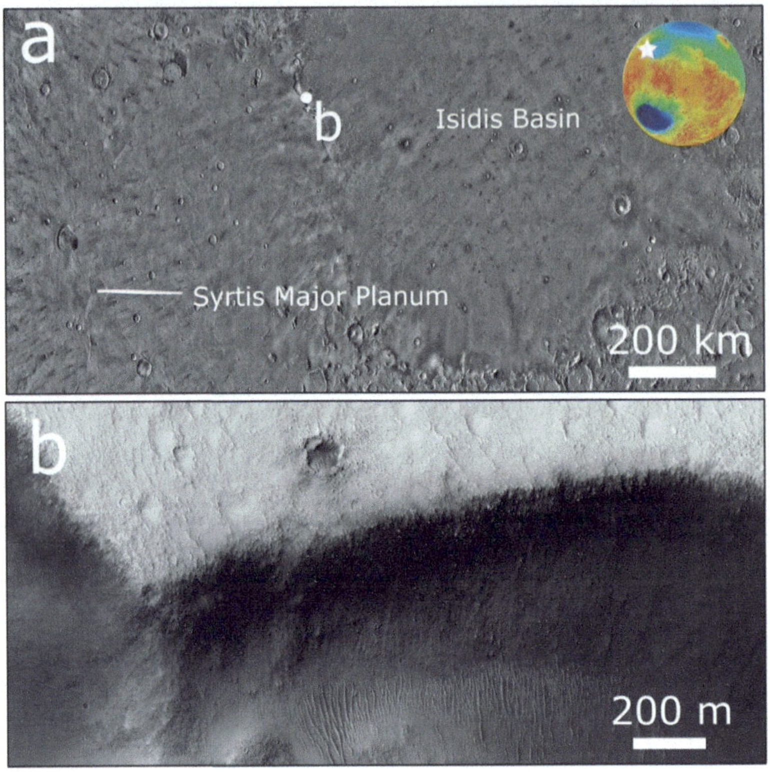

Fig. 18 **a** THEMIS mosaic context image showing the area located between Syrtis Major and the western Isidis Planitia region, dark wind streaks characterize the majority of the impact craters in the area, the direction of the wind indicated by the dark streaks is from east to west-northwest; the letter b indicates the location of a putative pyroclastic deposit within a crater distant ~750 km east of Syrtis Major against the direction of the wind. **b** Detail of the HiRISE image ESP_036196_1945 showing a putative pyroclastic deposit, it is well visible in the image the friable nature and the dark colour of the material with boulders fallen downhill. Sand ripples are also visible downhill along the east–west direction, north is towards the top of the image

circular outline in plan view, show fine-scale layering and cross-bedding at in-situ scale [108]. One of the areas containing bright deposits is called "Home Plate" (Fig. 19b). This is a ~90 m wide approximately circular-shaped low plateau about 2–4 m high, in which the presence of inward-dipping layered rocks were proposed to be formed by lapilli and volcanic ash later reworked by eolian processes [245]. It was even proposed that this material resulted from hydrovolcanic activity that occurred within Gusev crater leading to the formation of tuff ring or maar, which was later eroded [108]. A candidate volcanic source vent dubbed "Goddard" was discovered ~170 m south of Home Plate (Fig. 19b), but not explored by the rover because the fine particulate regolith precluded further driving. However, given the distances involved in explosive eruptions it is more likely that it resulted from materials transported from Apollinaris Patera that lies ~280 km north of the Gusev crater rim [28, 245]. Also, olivine-rich pyroclastic rocks were observed in some outcrops within Gusev crater supporting the theory that volcanism was active inside the crater [359]. The observation of a possible volcanic bomb-sag (Fig. 19c) and the grain sorting around Goddard within the Home Plate suggests that the deposits originate from hydrovolcanic activity within the Gusev crater itself [360] or from an impact in which a volcanic bomb would not be expected to travel more than 350 m from its source [361].

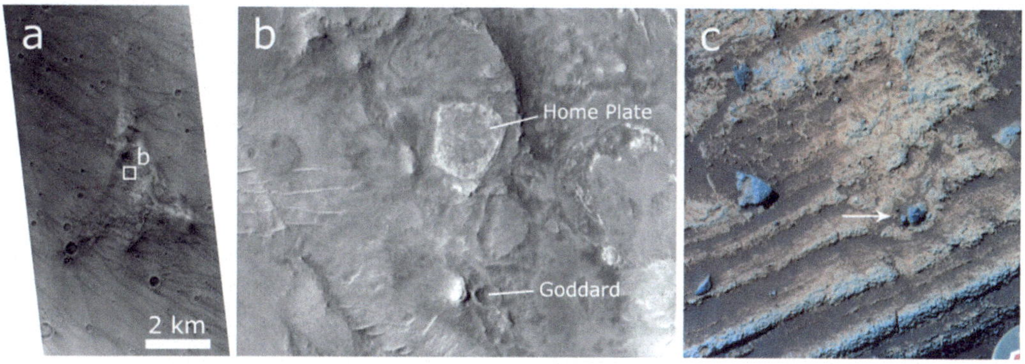

Fig. 19 Evidence of explosive volcanism as observed within Gusev crater. Panel **a** shows Columbia Hills on the floor Gusev crater investigated in-situ by the Spirit rover, in which bright semi-circular outcrops were observed. One of these areas containing bright deposits is a Home Plate (**b**), from which putative volcanic vent named Goddard was discovered. Panels a-c are based on HiRISE image PSP_005456_1650 and north is up. Panel **c** shows a putative volcanic bomb, which landed into the stratas photographed by the Spirit rover

8 Conclusions

As has been laid out in this chapter, a broad spectrum of observations indicate that signs of explosive volcanism are still visible on the surface of Mars. The most spectacular are the pyroclastic deposits like the MFF or some calderas even though the latter not always are associated to explosive volcanism. However, the sparsity of explosive edifices compared to effusive edifices implies that explosive volcanism was unevenly distributed in time and space. The more common occurrence of ancient martian explosive constructs hints that magmas ascending to the surface of Mars in the Noachian had a higher volatile content than Hesperian magmas and/or that the interaction between magma/lava and water was much more frequent in the Noachian than later in Mars' history. This would point to a transition from early frequent explosive volcanism to a younger, dominantly effusive volcanism. However, the postulated thicker atmosphere of early Mars would inhibit explosive volcanism with respect to a thinner one. It remains unclear, even after decades of Martian exploration, why such a transition in volcanic styles occurred. It is possible that the Martian mantle may have contained some water in the past, but got drier as volatiles were released by volcanism and never recycled due to the absence of plate tectonics [91] or available conditions on the surface to maintain the stability of water [81]. This limited the ability of the planet to replace the gases lost from the atmosphere to space [70] or those buried underneath the Martian surface [71]. Therefore, the evolution of the entire planet has been affected by the decline of its internal heat and, as a consequence, of its volcanism.

References

1. McCauley JF, Carr MH, Cutts JA, Hartmann WK, Masursky H, Milton DJ et al (1972) Preliminary mariner 9 report on the geology of Mars. Icarus 17:289–327
2. Greeley R, Spudis PD (1981) Volcanism on Mars. Rev Geophys [Internet] [cited 2018 Sep 23] 19 (1):13–41. http://doi.wiley.com/10.1029/RG019i001p00013
3. Grott M, Baratoux D, Hauber E, Sautter V, Mustard J, Gasnault O, et al (2013) Long-term evolution of the Martian crust-mantle system. Space Sci Rev [Internet] [cited 2018 Nov 6] 174(1–4):49–111. http://link.springer.com/10.1007/s11214-012-9948-3
4. Scott D, Tanaka K (1978) Geologic map of Mars [Internet]. Reston [cited 2019 Dec 7]. https://pubs.er.usgs.gov/publication/i1083
5. Moore HJ (2001) Geologic map of the Tempe-Mareotis region of Mars [Internet]. Reston, Virginia [cited 2019 Jul 19]. https://pubs.usgs.gov/imap/i2727/

6. Scott DH, Tanaka KL (1986) Geologic map of the western equatorial region of Mars [Internet]. IMAP 1802 [cited 2019 Apr 6]. https://pubs.er.usgs.gov/publication/i1802A
7. Tanaka KL, Scott DH (1987) Geologic map of the polar regions of Mars. USGS Miscellaneous Investigations Series Map I–1802–C
8. Hodges CA, Moore HJ (1994) Atlas of volcanic landforms on Mars [Internet]. Professional Paper 1534 [cited 2019 Jul 19]. https://pubs.er.usgs.gov/publication/pp1534
9. Carr MH (1973) Volcanism on Mars. 78(20):4049–4062.
10. Scott DH, Schaber GG, Horstman KC, Dial ALJ (1980) Lava flow maps of the Tharsis province on Mars [Internet]. Reports of planetary geology program, 1979–1980, pp 179–180 [cited 2019 Mar 27]. http://adsabs.harvard.edu/abs/1980rpgp.rept..179S
11. Schaber GG, Horstman KC, Dial Jr AL (1978) Lava flow materials in the Tharsis region of Mars. In: Lunar and planetary science conference, 9th [Internet]. Houston, Tex., March 13–17. Pergamon Press, Inc., New York, [cited 2020 Jan 16], pp 3433–3458. https://ui.adsabs.harvard.edu/abs/1978LPSC....9.3433S/abstract
12. Greeley R, Bridges NT, Crown DA, Crumpler L, Fagents SA, Mouginis-Mark PJ, et al (2000) Volcanism on the Red planet: Mars. In: Environmental effects on volcanic eruptions. Springer, US, pp 75–112
13. Mouginis-Mark P, Yoshioka MT (1998) The long lava flows of Elysium Planita, Mars. J Geophys Res Planets [Internet] [cited 2019 Mar 27] 103 (E8):19389–19400. http://doi.wiley.com/10.1029/98JE01126
14. Zimbelman JR, Garry WB, Bleacher JE, Crown DA (2015) Volcanism on Mars. In: The Encyclopedia of volcanoes [Internet]. Academic Press [cited 2019 Jul 21], pp 717–728. https://www.sciencedirect.com/science/article/pii/B9780123859389000419
15. Leone G (2014) A network of lava tubes as the origin of Labyrinthus Noctis and Valles Marineris on Mars. J Volcanol Geotherm Res 277:1–8
16. Leone G (2016) Alignments of volcanic features in the southern hemisphere of Mars produced by migrating mantle plumes. J Volcanol Geotherm Res 309:78–95
17. Leverington DW (2019) Formation of Ares Vallis (Mars) by effusions of low-viscosity lava within multiple regions of chaotic terrain. Geomorphology [Internet] [cited 2019 Aug 23] 345:106828. https://www.sciencedirect.com/science/article/pii/S0169555X19303009
18. Hopper JP, Leverington DW (2014) Formation of Hrad Vallis (Mars) by low viscosity lava flows. Geomorphology [Internet] [cited 2018 Oct 25] 207:96–113. https://www.sciencedirect.com/science/article/pii/S0169555X13005552
19. Koeppen WC, Hamilton VE (2008) Global distribution, composition, and abundance of olivine on the surface of Mars from thermal infrared data. J Geophys Res E Planets 113(5):E05001
20. Hiesinger H (2004) The Syrtis major volcanic province, Mars: synthesis from Mars global surveyor data. J Geophys Res 109:E01004, https://doi.org/10.1029/2003JE002143
21. Bramble MS, Mustard JF, Salvatore MR (2017) The geological history of Northeast Syrtis Major, Mars. Icarus [Internet] [cited 2019 Aug 18] 293:66–93. https://www.sciencedirect.com/science/article/pii/S0019103516303499?via%3Dihub
22. Peterson JE (1978) Volcanism in the Noachis-Hellas region of Mars, 2. Lunar Planet Sci Conf Proc [Internet] [cited 2019 Sep 6] 3:3411–3432. https://ui.adsabs.harvard.edu/abs/1978LPSC....9.3411P/abstract
23. Greeley R, Crown DA (1990) Volcanic geology of Tyrrhena Patera, Mars. J Geophys Res [Internet] [cited 2018 Sep 19] 95(B5):7133. http://doi.wiley.com/10.1029/JB095iB05p07133
24. Crown DA, Greeley R (1993) Volcanic geology of Hadriaca Patera and the eastern Hellas region of Mars. J Geophys Res Planets [Internet] [cited 2018 Sep 19] 98(E2):3431–3451. https://agupubs.onlinelibrary.wiley.com/doi/abs/10.1029/92JE02804
25. Williams DA, Greeley R, Werner SC, Michael G, Crown DA, Neukum G, et al (2008) Tyrrhena Patera: geologic history derived from Mars express high resolution stereo camera. J Geophys Res [Internet] [cited 2018 Nov 5] 113(E11):E11005. http://doi.wiley.com/10.1029/2008JE003104
26. Michalski JR, Bleacher JE (2013) Supervolcanoes within an ancient volcanic province in Arabia Terra, Mars. Nature 502(7469):47–52
27. Edgett KS (1997) Aeolian dunes as evidence for explosive volcanism in the tharsis region of Mars. Icarus 130(1):96–114
28. Kerber L, Head JW, Madeleine JB, Forget F, Wilson L (2011) The dispersal of pyroclasts from Apollinaris Patera, Mars: implications for the origin of the Medusae Fossae formation. Icarus 216 (1):212–220
29. Brož P, Hauber E (2012) A unique volcanic field in Tharsis, Mars: pyroclastic cones as evidence for explosive eruptions. Icarus [Internet] [cited 2018 Nov 6] 218(1):88–99. https://www.sciencedirect.com/science/article/pii/S001910351100457X
30. Wilson L, Head JW (1994) Mars: review and analysis of volcanic eruption theory and relationships to observed landforms. Rev Geophys [Internet] [cited 2018 Sep 22] 32(3):221. http://doi.wiley.com/10.1029/94RG01113
31. Wilson L (1980) Relationships between pressure, volatile content and ejecta velocity in three types of volcanic explosion. J Volcanol Geotherm Res 8(2–4):297–313

32. Stix J, Phillips JC (2012) An analog investigation of magma fragmentation and degassing: effects of pressure, volatile content, and decompression rate. J Volcanol Geotherm Res 15(211–212):12–23
33. Guzewich SD, Newman CE, de la Torre Juárez M, Wilson RJ, Lemmon M, Smith MD, et al (2016) Atmospheric tides in Gale Crater, Mars. Icarus [Internet] [cited 2019 Apr 5] 268:37–49. https://www.sciencedirect.com/science/article/pii/S0019103515005850
34. Wilson L (2009) Volcanism in the solar system. Nat Geosci 2:389–397. https://doi.org/10.1038/ngeo529
35. Wood CA (1979) Monogenetic volcanoes of the terrestrial planets. In: Lunar Planet Sci Conf 10th, Houston, Tex, March 19–23, 1979, Proceedings, vol 3 (A80-23677 08-91) New York, Pergamon Press Inc, pp 2815–2840 [Internet] [cited 2018 Sep 23] 10:2815–2840. http://adsabs.harvard.edu/abs/1979LPSC...10.2815W
36. Kereszturi G, Nemeth K (2012) Monogenetic basaltic volcanoes: genetic classification, growth, geomorphology and degradation. In: Updates in volcanology—new advances in understanding volcanic systems [Internet]. InTech [cited 2020 Aug 8]. http://www.intechopen.com/books/updates-in-volcanology-new-advances-in-understanding-volcanic-systems/monogenetic-basaltic-volcanoes-genetic-classification-growth-geomorphology-and-degradation
37. Brož P, Hauber E (2013) Hydrovolcanic tuff rings and cones as indicators for phreatomagmatic explosive eruptions on Mars. J Geophys Res Planets [Internet] [cited 2019 Jul 22] 118(8):1656–1675. http://doi.wiley.com/10.1002/jgre.20120
38. Wänke H, Dreibus G (1994) Chemistry and accretion history of Mars. Philos Trans R Soc Lond Ser A Phys Eng Sci [Internet] [cited 2019 Oct 10] 349(1690):285–293. http://www.royalsocietypublishing.org/doi/10.1098/rsta.1994.0132
39. Carr MH, Wänke H (1992) Earth and Mars: water inventories as clues to accretional histories. Icarus 98(1):61–71
40. Grott M, Breuer D (2009) Implications of large elastic thicknesses for the composition and current thermal state of Mars. Icarus 201(2):540–548
41. Hauck SA, Phillips RJ (2002) Thermal and crustal evolution of Mars. J Geophys Res [Internet] [cited 2018 Oct 18] 107(E7):5052. http://doi.wiley.com/10.1029/2001JE001801
42. Breuer D, Spohn T (2006) Viscosity of the Martian mantle and its initial temperature: constraints from crust formation history and the evolution of the magnetic field. Planet Space Sci 54(2):153–169
43. White DSM, Dalton HA, Kiefer WS, Treiman AH (2006) Experimental petrology of the basaltic shergottite Yamato-980459: implications for the thermal structure of the Martian mantle. Meteorit Planet Sci [Internet] [cited 2020 Aug 8] 41 (9):1271–1290. https://onlinelibrary.wiley.com/doi/full/10.1111/j.1945-5100.2006.tb00521.x
44. Fraeman AA, Korenaga J (2010) The influence of mantle melting on the evolution of Mars. Icarus 210 (1):43–57
45. Grott M, Morschhauser A, Breuer D, Hauber E (2011) Volcanic outgassing of CO2 and H2O on Mars. Earth Planet Sci Lett 308(3–4):391–400
46. Morschhauser A, Grott M, Breuer D (2011) Crustal recycling, mantle dehydration, and the thermal evolution of Mars. Icarus 212(2):541–558
47. Ogawa M, Yanagisawa T (2011) Numerical models of Martian mantle evolution induced by magmatism and solid-state convection beneath stagnant lithosphere. J Geophys Res [Internet] [cited 2020 Aug 8] 116(E8):E08008. http://doi.wiley.com/10.1029/2010JE003777
48. Ruedas T, Tackley PJ, Solomon SC (2013) Thermal and compositional evolution of the martian mantle: effects of phase transitions and melting. Phys Earth Planet Inter [Internet] [cited 2019 Aug 19] 216:32–58. https://www.sciencedirect.com/science/article/pii/S0031920112002166
49. Plesa AC, Breuer D (2014) Partial melting in one-plate planets: implications for thermo-chemical and atmospheric evolution. Planet Space Sci 1(98):50–65
50. Sekhar P, King SD (2014) 3D spherical models of Martian mantle convection constrained by melting history. Earth Planet Sci Lett 15(388):27–37
51. Kiefer WS, Li Q (2016) Water undersaturated mantle plume volcanism on present-day Mars. Meteorit Planet Sci [Internet] [cited 2020 Aug 8] 51(11):1993–2010. http://doi.wiley.com/10.1111/maps.12720
52. Filiberto J, Baratoux D, Beaty D, Breuer D, Farcy BJ, Grott M, et al (2016) A review of volatiles in the Martian interior. Meteorit Planet Sci [Internet] [cited 2020 Feb 10] 51(11):1935–1958. http://doi.wiley.com/10.1111/maps.12680
53. Bargery AS, Wilson L (2010) Dynamics of the ascent and eruption of water containing dissolved CO2 on Mars. J Geophys Res [Internet] [cited 2019 Apr 22] 115:5008. https://agupubs.onlinelibrary.wiley.com/doi/pdf/10.1029/2009JE003403
54. Tanaka KL, Robbins SJ, Fortezzo CM, Skinner JA, Hare TM (2014) The digital global geologic map of Mars: chronostratigraphic ages, topographic and crater morphologic characteristics, and updated resurfacing history. Planet Space Sci [Internet] [cited 2019 Mar 26] 95:11–24. https://www.sciencedirect.com/science/article/pii/S0032063313000652
55. Brož P, Čadek O, Hauber E, Rossi AP (2014) Shape of scoria cones on Mars: insights from numerical modeling of ballistic pathways. Earth Planet Sci Lett 1(406):14–23
56. Heap MJ, Byrne PK, Mikhail S (2017) Low surface gravitational acceleration of Mars results in a thick

and weak lithosphere: implications for topography, volcanism, and hydrology. Icarus 1(281):103–114
57. Brož P, Hauber E, Platz T, Balme M (2015) Evidence for Amazonian highly viscous lavas in the southern highlands on Mars. Earth Planet Sci Lett 1 (415):200–212
58. Wilson L, Head JW (2007) Explosive volcanic eruptions on Mars: Tephra and accretionary lapilli formation, dispersal and recognition in the geologic record. J Volcanol Geotherm Res 163(1–4):83–97
59. Wilson L (2004) Evidence for a massive phreatomagmatic eruption in the initial stages of formation of the Mangala Valles outflow channel, Mars. Geophys Res Lett [Internet] [cited 2020 Aug 19] 31(15):L15701. http://doi.wiley.com/10.1029/2004GL020322
60. Glaze LS, Baloga SM (2002) Volcanic plume heights on Mars: limits of validity for convective models. J Geophys Res E Planets [Internet] [cited 2020 Aug 15] 107(10):16–1. https://agupubs.onlinelibrary.wiley.com/doi/full/10.1029/2001JE001830
61. Glaze LS, Baloga SM, Wimert J (2011) Explosive volcanic eruptions from linear vents on Earth, Venus, and Mars: comparisons with circular vent eruptions. J Geophys Res [Internet] [cited 2020 Nov 12] 116(E1):E01011. http://doi.wiley.com/10.1029/2010JE003577
62. Kerber L, Head JW, Madeleine J-B, Forget F, Wilson L (2012) The dispersal of pyroclasts from ancient explosive volcanoes on Mars: implications for the friable layered deposits. Icarus [Internet] [cited 2019 Aug 13] 219(1):358–381. https://www.sciencedirect.com/science/article/pii/S0019103512001091
63. Kerber L, Forget F, Madeleine J-B, Wordsworth R, Head JW, Wilson L (2013) The effect of atmospheric pressure on the dispersal of pyroclasts from martian volcanoes. Icarus [Internet] [cited 2018 Sep 22] 223(1):149–156. https://www.sciencedirect.com/science/article/pii/S0019103512004915
64. Riedel C, Ernst GGJ, Riley M (2003) Controls on the growth and geometry of pyroclastic constructs. J Volcanol Geotherm Res 127(1–2):121–152
65. Kleinhans MG, Markies H, De Vet SJ, In't Veld AC, Postema FN (2011) Static and dynamic angles of repose in loose granular materials under reduced gravity. J Geophys Res E Planets [Internet] [cited 2020 Sep 15] 116(11). https://agupubs.onlinelibrary.wiley.com/doi/full/10.1029/2011JE003865
66. Atwood-Stone C, McEwen AS (2013) Avalanche slope angles in low-gravity environments from active Martian sand dunes. Geophys Res Lett [Internet] [cited 2020 Sep 15] 40(12):2929–2934. http://doi.wiley.com/10.1002/grl.50586
67. McElroy MB, Yung YL, Nier AO (1976) Isotopic composition of nitrogen: implications for the past history of Mars' atmosphere. Science (80) [Internet] [cited 2020 Sep 16] 194(4260):70–72. https://science.sciencemag.org/content/194/4260/70
68. Brain DA, Jakosky BM (1998) Atmospheric loss since the onset of the Martian geologic record: combined role of impact erosion and sputtering. J Geophys Res Planets [Internet] [cited 2019 May 24] 103(E10):22689–22694. http://doi.wiley.com/10.1029/98JE02074
69. Anders E, Owen T (1977) Mars and earth: origin and abundance of volatiles. Science (80) [Internet] [cited 2020 Sep 16] 198(4316):453–465. https://science.sciencemag.org/content/198/4316/453
70. Jakosky BM, Brain D, Chaffin M, Curry S, Deighan J, Grebowsky J et al (2018) Loss of the Martian atmosphere to space: present-day loss rates determined from MAVEN observations and integrated loss through time. Icarus 15(315):146–157
71. Jakosky BM (2019) The CO2 inventory on Mars. Planet Space Sci 1(175):52–59
72. Lillis RJ, Frey HV, Manga M, Mitchell DL, Lin RP, Acuña MH et al (2008) An improved crustal magnetic field map of Mars from electron reflectometry: highland volcano magmatic history and the end of the martian dynamo. Icarus 194(2):575–596
73. Leone G, Tackley PJ, Gerya TV, May DA, Zhu G (2014) Three-dimensional simulations of the southern polar giant impact hypothesis for the origin of the Martian dichotomy. Geophys Res Lett 41 (24):8736–8743
74. Gillmann C, Lognonné P, Moreira M (2011) Volatiles in the atmosphere of Mars: the effects of volcanism and escape constrained by isotopic data. Earth Planet Sci Lett [Internet] [cited 2019 Apr 5] 303(3–4):299–309. https://www.sciencedirect.com/science/article/pii/S0012821X11000239
75. Kurokawa H, Sato M, Ushioda M, Matsuyama T, Moriwaki R, Dohm JM et al (2014) Evolution of water reservoirs on Mars: constraints from hydrogen isotopes in Martian meteorites. Earth Planet Sci Lett 394:179–185
76. Krasnopolsky VA (2015) Variations of the HDO/H2O ratio in the Martian atmosphere and loss of water from Mars. Icarus [Internet] 257:377–386. https://doi.org/10.1016/j.icarus.2015.05.021
77. Villanueva GL, Mumma MJ, Novak RE, Käufl HU, Hartogh P, Encrenaz T et al (2015) Strong water isotopic anomalies in the martian atmosphere: probing current and ancient reservoirs. Science 348(6231):218–221
78. Gillmann C, Lognonné P, Chassefière E, Moreira M (2009) The present-day atmosphere of Mars: where does it come from? Earth Planet Sci Lett 277(3–4):384–393
79. Greeley R (1987) Release of juvenile water on Mars: estimated amounts and timing associated with volcanism. Science [Internet] [cited 2019 Oct 13] 236(4809):1653–1654. http://www.ncbi.nlm.nih.gov/pubmed/17754317
80. Carr MH, Head JW (2015) Martian surface/near-surface water inventory: sources, sinks, and changes

with time. Geophys Res Lett [Internet] [cited 2019 May 23] 42(3):726–732. http://doi.wiley.com/10.1002/2014GL062464
81. Leone G (2020) The absence of an ocean and the fate of water all over the Martian history. Earth Sp Sci [Internet] 7(4):e2019EA001031. https://doi.org/10.1029/2019EA001031
82. Wallace P, Plank T, Edmons M, Hauri E (2018) Volatiles in magmas. In: Carroll MR, Holloway JR (eds). Volatiles in magmas, vol 30. De Gruyter Mouton, 1–517 p
83. Ushioda M, Takahashi E, Hamada M, Suzuki T (2014) Water content in arc basaltic magma in the Northeast Japan and Izu arcs: an estimate from Ca/Na partitioning between plagioclase and melt. Earth Planets Sp [Internet] [cited 2019 Apr 6] 66 (1):127. http://earth-planets-space.springeropen.com/articles/10.1186/1880-5981-66-127
84. Mouginis-Mark PJ, Wilson L, Head JW (1982) Explosive volcanism on Hecates Tholus, Mars: investigation of eruption conditions. J Geophys Res [Internet] [cited 2018 Nov 6] 87(B12):890–904. http://doi.wiley.com/10.1029/JB087iB12p09890
85. Plescia JB (1993) An assessment of volatile release from recent volcanism in Elysium, Mars. Icarus [Internet] [cited 2019 Mar 26] 104(1):20–32. https://www.sciencedirect.com/science/article/pii/S0019103583710791?via%3Dihub
86. O'Neill C, Lenardic A, Jellinek AM, Kiefer WS (2007) Melt propagation and volcanism in mantle convection simulations, with applications for Martian volcanic and atmospheric evolution. J Geophys Res E Planets 112(E07003):1–17
87. Grott M, Hauber E, Werner SC, Kronberg P, Neukum G (2005) High heat flux on ancient Mars: evidence from rift flank uplift at Coracis Fossae. Geophys Res Lett [Internet] [cited 2018 Nov 6] 32 (21):L21201. http://doi.wiley.com/10.1029/2005GL023894
88. Hirschmann MM, Withers AC (2008 Jun 15) Ventilation of CO2 from a reduced mantle and consequences for the early Martian greenhouse. Earth Planet Sci Lett 270(1–2):147–155
89. Johnson MC, Rutherford MJ, Hess PC (1991 Jan 1) Chassigny petrogenesis: melt compositions, intensive parameters and water contents of Martian (?) magmas. Geochim Cosmochim Acta 55(1):349–366
90. Guest A, Smrekar SE (2007 Oct 15) New constraints on the thermal and volatile evolution of Mars. Phys Earth Planet Inter 164(3–4):161–176
91. Balta JB, McSween HY (2013) Water and the composition of Martian magmas. Geology [Internet] [cited 2019 Oct 1] 41(10):1115–1118. https://pubs.geoscienceworld.org/geology/article/41/10/1115-1118/131070
92. Craddock RA, Greeley R (2009) Minimum estimates of the amount and timing of gases released into the martian atmosphere from volcanic eruptions. Icarus 204(2):512–526
93. Ramirez RM, Craddock RA (2018) The geological and climatological case for a warmer and wetter early Mars [Internet], vol 11, Nature Geoscience. Nature Publishing Group [cited 2020 Sep 16], pp 230–237. https://doi.org/10.1038/s41561-018-0093-9
94. Wordsworth RD (2016) The climate of early Mars. Annu Rev Earth Planet Sci [Internet] [cited 2019 Aug 16] 44(1):381–408. http://www.annualreviews.org/doi/10.1146/annurev-earth-060115-012355
95. Postawko SE, Kuhn WR (1986) Effect of the greenhouse gases (CO2, H2O, SO2) on Martian paleoclimate. J Geophys Res Solid Earth [Internet] [cited 2020 Sep 16] 91(B4):431–438. https://agupubs.onlinelibrary.wiley.com/doi/full/10.1029/JB091iB04p0D431%4010.1002/%28ISSN%292169-9356.LPSC16
96. Mischna MA, Baker V, Milliken R, Richardson M, Lee C (2013) Effects of obliquity and water vapor/trace gas greenhouses in the early martian climate. J Geophys Res Planets [Internet] [cited 2020 Sep 16] 118(3):560–576. http://doi.wiley.com/10.1002/jgre.20054
97. Halevy I, Head JW (2014) Episodic warming of early Mars by punctuated volcanism. Nat Geosci 7 (12):865–868
98. Kerber L, Forget F, Wordsworth R (2015) Sulfur in the early martian atmosphere revisited: experiments with a 3-D global climate model. Icarus 15 (261):133–148
99. Wordsworth R, Kalugina Y, Lokshtanov S, Vigasin A, Ehlmann B, Head J, et al (2017) Transient reducing greenhouse warming on early Mars. Geophys Res Lett [Internet] [cited 2020 Sep 16] 44(2):665–671. https://agupubs.onlinelibrary.wiley.com/doi/full/10.1002/2016GL071766
100. Tanaka KL (2000) Dust and ice deposition in the Martian geologic record. Icarus 144(2):254–266
101. Hynek BM, Phillips RJ, Arvidson RE (2003) Explosive volcanism in the Tharsis region: global evidence in the Martian geologic record. J Geophys Res [Internet] [cited 2019 Aug 8] 108(E9):5111. http://doi.wiley.com/10.1029/2003JE002062
102. Moore JM (1990) Nature of the mantling deposit in the heavily cratered terrain of northeastern Arabia, Mars. J Geophys Res [Internet] [cited 2020 Aug 15] 95(B9):14279–14289. https://agupubs.onlinelibrary.wiley.com/doi/full/10.1029/JB095iB09p14279
103. Grant JA, Schultz PH (1990) Gradational epochs on Mars: evidence from West-Northwest of Isidis Basin and Electris. Icarus 84(1):166–195
104. Wendt L, Bishop JL, Neukum G (2013) Knob fields in the Terra Cimmeria/Terra Sirenum region of Mars: stratigraphy, mineralogy and morphology. Icarus 225(1):200–215
105. Mandt KE, de Silva SL, Zimbelman JR, Crown DA (2008) Origin of the Medusae Fossae formation, Mars: insights from a synoptic approach. J Geophys Res E Planets [Internet] [cited 2020 Aug 15] 113

(12). https://agupubs.onlinelibrary.wiley.com/doi/full/10.1029/2008JE003076
106. Scott DH, Tanaka KL (1982) Ignimbrites of Amazonis Planitia region of Mars. J Geophys Res Solid Earth [Internet] [cited 2019 Aug 8] 87(B2):1179–1190. http://doi.wiley.com/10.1029/JB087iB02p01179
107. Bradley BA, Sakimoto SEH, Frey H, Zimbelman JR (2002) Medusae Fossae formation: new perspectives from Mars global surveyor. J Geophys Res E Planets [Internet] [cited 2020 Aug 20] 107(8):2–1. https://agupubs.onlinelibrary.wiley.com/doi/full/10.1029/2001JE001537
108. Squyres SW, Aharonson O, Clark BC, Cohen BA, Crumpler L, De Souza PA, et al (2007) Pyroclastic activity at home plate in Gusev crater, Mars. Science (80) [Internet] [cited 2020 Aug 25] 316 (5825):738–742. www.sciencemag.org/cgi/content/full/316/5825/736/DC1
109. Mandt K, Silva S de, Zimbelman J, Wyrick D (2009) Distinct erosional progressions in the Medusae Fossae formation, Mars, indicate contrasting environmental conditions. Icarus [Internet] [cited 2019 Aug 14] 204(2):471–477. https://www.sciencedirect.com/science/article/pii/S0019103509002772
110. Zimbelman JR, Scheidt SP (2012) Hesperian age for western medusae fossae formation, Mars. Science (80) [Internet] [cited 2020 Aug 25] 336 (6089):1683. http://science.sciencemag.org/content/336/6089/1683
111. Ojha L, Lewis K (2018) The density of the Medusae Fossae formation: implications for its composition, origin, and importance in Martian history. J Geophys Res Planets [Internet] [cited 2020 Aug 25] 123 (6):1368–1379. http://doi.wiley.com/10.1029/2018JE005565
112. Kress AM, Head JW (2015) Late Noachian and early Hesperian ridge systems in the south circumpolar Dorsa Argentea formation, Mars: evidence for two stages of melting of an extensive late Noachian ice sheet. Planet Space Sci [Internet] [cited 2019 Apr 13] 109–110:1–20. https://www.sciencedirect.com/science/article/pii/S0032063314003869
113. Whitten JL, Campbell BA, Plaut JJ (2018) Radar properties of the Dorsa Argentea formation, Mars determined from SHARAD and MARSIS data. AGUFM [Internet] [cited 2020 Aug 31] 2018: P51G-2961. https://ui.adsabs.harvard.edu/abs/2018AGUFM.P51G2961W/abstract
114. Tanaka KL, Kargel JS, MacKinnon DJ, Hare TM, Hoffman N (2002) Catastrophic erosion of Hellas basin rim on Mars induced by magmatic intrusion into volatile-rich rocks. Geophys Res Lett [Internet] [cited 2020 Aug 12] 29(8):37-1–37-4. http://doi.wiley.com/10.1029/2001GL013885
115. Tanaka KL, Leonard GJ (1995) Geology and landscape evolution of the Hellas region of Mars. J Geophys Res [Internet] [cited 2020 Aug 12] 100 (E3):5407. http://doi.wiley.com/10.1029/94JE02804
116. Williams DA, Greeley R, Fergason RL, Kuzmin R, McCord TB, Combe JP et al (2009) The Circum-Hellas Volcanic Province, Mars: overview. Planet Space Sci 57:895–916
117. Bernhardt H, Hiesinger H, Ivanov MA, Ruesch O, Erkeling G, Reiss D (2016) Photogeologic mapping and the geologic history of the Hellas basin floor, Mars. Icarus 15(264):407–442
118. Williams J-P, Dohm JM, Soare RJ, Flahaut J, Lopes RMC, Pathare AV, et al (2017) Long-lived volcanism within Argyre basin, Mars. Icarus [Internet] [cited 2019 Aug 13] 293:8–26. https://www.sciencedirect.com/science/article/pii/S0019103516307254
119. Gregg TKP, Crown DA, Greeley R (1998) Geologic map of part of the Tyrrhena Patera region of Mars (MTM Quadrangle-20252) [Internet]. IMAP [cited 2020 Aug 12]. http://pubs.er.usgs.gov/publication/i2556
120. Gregg TKP, Farley MA (2006) Mafic pyroclastic flows at Tyrrhena Patera, Mars: constraints from observations and models. J Volcanol Geotherm Res [Internet] [cited 2018 Nov 5] 155(1–2):81–89. https://www.sciencedirect.com/science/article/pii/S0377027306000606
121. Ghent RR, Anderson SW, Pithawala TM (2012) The formation of small cones in Isidis Planitia, Mars through mobilization of pyroclastic surge deposits. Icarus 217(1):169–183
122. Ivanov MA, Hiesinger H, Erkeling G, Hielscher FJ, Reiss D (2012) Major episodes of geologic history of Isidis Planitia on Mars. Icarus 218:24–46
123. Smith ML, Claire MW, Catling DC, Zahnle KJ (2014) The formation of sulfate, nitrate and perchlorate salts in the Martian atmosphere. Icarus 1 (231):51–64
124. Burns RG, Fisher DS (1990) Iron-sulfur mineralogy of Mars: magmatic evolution and chemical weathering products. J Geophys Res Solid Earth [Internet] [cited 2020 Sep 16] 95(B9):14415–14421. https://agupubs.onlinelibrary.wiley.com/doi/full/10.1029/JB095iB09p14415%4010.1002/%28ISSN%292169-9356.MARS4
125. Banin A, Han FX, Kan I, Cicelsky A (1997) Acidic volatiles and the Mars soil. J Geophys Res E Planets [Internet] [cited 2020 Sep 16] 102(E6):13341–13356. https://agupubs.onlinelibrary.wiley.com/doi/full/10.1029/97JE01160
126. Quinn RC, Zent AP, Grunthaner FJ, Ehrenfreund P, Taylor CL, Garry JRC (2005) Detection and characterization of oxidizing acids in the Atacama desert using the Mars oxidation instrument. Planet Space Sci 53(13):1376–1388
127. Schiffman P, Zierenberg R, Marks N, Bishop JL, Darby DM (2006 Nov 1) Acid-fog deposition of Kilauea volcano: a possible mechanism for the formation of siliceous-sulfate rock coatings on Mars. Geology 34(11):921–924
128. Hutchins KS, Jakosky BM (1996) Evolution of Martian atmospheric argon: implications for sources

of volatiles. J Geophys Res Planets [Internet] [cited 2020 Feb 10] 101(E6):14933–14949. http://doi.wiley.com/10.1029/96JE00860
129. Slipski M, Jakosky BM (2016) Argon isotopes as tracers for martian atmospheric loss. Icarus 1 (272):212–227
130. Cassidy M, Manga M, Cashman K, Bachmann O (2018) Controls on explosive-effusive volcanic eruption styles [Internet], vol 9. Nature communications. Nature Publishing Group, [cited 2020 Aug 8], pp 1–16. www.nature.com/naturecommunications
131. Castro JM, Gardner JE (2008) Did magma ascent rate control the explosive-effusive transition at the Inyo volcanic chain, California? Geology 36 (4):279–282
132. Edmonds M, Herd RA (2007) A volcanic degassing event at the explosive-effusive transition. Geophys Res Lett [Internet] [cited 2020 Aug 8] 34(21): L21310. http://doi.wiley.com/10.1029/2007GL031379
133. Gonnermann HM, Manga M (2003) Explosive volcanism may not be an inevitable consequence of magma fragmentation. Nature [Internet] [cited 2020 Aug 8] 426(6965):432–435. www.nature.com/nature
134. Walker GPL, Croasdale R (1971) Characteristics of some basaltic pyroclastics. Bull Volcanol [Internet] [cited 2020 Aug 8] 35(2):303–317. https://link.springer.com/article/10.1007/BF02596957
135. Walker GPL (1973) Explosive volcanic eruptions—a new classification scheme. Geol Rundschau [Internet] [cited 2020 Aug 8] 62(2):431–446. https://link.springer.com/article/10.1007/BF01840108
136. Houghton BF, Nairn IA (1991) The 1976–1982 Strombolian and phreatomagmatic eruptions of White Island, New Zealand: eruptive and depositional mechanisms at a "wet" volcano. Bull Volcanol [Internet] [cited 2020 Aug 8] 54(1):25–49. https://link.springer.com/article/10.1007/BF00278204
137. Houghton BF, Wilson CJN, Smith IEM (1999) Shallow-seated controls on styles of explosive basaltic volcanism: a case study from New Zealand. J Volcanol Geotherm Res 91(1–2):97–120
138. Mastin LG, Christiansen RL, Thornber C, Lowenstern J, Beeson M (2004) What makes hydromagmatic eruptions violent? Some insights from the Keanakāko'i Ash, Kīlauea Volcano, Hawai'i. J Volcanol Geotherm Res 137(1–3 SPEC. ISS.):15–31
139. Németh K, Cronin SJ (2011 Apr 15) Drivers of explosivity and elevated hazard in basaltic fissure eruptions: the 1913 eruption of Ambrym Volcano, Vanuatu (SW-Pacific). J Volcanol Geotherm Res 201(1–4):194–209
140. Németh K, Cronin SJ, Smith IEM, Agustin Flores J (2012) Amplified hazard of small-volume monogenetic eruptions due to environmental controls, Orakei Basin, Auckland Volcanic Field, New Zealand. Bull Volcanol [Internet] [cited 2020 Aug 8] 74(9):2121–2137. https://link.springer.com/article/10.1007/s00445-012-0653-6
141. Liu EJ, Cashman KV, Rust AC, Höskuldsson A (2017) Contrasting mechanisms of magma fragmentation during coeval magmatic and hydromagmatic activity: the Hverfjall Fires fissure eruption, Iceland. Bull Volcanol [Internet] [cited 2020 Aug 8] 79(10):1–26. https://link.springer.com/article/10.1007/s00445-017-1150-8
142. Mura A, Adriani A, Tosi F, Lopes RMC, Sindoni G, Filacchione G, et al (2020) Infrared observations of Io from Juno. Icarus 341:113607
143. Roth L, Boissier J, Moullet A, Sánchez-Monge Á, de Kleer K, Yoneda M, et al (2020) An attempt to detect transient changes in Io's SO2 and NaCl atmosphere. Icarus 350:113925
144. Kahn R (1985) The evolution of CO2 on Mars. Icarus 62(2):175–190
145. Malin MC, Caplinger MA, Davis SD (2001) Observational evidence for an active surface reservoir of solid carbon dioxide on Mars. Science (80) [Internet] [cited 2020 Aug 11] 294 (5549):2146–2148. www.sciencemag.org
146. Dundas CM, McEwen AS, Chojnacki M, Milazzo MP, Byrne S, McElwaine JN, et al (2017) Granular flows at recurring slope lineae on Mars indicate a limited role for liquid water. Nat Geosci [Internet] [cited 2019 May 26] 10(12):903–907. http://www.nature.com/articles/s41561-017-0012-5
147. Leighton RB, Murray BC, Sharp RP, Allen JD, Sloan RK (1965) Mariner IV photography of Mars: initial results. Science [Internet] [cited 2019 May 25] 149(3684):627–630. http://www.ncbi.nlm.nih.gov/pubmed/17747569
148. Trokhimovskiy A, Fedorova A, Korablev O, Montmessin F, Bertaux J-L, Rodin A, et al (2015) Mars' water vapor mapping by the SPICAM IR spectrometer: five Martian years of observations. Icarus [Internet] [cited 2019 Apr 5] 251:50–64. https://www.sciencedirect.com/science/article/pii/S0019103514005466
149. Gonnermann HM (2015) Magma fragmentation. Annu Rev Earth Planet Sci [Internet] [cited 2020 Aug 10] 43(1):431–458. http://www.annualreviews.org/doi/10.1146/annurev-earth-060614-105206
150. Sheridan MF, Wohletz KH (1983) Hydrovolcanism: basic considerations and review. J Volcanol Geotherm Res [Internet] [cited 2018 Sep 22] 17 (1–4):1–29. https://www.sciencedirect.com/science/article/pii/0377027383900604
151. Leone G, Wilson L (2001) Density structure of Io and the migration of magma through its lithosphere. J Geophys Res Planets 106(E12):32983–32995
152. Davies AGD (2001) Volcanism on Io: the view from Galileo. Astron Geophy 42(2):10–15. https://doi.org/10.1046/j.1468-4004.2001.42210.x
153. Basaltic Volcanism Study Project (U.S.), Lunar and Planetary Institute (1981) Basaltic volcanism on the terrestrial planets [Internet]. Pergamon Press, [cited

2019 Jan 19], 1286 p. http://ads.harvard.edu/books/bvtp/
154. Nairn IA, Self S (1978) Explosive eruptions and pyroclastic avalanches from Ngauruhoe in February 1975. J Volcanol Geotherm Res 3(1–2):39–60
155. Fitzgerald RH, Tsunematsu K, Kennedy BM, Breard ECP, Lube G, Wilson TM et al (2014) The application of a calibrated 3D ballistic trajectory model to ballistic hazard assessments at Upper Te Maari, Tongariro. J Volcanol Geotherm Res 1 (286):248–262
156. Vanderkluysen L, Harris AJL, Kelfoun K, Bonadonna C, Ripepe M (2012) Bombs behaving badly: unexpected trajectories and cooling of volcanic projectiles. Bull Volcanol [Internet] [cited 2020 Aug 10] 74(8):1849–1858. https://link.springer.com/article/10.1007/s00445-012-0635-8
157. Cataldo E, Wilson L, Lane S, Gilbert J (2002) A model for large-scale volcanic plumes on Io: implications for eruption rates and interactions between magmas and near-surface volatiles. J Geophys Res Planets [Internet] [cited 2020 Aug 10] 107 (E10):19–1. https://agupubs.onlinelibrary.wiley.com/doi/full/10.1029/2001JE001513-1
158. Wilson L, Head JW (1983) A comparison of volcanic eruption processes on Earth, Moon, Mars, Io and Venus. Nature 302:663–669
159. Davies AG, Davies RL, Veeder GJ, de Kleer K, de Pater I, Matson DL, et al (2018) Discovery of a powerful, transient, explosive thermal event at Marduk Fluctus, Io, in *Galileo* NIMS Data. Geophys Res Lett [Internet] [cited 2018 Sep 26] 45 (7):2926–2933. http://doi.wiley.com/10.1002/2018GL077477
160. Head JW, Wilson L (1989 Jul 1) Basaltic pyroclastic eruptions: Influence of gas-release patterns and volume fluxes on fountain structure, and the formation of cinder cones, spatter cones, rootless flows, lava ponds and lava flows. J Volcanol Geotherm Res 37(3–4):261–271
161. Ollila AM, Newsom HE, Clark B, Wiens RC, Cousin A, Blank JG et al (2014) Trace element geochemistry (Li, Ba, Sr, and Rb) using Curiosity's ChemCam: early results for Gale crater from Bradbury Landing site to Rocknest. J Geophys Res E Planets 119:255–285
162. Sautter V, Fabre C, Forni O, Toplis MJ, Cousin A, Ollila AM, et al (2014) Igneous mineralogy at Bradbury rise: the first ChemCam campaign at Gale crater. J Geophys Res Planets [Internet] [cited 2019 Mar 27] 119(1):30–46. http://doi.wiley.com/10.1002/2013JE004472
163. Schmidt ME, Campbell JL, Gellert R, Perrett GM, Treiman AH, Blaney DL, et al (2014) Geochemical diversity in first rocks examined by the curiosity rover in gale crater: evidence for and significance of an alkali and volatile-rich igneous source. J Geophys Res Planets [Internet] [cited 2018 Oct 18] 119 (1):64–81. http://doi.wiley.com/10.1002/2013JE004481
164. Gasparri D, Leone G, Cataldo V, Punjabi V, Nandakumar S (2020) Lava filling of Gale crater from Tyrrhenus Mons on Mars. J Volcanol Geotherm Res 389:106743
165. Liu EJ, Cashman KV, Rust AC, Edmonds M (2018) Insights into the dynamics of mafic magmatic-hydromagmatic eruptions from volatile degassing behaviour: the Hverfjall Fires, Iceland. J Volcanol Geotherm Res 1(358):228–240
166. Saucedo R, Macías JL, Ocampo-Díaz YZE, Gómez-Villa W, Rivera-Olguín E, Castro-Govea R, et al (2017) Mixed magmatic-phreatomagmatic explosions during the formation of the Joya Honda maar, San Luis Potosí, Mexico. In: Geological society special publication [Internet]. Geological Society of London [cited 2020 Aug 11], pp 255–279. https://sp.lyellcollection.org/content/446/1/255
167. Nunns AG, Hochstein MP (2019) Geophysical constraints on the structure and formation of Onepoto, Orakei, Pupuke and Tank Farm maar volcanoes, Auckland volcanic field. New Zeal J Geol Geophys [Internet] [cited 2020 Aug 11] 62 (3):341–356. https://www.tandfonline.com/doi/abs/10.1080/00288306.2019.1581239
168. Graettinger AH (2018 May) Trends in Maar crater size and shape using the global Maar Volcano location and shape (MaarVLS) database. J Volcanol Geotherm Res 15(357):1–13
169. Allen CC (1979 Dec 30) Volcano-ice interactions on Mars. J Geophys Res Solid Earth 84(B14):8048–8059
170. Thordarson T, Self S (1993) The Laki (Skaftár Fires) and Grímsvötn eruptions in 1783–1785. Bull Volcanol [Internet] [cited 2020 Aug 11] 55(4):233–263. https://link.springer.com/article/10.1007/BF00624353
171. Keszthelyi LP, Jaeger WL, Dundas CM, Martínez-Alonso S, McEwen AS, Milazzo MP (2010) Hydrovolcanic features on Mars: preliminary observations from the first Mars year of HiRISE imaging. Icarus [Internet] [cited 2018 Oct 18] 205(1):211–229. https://www.sciencedirect.com/science/article/pii/S0019103509003443
172. Hamilton CW, Fagents SA, Thordarson T (2011) Lava–ground ice interactions in Elysium Planitia, Mars: geomorphological and geospatial analysis of the Tartarus Colles cone groups. J Geophys Res [Internet] [cited 2019 Aug 20] 116(E3):E03004. http://doi.wiley.com/10.1029/2010JE003657
173. Dundas CM (2017) Effects of lava heating on volatile-rich slopes on Io. J Geophys Res Planets [Internet] [cited 2020 Aug 11] 122(3):546–559. http://doi.wiley.com/10.1002/2016JE005177
174. Snyder CW (1979) The planet Mars as seen at the end of the Viking mission. J Geophys Res Solid Earth 84(B14):8487–8519
175. Ingersoll AP (1970) Mars: occurrence of liquid water. Science [Internet] [cited 2019 Apr 17] 168 (3934):972–973. http://www.ncbi.nlm.nih.gov/pubmed/17844187

176. Leone G (2017) Mangala Valles, Mars: a reassessment of formation processes based on a new geomorphological and stratigraphic analysis of the geological units. J Volcanol Geotherm Res 337:62–80
177. Hauber E, Bleacher J, Gwinner K, Williams D, Greeley R (2009) The topography and morphology of low shields and associated landforms of plains volcanism in the Tharsis region of Mars. J Volcanol Geotherm Res 185(1–2):69–95
178. Mouginis-Mark PJ (2005) Prodigious ash deposits near the summit of Arsia Mons volcano, Mars. Geophys Res Lett [Internet] [cited 2019 Apr 5] 29 (16):15-1–15-4. http://doi.wiley.com/10.1029/2002GL015296
179. Hauber E, Brož P, Jagert F, Jodowski P, Platz T (2011) Very recent and wide-spread basaltic volcanism on Mars. Geophys Res Lett [Internet] [cited 2020 Aug 11] 38(10):L10201 https://agupubs.onlinelibrary.wiley.com/doi/full/10.1029/2011GL047310
180. Bouley S, Baratoux D, Paulien N, Missenard Y, Saint- B (2018) The revised tectonic history of Tharsis. Earth Planet Sci Lett 15(488):126–133
181. Plescia JB (2004) Morphometric properties of Martian volcanoes. J Geophys Res [Internet] [cited 2018 Sep 22] 109(E3):E03003. http://doi.wiley.com/10.1029/2002JE002031
182. Williams DA, Greeley R, Zuschneid W, Werner SC, Neukum G, Crown DA, et al (2007) Hadriaca Patera: insights into its volcanic history from Mars express high resolution stereo camera. J Geophys Res [Internet] [cited 2020 Aug 13] 112 (E10):E10004. http://doi.wiley.com/10.1029/2007JE002924
183. Williams DA, Manfredi L, Fergason RL, Combe J-P, Poulet F, Rosemberg C, et al (2010) Surface-compositional properties of the Malea Planum region of the Circum-Hellas Volcanic Province, Mars. Earth Planet Sci Lett [Internet] [cited 2019 Sep 6] 294(3–4):451–465. https://www.sciencedirect.com/science/article/pii/S0012821X09006669
184. Williams DA, Greeley R, Manfredi L, Raitala J, Neukum G (2010) The Circum-Hellas Volcanic Province, Mars: assessment of wrinkle-ridged plains. Earth Planet Sci Lett 294:492–505
185. Leonard GJ, Tanaka KL (2001) Geologic map of the Hellas region of Mars. US Department of the Interior, US Geological Survey, Reston, Virginia
186. Frey H (2008) Ages of very large impact basins on Mars: implications for the late heavy bombardment in the inner solar system. Geophys Res Lett 35(13): L13203
187. Lillis RJ, Manga M, Mitchell DL, Lin RP, Acuna MH (2006) Unusual magnetic signature of the Hadriaca Patera Volcano: implications for early Mars. Geophys Res Lett [Internet] [cited 2020 Aug 12] 33(3):L03202. http://doi.wiley.com/10.1029/2005GL024905
188. Lopes RMC, Kilburn CRJ (1990) Emplacement of lava flow fields: application of terrestrial studies to Alba Patera, Mars. J Geophys Res [Internet] [cited 2019 Mar 27] 95(B9):14383. http://doi.wiley.com/10.1029/JB095iB09p14383
189. Ivanov MA, Head JW (2006) Alba Patera, Mars: topography, structure, and evolution of a unique late Hesperian-early Amazonian shield volcano. J Geophys Res E Planets 111(E9):E09003
190. Cattermole P (1990) Volcanic flow development at Alba Patera, Mars. Icarus [Internet] [cited 2019 Mar 27] 83(2):453–493. https://www.sciencedirect.com/science/article/pii/001910359090079O?via%3Dihub
191. Schneeberger DM, Pieri DC (1991) Geomorphology and stratigraphy of Alba Patera, Mars. J Geophys Res [Internet] [cited 2020 Aug 13] 96 (B2):1907. http://doi.wiley.com/10.1029/90JB01662
192. Mouginis-Mark PJ, Wilson L, Zimbelman JR (1988) Polygenic eruptions on Alba Patera, Mars. Bull Volcanol [Internet] [cited 2020 Aug 13] 50 (6):361–379. https://link.springer.com/article/10.1007/BF01050636
193. Leone G (2015) Transition topography (Mars). In: Encyclopedia of planetary landforms [Internet]. Springer, New York, NY, pp 1–6. https://doi.org/10.1007/978-1-4614-9213-9_650-1
194. Robinson MS, Mouginis PJ, Zimbelman JR, Wu SSC, Ablin KK, Howington AE (1993) Chronology, eruption duration, and atmospheric contribution of the Martian volcano Apollinaris patera. Icarus 104(2):301–323
195. El Maarry MR, Dohm JM, Marzo GA, Fergason R, Goetz W, Heggy E, et al (2012) Searching for evidence of hydrothermal activity at Apollinaris Mons, Mars. Icarus [Internet] [cited 2018 Sep 22] 217(1):297–314. https://www.sciencedirect.com/science/article/pii/S0019103511004155
196. Werner SC (2009) The global Martian volcanic evolutionary history. Icarus 201:44–68
197. Malin MC, Edgett KS (2000) Sedimentary rocks of early Mars. Science (80) [Internet] [cited 2020 Aug 13] 290(5498):1927–1937. www.sciencemag.org
198. Edgett KS, Malin MC (2002) Martian sedimentary rock stratigraphy: outcrops and interbedded craters of northwest Sinus Meridiani and southwest Arabia Terra. Geophys Res Lett [Internet] [cited 2020 Aug 13] 29(24):32-1–32-4. http://doi.wiley.com/10.1029/2002GL016515
199. Bishop JL, Rampe EB (2016) Evidence for a changing Martian climate from the mineralogy at Mawrth Vallis. Earth Planet Sci Lett 15(448):42–48
200. Wada K (1978) Allophane and imogolite. Dev Sedimentol 26(C):147–187

201. Levard C, Basile-Doelsch I (2016) Geology and mineralogy of imogolite-type materials. In: Developments in clay science. Elsevier B.V., pp 49–65
202. Rogers AD, Warner NH, Golombek MP, Head JW, Cowart JC (2018) Areally extensive surface bedrock exposures on Mars: many are clastic rocks, not Lavas. Geophys Res Lett [Internet] [cited 2019 Sep 11] 45(4):1767–1777. http://doi.wiley.com/10.1002/2018GL077030
203. Mandon L, Quantin-Nataf C, Thollot P, Mangold N, Lozac'h L, Dromart G, et al (2020) Refining the age, emplacement and alteration scenarios of the olivine-rich unit in the Nili Fossae region, Mars. Icarus 336:113436
204. Kremer CH, Mustard JF, Bramble MS (2019) A widespread olivine-rich ash deposit on Mars. Geology 47(7):677–681
205. Fawdon P, Skok JR, Balme MR, Vye-Brown CL, Rothery DA, Jordan CJ (2015) The geological history of Nili Patera, Mars. J Geophys Res Planets [Internet] [cited 2020 Aug 15] 120(5):951–977. http://doi.wiley.com/10.1002/2015JE004795
206. Robbins SJ, Achille G Di, Hynek BM. The volcanic history of Mars: high-resolution crater-based studies of the calderas of 20 volcanoes. Icarus [Internet] [cited 2018 Nov 13] 211(2):1179–1203. https://www.sciencedirect.com/science/article/pii/S0019103510004318
207. Plescia JB (2000) Geology of the Uranius Group Volcanic constructs: Uranius Patera, Ceraunius Tholus, and Uranius Tholus. Icarus [Internet] [cited 2019 May 17] 143(2):376–396. https://www.sciencedirect.com/science/article/abs/pii/S0019103599962590
208. Reimers CE, Komar PD (1979) Evidence for explosive volcanic density currents on certain Martian volcanoes. Icarus [Internet] [cited 2018 Sep 22] 39(1):88–110. https://www.sciencedirect.com/science/article/pii/0019103579901039
209. Plescia JB (2003) Tharsis Tholus: an unusual Martian volcano. Icarus 165(2):223–241
210. Platz T, Münn S, Walter TR, Procter JN, McGuire PC, Dumke A et al (2011) Vertical and lateral collapse of Tharsis Tholus, Mars. Earth Planet Sci Lett. 305(3–4):445–455
211. Mouginis-Mark PJ, Christensen PR (2005) New observations of volcanic features on Mars from the THEMIS instrument. J Geophys Res Planets [Internet] [cited 2019 Jul 30];110(E8). http://doi.wiley.com/10.1029/2005JE002421
212. Hauber E, Van Gasselt S, Ivanov B, Werner S, Head JW, Neukum G, et al (2005) Discovery of a flank caldera and very young glacial activity at Hecates Tholus, Mars. Nature [Internet] [cited 2020 Aug 14] 434(7031):356–61. www.nature.com/nature
213. De Pablo MA, Michael GG, Centeno JD (2013) Age and evolution of the lower NW flank of the Hecates Tholus volcano, Mars, based on crater size-frequency distribution on CTX images. Icarus 226 (1):455–469
214. de Pablo Hernández MÁ, Carrillo JDC (2012) Geomorphological map of the lower NW flank of the Hecates Tholus volcano, Mars (scale 1:100,000). J Maps [Internet] [cited 2020 Aug 14] 8(3):208–214. https://www.tandfonline.com/doi/full/10.1080/17445647.2012.703902
215. Stewart EM, Head JW (2001) Ancient Martian volcanoes in the Aeolis region: new evidence from MOLA data, JGeophys Res 106(E8):17505–17513. https://doi.org/10.1029/2000JE001322
216. Bleacher JE, Greeley R, Williams DA, Cave SR, Neukum G (2007) Trends in effusive style at the Tharsis Montes, Mars, and implications for the development of the Tharsis province. J Geophys Res E Planets 112:E09005, https://doi.org/10.1029/2006JE002873
217. Moyer EJ, Irion FW, Yung YL, Gunson MR (1996) ATMOS stratospheric deuterated water and implications for troposphere-stratosphere transport. Geophys Res Lett [Internet] [cited 2019 Apr 5] 23 (17):2385–2388. https://agupubs.onlinelibrary.wiley.com/doi/abs/10.1029/96GL01489%4010.1002/%28ISSN%291944-8007.ATLASS1
218. Peters SI, Christensen PR (2017) Flank vents and graben as indicators of Late Amazonian volcanotectonic activity on Olympus Mons. J Geophys Res Planets 122:501–523
219. Carr MH, Greeley R, Blasius KR, Guest JE, Murray JB (1977) Some Martian volcanic features as viewed from the Viking orbiters. J Geophys Res [Internet] [cited 2020 Aug 17] 82(28):3985–4015. https://agupubs.onlinelibrary.wiley.com/doi/full/10.1029/JS082i028p03985
220. Frey H, Jarosewich M (1982) Subkilometer Martian volcanoes: properties and possible terrestrial analogs. J Geophys Res [Internet] [cited 2018 Nov 6] 87(B12):9867–9879. http://doi.wiley.com/10.1029/JB087iB12p09867
221. Plescia JB (1994) Geology of the small Tharsis Volcanoes: Jovis Tholus, Ulysses Patera, BibIls Patera, Mars. Icarus 111(1):246–269
222. West M (1974) Martian volcanism: additional observations and evidence for pyroclastic activity. Icarus 21(1):1–11
223. Keszthelyi L, Jaeger W, McEwen A, Tornabene L, Beyer RA, Dundas C et al (2008) High resolution imaging science experiment (HiRISE) images of volcanic terrains from the first 6 months of the Mars reconnaissance orbiter primary science phase. J Geophys Res E Planets 113:E04005, https://doi.org/10.1029/2007JE002968
224. Harrison KP, Chapman MG (2008) Evidence for ponding and catastrophic floods in central Valles Marineris, Mars. Icarus 198(2):351–364
225. Brož P, Hauber E, Wray JJ, Michael G (2017) Amazonian volcanism inside Valles Marineris on Mars. Earth Planet Sci Lett 1(473):122–130

226. Meresse S, Costard F, Mangold N, Masson P, Neukum G (2008) Formation and evolution of the chaotic terrains by subsidence and magmatism: hydraotes Chaos, Mars. Icarus 194(2):487–500
227. Korteniemi J, Kukkonen S (2018) Volcanic structures within Niger and Dao Valles, Mars, and implications for outflow channel evolution and Hellas basin rim development. Geophys Res Lett [Internet] [cited 2020 Aug 17] 45(7):2934–2944. https://onlinelibrary.wiley.com/doi/abs/10.1002/2018GL077067
228. Lanz JK, Saric MB (2009) Cone fields in SW Elysium Planitia: hydrothermal venting on Mars? J Geophys Res [Internet] [cited 2018 Oct 18] 114 (E2):E02008. http://doi.wiley.com/10.1029/2008JE003209
229. Lanz JK, Wagner R, Wolf U, Kröchert J, Neukum G (2010) Rift zone volcanism and associated cinder cone field in Utopia Planitia, Mars. J Geophys Res [Internet] [cited 2020 Aug 18] 115 (E12):E12019. http://doi.wiley.com/10.1029/2010JE003578
230. Skinner JA, Tanaka KL (2007 Jan 1) Evidence for and implications of sedimentary diapirism and mud volcanism in the southern Utopia highland-lowland boundary plain, Mars. Icarus 186(1):41–59
231. Okubo CH (2016) Morphologic evidence of subsurface sediment mobilization and mud volcanism in Candor and Coprates Chasmata, Valles Marineris, Mars. Icarus 1(269):23–37
232. Wheatley DF, Chan MA, Okubo CH (2019) Clastic pipes and mud volcanism across Mars: terrestrial analog evidence of past Martian groundwater and subsurface fluid mobilization. Icarus 1(328):141–151
233. Senthil Kumar P, Krishna N, Prasanna Lakshmi KJ, Raghukanth STG, Dhabu A, Platz T (Jan) Recent seismicity in Valles Marineris, Mars: insights from young faults, landslides, boulder falls and possible mud volcanoes. Earth Planet Sci Lett 1(505):51–64
234. Fagents SA, Lanagan P, Greeley R (2002) Rootless cones on Mars: a consequence of lava-ground ice interaction. Geol Soc Lond Spec Publ [Internet] [cited 2019 Aug 20] 202(1):295–317. http://sp.lyellcollection.org/lookup/doi/10.1144/GSL.SP.2002.202.01.15
235. Reynolds P, Brown RJ, Thordarson T, Llewellin EW, Fielding K (2015) Rootless cone eruption processes informed by dissected tephra deposits and conduits. Bull Volcanol [Internet] [cited 2020 Aug 18] 77(9):1–17. http://imagej.nih
236. Cattermole P (1986) Linear volcanic features at Alba Patera, Mars—PRobable spatter ridges. J Geophys Res [Internet] [cited 2020 Aug 18] 91(B13): E159. https://agupubs.onlinelibrary.wiley.com/doi/full/10.1029/JB091iB13p0E159
237. Greeley R, Fagents SA (2001) Icelandic pseudocraters as analogs to some volcanic cones on Mars. J Geophys Res Planets [Internet] [cited 2019 Aug 20] 106(E9):20527–20546. http://doi.wiley.com/10.1029/2000JE001378
238. Baker VR (2001) Water and the martian landscape. Nature [Internet] [cited 2019 Apr 11] 412 (6843):228–236. http://www.nature.com/articles/35084172
239. Feldman WC, Prettyman TH, Maurice S, Plaut JJ, Bish DL, Vaniman DT, et al (2004) Global distribution of near-surface hydrogen on Mars. J Geophys Res E Planets [Internet] [cited 2020 Aug 18] 109(9). https://agupubs.onlinelibrary.wiley.com/doi/full/10.1029/2003JE002160
240. Christensen PR (2006) Water at the Poles and in Permafrost regions of Mars. Elements [Internet] [cited 2019 Apr 5] 2(3):151–155. https://pubs.geoscienceworld.org/elements/article/2/3/151-155/137698
241. Byrne S, Dundas CM, Kennedy MR, Mellon MT, McEwen AS, Cull SC, et al (2009) Distribution of mid-latitude ground ice on mars from new impact craters. Science (80) [Internet] [cited 2020 Aug 18] 325(5948):1674–1676. http://science.sciencemag.org/
242. Smith PH, Tamppari LK, Arvidson RE, Bass D, Blaney D, Boynton W V, et al (2009) H2O at the Phoenix landing site. Science [Internet] [cited 2019 Apr 5] 325(5936):58–61. http://www.ncbi.nlm.nih.gov/pubmed/19574383
243. Vincendon M, Forget F, Mustard J (2010) Water ice at low to midlatitudes on Mars. J Geophys Res [Internet] [cited 2020 Aug 18] 115(E10):E10001. http://doi.wiley.com/10.1029/2010JE003584
244. Rossbacher LA, Judson S (1981) Ground ice on Mars: inventory, distribution, and resulting landforms. Icarus [Internet] [cited 2019 Aug 8] 45 (1):39–59. https://www.sciencedirect.com/science/article/pii/0019103581900051
245. Head JW, Wilson L (2007) Heat transfer in volcano-ice interactions on Mars: synthesis of environments and implications for processes and landforms. In: Annals of glaciology
246. Wilson L, Mouginis-Mark PJ (2003) Phreatomagmatic dike-cryosphere interactions as the origin of small ridges north of Olympus Mons, Mars. Icarus 165(2):242–252
247. Chapman MG, Tanaka KL (2001) Interior trough deposits on Mars: Subice volcanoes? J Geophys Res Planets [Internet] [cited 2020 Aug 19] 106 (E5):10087–10100. http://doi.wiley.com/10.1029/2000JE001303
248. Ghatan GJ, Head JW III (2002) Candidate subglacial volcanoes in the south polar region of Mars: morphology, morphometry, and eruption conditions. J Geophys Res Planets 107(E7):5048–5071
249. Chapman MG, Allen CC, Gudmundsson MT, Gulick VC, Jakobsson SP, Lucchitta BK, et al (2000) Volcanism and ice interactions on Earth and Mars. In: Environmental effects on volcanic eruptions [Internet]. Springer, US [cited 2020 Aug 19],

pp 39–73. https://link.springer.com/chapter/10.1007/978-1-4615-4151-6_3
250. Christiansen EH (1989) Lahars in the Elysium region of Mars. Geology [Internet] [cited 2018 Oct 22] 17(3):203. https://pubs.geoscienceworld.org/geology/article/17/3/203-206/204846
251. Russell PS, Head JW (2003) Elysium-Utopia flows as mega-lahars: a model of dike intrusion, cryosphere cracking, and water-sediment release. J Geophys Res E Planets [Internet] [cited 2020 Aug 18] 108(6). https://agupubs.onlinelibrary.wiley.com/doi/full/10.1029/2002JE001995
252. Wilson L, Mouginis-Mark PJ (2003) Phreatomagmatic explosive origin of Hrad Vallis, Mars. J Geophys Res E Planets [Internet] [cited 2020 Aug 18] 108(8):1–1. https://agupubs.onlinelibrary.wiley.com/doi/full/10.1029/2002JE001927
253. Pedersen GBM (2013) Frozen Martian lahars? Evaluation of morphology, degradation and geologic development in the Utopia-Elysium transition zone. Planet Space Sci 85:59–77
254. Lavigne F, Thouret JC, Voight B, Suwa H, Sumaryono A (2000) Lahars at Merapi volcano, central Java: an overview. J Volcanol Geotherm Res 100(1–4):423–456
255. Major JJ, Newhall CG (1989) Snow and ice perturbation during historical volcanic eruptions and the formation of lahars and floods—a global review [Internet], vol 52. Bulletin of volcanology. Springer [cited 2020 Aug 18], pp 1–27. https://link.springer.com/article/10.1007/BF00641384
256. Leverington DW (2007) Was the Mangala Valles system incised by volcanic flows? J Geophys Res [Internet] [cited 2018 Oct 25] 112(E11):E11005. http://doi.wiley.com/10.1029/2007JE002896
257. Leverington DW (2004) Volcanic rilles, streamlined islands, and the origin of outflow channels on Mars. J Geophys Res [Internet] [cited 2018 Oct 25] 109 (E10):E10011. http://doi.wiley.com/10.1029/2004JE002311
258. Leverington DW (2006) Volcanic processes as alternative mechanisms of landform development at a candidate crater-lake site near Tyrrhena Patera, Mars. J Geophys Res E Planets 111(E11):E11002
259. Leverington DW (2011) A volcanic origin for the outflow channels of Mars: key evidence and major implications. Geomorphology 132:51–75
260. Gudmundsson MT, Sigmundsson F, Björnsson H (1997) Ice-volcano interaction of the 1996 Gjalp subglacial eruption, Vatnajokull, Iceland. Nature [Internet] [cited 2020 Aug 19] 389(6654):954–957. https://www.nature.com/articles/40122
261. Honnorez J, Kirst P (1975) Submarine basaltic volcanism: morphometric parameters for discriminating hyaloclastites from hyalotuffs. Bull Volcanol [Internet] [cited 2020 Aug 19] 39(3):441–465. https://link.springer.com/article/10.1007/BF02597266
262. Skilling IP (1994) Evolution of an englacial volcano: Brown Bluff, Antarctica. Bull Volcanol [Internet] [cited 2020 Aug 19] 56(6–7):573–591. https://link.springer.com/article/10.1007/BF00302837
263. Chapman MG (1994 Jun 1) Evidence, age, and thickness of a Frozen Paleolake in Utopia Planitia, Mars. Icarus 109(2):393–406
264. Chapman MG (2003 Feb 1) Sub-ice volcanoes and ancient oceans/lakes: a Martian challenge. Glob Planet Change 35(3–4):185–198
265. Martínez-Alonso S, Mellon MT, Banks ME, Keszthelyi LP, McEwen AS (2011) Evidence of volcanic and glacial activity in Chryse and Acidalia Planitiae, Mars. Icarus 212(2):597–621
266. Scanlon KE, Head JW, Wilson L, Marchant DR (2014) Volcano–ice interactions in the Arsia Mons tropical mountain glacier deposits. Icarus [Internet] [cited 2019 Apr 5] 237:315–339. https://www.sciencedirect.com/science/article/pii/S0019103514002164
267. Ann Hodges C, Moore HJ (1979) The subglacial birth of Olympus Mons and its aureoles. J Geophys Res [Internet] [cited 2020 Aug 19] 84(B14):8061. https://agupubs.onlinelibrary.wiley.com/doi/full/10.1029/JB084iB14p08061
268. Frey H, Lowry BL, Chase SA (1979) Pseudocraters on Mars. J Geophys Res [Internet] [cited 2019 Aug 20] 84(B14):8075. http://doi.wiley.com/10.1029/JB084iB14p08075
269. Mouginis PJ (1985 Nov 1) Volcano/ground ice interactions in Elysium Planitia, Mars. Icarus 64 (2):265–284
270. Malin MC (1992) Mars observer camera. J Geophys Res [Internet] [cited 2020 Aug 20] 97(E5):7699–7718. https://agupubs.onlinelibrary.wiley.com/doi/full/10.1029/92JE00340
271. Tanaka KL, Fortezzo CM, Hayward RK, Rodriguez JAP, Skinner JA (2011) History of plains resurfacing in the Scandia region of Mars. Planet Space Sci 59(11–12):1128–1142
272. Greeley R, Guest JE (1987) Geologic map of the eastern equatorial region of Mars [Internet] IMAP [cited 2019 Jul 27]. https://pubs.er.usgs.gov/publication/i1802B
273. Bruno BC, Fagents SA, Thordarson T, Baloga SM, Pilger E (2004) Clustering within rootless cone groups on Iceland and Mars: Effect of nonrandom processes. J Geophys Res [Internet] [cited 2019 Aug 20] 109(E7):E07009. http://doi.wiley.com/10.1029/2004JE002273
274. Bruno BC, Fagents SA, Hamilton CW, Burr DM, Baloga SM (2006) Identification of volcanic rootless cones, ice mounds, and impact craters on Earth and Mars: using spatial distribution as a remote sensing tool. J Geophys Res [Internet] [cited 2020 Aug 20] 111(E6):E06017. http://doi.wiley.com/10.1029/2005JE002510
275. Noguchi R, Kurita K (2015) Unique characteristics of cones in Central Elysium Planitia, Mars. Planet Space Sci 111(1):44–54

276. Dundas CM, Keszthelyi LP (2013) Modeling steam pressure under martian lava flows. Icarus 226 (1):1058–1067
277. Hamilton CW, Fagents SA, Wilson L (2010) Explosive lava-water interactions in Elysium Planitia, Mars: geologic and thermodynamic constraints on the formation of the Tartarus Colles cone groups. J Geophys Res [Internet] [cited 2020 Aug 20] 115 (E9):E09006. http://doi.wiley.com/10.1029/2009JE003546
278. Bertagnini A, Landi P, Santacroce R, Sbrana A (1991) The 1906 eruption of Vesuvius: from magmatic to phreatomagmatic activity through the flashing of a shallow depth hydrothermal system. Bull Volcanol [Internet] [cited 2020 Aug 19] 53 (7):517–532. https://link.springer.com/article/10.1007/BF00298153
279. Clarke H, Troll VR, Carracedo JC (2009) Phreatomagmatic to Strombolian eruptive activity of basaltic cinder cones: Montaña Los Erales, Tenerife Canary Islands. J Volcanol Geotherm Res 180(2–4):225–245
280. Cioni R, Sbrana A, Vecci R (1992) Morphologic features of juvenile pyroclasts from magmatic and phreatomagmatic deposits of Vesuvius. J Volcanol Geotherm Res 51(1–2):61–78
281. Mouginis PJ, Wilson L (2019) Late-stage intrusive activity at Olympus Mons, Mars: summit inflation and giant dike formation. Icarus 1(319):459–469
282. Pedersen GBM, Head JW, Wilson L (2010) Formation, erosion and exposure of early Amazonian dikes, dike swarms and possible subglacial eruptions in the Elysium Rise/Utopia Basin Region, Mars. Earth Planet Sci Lett 294(3–4):424–439
283. Levy JS, Goudge TA, Head JW, Fassett CI (2017) Candidate volcanic and impact-induced ice depressions on Mars. Icarus 15(285):185–194
284. Scott ED, Wilson L (1999) Evidence for a sill emplacement event on the upper flanks of the Ascraeus Mons shield volcano, Mars. J Geophys Res Planets [Internet] [cited 2020 Aug 19] 104 (E11):27079–27089. http://doi.wiley.com/10.1029/1999JE001049
285. Morris AR, Mouginis PJ (2006) Thermally distinct craters near Hrad Vallis, Elysium Planitia, Mars. Icarus 180(2):335–347
286. Pajola M, Rossato S, Baratti E, Mangili C, Mancarella F, McBride K, et al (2016) The Simud–Tiu Valles hydrologic system: a multidisciplinary study of a possible site for future Mars on-site exploration. Icarus [Internet] [cited 2019 Jul 7] 268:355–381. https://www.sciencedirect.com/science/article/pii/S0019103516000038
287. Picardi G, Plaut JJ, Biccari D, Bombaci O, Calabrese D, Cartacci M, et al (2005) Planetary science: Radar soundings of the subsurface of Mars. Science (80) [Internet] [cited 2020 Aug 24] 310(5756):1925–1928. http://science.sciencemag.org/
288. Watters TR, Campbell B, Carter L, Leuschen CJ, Plaut JJ, Picardi G, et al (2007) Radar sounding of the Medusae Fossae formation Mars: equatorial ice or dry, low-density deposits? Science [Internet] [cited 2019 Apr 6] 318(5853):1125–1128. http://www.ncbi.nlm.nih.gov/pubmed/17975034
289. Orosei R, Rossi AP, Cantini F, Caprarelli G, Carter LM, Papiano I, et al (2017) Radar sounding of Lucus Planum, Mars, by MARSIS. J Geophys Res Planets [Internet] [cited 2020 Aug 24] 122 (7):1405–1418. http://doi.wiley.com/10.1002/2016JE005232
290. Ganesh I, Carter LM, Smith IB (2020) SHARAD mapping of Arsia Mons caldera. J Volcanol Geotherm Res 390:106748
291. Fassett CI, Head JW (2007) Valley formation on martian volcanoes in the Hesperian: evidence for melting of summit snowpack, caldera lake formation, drainage and erosion on Ceraunius Tholus. Icarus [Internet] [cited 2019 Jul 20] 189(1):118–135. https://www.sciencedirect.com/science/article/pii/S001910350700022X
292. Fergason RL, Christensen PR (2008) Formation and erosion of layered materials: geologic and dust cycle history of eastern Arabia Terra, Mars. J Geophys Res E Planets [Internet] [cited 2020 Aug 24] 113(12). https://agupubs.onlinelibrary.wiley.com/doi/full/10.1029/2007JE002973
293. Lewis KW, Aharonson O (2014) Occurrence and origin of rhythmic sedimentary rocks on Mars. J Geophys Res Planets [Internet] [cited 2020 Aug 24] 119(6):1432–1457. http://doi.wiley.com/10.1002/2013JE004404
294. Annex AM, Lewis KW (2020) Regional correlations in the layered deposits of Arabia Terra, Mars. J Geophys Res Planets [Internet] [cited 2020 Aug 24] 125(6). https://onlinelibrary.wiley.com/doi/abs/10.1029/2019JE006188
295. Sullivan R, Arvidson R, Bell JF, Gellert R, Golombek M, Greeley R, et al (2008) Wind-driven particle mobility on Mars: insights from Mars exploration Rover observations at "El Dorado" and surroundings at Gusev Crater. J Geophys Res [Internet] [cited 2020 Aug 25] 113(E6). https://agupubs.onlinelibrary.wiley.com/doi/full/10.1029/2008JE003101
296. Adeli S, Hauber E, Kleinhans M, Le Deit L, Platz T, Fawdon P, et al (2016) Amazonian-aged fluvial system and associated ice-related features in Terra Cimmeria, Mars. Icarus [Internet] [cited 2019 Jul 7] 277:286–299. https://www.sciencedirect.com/science/article/abs/pii/S0019103516301786
297. Michalski JR, Dobrea EZN, Niles PB, Cuadros J (2017) Ancient hydrothermal seafloor deposits in Eridania basin on Mars. Nat Commun [Internet]

[cited 2020 Aug 24] 8(1):1–10. www.nature.com/naturecommunications
298. Grant JA, Wilson SA, Noe Dobrea E, Fergason RL, Griffes JL, Moore JM et al (2010) HiRISE views enigmatic deposits in the Sirenum Fossae region of Mars. Icarus 205(1):53–63
299. Adeli S, Hauber E, Le Deit L, Jaumann R (2015) Geologic evolution of the eastern Eridania basin: implications for aqueous processes in the southern highlands of Mars. J Geophys Res Planets [Internet] [cited 2020 Aug 24] 120(11):1774–1799. http://doi.wiley.com/10.1002/2015JE004898
300. Ehlmann BL, Mustard JF, Murchie SL (2010) Geologic setting of serpentine deposits on Mars. Geophys Res Lett 37:L06201
301. Che C, Glotch TD (2014) Thermal alteration: a possible reason for the inconsistency between OMEGA/CRISM and TES detections of phyllosilicates on Mars? Geophys Res Lett 41:321–327
302. Niles PB, Michalski J, Ming DW, Golden DC (2017) Elevated olivine weathering rates and sulfate formation at cryogenic temperatures on Mars. Nat Commun [Internet] [cited 2020 Aug 28] 8(1):1–5. www.nature.com/naturecommunications
303. Leverington DW (2020) Incision of Ma'adim Vallis (Mars) by dry volcanic megafloods effused from multiple highland sources. Planet Space Sci 191:105021
304. Leone G (2014) A network of lava tubes as the origin of Labyrinthus Noctis and Valles Marineris on Mars. J Volcanol Geotherm Res 277:1–8
305. Sakimoto SEH, Frey HV, Garvin JB, Roark JH (1999) Topography, roughness, layering, and slope properties of the Medusae Fossae formation from Mars Orbiter Laser Altimeter (MOLA) and Mars Orbiter Camera (MOC) data. J Geophys Res Planets [Internet] [cited 2020 Aug 25] 104(E10):24141–24154. http://doi.wiley.com/10.1029/1999JE001044
306. Kerber L, Head JW (2010) The age of the Medusae Fossae formation: evidence of Hesperian emplacement from crater morphology, stratigraphy, and ancient lava contacts. Icarus 206(2):669–684
307. Harrison SK, Balme MR, Hagermann A, Murray JB, Muller JP (2010) Mapping Medusae Fossae formation materials in the southern highlands of Mars. Icarus 209(2):405–415
308. de Silva SL, Bailey JE, Mandt KE, Viramonte JM (2010) Yardangs in terrestrial ignimbrites: synergistic remote and field observations on Earth with applications to Mars. Planet Space Sci [Internet] [cited 2019 Aug 14] 58(4):459–471. https://www.sciencedirect.com/science/article/pii/S0032063309003018
309. Zimbelman JR, Griffin LJ (2010) HiRISE images of yardangs and sinuous ridges in the lower member of the Medusae Fossae formation, Mars. Icarus 205:198–210
310. Kerber L, Head JW (2012) A progression of induration in Medusae Fossae Formation transverse aeolian ridges: evidence for ancient aeolian bedforms and extensive reworking. Earth Surf Process Landforms [Internet] [cited 2020 Aug 25] 37 (4):422–433. http://doi.wiley.com/10.1002/esp.2259
311. Liu J, Di K, Gou S, Yue Z, Liu B, Xiao J, et al (2020) Mapping and spatial statistical analysis of Mars Yardangs. Planet Space Sci 192:105035
312. Schultz PH, Lutz AB (1988) Polar wandering of Mars. Icarus 73(1):91–141
313. Forsythe RD, Zimbelman JR (1988) Is the Gordii Dorsum escarpment on Mars an exhumed transcurrent fault? Nature 336(6195):143–146
314. Mouginis-Mark PJ, Zimbelman JR (2020) Rafted pumice: a new model for the formation of the Medusae Fossae formation, Mars. Icarus 343:113684
315. Bandfield JL (2002) Global mineral distributions on Mars. J Geophys Res 107(E6). doi:10.1029/2001JE001510
316. Ruff SW, Christensen PR (2002) Bright and dark regions on Mars: particle size and mineralogical characteristics based on thermal emission spectrometer data. J Geophys Res E Planets [Internet] [cited 2020 Aug 26] 107(12):2–1. https://agupubs.onlinelibrary.wiley.com/doi/full/10.1029/2001JE001580
317. Ruff SW, Christensen PR (2007) Basaltic andesite, altered basalt, and a TES-based search for smectite clay minerals on Mars. Geophys Res Lett [Internet] [cited 2020 Aug 26] 34(10):L10204. http://doi.wiley.com/10.1029/2007GL029602
318. Ody A, Poulet F, Langevin Y, Bibring JP, Bellucci G, Altieri F, et al (2012) Global maps of anhydrous minerals at the surface of Mars from OMEGA/MEx. J Geophys Res E Planets [Internet] [cited 2020 Mar 5];117(9). http://doi.wiley.com/10.1029/2012JE004117
319. Ody A, Poulet F, Bibring JP, Loizeau D, Carter J, Gondet B, et al (2013) Global investigation of olivine on Mars: insights into crust and mantle compositions. J Geophys Res E Planets [Internet] [cited 2020 Mar 5] 118(2):234–262. http://doi.wiley.com/10.1029/2012JE004149
320. Harmon JK, Arvidson RE, Guinness EA, Campbell BA, Slade MA (1999) Mars mapping with delay-Doppler radar. J Geophys Res Planets [Internet] [cited 2020 Aug 26] 104(E6):14065–14089. http://doi.wiley.com/10.1029/1998JE900042
321. Carter LM, Campbell BA, Watters TR, Phillips RJ, Putzig NE, Safaeinili A et al (2009) Shallow radar (SHARAD) sounding observations of the Medusae Fossae formation, Mars. Icarus 199(2):295–302
322. Hood LL, Harrison KP, Langlais B, Lillis RJ, Poulet F, Williams DA (2010) Magnetic anomalies near Apollinaris Patera and the Medusae Fossae formation in Lucus Planum, Mars. Icarus 208 (1):118–131
323. Wilson L, Head JW (2009) Tephra deposition on glaciers and ice sheets on Mars: influence on ice

survival, debris content and flow behavior. J Volcanol Geotherm Res [Internet] [cited 2019 Apr 6] 185(4):290–297. https://www.sciencedirect.com/science/article/pii/S037702730800543X
324. Banfield D, Spiga A, Newman C, Forget F, Lemmon M, Lorenz R, et al (2020) The atmosphere of Mars as observed by InSight. Nat Geosci [Internet] [cited 2020 Aug 28] 13:190–198.https://doi.org/10.1038/s41561-020-0534-0
325. Lellouch E, Rosenqvist J, Goldstein JJ, Bougher SW, Paubert G (1991) First absolute wind measurements in the middle atmosphere of Mars. Astrophys J [Internet] [cited 2020 Aug 28] 383:401. https://ui.adsabs.harvard.edu/abs/1991ApJ...383..401L/abstract
326. Moreno R, Lellouch E, Forget F, Encrenaz T, Guilloteau S, Millour E (2009) Wind measurements in Mars' middle atmosphere: IRAM Plateau de Bure interferometric CO observations. Icarus 201 (2):549–563
327. Lee SW, Thomas PC, Veverka J (1982) Wind streaks in Tharsis and Elysium: implications for sediment transport by slope winds. J Geophys Res Solid Earth [Internet] [cited 2020 Aug 28] 87 (B12):10025–10041. https://agupubs.onlinelibrary.wiley.com/doi/full/10.1029/JB087iB12p10025%4010.1002/%28ISSN%292169-9356.MARS3
328. Muhleman DO, Butler BJ, Grossman AW, Slade MA (1991) Radar images of mars. Science (80) [Internet] [cited 2020 Aug 28] 253 (5027):1508–1513. https://science.sciencemag.org/content/253/5027/1508
329. Edgett KS, Butler BJ, Zimbelman JR, Hamilton VE (1997) Geologic context of the Mars radar "Stealth" region in southwestern Tharsis. J Geophys Res Planets [Internet] [cited 2020 Aug 28] 102 (E9):21545–21567. http://doi.wiley.com/10.1029/97JE01685
330. Ojha L, Lewis K, Karunatillake S, Schmidt M (2018) The Medusae Fossae formation as the single largest source of dust on Mars. Nat Commun [Internet] [cited 2020 Aug 29] 9(1):1–7. www.nature.com/naturecommunications
331. Boynton WV, Taylor GJ, Evans LG, Reedy RC, Starr R, Janes DM, et al (2007) Concentration of H, Si, Cl, K, Fe, and Th in the low- and mid-latitude regions of Mars. J Geophys Res 112(E12):E12S99
332. Tanaka KL, Kolb EJ (2001) Geologic history of the polar regions of Mars based on Mars global survey data I. Noachian and Hesperian Periods. Icarus 154 (1):3–21
333. Bleacher JE, Sakimoto SEH, Garvin JB, Wong M (2003) Deflation/erosion rates for the Parva Member, Dorsa Argentea Formation and implications for the south polar region of Mars. J Geophys Res E Planets [Internet] [cited 2020 Aug 31] 108(7). https://agupubs.onlinelibrary.wiley.com/doi/full/10.1029/2001JE001535
334. Ackiss S, Horgan B, Seelos F, Farrand W, Wray J (2018) Mineralogic evidence for subglacial volcanism in the Sisyphi Montes region of Mars. Icarus 1(311):357–370
335. Wright HMN, Lesti C, Cas RAF, Porreca M, Viramonte JG, Folkes CB, et al (2011) Columnar jointing in vapor-phase-altered, non-welded Cerro Galán Ignimbrite, Paycuqui, Argentina. Bull Volcanol [Internet] [cited 2020 Aug 31] 73(10):1567–1582. https://link.springer.com/article/10.1007/s00445-011-0524-6
336. Lim C, Huh M, Yi K, Lee C (2015) Genesis of the columnar joints from welded Tuff in Mount Mudeung National Geopark, Republic of Korea. Earth Planets Sp [Internet] [cited 2020 Aug 31] 67(1):152. http://www.earth-planets-space.com/content/67/1/152
337. Carter J, Poulet F, Bibring J-P, Mangold N, Murchie S (2013) Hydrous minerals on Mars as seen by the CRISM and OMEGA imaging spectrometers: updated global view. J. Geophys. Res. Planets 118:831–858. https://doi.org/10.1029/2012JE004145
338. Hiesinger H, Head III JW (2002) Topography and morphology of the Argyre Basin, Mars: implications for its geologic and hydrologic history. Planet Space Sci [Internet] [cited 2019 Jul 7] 50(10–11):939–981. https://www.sciencedirect.com/science/article/pii/S0032063302000545
339. Bernhardt H, Hiesinger H, Reiss D, Ivanov M, Erkeling G (2013) Putative eskers and new insights into glacio-fluvial depositional settings in southern Argyre Planitia, Mars. Planet Space Sci [Internet] [cited 2019 Jul 7] 85:261–278. https://www.sciencedirect.com/science/article/pii/S0032063313001633
340. Kargel JS, Strom RG (1992) Ancient glaciation on Mars. Geology 20(1):3–7
341. Zimbelman J, Edgett K (1992) The Tharsis Montes, Mars: comparison of volcanic and modified landforms. In: Proceedings of lunar and planetary science [Internet] [cited 2018 Nov 17] 31–44. https://repository.si.edu/bitstream/handle/10088/2735/199205.pdf
342. Leverington DW (2009) Reconciling channel formation processes with the nature of elevated outflow systems at Ophir and Aurorae Plana, Mars. J Geophys Res Planets [Internet] [cited 2018 Oct 25] 114(E10). http://doi.wiley.com/10.1029/2009JE003398
343. Michalski J, Niles PB (2012) Atmospheric origin of martian interior layered deposits: links to climate change and the global sulfur cycle. Geology 40 (5):419–422
344. Herkenhoff KE, Vasavada AR (1999) Dark material in the polar layered deposits and dunes on Mars. J Geophys Res Planets [Internet] [cited 2020 Sep 8] 104(E7):16487–16500. http://doi.wiley.com/10.1029/1998JE000589
345. Paige DA, Bachman JE, Keegan KD (1994) Thermal and albedo mapping of the polar regions of Mars using Viking thermal mapper observations:

1. North polar region. J Geophys Res [Internet] [cited 2020 Sep 14] 99(E12):25959–25991. https://agupubs.onlinelibrary.wiley.com/doi/full/10.1029/93JE03428
346. Byrne S, Murray BC (2002) North polar stratigraphy and the paleo-erg of Mars. J Geophys Res E Planets [Internet] [cited 2020 Sep 14] 107(6):11–1. https://agupubs.pericles-prod.literatumonline.com/doi/full/10.1029/2001JE001615
347. Tsoar H, Greeley R, Peterfreund AR (1979) MARS: the North Polar sand sea and related wind patterns. J Geophys Res Solid Earth [Internet] [cited 2020 Sep 14] 84(B14):8167–8180. https://agupubs.onlinelibrary.wiley.com/doi/full/10.1029/JB084iB14p08167%4010.1002/%28ISSN%292169-9356.MARS2
348. Garvin JB, Sakimoto SEH, Frawley JJ, Schnetzler CC, Wright HM (2000) Topographic evidence for geologically recent near-polar volcanism on Mars. Icarus 145(2):648–652
349. Ghatan GJ (2003) Cavi Angusti, Mars: characterization and assessment of possible formation mechanisms. J Geophys Res 108(E5):5045–5063
350. Kolb EJ, Tanaka KL (2001) Geologic history of the polar regions of Mars based on Mars global surveyor data II. Amazonian period. Icarus 154 (1):22–39
351. Brothers TC, Holt JW, Spiga A (2013) Orbital radar, imagery, and atmospheric modeling reveal an aeolian origin for Abalos Mensa, Mars. Geophys Res Lett [Internet] [cited 2020 Sep 15] 40(7):1334–1339. http://doi.wiley.com/10.1002/grl.50293
352. Brothers TC, Holt JW, Spiga A (2015) Planum Boreum basal unit topography, Mars: irregularities and insights from SHARAD. J Geophys Res Planets [Internet] [cited 2020 Sep 14] 120(7):1357–1375. http://doi.wiley.com/10.1002/2015JE004830
353. Brothers SC, Kocurek G, Holt JW (2018) Sequence architecture of the cavi unit, Chasma Boreale, Mars. Icarus 1(308):42–60
354. Fishbaugh KE, Head JW (2005) Origin and characteristics of the Mars north polar basal unit and implications for polar geologic history. Icarus 174(2 SPEC. ISS.):444–474
355. Tanaka KL, Rodriguez JAP, Skinner JA, Bourke MC, Fortezzo CM, Herkenhoff KE et al (2008) North polar region of Mars: advances in stratigraphy, structure, and erosional modification. Icarus 196(2):318–358
356. Squyres SW, Arvidson RE, Blaney DL, Clark BC, Crumpler L, Farrand WH, et al (2006) Rocks of the Columbia Hills. J Geophys Res Planets [Internet] [cited 2019 Dec 9] 111(E2):E02S11. http://doi.wiley.com/10.1029/2005JE002562
357. Farrand WH, Bell JF, Johnson JR, Squyres SW, Soderblom J, Ming DW (2006) Spectral variability among rocks in visible and near-infrared multispectral Pancam data collected at Gusev crater: examinations using spectral mixture analysis and related techniques. J Geophys Res Planets 111:E02S15, http://doi.wiley.com/10.1029/2005JE002495
358. Herkenhoff KE, Squyres SW, Anderson R, Archinal BA, Arvidson RE, Barrett JM, et al (2006) Overview of the microscopic imager investigation during spirit's first 450 sols in Gusev crater. J Geophys Res Planets [Internet] [cited 2020 Sep 15] 111(E2). http://doi.wiley.com/10.1029/2005JE002574
359. Steven WR, Paul BN, Fabrizio A, Amanda BC (2014) Evidence for a Noachian-aged ephemeral lake in Gusevcrater, Mars. Geology 2014 42 (4):359–362. https://doi.org/10.1130/G35508.1
360. Lewis KW, Aharonson O, Grotzinger JP, Squyres SW, Bell JF, Crumpler LS, et al (2008) Structure and stratigraphy of home plate from the spirit mars exploration rover. J Geophys Res [Internet] [cited 2020 Sep 15] 113(E12):E12S36. http://doi.wiley.com/10.1029/2007JE003025
361. Manga M, Patel A, Dufek J, Kite ES (2012) Wet surface and dense atmosphere on early Mars suggested by the bomb sag at Home Plate, Mars. Geophys Res Lett [Internet] [cited 2020 Sep 15] 39 (1). http://doi.wiley.com/10.1029/2011GL050192

The Origin and Evolution of Volcanism at Martian Highland Paterae: A Review of the Current State of Knowledge

Nicholas P. Lang, Caitlin Ahrens, and Alexandra K. Farrell

Abstract

Highland paterae are broad, low-sloped volcanoes on Mars that have a large summit caldera and channeled flanks. Crater-modeled ages suggest they may represent some of the oldest central vent volcanoes on the planet. The presence of channeled flanks that surround a younger summit caldera with topographically smooth floor material has led to the interpretation that highland paterae record a transition from explosive volcanic eruptions to more effusive eruptive styles. In this chapter we describe the geology of seven highland paterae (Amphitrites Patera, Peneus Patera, Malea Patera, Pityusa Patera, Tyrrhena Patera, Hadriaca Patera, and Apollinaris Mons), discuss what that reveals about their history and origin, and then discuss possible avenues of future research on these intriguing Martian landforms.

N. P. Lang (✉)
Department of Geology, Mercyhurst University, 501 E. 38th Street, Erie, PA 16546, USA
e-mail: nlang@mercyhurst.edu

C. Ahrens
NASA Goddard Space Flight Center, Greenbelt, MD, USA

A. K. Farrell
Alaska Earthquake Center, University of Alaska, Fairbanks, AK 99775, USA

G. Leone (ed.), *Mars: A Volcanic World*,
https://doi.org/10.1007/978-3-030-84103-4_9

1 Introduction

Highland paterae [1–3] represent one of the most distinctive landforms on the Martian surface. The term patera (paterae if plural) was originally applied to any large, flat-floored depression with irregular margins identified in Mariner 9 and Viking imagery and included volcanoes as well as impact craters [4; see Chap. 8]. However, highland paterae represent an individual class of patera that is volcanic in origin and concentrated mostly around the Hellas basin region [1–5] (Fig. 1). This class of patera has typically included Tyrrhena, Hadriaca, Amphitrites, and Peneus Paterae, which are all characterized by very low slopes of $<1°$ to $\sim 1.5°$ that, in turn, are incised by channel complexes radial to a prominent central caldera [1, 6, 7]. The similar morphologies and characteristics of these four paterae suggest a common origin that, based on the presence of abundant channels that cut into an apparently easily erodible surface, may have included explosive activity with effusive eruptions likely occurring later in their eruptive history [3, 5–9; see Chap. 8 *for further descriptions of explosive eruption styles*]. They are among the oldest central vent volcanoes on Mars [5, 9–12] and may represent a global transition from flood lava eruptions, which dominated the early volcanic history of Mars, to more localized eruptions [6].

The objective of this chapter is to outline our current understanding of the processes recorded

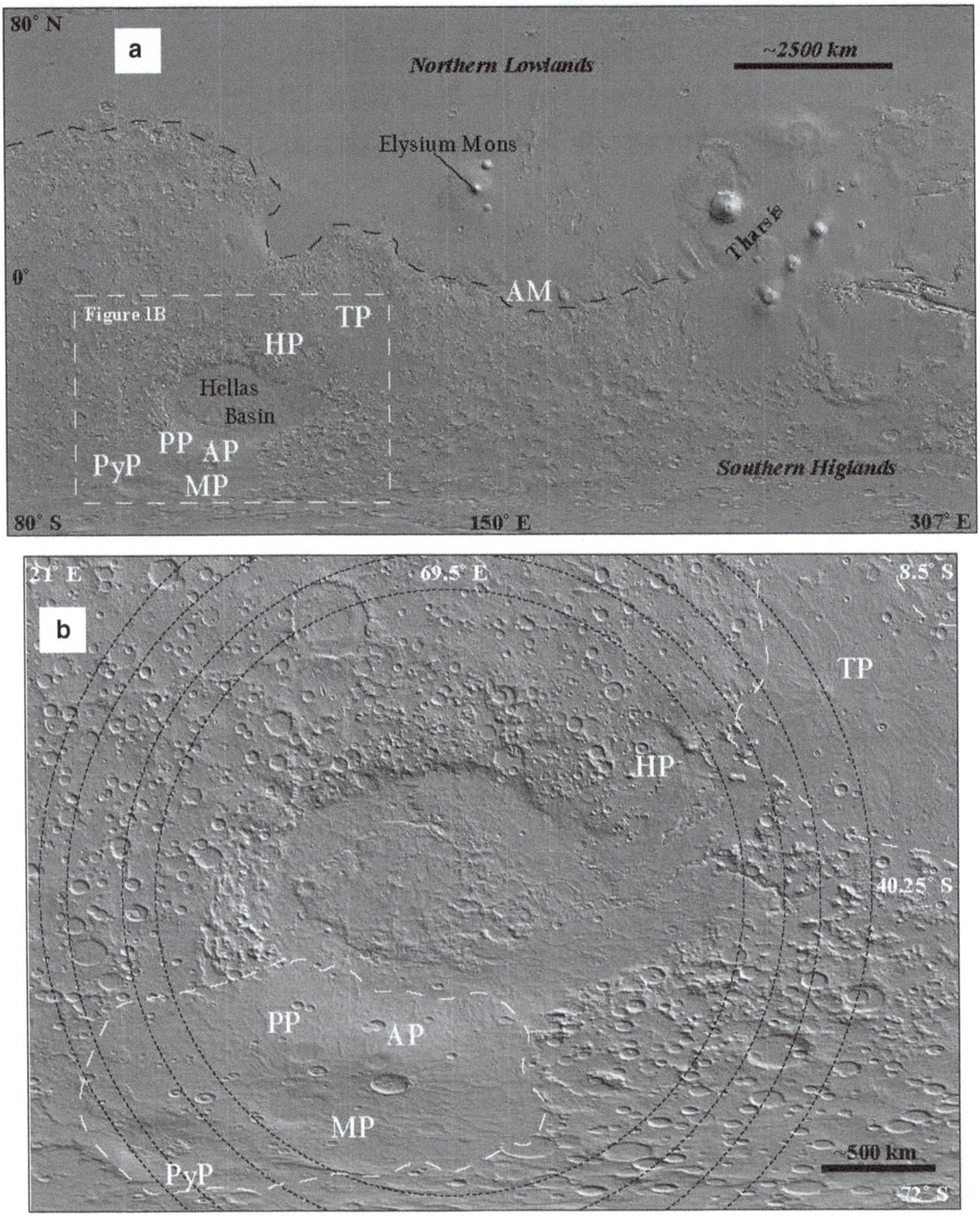

Fig. 1 Shaded relief images highlighting the locations of the highland paterae discussed in this chapter. **a** Regional Mars Orbiter Laser Altimeter (MOLA) shaded relief image of Mars highlighting the locations of the seven highland paterae (white text) discussed in this chapter. For reference, the locations of the Tharsis volcanic province, Elysium Mons, and the Hellas Basin are also highlighted. **b** MOLA shaded relief image of the Hellas Basin region showing the locations of the six highland paterae located in this area. Dashed white lines denote the approximate boundaries of volcanic provinces in which the paterae occur. Hesperian Planum along the northeast corner of Hellas Basin and Malea Planum is along the southwest corner of the basin. The dotted black circles represent the approximate locations of crater rings as indicated by previous workers [1]. AM = Apollinaris Mons, AP = Amphitrites Patera; HP = Hadriaca Patera, MP = Malea Patera, PP = Peneus Patera, PyP = Pityusa Patera, TP = Tyrrhena Patera

at Martian highland paterae. These include the style(s) of volcanic eruptions, processes involved in caldera formation, and the construction and subsequent erosion of the main edifice. In addition to the four traditionally recognized highland paterae, this chapter will also address Apollinaris Mons, Pityusa Patera, and Malea Patera. Although Apollinaris Mons, which is located along the northern lowland-southern highland transition, was not originally classified as a highland patera [see 4, 13, 14], it contains many similar morphologic aspects including a large central caldera and extensively channeled and dissected flanks [13; see also, 3]. Thus, it may have an origin similar to the other four highland paterae and is therefore included in this chapter. Pityusa and Malea Paterae are spatially associated with Amphitrites and Peneus Patera, have broadly similar morphologies, and may be part of the same magmatic province [5]; therefore they are included here as well. It is our hope that this chapter can provide a starting point for those looking to learn more about highland paterae and/or continue to address outstanding questions about this distinct type of Martian volcano.

1.1 Scientific Rationale

The broad, low-sloped, heavily channeled flanks of highland paterae make them an intriguing class of central vent volcanoes on Mars [4]. Based on crater modeled ages ranging from ~3.2 to ~3.9 Ga [5, 9, 10, 13, 15–17; see also, 18, 19] (Fig. 2), they also represent some of the oldest recognized central vent volcanoes on the planet where they straddle the transition from the Noachian to the Hesperian. Therefore, studying highland paterae is critical because they further our understanding of Martian volcanic systems by providing a glimpse into volcanic processes on early Mars. In addition, because the constructs have existed for a majority of the planet's history, they undoubtedly must have recorded surface processes that have operated upon them since their formation. This includes wind processes as well as any possible interactions with volatile-rich materials that may have occurred. Thus, highland paterae also provide a window into other surface processes that have operated through time and can potentially help further constrain the planet's volatile history. Increasing our understanding of early Martian volcanism and volatile history is essential to unlock and document the evolution of the Martian surface. To elaborate, work by [6] and [7] on Tyrrhena and Hadriaca Paterae, respectively, suggests that highland paterae were built from a series of explosive eruptions that temporally transitioned into more localized, effusive eruptions. Robinson et al. [13] came to similar conclusions for Apollinaris Mons [see also, 19, 20]. The occurrence of possibly two distinct styles of eruption (explosive and effusive) raises several critical questions regarding highland paterae. Specifically, these are:

(1) *What triggered the explosive eruptions and why the transition to effusive eruptions?* Based on the presence of Hadriaca and Tyrrhena near the eastern edge of the Hellas basin, [6] and [7] suggested explosive eruptions may have been triggered by interaction of magma with a near-surface reservoir of ice and/or water. However, they also suggested that an alternative explanation may be the volatile content of the magma itself [see also, 21, 22 and this chapter]. Determining between these two eruption triggers has implications for early Mars' near-surface volatile content and magma compositions. To illustrate, if the explosive nature of the eruptions was triggered by the interaction of magma with a near-surface volatile then highland paterae would be indicators of hydrovolcanism, which then makes distinct predictions as to facies that should be observed within the deposits (assuming high enough effective resolution imagery [e.g., 23]. If true, then the change to effusive volcanism could then be recording the loss of volatiles in the region [6, 7]; once volatiles disappeared, explosive eruptions end and effusive eruptions begin [7; see also, 6; *see this chapter for further discussion*]. However, if explosive eruptions were triggered by the composition of the magma or some other physical characteristic of the volcano (e.g.,

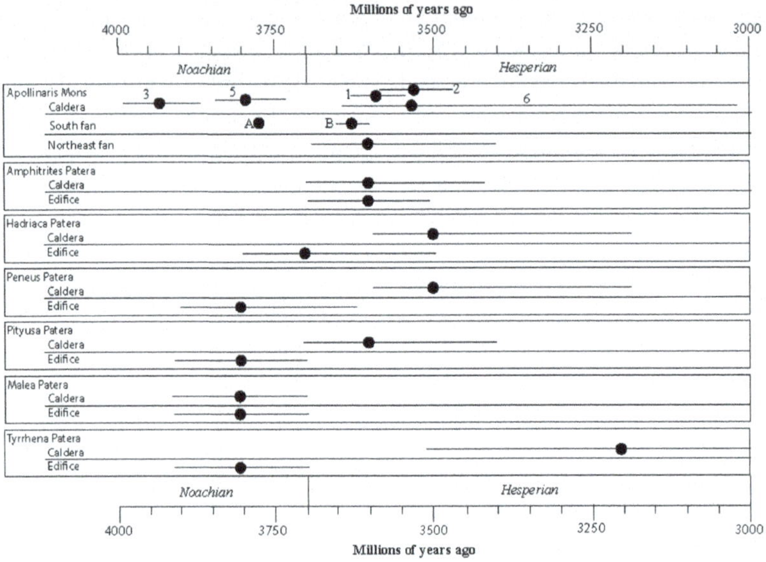

Fig. 2 Crater modeled ages for the seven highland paterae discussed in this chapter. Black dot represents the reported crater modeled age and thin solid lines extending from the circles represent the reported error for the crater model age; these lines are not meant to represent the length of time the volcanoes may have been active. With the exception of Apollinaris Mons, ages for all volcanoes are from [5]. Caldera ages for Apollinaris Mons are from [13]; caldera ages for Apollinaris are offset vertically only for clarity. Numbers next to each reported age for the Apollinaris caldera refers to specific surfaces identified and dated by [13]; surface 1 is the same as our identified inner caldera and surfaces 2–6 are the same as our identified outer caldera. The south fan age is from [16], who did not report error bars with this age. The northeast fan age is from [17] and is their identified unit Amfu

conduit) or magma (e.g., degree of crystallinity or volatile content), then the presence of volatiles in the near surface are not required (also see Chap. 1). If explosivity is related to magma composition (i.e., felsic vs. mafic magma), then the transition to effusive eruptions may reflect a change to more mafic magma compositions. If magma composition is the trigger for explosive eruptions, then this raises even more questions such as why the transition from silicic to more mafic volcanism?

(2) *What were the processes involved in caldera formation?* Calderas are depressions that form due to the removal of magmatic support from the crust overlying a magma chamber [24, 25]. Thus, they can provide critical clues regarding eruptive and magma chamber processes [e.g., 26, 27]. Each highland patera contains a single caldera that is almost more pronounced than the edifices themselves [14], though Apollinaris does contain a second, smaller caldera inset into a larger, older caldera [13]. However, the number of eruptive events they record and the processes leading to their formation are unknown. For example, it is unclear if highland paterae calderas formed from a single eruptive event or if they formed gradually due to long term draining of the chamber. Constraining the processes involved in caldera collapse at highland paterae can therefore provide useful information regarding the internal history of the volcanoes as well as help constrain the eruptive history and style [e.g., 26].

(3) *What processes have operated on the flanks?* All highland paterae exhibit numerous channels on their flanks. However, the origin of these channels has been unclear [e.g., 28]. The occurrence of channels within possible pyroclastic deposits suggests an origin involving volcanic density currents [28], but the v-shaped nature of some of the channels [28] has led to postulations of modification by water or

lava [see 29, 30, *see also Chap.* 1]. Similar to question #1 above, differentiating between these mechanisms of channel formation has implications for volcanic and related (e.g., erosional) processes operating at highland paterae. To illustrate, if the channels are volcanic in origin, then this implies that volcanic materials were emplaced by traveling along the ground as either lava flows and/or ground-hugging pyroclastic flows. However, if the channels were carved or even modified by a substance other than lava or pyroclastic flows, then other erosional processes must have operated on these volcanoes. Further, if the channels are purely volcanic in origin, then erosion of the volcano may be closely tied to individual eruptive episodes.

(4) *What was the duration of volcanic activity?* Do highland paterae represent long-lived volcanic systems or are they the result of short-lived catastrophic-like events that resulted from the interaction of magma with a near-surface reservoir of volatiles? Crater counts on Hadriaca Patera and Apollinaris Mons suggest that these volcanoes are long-lived systems lasting for periods of $\sim 10^7$ years [5, 9, 10, 13, 15, 16]. Such long-lived systems suggest a stable magmatic source (perhaps a plume?), which in turn has implications for the internal processes and evolution (including the thermal structure and mechanisms of heat loss) of early Mars.

These four questions make highland paterae interesting landforms to continue to investigate. In the next section we will describe the general geology and regional context of seven highland paterae. We will follow those descriptions with another visit to the four outstanding questions discussed above.

1.2 Highland Paterae Context and Descriptions

The highland paterae addressed in this chapter include: Amphitrites Patera, Peneus Patera, Malea Patera, Pityusa Patera, Tyrrhena Patera, Hadriaca Patera, and Apollinaris Mons. In this section we describe the morphologic characteristics of these seven highland paterae and outline a broad geologic history for each. However, prior to describing each center, it is worthwhile to consider their geologic context.

Six of the highland paterae occur around the rim of Hellas basin, a ~2300 km diameter impact basin centered at ~41° S, 67° E (Fig. 1 b). The northeastern and southern edges of the basin have displayed extensive evidence for volcanism [5; see also, 31], which can be denoted in shaded relief images (e.g., Fig. 1) as topographically smooth terrains that stand out in contrast to the surrounding southern highlands. This has led to this area being labeled the Circum-Hellas Volcanic Province [5]. These volcanic centers have been noted to correspond with ring fractures associated with Hellas basin [1], which may have facilitated magma ascent and eruption [1]. Amphitrites (59° S, 61.1° E), Peneus (58.1° S, 52.5° E), Malea (63.8° S, 52° E), and Pityusa (67° S, 38.5° E) Paterae are located along the southwest portion of this volcanic province where they define an area called Malea Planum, which is interpreted as a region that experienced flood volcanism [30]. Hadriaca (92.12° E, 21° S) and Tyrrhena (106.44° E, 21° S) Paterae occur along the northeast edge of the Hellas basin, a region that includes Hesperia Planum which, similar to Malea Planum, is interpreted as a region of flood volcanism [32]. Abundant evidence for erosion is also present along the Hellas rim and includes the occurrence of Dao, Niger, and Harmakhis Valles in the northeast corner of the basin [33] and in the Malea Planum region with multiple integrated sinuous channel networks that in some cases may have been inverted into ridges [34].

Apollinaris Mons is geographically distinct from the other six highland paterae, but, similar to them, does occur along a topographic transition (Fig. 1). This topographic transition is the boundary between the northern lowlands and southern highlands and may be part of a larger volcanic province [35] that includes alkalic flood basalts that erupted onto the floor of Gusev Crater to the south [16], which has shown evidence for phreatomagmatic activity [36]. In addition, surrounding Apollinaris Mons is chaotic terrain [13], which consists of elevated knobs occurring in groups or as single mesas. The presence of this terrain has been interpreted to be the result of ice-related processes [37], hydrothermal vent-related activity [38], or volcanic erosion (Leone, 2014, A network of lava tubes as origin of Labyrinthus Noctis and Valles Marineris on Mars).

1.2.1 Amphitrites Patera

Amphitrites Patera (Fig. 3a, b) is ~280 km in diameter with channelized and eroded flanks (Fig. 4a) that slope from ~1.2° to ~1.5° and that reach up to an elevation of ~1–1.3 km above the surrounding plains, forming a broad shield-shaped structure (Fig. 3b). It is topped by a ~120 km diameter poorly defined caldera [39, 40]. The caldera's depth ranges from ~150–800 m and appears to be defined by discontinuous graben (Fig. 4b); a ~10 km wide ridge also marks the rim of the caldera on all but the southern edge. The lack of well-defined margins to this caldera suggest possible formation through down sagging, or a later (volcanic overflow?) deposit has obscured the existence of fractures and/or graben. The caldera floor hosts a suite of northeast-trending contractional ridges (mare-like ridges [5]) that extend onto the northeast and southwest parts of the volcano's flanks. In the center of the caldera (~58.9° S, 60.6° E) is a ~40 km wide depression, which could represent another, younger caldera (Fig. 3 b). The flanks are incised by ~< 1 to ~5 km wide channels that radiate from the caldera. Mars Orbital Camera (MOC) views of these channel walls indicate they do not contain geologic layering, at least at the resolution of MOC. The abundant presence of channels along the flanks indicate they are composed of loose, easily erodible materials that have been interpreted as pyroclastic deposits; ridges cut across the channels and seemingly represent tectonic activity that post-dated channel formation. Evidence for effusive volcanism having occurred at Amphitrites is limited and includes the seemingly morphologically smooth caldera floors that have been subsequently deformed by contractional structures. However, the extent to which this smoothness is related to erupted materials versus surficial deposits [5, 15] is not clear (Fig. 4d?). For example, [40] interpreted some ridges on the caldera floor as flow fronts, but a surficial ice-rich deposit termed "swiss cheese" terrain [15] has been proposed to cover the deposits [see also, 40] and could also be responsible for obscuring the edges of the caldera making volcanic unit identification difficult. Further, at least one channel on the northwest flank of Amphitrites contains an elongated chain of craters that has been interpreted as collapsed sections of lava tubes [5] (Fig. 4e). Dark material noted on the caldera floor and parts of the flanks suggest a basaltic composition (pyroxene and olivine) of materials at this volcano, but the degree to which these materials represent in-place volcanic deposits versus wind-derived deposits is not clear [5]. If the identification of a possible second caldera in the center of Amphitrites is correct, then there may have been multiple episodes of magma rising and falling beneath this edifice. It was noted the presence of a strong positive gravity anomaly at Amphitrites that could be indicative of a dense subsurface magma body [5]. Crater modeled ages place both the flanks of the volcano and its caldera at 3.6 Ga [5], which is similar to ages for other highland paterae in the vicinity of Hellas Basin.

1.2.2 Peneus Patera

Peneus Patera (Fig. 5a, b) is ~290 km in diameter with little topographic relief compared to Amphitrites Patera. Similar to Amphitrites Patera, Peneus has flanks that slope from ~1.2° to ~1.5° and that reach up to an elevation of ~0.7 km above the surrounding terrain (Fig. 5 b). The top of the volcano contains a ~125 km in diameter caldera that is bound by a ~25 km wide band of concentric graben [39]. A possible dome or thick flow appears to locally cover the graben along the southwest corner of the caldera (Fig. 6). Channels occur along the northern flank and radiate north from the caldera, though the channels are not as well defined as they are on Amphitrites; morphologically these channels

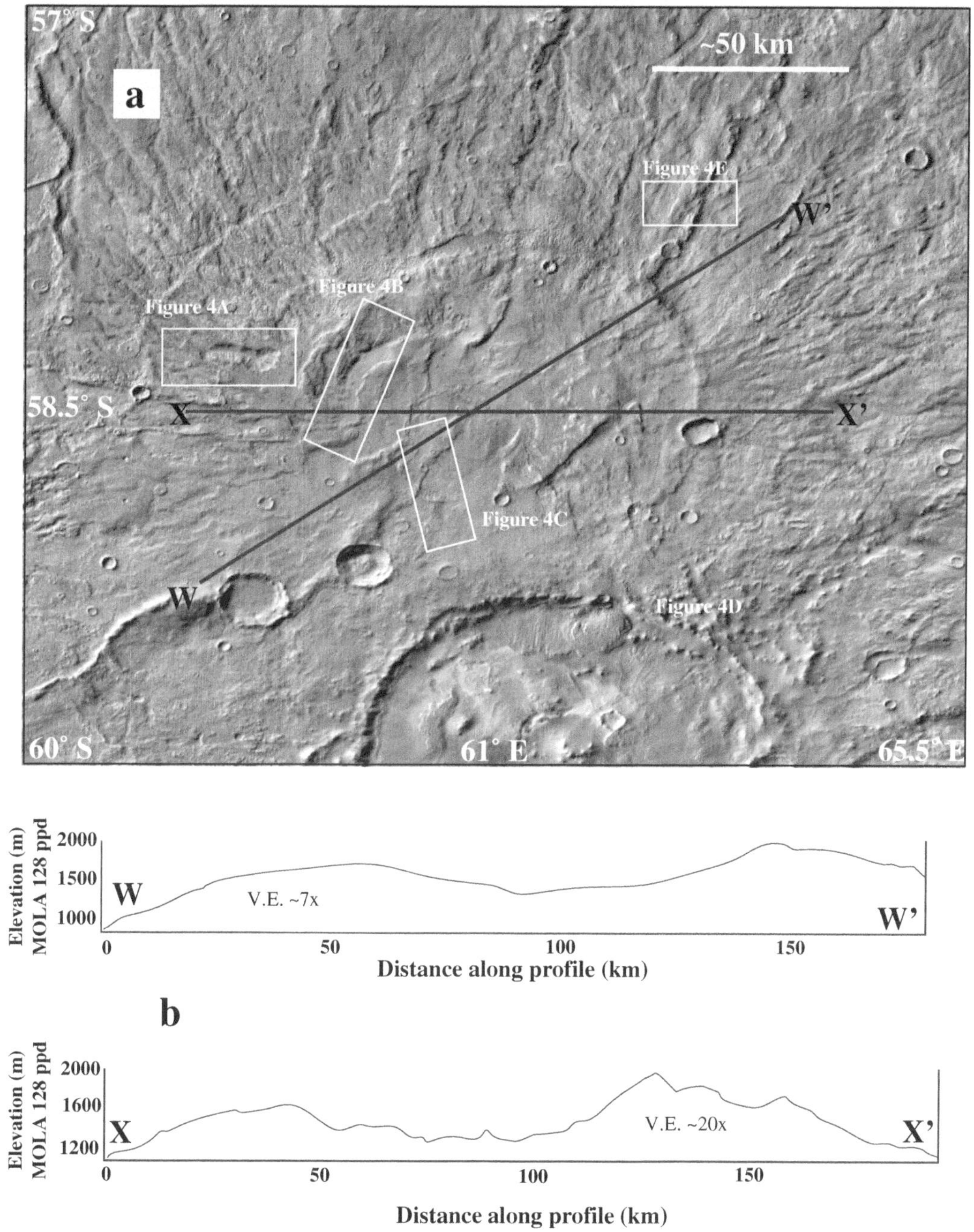

Fig. 3 Image and morphologic characteristics of Amphitrites Patera. **a** THermal EMission Imaging System (THEMIS) daytime infrared (IR) image of Amphitrites Patera illustrating its overall appearance. White boxes note the locations of images highlighted in Fig. 4. Solid black lines of W-W′ and X-X′ show the locations of topographic profiles shown in (**b**). **b** Topographic profiles of Amphitrites Patera. V.E. = Vertical Exaggeration

appear broadly similar to those on Amphitrites' flanks, suggesting that Peneus' northern flanks may be comprised of similar material. The floor of the caldera is relatively smooth, though wrinkle ridges occur locally. Similar to Amphitrites Patera, crater modeled ages of Peneus'

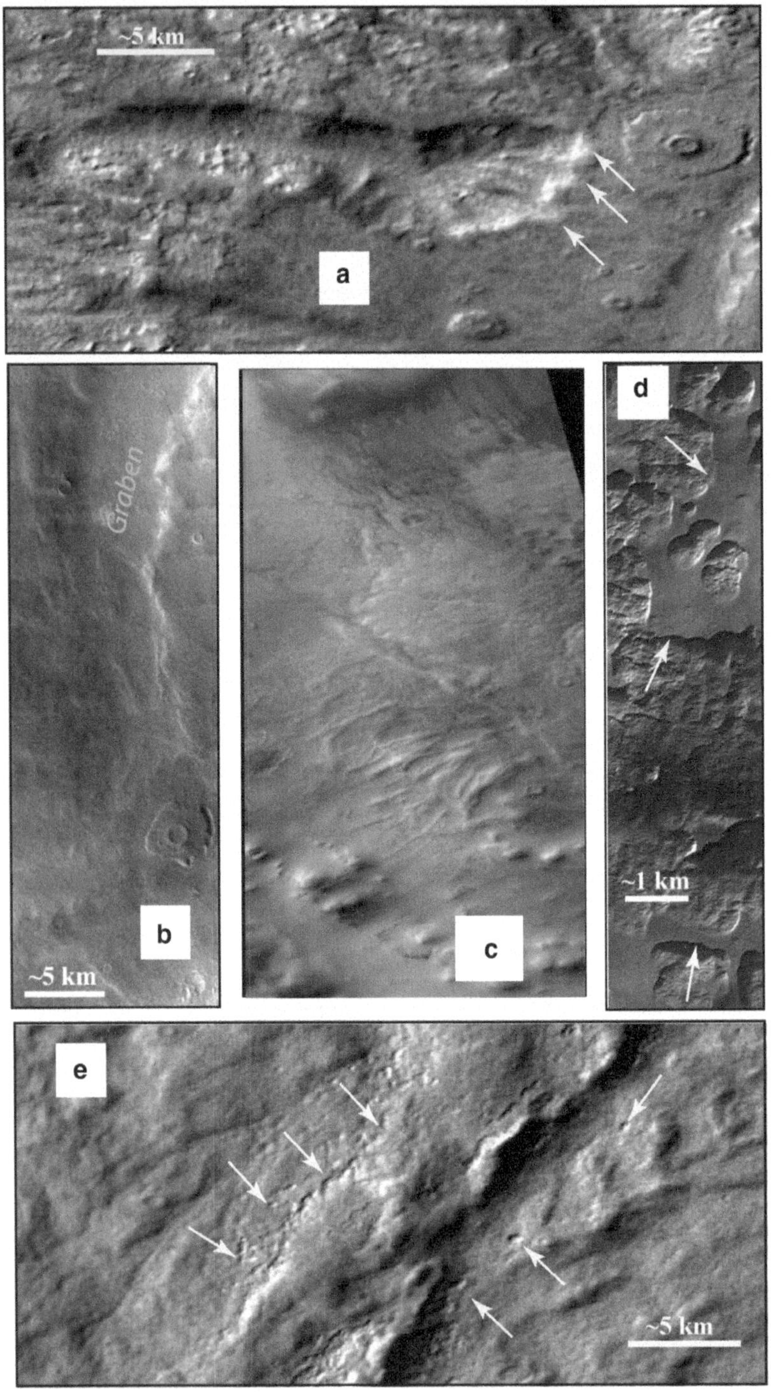
~5 km
a
Graben
b
~5 km
c
d
~1 km
e
~5 km

Fig. 4 Views of the Amphitrites Patera flanks and caldera floor. **a** Portion of THEMIS daytime IR image highlighting a series of mass wasting scarps (white arrows) and associated channels on the western flank. **b** Portion of THEMIS VIS image V07351006 showing the western edge of the summit caldera highlighting a poorly defined (highly eroded?) graben. **c** Portion of CTX image B12_014387_1200_XN_60S299W showing characteristics of the floor of what may be a smaller, secondary inner caldera within Amphitrites. The floor of both the inner and outer calderas show morphologic evidence for easily erodible deposits similar to those observed at highland paterae north of Hella (bottom of image near '**c**'); following previous workers' interpretations, we interpret these deposits as possible pyroclastic materials. Possible wind-derived deposits are located near the top of the image (dark tones). Image is ~25 km across. **d** Portion of MOC image M2101239 showing characteristics of the south flank of Amphitrites Patera. Most of the floor here is covered by possible volatile-rich materials (white arrows). These deposits obscure much of the underlying geology making identification of distinct volcanic deposits difficult. **e** Portion of THEMIS daytime IR image highlighting multiple series of aligned pits (white arrows) that may represent lava tubes [5], though an origin related to the surficial volatile-rich deposit cannot be excluded

caldera indicate it is ~3.5 Ga [15], whereas crater modeled ages of its flanks indicate they are ~3.8 Ga suggesting activity here began before Amphitrites [5]. Based on the smoothness of the caldera floor and the presence of a dome-like feature in the caldera's southwest corner, effusive activity appears to characterize the caldera's interior.

1.2.3 Malea and Pityusa Paterae

Malea and Pityusa Paterae are two irregularly-shaped depressions located south-southwest of Peneus and Amphitrites Paterae (Fig. 7). Both are broad, poorly-defined structures when observed in high resolution images (e.g., MOC) and MOLA data [5, 40]. They are also both ~400 km in diameter [40]. The floor of Malea Patera resides at ~−10 m in elevation, but is surrounded by topographic highs that rise more than 1 km above the floor. Wrinkle ridges cut the floor of this depression, which also contains pedestal craters, dark patches of basaltic material, and a mesa of layered material. It was also noted the presence of a possible layer that appears in MARSIS data that could represent a buried lava flow, pyroclastic deposit, or even ice [5]. Crater counts of Malea Patera indicate a crater-model age of ~3.8 Ga. Similar to Malea, Pityusa Patera is a depression surrounded by a pronounced rim. The floor of the depression is ~170 × 200 km in size and is deformed by a series of wrinkle ridges. The floor also contains deposits of layered material, dark (basaltic) sand dune materials, and a bright and smooth plateau that hosts impact craters. Similar to Malea Patera, Pityusa also has a crater-modeled age of ~3.8 Ga. Initial work on these paterae suggested they are not volcanic in origin, but possibly rather old, filled impact craters [40] or even the surface expression of dense intrusions [41]. However, Williams et al. [5], in a subsequent examination, postulated that these two depressions likely are volcanic in origin where the layered mesa and plateau deposits likely represent ignimbrites. Ultimately, the morphology of these two structures is distinctly different from Amphitrites and Peneus Paterae implying that if they are volcanic, then volcanism there may have occurred in a style distinct from the other two paterae. Consistent with a volcanic interpretation is the identification of relatively weak gravity signatures [5, 42] under both of these features, which could be related to collapse of the basin floor into an empty magma chamber after an eruption [5].

1.2.4 Hadriaca Patera

Hadriaca Patera (−30.2° N, 92.79° E) is a 330 by 550 km depression with a maximum relief of 1.1 km [43] located northeast of the Hellas Basin (Fig. 8a). The effects of contemporaneous volcanic and tectonic activity of the Hadriaca region further implies the complexity of its environment and evolution [7, 9, 33].

Overall Morphology

Hadriaca was estimated Hesperian in age [7, 44, 45], and the overall morphology is consistent with formation by pyroclastic flows [44]. Crown and Greeley [7] suggests the explosive eruption style inferred at this patera is attributed to the interaction of magma with volatiles (*see also this*

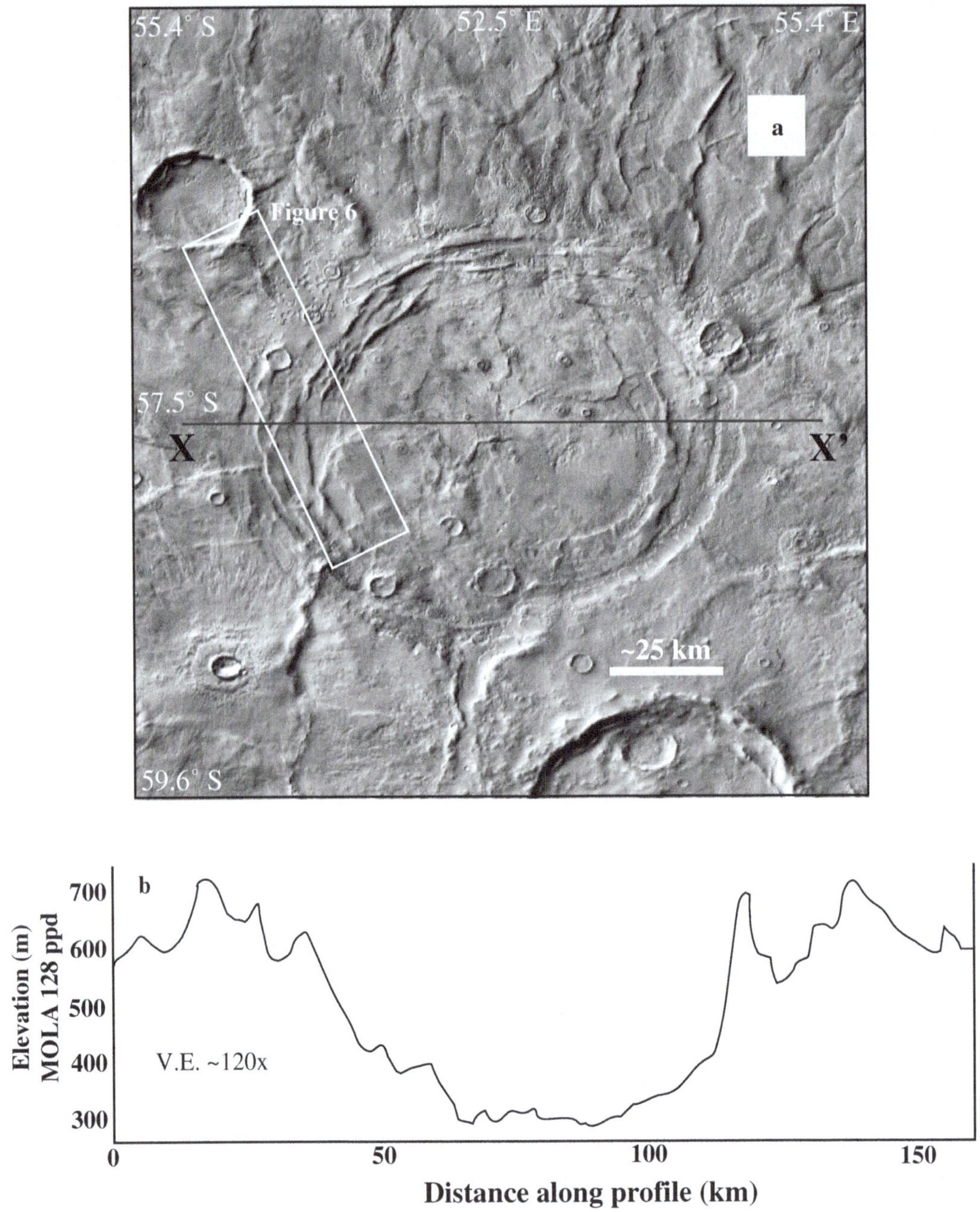

Fig. 5 Characteristics of Peneus Patera. **a** THEMIS daytime IR image highlighting Peneus' caldera and flanks. Channels occur on the northern flank, but are seemingly absent on the rest of the volcano. White box denotes the location of Fig. 6. Black X-X′ line notes the location of the topographic profile shown in (**b**). **b** Topographic profile of Peneus Patera. V.E. = Vertical Exaggeration

chapter). Extensive volcanic material deposits are observed as visible layers in the depressions and extend to a depth of 1–2 km [46], consisting of layered, heavily dissected deposits and suggesting a low-permeability ash mantle from sapping valleys at the flanks [7, 28, 45].

The formation of the Hadriaca Patera edifice took place between 3.9–3.7 Ga [9]. Crater counts

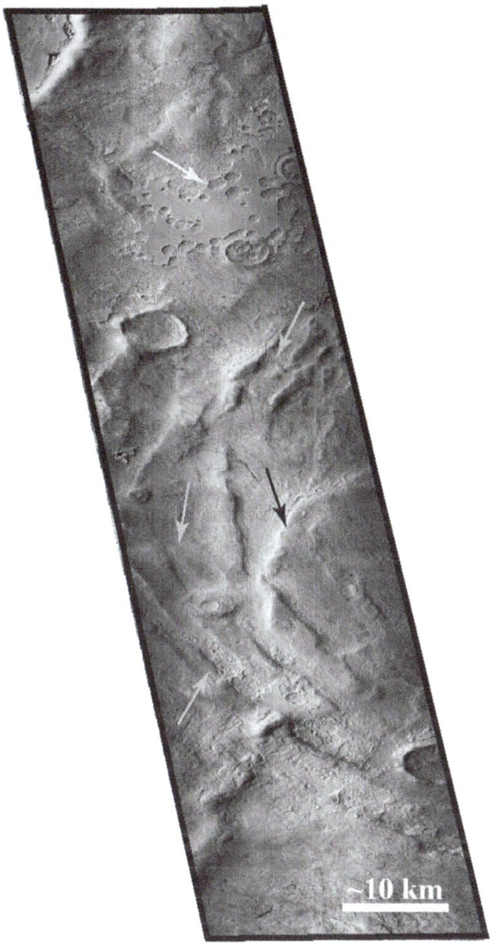

Fig. 6 Portion of Mars Reconnaissance Orbiter (MRO) Context (CTX) Camera image B11_013728_1222_XN_57S309W over the southwestern part of Peneus Patera. White arrow highlights possible volatile-rich materials. These deposits obscure much of the underlying geology making identification of distinct volcanic deposits difficult. Gray arrows highlight the graben that mark the edge of this caldera. Black arrow highlights the possible volcanic dome that occurs within the caldera's interior

and modeled ages suggest early shield-building volcanic eruptions during this time, with caldera formation no later than ~3.5 Ga [9]. A variety of geologic activity occurred in the caldera and northern flank at ~3.3–3.5 Ga, likely including pyroclastic flows and differential erosion/deposition by volatile (or wind) activity [9]. Williams et al. [9] also stated that resurfacing events in the caldera and on the eastern flank occurred at ~2.4–2.6 Ga, during which the eastern flank's morphology suggests fluvial erosional activity. As recent as ~1.5 Ga, the Hadriaca Patera caldera was resurfaced by some process (suggested to be differential wind deposition or erosion) coextensively with the resurfacing of the heavily channeled southern flank in the Amazonian Period. Unlike the Tharsis shields, it was estimated that major activity ended at Hadriaca over a billion years ago [9].

To the north of Hadriaca Patera are dissected plateau materials that embay the highlands, and with the heavily cratered Noachian highland and mountainous terrains to the west, together appear to have served as topographic barriers to the emplacement of the materials forming the flanks of Hadriaca [7]. The plateau is bounded by scarps and extends around the western wall of the caldera, and it can be distinguished from the rest of the channeled flanks due to its different morphology [7, 33]. To the southeast, Hadriaca Patera is cut by the Dao Vallis outflow system [7], where crosscutting relationships between outflow channels and channeled plains imply a complex period of volatile-driven deposition and erosion [7, 47]. Dao, Harmahkis, and Reull Valles are interpreted to be collapsed regions of volcanic and sedimentary plains which have been eroded by surface/subsurface volatile processes [47]. Like all the other outflow channels, also these ones may have been directly carved by lava flows as an alternative explanation [41].

Caldera

The nearly-circular summit caldera (Fig. 8b) has a diameter of 77 km and is observed to be well-defined [7, 44], probably indicative that the magma chamber below remained at its current position over a prolonged period of time [45]. Something to note is the morphological similarities between Hadriaca and Tyrrhena paterae, suggesting a similar volcanic history with an early pyroclastic-dominant phase, followed by effusive eruptions at their summit calderas [7; see also chapter "Explosive Volcanism on Mars"].

The rim of the caldera is variable. The northern and eastern caldera walls are indistinct, possibly displaying the effects of overflow or mantling by

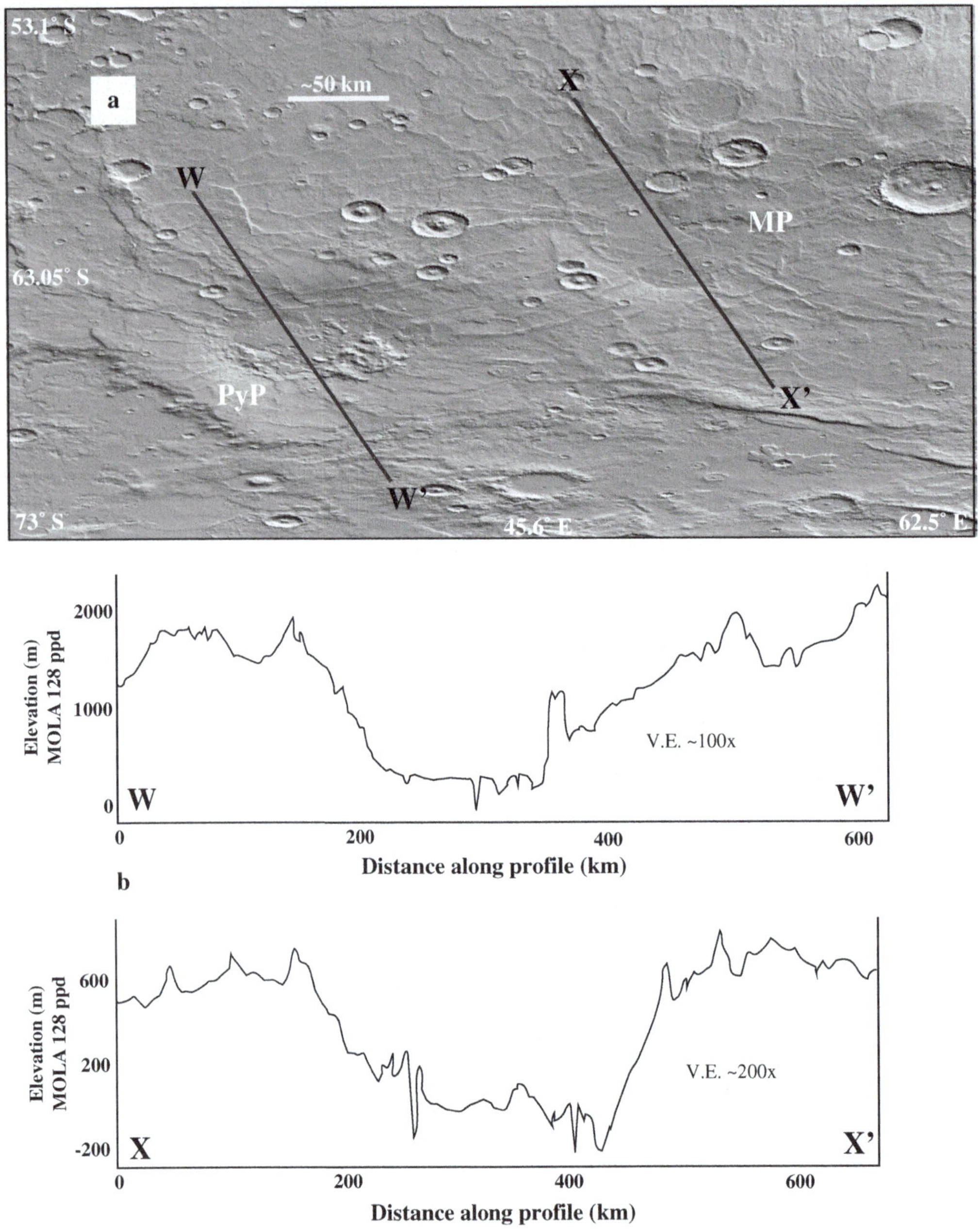

Fig. 7 Characteristics of Pityusa and Malea Planum. **a** MOLA shaded relief image of the southern portion of Malea Planum highlighting the morphologic characteristics of Pityusa and Malea Paterae. PyP = Pityusa Patera; MP = Malea Patera. Black lines W-W′ and X-X′ denote the locations of topographic profiles highlighted in (**b**). **b** Topographic profiles of Pityusa Patera (W-W′) and Malea Patera (X-X′). V.E. = Vertical Exaggeration

volcanic materials [33]. The steplike form to the west is revealing of multiple episodes of collapse. A curvilinear wrinkle ridge separates the eastern region of the caldera from the main floor [7]. This area contains numerous small (<2 km), domelike features on an elevated surface. From Williams et al. [9], the position of the small domes suggest that they may be the youngest deposits associated

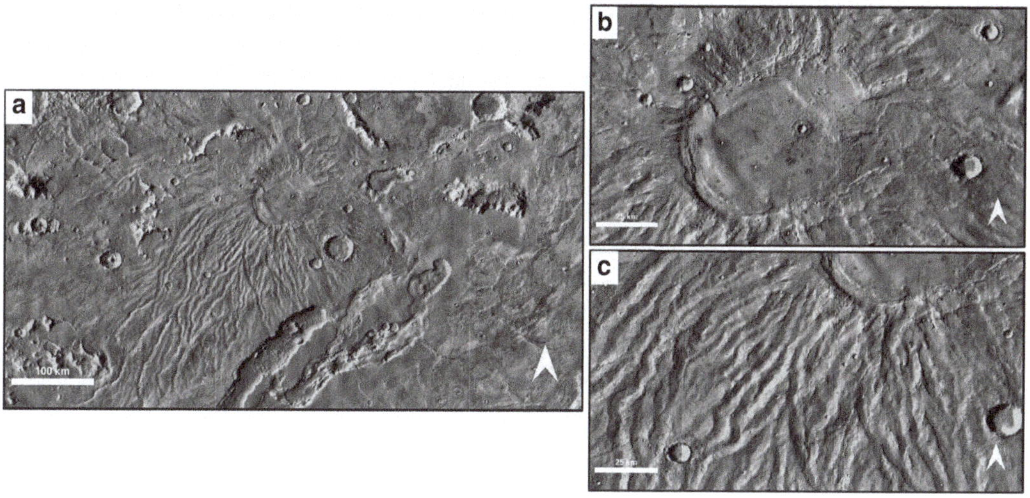

Fig. 8 Morphological overview of Hadriaca Patera. **a** THEMIS daytime IR image of Hadriaca Patera. **b** Closeup of the Hadriaca Patera caldera and arcuate rim features. **c** Closeup view of the southwestern dissected flanks

with Hadriaca Patera. These domes may be interpreted as either constructional volcanic features (i.e., lava domes or cinder cones), or as erosional remnants of a pyroclastic deposit or lava flow. Scarp morphologies are observed on the caldera floor and extend across the rim into the surrounding flank materials [7], where most of the scarps appear eroded. The north rim of the caldera exhibits circumferential fractures with well-defined edges [9], consistent with late-stage sagging of the summit area.

The variability of the caldera rim, the scarp morphology, and the lack of any observable primary lava flow features are consistent with eroded deposits emplaced by explosive eruptive activity, though a combination of eruptive styles is indeed plausible [7, 48]. Understanding the nature of these multiple geologic features in and around the caldera is important for understanding the types of processes that formed Hadriaca and its geologic history, providing insight into both aggradational and degradational events [9].

Unlike the nested calderas in the Tharsis region, where there are discrete collapse events separated by caldera floor resurfacing, the Hadriaca caldera is not as clear [49, 50]. The lack of nested craters and the simplicity of the caldera morphology suggest that one large episode of volcanism occurred, or if multiple events did occur, then the magma chamber remained relatively uniform, or at the very least the latest episode of volcanism was the largest in magnitude and essentially erased signs of previous summit activity [7, 49]. However, from MOLA data showing non-uniform topography in the caldera, a study by Williams et al. [9] pursues the question of where the resurfacing processes likely occurred in the caldera. From this observation, the western central caldera and a trough along the southwest rim are the lowest elevation regions separated by a raised ridge. The southwest trough and raised ridge are a part of a relatively smoother terrain compared to the west central region that has a rougher, more heavily cratered texture. These features suggest that the trough and ridge have been resurfaced more recently.

Flank Channels

With Hadriaca Patera elongated to the southwest, the channeled flanks that surround the caldera are asymmetric (Fig. 8c) [7, 44]. Flank slopes range from 0.10° to the northeast to 0.47° in the south and west [7], where for reference the average flank slope for Martian montes is 3.92°, and 2.78° for Martian paterae [51]. Surrounding the summit caldera are channeled flanks comprised of extensively layered and friable deposits [7, 32, 47, 52, 53]. The morphology of these channels may suggest erosional processes by groundwater sapping and surface runoff, but lava overflows might be more plausible in the low-pressure Martian environment where gravity-driven flow models [9, 47] demonstrate that the distribution of flank materials can be ascribed to the emplacement of pyroclastic flows. The layering pattern observed within Hadriaca Patera could be due to the welding of pyroclastic flows, which would control its vulnerability to erosion [54]. The flank valleys on the south side of Hadriaca Patera are shallow, scarp-bounded troughs, some with inner V-shaped interior valleys [9]. These morphologies, along with the theater-headed and erosional remnant mesas of the upper flank materials, have also been attributed to sapping and surface runoff [6, 7, 27]. However, also in this case lava flows can be a plausible alternative explanation. There is a higher density of channels and of narrow V-shaped valleys, compared to Tyrrhena Patera [7]. According to [7], differences in channel density could indicate the relative ages of the volcanoes, though crater frequencies and channel morphology suggest that Hadriaca Patera is younger than Tyrrhena [27, 55].

To the south and west of the caldera are degraded, trough-shaped valleys that form a radial pattern [27]. The channels are continuous for large distances and lack tributaries. Some theater-type channel heads are also observed, particularly near the caldera. Scarps within the channel walls suggest that enlargement of the channels may have occurred [9]. Many of the channels are a typical V shape, and deeper dissection is observed in the southwest flanks of Hadriaca.

Hadriaca Patera is observed to have a population of ~180 dikes, ridges and fractures extending some hundreds of kilometers from the caldera rim, some as much as 450 km [56]. This dike grouping is similar to grabens studied in the Tharsis Montes region [50, 57, 58]. Korteniemi et al. [56] observed that the dike groups directly south of Hadriaca were probably fed by a magma chamber related to the volcano, which may have caused a roughly radial dike swarm extending up to 550 km south from the volcano center. The overall orientation of the dike population is from northwest-southeast to north–south, possibly controlled by a concentric fracture system around the Hellas basin [9, 56]. Southeast of the volcano, the rest of the dikes occur in several isolated groups.

It should be noted that the distribution and orientation of the channels at Hadriaca do not reflect regional or local structural influence, as is common in terrestrial sapping environments [7, 27, 59, 60]. Crown and Greeley [7] noted that the continuous form of the channels and their extensivity are not consistent with the complex surface morphology of a typical lava flow field. Differences in channel morphology could also be attributed to different material properties and the presence of subsurface mechanisms (e.g., variations in permeability, distribution of dikes, structural discontinuities, etc.).

Gravity and Magnetic Anomalies

Several studies have identified a positive gravity anomaly at Hadriaca and Tyrrhena Patera, which could indicate solidified, dense mafic magma bodies underneath [5, 61, 62]. Typically, Martian volcanoes show either a localized magnetic minimum or a total absence of crustal magnetism [62]. Interestingly, Hadriaca Patera is the only Martian volcano with a clear positive magnetic anomaly directly over the volcanic edifice, as observed by using electron reflection magnetometry [63–65].

Lillis et al. [62] suggests three possible scenarios for the unusual magnetic signature at Hadriaca Patera: (i) the magma body and volcano were emplaced after the Hellas impact (and therefore after the termination of an early dynamo), though this may not have been the case if a volcano formed at a location with a pre-existing magnetic anomaly; (ii) the Hadriaca magma chamber (pre-edifice) was emplaced before the Hellas impact and was magnetized by thermoremanent magnetization (TRM) in an early pre-Hellas dynamo magnetic field; and (iii) the bulk of Hadriaca was emplaced after the Hellas impact in an area not strongly magnetized, where the dikes and sills that formed beneath the volcano cooled and acquired TRM in a post-Hellas global magnetic field [66]. This last hypothesis is consistent with the interpretations of [67] and [68].

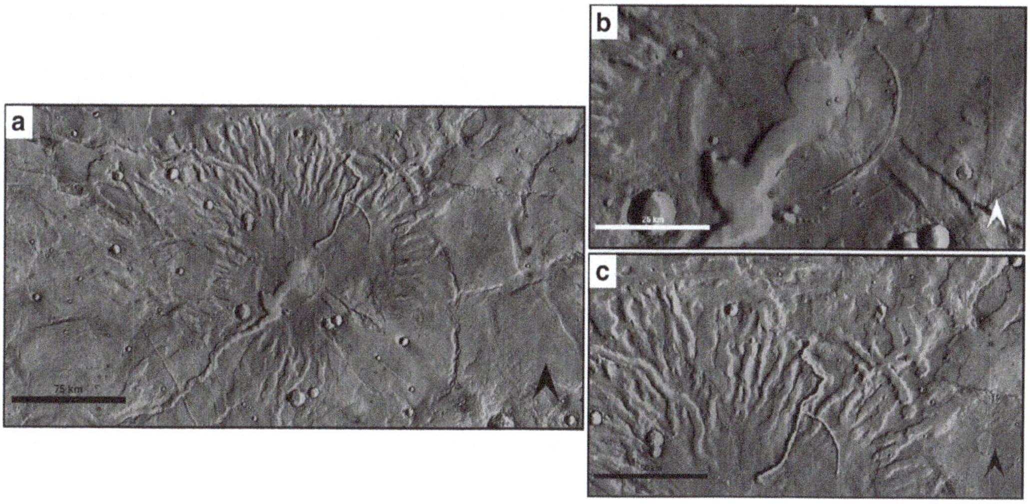

Fig. 9 THEMIS daytime IR image of Tyrrhena Patera. **a** Overall morphological image of the Tyrrhena Patera region and associated flanks. **b** Closeup of the asymmetric Tyrrhena caldera and nearby thin ring fractures. **c** Closeup of the northeastern flank channels and arcuate fractures

1.2.5 Tyrrhena Patera

Tyrrhena Patera (21.36° S, 106.5° E) is a low-relief central vent volcano (Fig. 9a), northeast of the Hellas basin, with a summit at 1.5 km above the surrounding plains [1, 2, 6, 69]. Tyrrhena Patera is observed to be the earliest central vent volcanic system identified on Mars [6]. This highland volcano is associated with a very-localized gravity anomaly that may have formed in the presence of a global magnetic field, and the gravity and topography studies at this site are consistent with the absence of subsurface loading [63, 70, 71]. The volcano possesses an irregularly shaped summit caldera, shallow flank slopes (~0.09–0.37°), and rilles interpreted to be lava channels [6, 69, 72] which cut through the caldera and flanks. The erosional morphology of the shield's flanks suggest that it is composed of friable material, with the flank slopes also more consistent with ash-flow sheets and pyroclastic materials than constructional basaltic flows [6, 70, 72].

Caldera and Flow Deposits

The summit caldera complex (Fig. 9b) is associated with a large lava flow field (~1000 km by 200 km) and a tectonic rille on the southwestern portion of the volcano flank [6, 8]. Formation of a caldera rille and modification of the main edifice was estimated to start in the late Hesperian–Early Amazonian, with a caldera 3.3 Ga old from crater retention age [70, 73–75]. Activity was proposed to continue on through the late Amazonian between 0.8 and 1.4 Ga [76], with the most recent resurfacing attributed to wind erosion and deposition [70, 76], although this relatively young age is not supported by available thermal models (see chapters "History of Scientific Studies and Current Views of Mars" and "The Formation of Mars and the Origin of Its Volcanic Provinces").

The Tyrrhena basal extent is estimated to be from 200–300 km, up to 600 km [8, 69, 70, 73], based on the extent of associated lava flows [72]. The lava flows are the youngest materials at Tyrrhena [6, 33, 75]. The erodible nature of the shield materials combined with the extremely low flank slopes suggests these materials were emplaced via pyroclastic flows and have a mafic composition [6, 72, 77]. Previous mapping suggested that the pyroclastic materials from Tyrrhena could be traced as far as 600 km from the central caldera vent [6]. However, new higher resolution images show that pyroclastic material might be found in craters located downwind from the central caldera vent (see this chapter). Other than the young lava flows emanating to the southwest of the summit, no primary flow features (e.g., lobate flow margins, channels, pit-crater chains) are observed [6, 8]. The shield materials are instead characterized by a smooth, locally dune-covered surface bounded by steep convex scarps (Fig. 9c), indicating a more erosional rather than depositional origin [6]. The shield material layers are observed to be stair-like, where individual layers are between 50–325 m thick, and it is suggested that these layers may be cooling (i.e., flows cooled together) or eruptive (distinct flows that were emplaced and cooled independently) units [6, 8]. Similar to Hadriaca Patera, the channels dissecting Tyrrhena's flanks display morphologies interpreted as modification by groundwater sapping that is enhanced in permeable, pyroclastic materials [6, 27, 75]. However, the morphology of the flank valleys suggests that there may be buried lava flows [27].

Flow Activity

The Tyrrhena Patera lava flows are observed to extend from the central caldera to the southwest and cap the summit region [6]. The distribution and lengths of the observed deposits are consistent with an origin by the emplacement of gravity-driven ash flows, either by hydro-magmatic or magmatic explosive eruptions [6]. A hydro-magmatic origin for this volcano may suggest climatic changes and may explain the absence of this style of volcanism in later geologic times [6]. Otherwise, a magmatic origin implies an evolution in Martian magmas from Hesperian to Amazonian time, during which the prominent shields in the Tharsis region formed [6].

There are three key intervals of major activity at Tyrrhena Patera [70, 76]. This starts with the formation of the edifice in the Noachian Period (~3.7–4.0 Ga) by multiple eruptions, shortly following the Hellas impact (~4 Ga) and the formation of Hadriaca Patera (~3.9 Ga). Subsequently, the style of volcanism changed from dominantly explosive to effusive volcanism [8, 70]. The modification of the edifice and formation of the caldera rille and channels were estimated to occur in the Hesperian Period (with possible extension into the Amazonian Period). In the final stage, a modification of the edifice may have taken place in the Late Amazonian Epoch (~0.8–1.4 Ga), with the most recent event being modification on the upper shield, caldera rille, and channels at ~800 Ma [76]. But, again, this relatively young age is not supported by available thermal models (see chapters "History of Scientific Studies and Current Views of Mars" and "The Formation of Mars and the Origin of Its Volcanic Provinces").

Principal Units

The volcano is composed of several principal units, based on their morphology and erosional characteristics. These units are the combination of observations from [76] and [6] and consist of summit and basal shield materials, caldera rille floor, channel and flank flow materials, and smooth plains.

Summit Shield Materials

The summit shield materials can be separated into two subcategories: upper and lower summit shield materials. The upper unit is smooth and continuous, forming isolated mesas [75]. It consists of steep, amphitheater-shaped scarps oriented away from the summit, forming the walls of wide, flat-floored channels radiating to the summit. Observations from [75] also show that this unit contains crater chains, arcuate graben, and concentric ridges. This unit likely formed through pyroclastic deposition erupted from Tyrrhena, in which the arcuate graben and crater chains mark the collapse zone around the summit produced by magma withdrawal or eruption [76]. The erosional style and morphology suggest that it is composed of volcanic ash [48, 51, 78].

The lower summit shield unit is a discontinuous, smooth (displaying localized hummocky texture) surface, with isolated steep-sided mesas [75]. This sub-unit also contains narrow (<500 m wide), V-shaped channels facing away from the summit. This unit is interpreted to be erosionally modified [75].

Basal Shield Materials

The basal shield materials unit extends as far as 340 km to the south and over 300 km to the north-west from the central caldera. It consists of a smooth surface, though discontinuous around the summit units, and heavily dissected by ~5 km wide, flat-floored channels [75]. It occurs as isolated, steep-sided mesas. This unit consists of pyroclastic deposits, heavily modified by groundwater sapping or lava erosion, with slight variations from differential erosion.

Caldera Rille Floor
The rille-like caldera floor is the smoothest (least cratered) part of Tyrrhena Patera. This unit is interpreted to be late-stage lava flows that feed into a large lava flow field southwest of the volcano [6, 75, 79]. Two sets of ring fractures and three rille-like channels can be identified at the summit. The inner ring fractures are composed of discontinuous grabens [6].

Channel and Flank Flow Materials
The channel floor has a mottled surface locally cut by shallow, narrow channels within flat-floored, steep-walled channels that dissect the shield materials [75]. This unit is interpreted to be reworked summit and basal materials, which were transported and subsequently deposited [76]. The southwest flank flow unit is oriented toward the central caldera, and consists of tens to hundreds of individual lava flow lobes emplaced by channel flow mechanisms [6]. Several small channels are representative of collapsed lava tubes.

Smooth Plains
The smooth plains extend west and northwest from the basal shield unit, about 600 km from the central caldera and may be the earliest exposed eruption unit [6]. This unit, also noted as the "etched plains materials unit," is a broad, flat plain generally lacking in ridges, but with shallow scarps. It embays older, cratered terrain, hence the "etched" term to describe the eroded, diffuse texture. In some locations, it appears smooth deposits mantle the ridges. This unit is either thin lava flows or pyroclastic material, with the mantling deposits originating from reworked shield material [75, 76].

Ridged Plains
The ridged plains to the east of Tyrrhena Patera consist of flood lavas and are characterized by numerous mare ridges [6, 80]. Evidence for the control of fluvial channels by volcanic topography is suggested along the contact between the southwest flank flow unit and the ridged plains, as implied by Greeley and Crown [6].

1.2.6 Apollinaris Mons

Apollinaris Mons (8° S, 174° E) is a ~200 km diameter, ~5.5 km tall singular and isolated volcano located along the transition zone (Leone, 2015, Transition Topography, Encyclopedia of Planetary Landforms) between the northern lowlands and southern highlands ~200 km north of Gusev Crater (Figs. 1a and 10). The Medusae Fossae Formation (MFF) [e.g., 81] surrounds Apollinaris on all but its southern edge; deposits morphologically similar to the MFF occur on Apollinaris' summit (Fig. 11a), suggesting that the MFF may have covered Apollinaris at one time [82]. Further, Apollinaris Mons shows evidence of overlying and interfingering with the MFF (Figs. 11b, c), suggesting a complex relationship and temporal history between these two units. The origin of the MFF is uncertain; however, based on its distribution and textural appearance, it may represent volcanic deposits [81] sourced from Apollinaris Mons, which would suggest Apollinaris-related deposits may be more globally distributed [83].

Overall Morphology

Apollinaris Mons has been divided into four geomorphic units that include the main edifice, an outer caldera, an inner caldera, and fan materials on the south side of the edifice [13, 19] (Fig. 10). The main edifice is roughly circular in planform with slopes of ~1.5° towards the base which steepen to ~5° near the summit [69]. Much of the edifice is bounded by a ~2000 m high basal scarp that is covered on the south side of the edifice by the fan deposit. Crater modeled ages of the Apollinaris Mons flanks range from ~3.5 Ga [13] to ~3.6 Ga [17] to ~3.7 Ga [16], whereas the caldera may have experienced a more protracted history with crater modeled ages ranging from ~3.9 to ~3.6 Ga [10]. There is a positive magnetic signature that corresponds with Apollinaris' edifice, which is likely indicative of an extinct magma chamber [84]. However, the occurrence of this anomaly

Fig. 10 Overall geology of Apollinaris Mons. **a** THEMIS daytime IR image of Apollinaris Mons. White boxes denote the locations of Fig. 11a through h. Black line X-X′ shows the location of the topographic profile shown in (**b**). **b** Topographic profile of Apollinaris Mons. V.E. = Vertical Exaggeration. **c** Geologic map of Apollinaris Mons. **d** Sequence of Map Units to accompany the geologic map shown in (**c**)

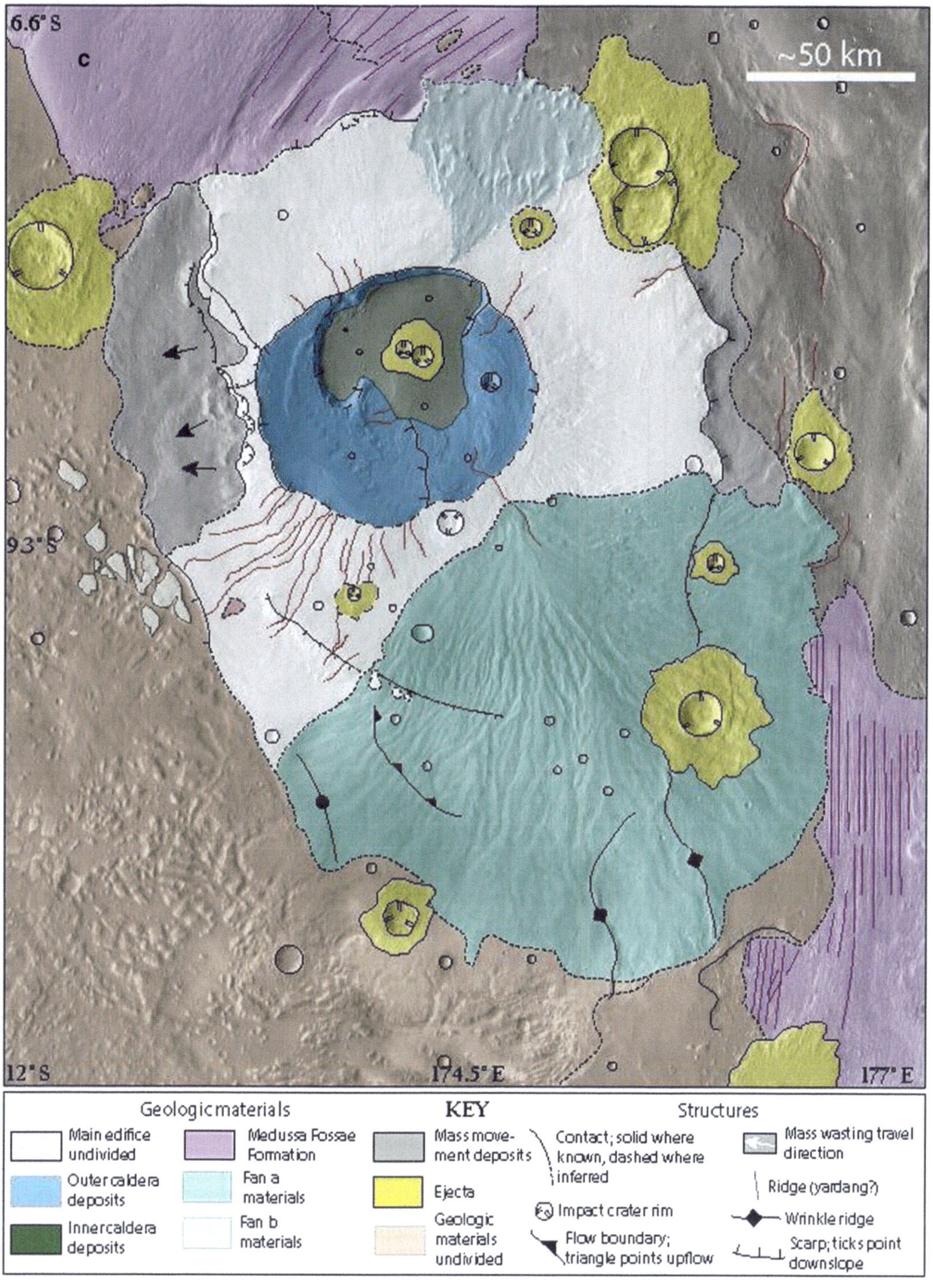

Fig. 10 (continued)

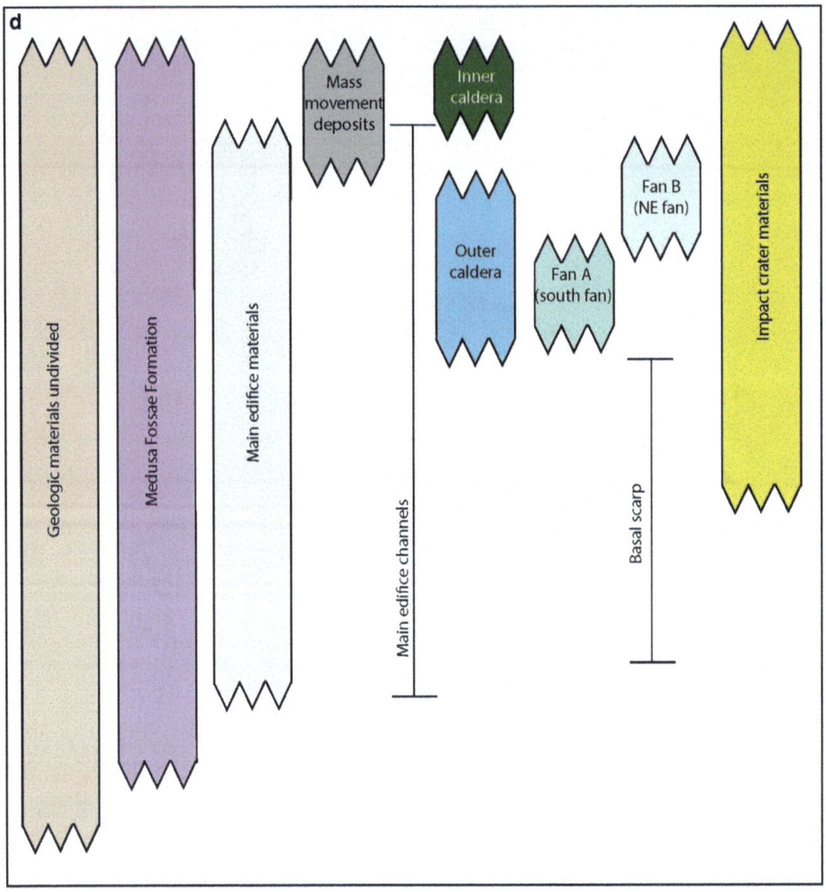

Fig. 10 (continued)

appears to be most consistent with iron-rich basalt that was present in an oxidized environment [84].

Main Edifice

Much of the main edifice consists of apparently amorphous materials that appear to lack large-scale internal structuring such as layering, jointing, and mechanical strength (Fig. 11d). Consequently, much of the main edifice has been lumped into one unit. However, the uppermost few meters of these deposits seem to be mechanically stronger than the underlying materials.

Present on the west side of the edifice are a series of slump blocks (Fig. 11d); the western side of the volcano has a slightly steeper slope than the eastern side (slope = 4.5°–5° on the western side vs. ~1–3° on the eastern side; Fig. 10b). It is possible the steeper nature of the western side of Apollinaris led to the slumping there compared to the east side. The slump blocks are isolated in their occurrence, but appear to post-date formation of both the basal scarp and the channels on the main edifice; thus, similar to the basal scarp, these blocks may have failed while Apollinaris was active. Cutting through the edifice's deposits are channels that are straight on the uppermost slopes near the summit caldera complex, but either dissipate or meander at shallower slopes. The channels lack distributaries, but some do have minor tributaries near the summit (Fig. 11a). The overall lack of extensive tributaries suggests they do not represent fluvial systems. At the termini of a few

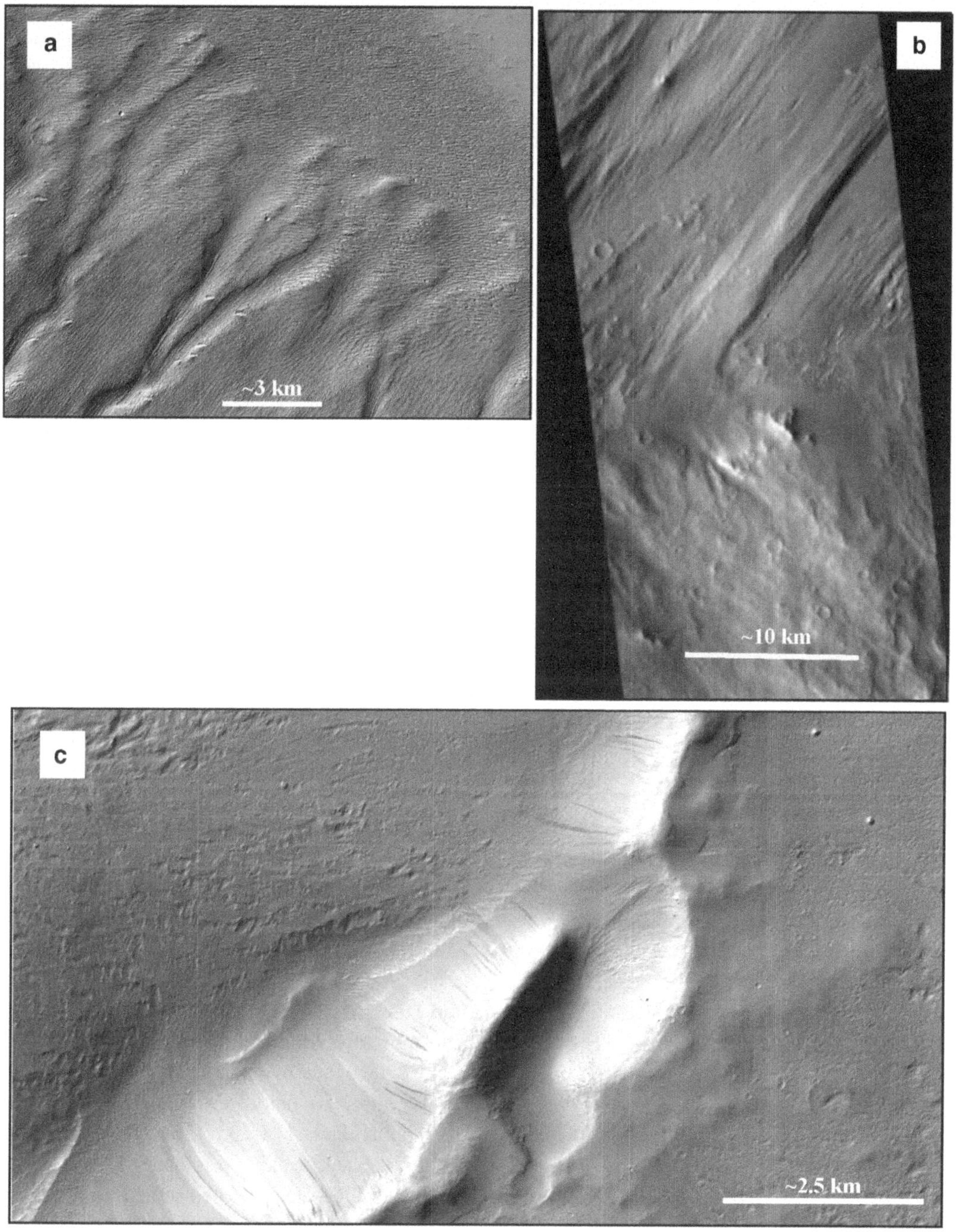
a
~3 km
b
~10 km
c
~2.5 km

◀ **Fig. 11** Highlights of the Apollinaris Mons edifice. **a** CTX image of the southwest edge of the boundary between the main edifice and summit caldera highlighting possible MFF deposits (dark gray layer) that cover Apollinaris Mons here and channels that appear to originate at the summit caldera. Channels exhibit limited tributaries. **b** CTX image of the northern boundary of Apollinaris Mons with the MFF. The stratigraphic relationship between the MFF and Apollinaris Mons here is not clear, but Apollinaris Mons appears to be either residing on top of the MFF here or may be representing the source of material that is comprising the MFF here. **c** CTX image of another segment of the northern boundary between the MFF and Apollinaris Mons. The stratigraphic relationship between the MFF and Apollinaris Mons is unclear here as well. It appears that some of the MFF may have accumulated against Apollinaris Mons, but this relationship is obscured by material that appears to have shed off the wall of Apollinaris Mons. Also note the lack of obvious layering in the exposed walls of Apollinaris Mons. **d** CTX image of the western flank of Apollinaris Mons highlighting localized slumping; white arrows highlight the head scarps of these localized mass wasting events. Note how flank channels are truncated along the edges of these scarps indicating these mass wasting events likely occurred after channel formation. **e** CTX image of channel termini at the bottom of the southwestern flank of Apollinaris Mons. White arrows highlight leveed materials that extend out of these channels. The type of material comprising these levees is not known, but could represent lava flows, debris flows, or pyroclastic deposits. **f** Portion of CTX image P18_007882_1703_XN_09S185W highlighting a part of the northeastern fan on Apollinaris Mons. White arrows highlight the location of layering and jointing within this fan deposit suggesting it may represent a pyroclastic deposit that was welded upon emplacement. **g** Portion of MOC image S0602038 highlighting well-developed layering (black arrows) within the south fan deposit. **h** Portion of CTX image P22_009464_1693_XN_10S183W highlighting where the toe of the south fan deposit has extended into yardangs of the MFF (white arrows) indicating that south fan deposits are younger than the MFF. Black arrow indicates where a yardang has been possibly buried by deposits of the south fan

Fig. 11 (continued)

channels on the western flank, leveed materials appear to have emerged; the presence of levees in these materials indicates the material has some mechanical strength, which would be consistent with them being either lava flows, pyroclastic flows, or perhaps even debris flows [13] (Fig. 11e). Although these deposits have travelled through these channels, it is not clear if these are the materials that carved the channels. No materials can be observed to have emerged from other Apollinaris channels.

Fan Deposits

The most discernible units on the flanks of the main edifice are two fan-shaped deposits—one on the northeast flank [85] (Fan B) and one on the south flank [13, 35, 86] (Fan A). Both deposits appear to have originated at or near the summit of the volcano. The northeast fan deposit has largely been eroded away and occurs predominantly as a series of erosional remnants, but likely represents a pyroclastic flow deposit [85]. Figure 11f highlights this fan deposit and shows it contains multiple layers that may be related to welding and jointing within an ignimbrite. The south fan appears to be more resistant, though it does contain a series of channels. The presence

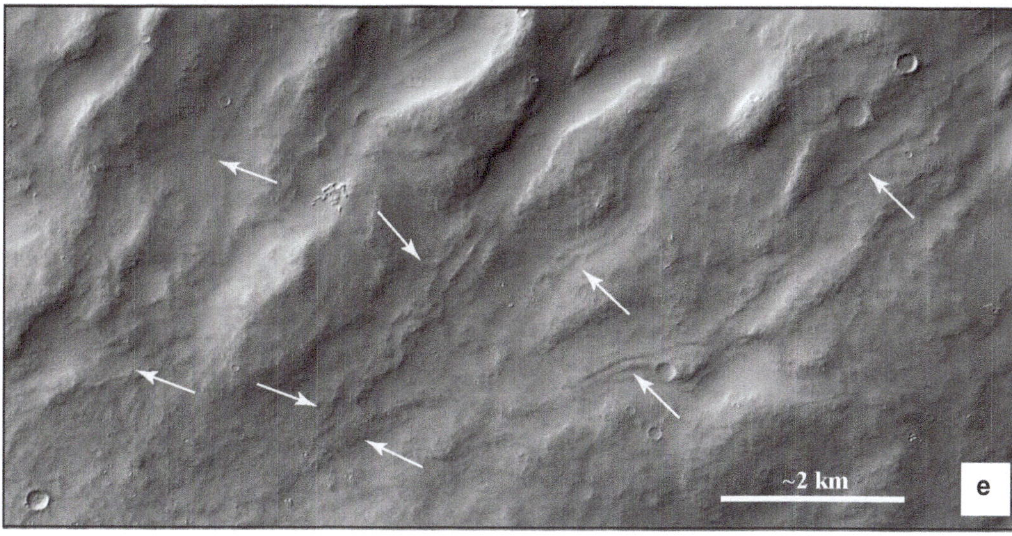

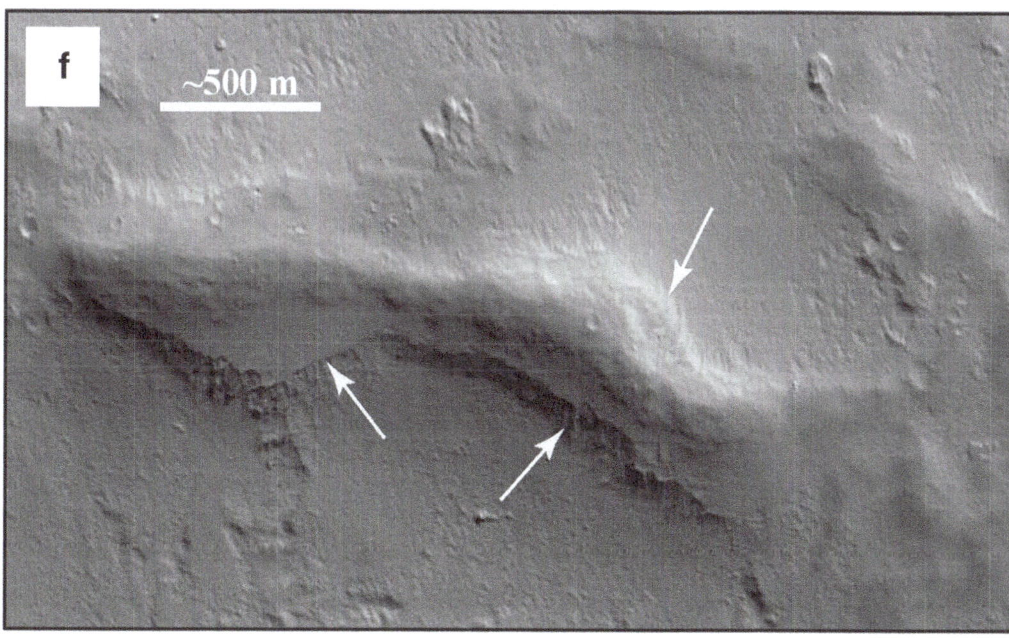

Fig. 11 (continued)

of these channels has led to the interpretation of this deposit as a series of pahoehoe flows [13] because (1) the valleys are not the same narrow, incised valleys seen in the main edifice, suggesting a different, mechanically stronger unit, (2) there are no high-relief flow lobes, as would be expected with an a'a flow, and (3) the unit spreads out into a fan, getting wider with distance from the outer caldera. However, the presence of layers in this deposit (as exposed in the walls of impact craters (Fig. 11g)) has led to alternative interpretations that it could represent pyroclastic [35, 86] or possibly debris flow deposits [86]. The toe of the south fan embays yardangs of the MFF (Fig. 11h), suggesting it is younger than the MFF here. Interestingly, topographic profiles along the channel that fed this fan show an increase in elevation in moving from the caldera

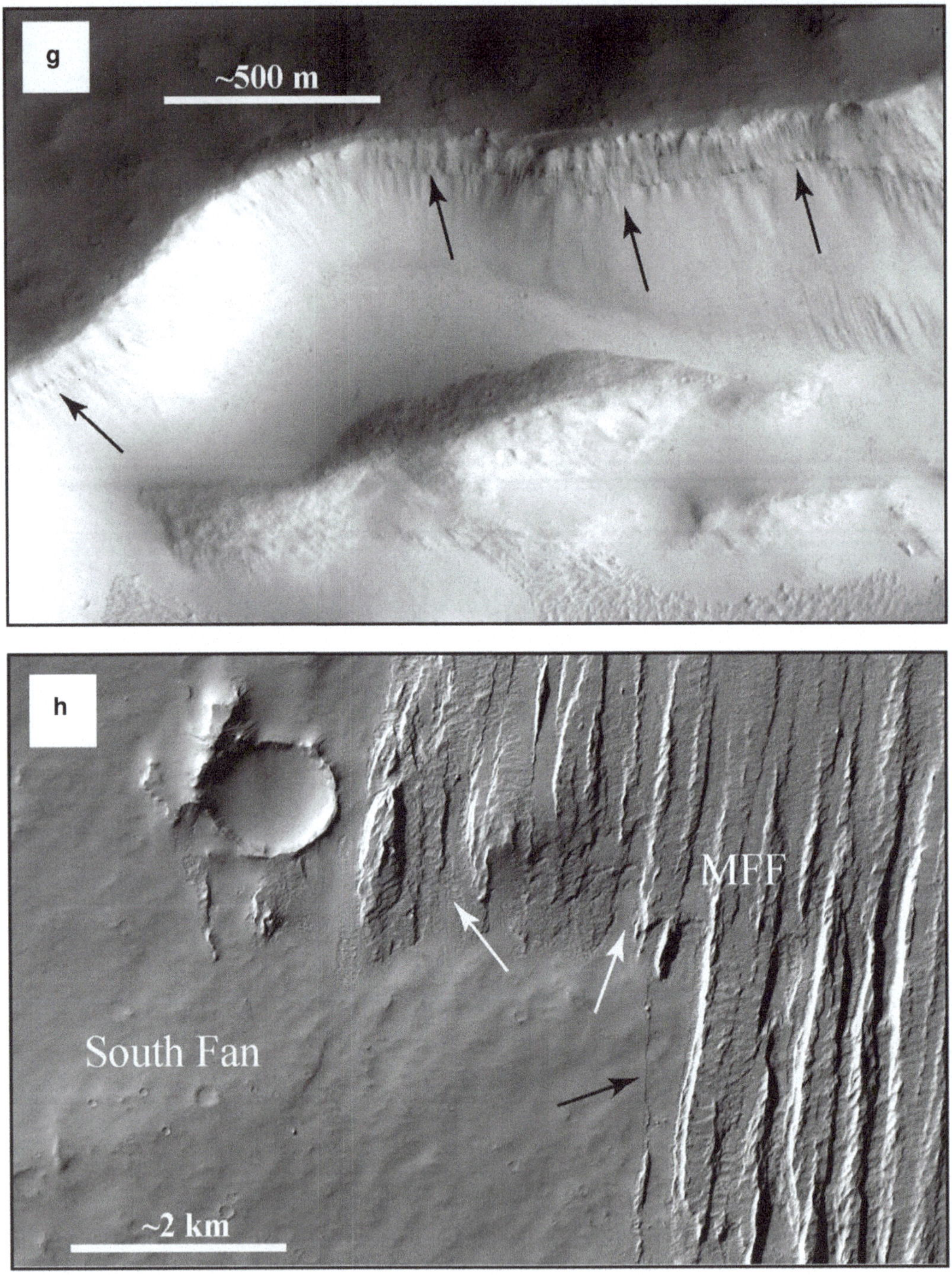

Fig. 11 (continued)

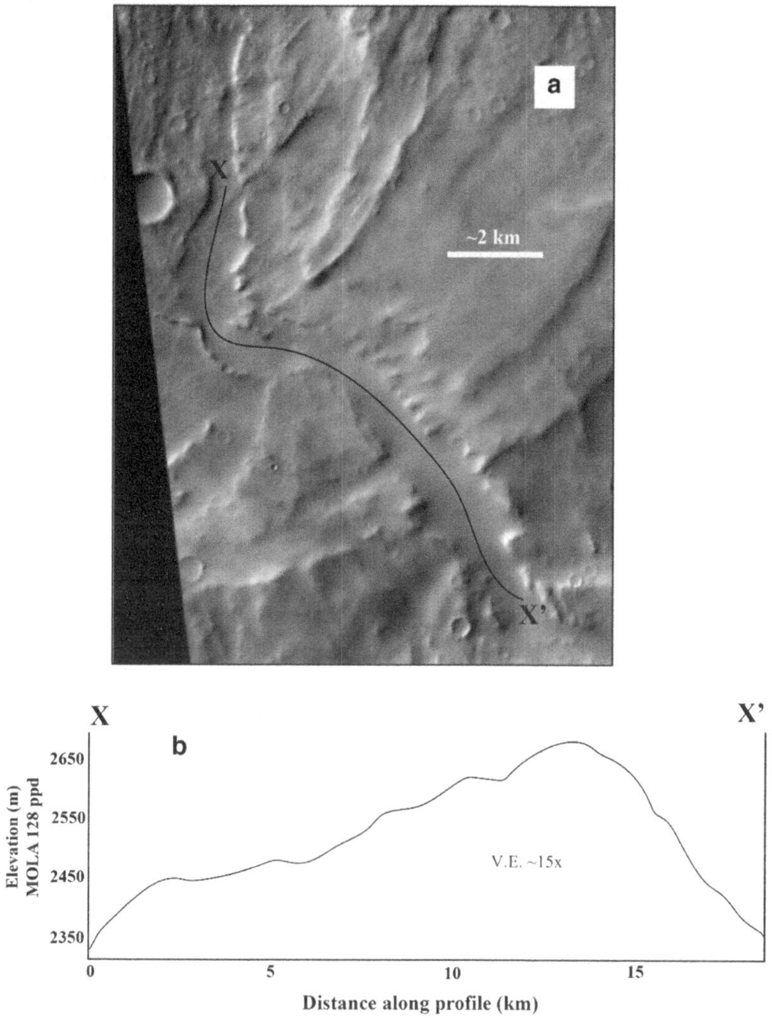

Fig. 12 Characteristics of the summit channel that feeds the south fan on Apollinaris Mons. **a** Portion of CTX image D02_027859_1708_XI_09S185W highlighting the channel that traverses from inside the outer caldera to the south fan. Black curved line X-X′ denotes the location of the topographic profile shown in (**b**). **b** Topographic profile of the channel that traverses from the outer caldera to the south fan. V.E. = Vertical Exaggeration

floor to the channel mouth (Fig. 12); this suggests that the caldera may have experienced subsidence subsequent to emplacement of the south fan. Crater modeled ages suggest the northeast fan could be younger than the south fan. [87] highlighted that the more pronounced layering and possible jointing in the fan deposits may represent higher degrees of welding compared to the flank and caldera pyroclastic deposits, which could be tied to higher eruption temperatures later in the volcano's history.

Summit Caldera

The summit caldera of Apollinaris Mons is ~80 km × 70 km and can be divided into an outer and inner caldera (Fig. 13). The outer caldera is ~80 km in diameter and is marked by a steep scarp along its outermost edge. Its floor is covered by deposits of etched, easily erodible materials that, following previous workers, we interpret as pyroclastic deposits [3, 13] (Fig. 14 a). Along the eastern and southern parts of the outer caldera, concentric troughs, which we interpret as graben modified by possible fluvial activity, cut these deposits (Fig. 14b). In turn, etched materials that we again interpret as pyroclastic deposits cover the graben in the western part of the outer caldera. Figure 13c shows a side view of these materials; they appear to be easily eroded, but do appear to contain some weak form of layering. The outer caldera has been broken into large (>25 km diameter) fault blocks that dip north into the inner caldera; the cause of the fractures on the outer caldera floor is unknown, but it is possible they could be the result of piecemeal caldera collapse, though their occurrence during formation of the inner caldera cannot be ruled out. The broken outer caldera floor is embayed by inner caldera deposits that appear to be mechanically stronger than outer caldera deposits. They locally host contractional ridges, which is consistent with previous interpretations that inner caldera materials are effusive deposits [13]; in fact, these deposits seem to reflect a lava lake. From cross-cutting relations, we envision the following history for Apollinaris Mons' summit caldera: (1) Two periods of explosive eruptions. As evidenced by graben cutting the pyroclastic deposits, the outer caldera either formed synchronously with these eruptions or post-dated them; it is possible that a second period of explosive volcanism resulted in the formation of the outer caldera. (2) Formation of the inner caldera, which occurred an unknown amount of time after outer caldera formation and was associated with filling by a lava lake. Based on the embayment of outer caldera materials by inner caldera materials, it appears that the inner caldera may have formed through a trap-door-like process. (3) Channel formation within Apollinaris Mons' summit caldera and graben modification, though it is possible that channel formation could have occurred throughout caldera evolution. Evidence for this relates to the channel that feeds the fan on the southern part of the edifice. As noted earlier, a MOLA-derived topographic profile taken along the length of the channel indicates that the channel slopes up from from inside the caldera to the rim (Fig. 12). The history recorded at this summit caldera is consistent with that recorded at Hadriaca and Tyrrhena Paterae.

1.3 Summary of Current Understanding of Highland Paterae

We opened this chapter with four broad questions that have motivated work on highland paterae. After spending the last section describing the morphology, topography, and ages of seven highland paterae, it is worthwhile to revisit those questions and see where we currently stand with answering them.

(1) *What triggered the explosive eruptions and why the transition to effusive eruptions?*

The answer to this question is still ambiguous. It seems unlikely that explosive eruptions would be due to a felsic magma composition. Although some felsic volcanic deposits have been identified on Mars near Syrtis Major [7, 88], the vast majority of volcanism on Mars has been shown to be basaltic in composition [89, 90]. Even intermediate compositions that were suggested to be possible in the northern lowlands [89] have been shown to possibly be weathered basalt [90]. In addition, although surficial deposits in the Malea Planum region cannot be definitively tied to originating in that location, their mafic mineralogies [5; see also 31] are also consistent with volcanism in that region being mafic in composition. The detection of possible water–ice in the subsurface of some regions of Mars [91–94] is consistent with original interpretations that explosive eruptions at highland paterae could be tied with magma-volatile interactions early in each volcano's evolution. In this scenario, initial interactions of magma with a volatile such as ground ice would have led to explosive eruptions; a subsequent loss of ground ice either through changing climatic conditions [e.g., 95, 96] or dissipation due to magma interaction could have led to subsequent effusive eruptions. However, alternative scenarios could also account for this explosive to effusive transition.

Specifically, the possibility that explosive eruptions could be related to magma degassing [97], rapid ascent of highly vesiculated magma [e.g., 98], conduit-related processes (e.g., conduit narrowing) [e.g., 99], and high groundmass crystallinity cannot be fully ruled out and testing between each of these scenarios is currently difficult [see Chap. 0 and Chap. 8 for further discussion]. Ultimately, any hypothesis that addresses the explosive to effusive transition would need to account for why this transition occurred at about the same time in each volcano's evolution.

(2) *What were the processes involved in caldera formation?*

Calderas at each of the highland paterae vary from having well-defined margins to more subtle transitions. Well-defined margins are typically noted by the presence of fractures and/or graben (i.e., Apollinaris, Peneus, Hadriaca, and Tyrhenna) that appear to have facilitated a coherent, piston-like drop of the caldera floor. Apollinaris underwent at least two caldera formation events and Hadriaca Patera may have undergone multiple events. Peneus, Hadriaca, and Tyrrhena Paterae appear to have undergone one caldera-forming event each that, based on the seemingly coherent nature of the caldera floors, may have been a piston-style caldera formation where the floor dropped along ring faults into an evacuating magma chamber. However, it is unclear if possible dropping of the floor occurred instantaneously or gradually. Based on analog experiments [100], large coherent collapse of caldera floors, such as what is observed at these multiple highland paterae, suggests a shallow magma chamber with a large diameter (low caldera-roof aspect ratio). The occurrence of possible concentric graben in the outer caldera at Apollinaris Mons suggests that the outer caldera may have collapsed in a more piecemeal fashion.

(3) *What processes have operated on the flanks?*

Processes that are suggested to have operated on highland paterae flanks appear to be a combination wind, mass wasting, and those that produce channels. Wind processes are apparent in the appearance of surficial deposits in the calderas of Malea Planum and on the flanks of Apollinaris Mons; possible yardangs are also present on the southwest flank of Apollinaris Mons. These are likely the current processes operating on these volcanic structures, where they are serving to transport and remove materials from the edifices, thus obscuring their surfaces and complicating interpretation of recorded volcanic histories at each site.

Mass wasting processes have occurred predominantly in highland paterae calderas as local isolated events, but they have occurred at a larger scale at Apollinaris Mons. The higher occurrence of mass wasting events at Apollinaris Mons may be due to the steeper flanks of the edifice compared to the other highland paterae described in this chapter. The most obvious evidence of mass wasting at Apollinaris Mons is the basal scarp that surrounds the edifice; the fan material on Apollinaris' south flank covers the scarp, indicating that this scarp must have occurred during the volcano's active history. The cause of the formation of the basal scarp is unknown and the absence of most materials that were removed (i.e., lack of deposits associated with this mass wasting event) is striking—were they possibly removed by wind? Isolated slumping occurred on Apollinaris' western flank and most likely post-dated formation of the basal scarp. The trigger for these slumping events is also not clear, but may be related to the steeper nature of the western flank of this volcano and the seemingly weak nature (e.g., lack of obvious layering and possible welding) of these materials.

The formation of channels on the flanks of the volcanoes may perhaps be the most ambiguous process as there is no clear evidence pointing to their origin. The lack of higher stream orders for any of the channels suggests they are not evolved fluvial channels. Possible lava tubes in channels at Amphitrites Patera [5] suggests volcanic processes played a role in channel formation, or at least took advantage of pre-existing channels. Most channels do not have obvious transported materials at their termini, further complicating this question. However, channels on the

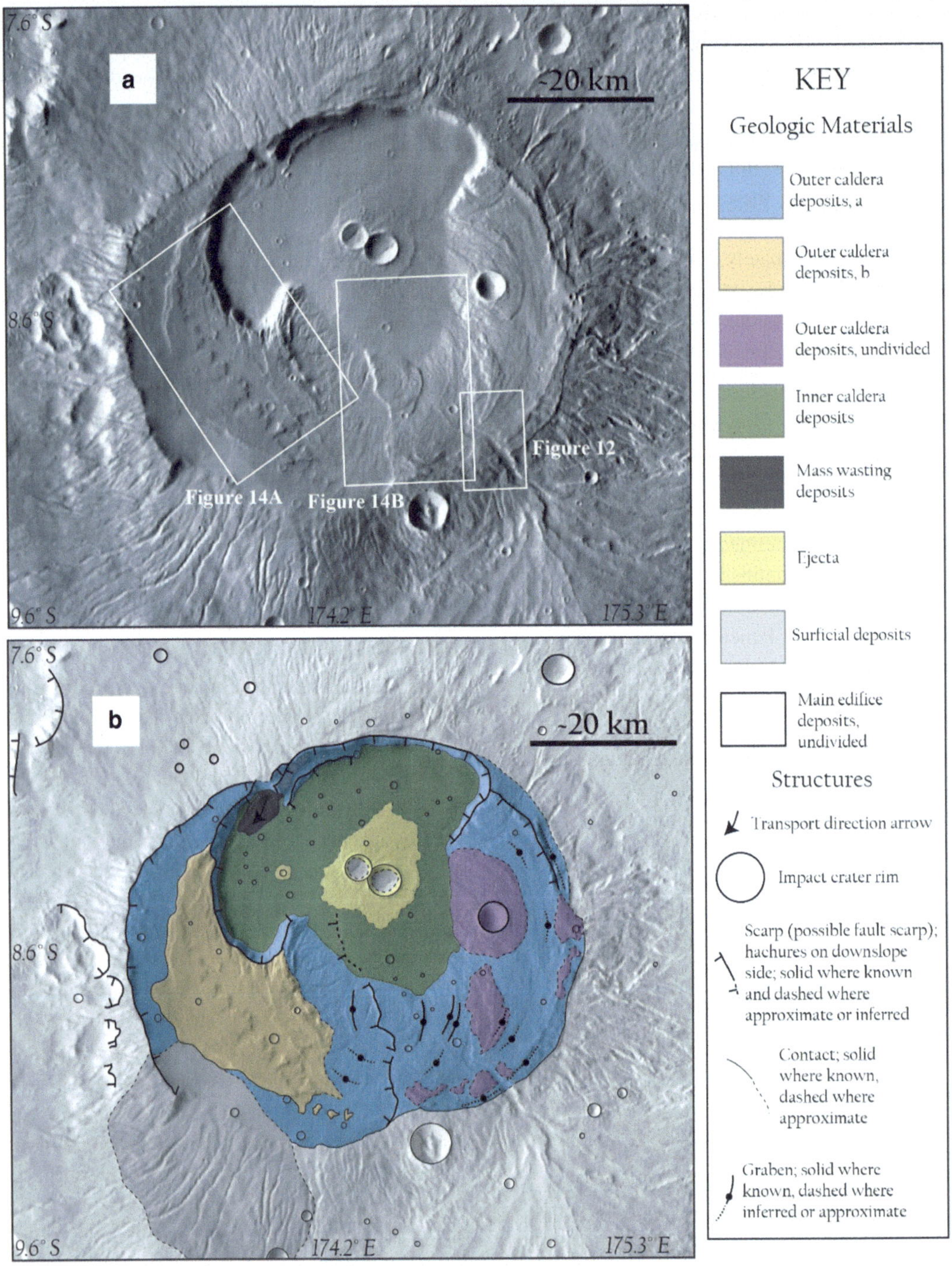

Fig. 13 Geology of the summit caldera of Apollinaris Mons. **a** THEMIS daytime IR image of the two calderas at the Apollinaris Mons summit. White boxes denote the locations of Figs. 12 and 14. **b** Geologic map of the two calderas at the Apollinaris Mons summit

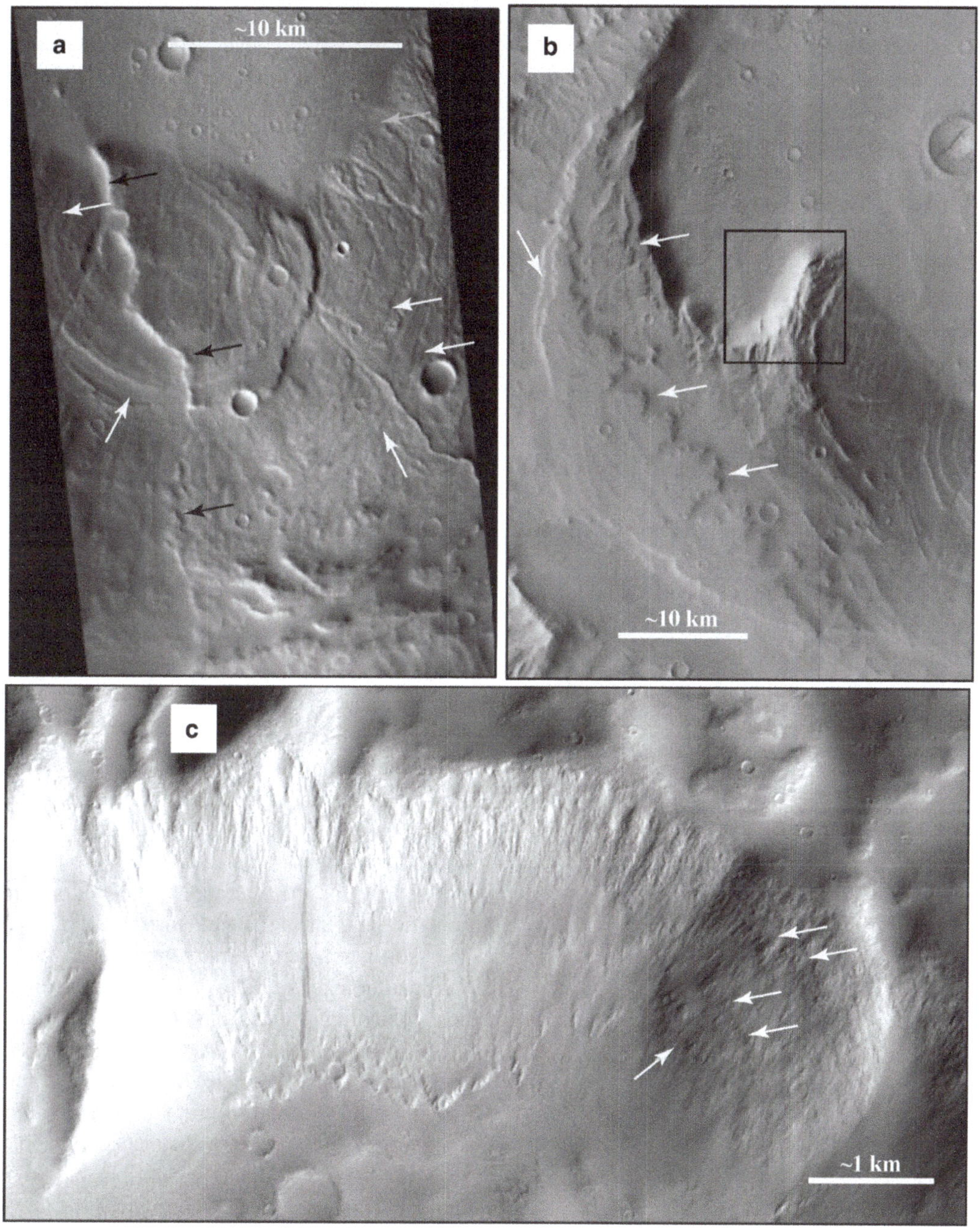

Fig. 14 Characteristics of various parts of Apollinaris Mons. **a** Portion of CTX image D10_031116_1708_XI_09S185W that covers parts of the outer and inner caldera of Apollinaris Mons. White arrows highlight possible graben in the outer caldera; some of these graben may have subsequently been modified by aqueous processes. Black arrows highlight a scarp that has broken part of the outer caldera floor; the scarp appears to extend under the smooth material comprising the inner caldera. Gray arrow highlights where the smooth materials comprising the inner caldera appear to have onlapped onto the floor of the outer caldera. **b** CTX image mosaic over the western edge of the Apollinaris Mons summit caldera; observed in this image is the outer caldera and inner caldera. White arrows highlight layering within the etched deposits found on the outer caldera floor; we interpret these as pyroclastic deposits. Black box denotes the location of Fig. 13c. **c** Portion of CTX image B01_010163_1712_XN_08S186W of the scarp that marks a boundary between the inner and outer calderas at Apollinaris Mons. White arrows highlight crude layering that is observed within these possible pyroclastic deposits

southwestern flank of Apollinaris Mons could provide some clues. The preserved materials that have emerged from these channels appear to contain levees, which indicates that these deposits may have behaved as a Bingham plastic or at least as a non-Newtonian substance [e.g., 101, 102]. The apparent viscous behavior of these materials does not uniquely identify them, but it does suggest they may represent either lava flows [e.g., 101], mass wasting materials [86], or even pyroclastic deposits [e.g., 103]. Given that the channels, when present, extend to the summit, whatever carved these channels must have originated at the calderas.

(4) *What was the duration of volcanic activity?*

Given the sizes and volumes of the paterae discussed here [e.g., 39], it seems reasonable to assume that each of these paterae were active over long periods of geologic time. This assumption is supported by crater modeled ages for each of these centers. Crater modeled ages suggest activity at highland paterae could have persisted for upwards of ~200–500 million years (Fig. 2), with flank activity appearing to be older than activity within the calderas. It is likely this age difference could in part be attributed to wind or other volcanic resurfacing events within the calderas [e.g., 9], which makes it difficult to more fully constrain how much younger caldera deposits may be than flank materials. Regardless of the duration, though, each highland patera shows evidence that as activity waned it likely became concentrated at the summit. The amount of time over which that occurred is still unclear.

1.4 Possible Future Research Directions

It is exciting, though also understandably frustrating, that after over 40 years of analysis, fundamental questions regarding highland paterae are still not resolved. Short of human missions and/or sample return missions from these volcanic centers, it is likely that we will not be able to fully resolve these questions. It is worthwhile, though, to still consider future directions work can be taken to better understand these intriguing features. We explore that here by considering direct analyses that could be performed and also by considering some terrestrial analogs to highland paterae.

1.4.1 Direct Analyses of Highland Paterae

There are multiple directions that future examinations of highland paterae can take. Recent studies of the Circum-Hellas Volcanic Province [e.g., 5] and Apollinaris Mons [e.g., 17, 38, 86], for example, have emphasized detailed mapping of certain parts of each highland patera. For example, Williams et al. [5] utilized a combination of MOLA, High-Resolution Stereo Camera (HRSC), Thermal Emission Imaging System (THEMIS), Mars Orbiter Camera (MOC), High-Resolution Imaging Science Experiment (HiRISE), Context Imager (CTX), and multispectral (HRSC, Observatoire pour la Mineralogie, l'Eau, les Glaces et l'Activite (OMEGA)) data to conduct morphologic analyses, compositional analyses, and impact crater counting for paterae around the Hellas Basin. Chuang et al. [17] used similar datasets and approaches on the northeast fan on Apollinaris Mons and were subsequently able to constrain surface ages and erosion rates for this part of the volcano. Future efforts in examining highland paterae may be the most successful if following similar approaches.

In addition to the above mentioned approaches, obtaining finer-scale topographic data would also be useful. Specifically, improved topographic data may help identify the extent (including thicknesses) of erupted products. For example, the presence of a distinct topographic break on the flank could be attributable to a lithologic (i.e., pyroclastic deposit vs. lava flow) change. Further, integrating topography will be critical for determining unit thicknesses; once unit thicknesses are constrained, they can be used with the aerial extent of exposures to calculate volume estimates. Estimating the distributions and volumes of erupted products will provide more insight into the extent to which volatiles were emplaced into early Mars' atmosphere through paterae volcanism. Combining altimetry with visible data sets could allow for the identification of marker beds. The identification of a marker bed would be useful because if a distinct layer can be identified and traced, the relationship of that layer to topography may become evident, which can provide clues as to its emplacement style (i.e., fall vs. ground-hugging). Similar to the above mentioned studies, tying topographic data together with high-resolution images could be another means of differentiating between deposits emplaced through explosive activity (i.e., pyroclastic deposits), which may have an 'etched' appearance [3, 7] versus effusive activity (i.e., lava flows), which may be more resistant to erosion.

1.4.2 Examination of Possible Terrestrial Analogs

Terrestrial analogs provide a means of contextualizing and furthering our understanding of planetary landforms that are otherwise inaccessible [e.g., 104–106]. Although no specific analog studies have been undertaken for highland paterae, several have been proposed. Traditionally, most analogs for volcanism on Mars have emphasized the broad basaltic shield volcanoes such as those in the Tharsis region [e.g., 106, 107] and have examined terrestrial basaltic shield volcanoes such as those on the Hawaiian islands [107]. However, it is not clear if this analog is the most valid for highland paterae [107]. Any analog for the highland paterae should take into account their low topographic profiles, eroded flank deposits, and large central calderas [e.g., 6, 7, 107] as well as their large summit calderas and late-stage effusive activity [3, 106]. To that end, ignimbrite shields of the altiplano region of the central Andes Mountains have been proposed by some workers as a potential terrestrial analog [7, 107]. Ignimbrite shields have slopes of $\sim 1°–3°$ [107] that dip outward from a central caldera and are comprised of ash-flow material [108] with large well-defined central calderas or a central lava dome complex. It is possible they represent non-resurgent calderas that are surrounded by ignimbrite aprons that were subsequently filled by later lava flows and tuff deposits [109]. Slopes on these volcanoes have also been eroded and channelized in a manner morphologically similar to highland paterae. Even though they do appear morphologically similar in satellite images [e.g., 7], there are differences between ignimbrite shields and highland paterae that should be considered if using ignimbrite shields as an analog. For example, the magma composition of ignimbrite shields is felsic and related to subduction zone magmatism [110]. Further, the erupted volume of ignimbrite shields varies from ~ 100 to ~ 3000 km^3 [111, 112], whereas the erupted volume of highland paterae appear to be an order of magnitude larger and range from 1.6×10^4 (Hadriaca Patera) to 2.1×10^4 (Tyrrhena Patera) to 7.3×10^4 km^3 (Apollinaris Mons) [69].

Another proposed analog for highland paterae has been the Masaya Volcano Complex in Nicaragua [3], which is a nested caldera structure in a low basaltic shield [113]. The caldera was originally thought to have formed through magma drawdown [113], but more recent work has demonstrated that caldera collapse was synchronous with eruption of basaltic ignimbrites [97]. These explosive eruptions may have been triggered during periods when magma was unable to de-gas and became supersaturated in volatiles [97]; when degassing does occur, effusive eruptions of lava are erupted instead. Although this volcano is also associated with subduction zone magmatism, the explosive eruption of basaltic materials makes it an intriguing analog. Specifically, since the explosive eruptions were not seemingly triggered by interactions with water, Masaya Volcano could provide an alternative means of explaining highland paterae eruptions without calling upon magma-volatile interactions.

Additional insight could also be gained through the study of eruptive histories recorded within terrestrial basaltic volcanic systems. Specifically, a comparison of the sequence of eruptive events could be helpful. For example, eruptive stratigraphies recorded in the Cima Volcanic Field in the Mojave Desert of southeastern California, USA, indicates that initial stages of individual cinder cone eruptions were hydrovolcanic due to the interaction of magma with shallow groundwater [114] these eruptions eventually transitioned into strombolian eruption styles to create cinder cones which, in some cases, transitioned into effusive activity with the eruption of lava flows [114]. Similar stratigraphies are also recorded in the Garrotxa Volcanic Field in the northeastern Iberian Peninsula where early phreatomagmatic eruptions transitioned into violent Strombolian-style eruptions [98]; violent Strombolian eruptions were likely the result of the ascent of crystal-poor, highly vesiculated magma under fast decompression rates [98]. At Mount Etna, Italy, conduit processes (i.e., a narrowing of the conduit) [98] and the presence of high groundmass crystallinity [115] have also been predicted to result in basaltic systems erupting in an explosive manner. Consequently, comparison of basaltic volcanic systems to highland paterae could provide meaningful context as to why shifting eruption styles could occur.

Ultimately, although it is likely that no perfect one-to-one terrestrial analog may exist for highland paterae, it is worthwhile to consider these described terrestrial volcanic systems in future highland paterae analyses. Whether these studies are strictly remote sensing in their approach [e.g., 7] or are more field based [107], terrestrial analogs may serve as an insightful pathway for constraining highland paterae evolution.

1.5 Conclusions

In this chapter we summarized the current state of knowledge on martian highland paterae. These are an intriguing class of volcanic landform on the Martian surface that consist of broad low slopes that may signify explosive eruptions and a large central caldera that contains effusive deposits. Temporally, these features indicate there was a transition from explosive to effusive activity, but the cause of that transition is still not clear. Despite over 40 years of analysis, many outstanding questions regarding highland paterae still exist and further addressing of those questions will likely require some combination of finer-scale resolution topographic and compositional data as well as possible terrestrial analog analysis. We hope the work presented in this chapter will provide an ideal jumping-off point from which future studies of this unique landform can occur.

Acknowledgements This work was supported in part by NASA's Mars Data Analysis Program (MDAP, Grant number NNX09AI39G). The authors wish to thank G. Leone for his work as editor on this book.

References

1. Peterson JE (1978) Volcanism in the Noachis-Hellas region of Mars, 2. In: Proceedings of the 9th lunar and planetary science conference, pp 3411–3432
2. Schultz PH (1984) Impact basin control of volcanic and tectonic provinces on Mars. Lunar Planet Sci 728–729
3. Gregg T (2014) Highland Patera. In: Encyclopedia of planetary landforms. Springer, New York
4. Greeley R, Spudis PD (1978) Volcanism in the cratered terrain hemisphere of Mars. Geophys Res Lett 5:453–455
5. Williams D, Greeley R, Fergason R, Kuzmin R, McCord T et al (2009) The Circum-Hellas volcanic province. Planet Space Sci 57:895–916
6. Greeley R, Crown DA (1990) Volcanic geology of Tyrrhena Patera. J Geophys Res 95:7133–7149
7. Crown DA, Greeley R (1993) Volcanic geology of Hadriaca Patera and the eastern Hellas region of Mars. J Geophys Res 98:3431–3451
8. Gregg T, Farley M (2006) Mafic pyroclastic flows at Tyrrhena Patera, Mars: constraints from observations and models. J Volc Geoth Res 155:81–89
9. Williams D, Greeley R, Zuschneid W, Werner S, Neukum G et al (2007) Hadriaca Patera: insights into its volcanic history from Mars Express High Resolution Stereo Camera. J Geophys Res 112. https://doi.org/10.1029/2007JE002924
10. Robbins S, Di Achille G, Hynek BM (2011) The volcanic history of Mars: high-resolution crater-based studies of the calderas of 20 volcanoes. Icarus 211:1179–1203
11. Scott D, Carr M (1978) Geologic map of Mars. U.S. Geol. Surv. Misc. Invest. Ser., Map I-1083, scale 1:25M
12. Greeley R, Guest J (1987) Geologic map of the eastern equatorial region of Mars. U.S. Geol. Surv. Misc. Invest. Ser., Map I-1802B, scale 1:15,000,000
13. Robinson M, Mouginis-Mark P, Zimbelman J, Wu S, Ablin K, Howington-Kraus A (1993) Chronology, eruption, duration, and atmospheric contribution of the Martian volcano Apollinaris Patera. Icarus 104:301–323

14. Crumpler L, Head J, Aubele J (1996) Calderas on Mars: characteristics, structure, and associated flank deformation. In: McGuire WJ, Jones AP, Neuberg J (eds) Volcano instability on the Earth and other planets, vol 110. Geological Society Special Publication, pp 307–348
15. Greeley R, Williams D, Fergason R, Neukum G, Baratoux D et al (2007) Amphitrites and Peneus: new insights into highland paterae. In: Lunar and planetary science conference, Abstract #1373
16. Greeley R, Bernard FH, McSween HY, Neukum G, Pinet P et al (2005) Fluid lava flows in Gusev crater, Mars. J Geophys Res 110:E05008. https://doi.org/1029/2005JE002401
17. Chuang, FC, Crown, DA, Berman, DC (2019) Geology of the northeastern flank of Apollinaris Mons, Mars: Constraints on the erosional history from morphology, topography, and crater populations. Icarus 333:385–403
18. Mouginis-Mark PJ, Wilson L, Zimbelman JR (1988) Polygenetic eruptions on Alba Patera, Mars. Bull Volc 50:361–379
19. Scott DH, Dohm JM, Applebee DJ (1993) Geologic map of science study area 8, Apollinaris Patera region of Mars. U.S. Geol. Surv. Misc. Invest. Ser., Map I-2351
20. Wilson L, Head JW (1994) Mars: review and analysis of volcanic eruption theory and relationships to observed landforms. J Geophys Res 32:221–263
21. Wilson L, Head JW (2007) Explosive volcanic eruptions on Mars: Tephra and accretionary lapilli formation, dispersal and recognition in the geologic record. J Volc Geoth Res 163:83–97
22. Zimbelman JR (2001) Image resolution and evaluation of genetic hypotheses for planetary landscapes. Geomorphology 37:179–199
23. McGuire WJ (1996) Volcano instability: a review of contemporary themes. In: McGuire WJ, Jones AP, Neuberg J (eds) Volcano instability on the Earth and other planets, vol 110. Geological Society Publication, pp 1–23
24. Cole JW, Milner DM, Spink KD (2005) Calderas and caldera structures: a review. Earth Sci Rev 69:1–26
25. Mouginis-Mark PJ, Robinson MS (1992) Evolution of the Olympus Mons caldera, Mars. Bull Volc 54:347–360
26. Mouginis-Mark PJ, Rowland SK (2001) The geomorphology of planetary calderas. Geomorphology 37:201–223
27. Gulick VC, Baker VR (1990) Origin and evolution of valleys on Martian volcanoes. J Geophys Res 95:14325–14344
28. Reimers CE, Komar PD (1979) Evidence for explosive density currents on certain Martian volcanoes. Icarus 39:88–110
29. Fassett CI, Head JW (2007) Valley formation on Martian volcanoes in the Hesperian: evidence for melting of summit snowpack, caldera lake formation, drainage and erosion on Ceraunius Tholus. Icarus 189:118–135
30. Zimbelman JR, Garry WB, Bleacher JE, Crown DA (2015) Volcanism on Mars. In: The encyclopedia of volcanoes, 2nd ed, pp 717–728
31. Rogers AD, Nazarian AH (2013) Evidence for Noachian flood volcanism in Noachis Terra, Mars, and the possible role of Hellas impact basin tectonics. J Geophys Res: Planets 118:1094–1113. https://doi.org/10.1002/jgre.20083
32. Leonard JG, Tanaka K (2001) USGS Misc. Map Series I-2694
33. Crown DA, Price KH, Greeley R (1992) Geologic evolution of the east rim of the Hellas basin Mars. Icarus 100(1):1–25
34. Bernhardt H, Williams DA, Clark JD (2019) Malea Planum: timing and scale of deposition and erosion on the oldest of Mars' large volcanic provinces. In: 50th lunar and planetary science conference, Abstract 1435.pdf
35. Lang NP, McSween HY, Tornabene LL, Hardgrove CH, Christensen PR (2010) Re-examining the relationship between Apollinaris Patera and the Gusev basalts. J Geophys Res 115:E04006. https://doi.org/10.1029/2009JE003397
36. Squyres S, Aharonson O, Clark B, Cohen B, Crumpler L et al (2007) Pyroclastic activity at Home Plate in Gusev Crater, Mars. Science 316:738–742
37. Carr MH, Schaber GG (1977) Martian permafrost features. J Geophys Res 82:4039–4054
38. El Maary MR, Dohm JM, Marzo GA, Fergason R, Goetz W et al (2012) Searching for evidence of hydrothermal activity at Apollinaris Mons, Mars. Icarus 217:297–314
39. Plescia JB (2003) Amphitrites-Peneus Paterae/Malea Planum geology. In: Lunar and planetary science conference, Abstract #1478
40. Larson SK (2007) The origins of four paterae of Malea Planum, Mars; Unpublished M.S. thesis, Brigham Young University, Provo, UT, 120 pp
41. Lemoine FG, Smith DE, Rowlands DD, Zuber MT, Neumann GA, Chinn DS, Pavlis DE (2001) An improved solution of the gravity field of Mars (GMM-2B) from Mars Global Surveyor. J Geophys Res 106(#E10):23,359–23,376
42. McGovern P, Solomon S, Smith D, Zuber M, Simons M et al (2002) Localized gravity/topography admittance and correlation spectra on Mars: implications for regional and global evolution. J Geophys Res Planets 107(E12):19-1–19-25
43. Plescia J (2002) Martian volcano morphology. Lunar Planet Sci XXXIII, Abstract 1854
44. Kiefer W (2003) Gravity evidence for extinct magma chambers on Mars: Tyrrhena Patera and Hadriaca Patera. Lunar Planet Sci XXXIV, Abstract 1234
45. Musiol S, Cailleau B, Platz T, Kneissl T, Dumke A, Neukum G (2011) Outflow activity near Hadriaca Patera, Mars: fluid-tectonic interaction investigated

with high resolution stereo camera stereo data and finite element modeling. J Geophys Res Planets 116 (E8):E08001
46. Kostama V-P, Ivanov M, Raitala J, Tormanen T, Korteniemi J, Neukum G (2010) Evidence for multiple ice deposition phases on the north-eastern rim of the Hellas basin, Mars. EPSL 294:321–331
47. Squyres S, Wilhelms D, Moosman A (1987) Large-scale volcano-ground ice interactions on Mars. Icarus 70:385–408
48. Greeley R, Spudis P (1981) Volcanism on Mars. Rev Geophys Space Phys 19:13–41
49. Mouginis-Mark PJ (1981) Late stage summit activity of Martian shield volcanoes. In: Proceedings of the 12th Lunar and planetary science conference, pp 1431–1447
50. Zimbelman J, Edgett K (1992) The Tharsis Montes, Mars—comparison of volcanic and modified landforms. LPSC XXI:31–44
51. Pike RJ (1978) Volcanoes on the inner planets: some preliminary comparisons of gross topography. In: Proceedings of the 9th lunar and planetary science conference, pp 3239–3273
52. Crown D, Greeley R (2007) Geologic map of MTM-30262 and -30267 quadrangles, Hadriaca Patera region on Mars. USGS Scientific Investigations Series, I-2936
53. Kerber L, Head J, Madeleine J-B, Forget F, Wilson L (2012) The dispersal of pyroclasts from ancient explosive volcanoes on Mars: implications for the friable layered deposits. Icarus 219(1):358–381
54. Tanaka K, Kargel J, MacKinnon D, Hare T, Hoffman N (2002) Catastrophic erosion of Hellas basin rim on Mars induced by magmatic intrusion into volatile-rich rocks. Geophys Res Lett 29 (8):1195
55. Plescia JB, Saunders RS (1979) The chronology of the Martian volcanoes. In: Proceedings of the 10th lunar and planetary science conference, pp 2841–2859
56. Korteniemi J, Raitala J, Aittola M, Ivanov M, Kostama V et al (2010) Dike indicators in the Hadriaca Patera-Promethei Terra region, Mars. EPSL 294(3–4):466–478
57. McGovern P, Solomon SB (1993) State of stress, faulting, and eruption characteristics of large volcanoes on Mars. J Geophys Res 98:23553–23579
58. Montesi LGJ (2001) Concentric dikes on the flanks of Pavonis Mons: implications for the evolution of Martian shield volcanoes and mantle plumes. In: Ernst RE, Buchan KL (eds) Mantle Plumes: their identification through time. Spec Pap Geol Soc Am 352:165–181
59. Laity J, Malin M (1985) Sapping processes and the development of theater-headed valley networks on the Colorado Plateau. GSA Bull 96(2):203–217
60. Kochel R, Piper J (1986) Morphology of large valleys on Hawaii: evidence for groundwater sapping and comparisons with Martian valleys. JGR Solid Earth E175–E192
61. Arkani-Hamed J (2004) Timing of the Martian core dynamo. J Geophys Res 109:E03006. https://doi.org/10.1029/2003JE002195
62. Lillis R, Manga M, Mitchell D, Lin R, Acuna M (2006) Unusual magnetic signature of the Hadriaca Patera Volcano; implications for early Mars. GRL 33(3):L03202
63. Lillis R, Frey H, Manga M, Mitchell D, Lin R, Acuna M, Bougher S (2008) An improved crustal magnetic field map of Mars from electron reflectometry: highland volcano magmatic history and the end of the martian dynamo. Icarus 194(2):575–596
64. Lillis RJ, Mitchell DL, Lin RP, Connerney JEP, Acuna MH (2004) Mapping crustal magnetic fields at Mars using electron reflectometer. Geophys Res Lett 31:L15702. https://doi.org/10.1029/2004GL020189
65. Mitchell D, Lillis R, Lin R, Connernery J, Acuna M (2007) A global map of Mars' crustal magnetic field based on electron reflectometry. J Geophys Res 112 (E1):E11005
66. Kiefer WS (2004) Gravity evidence for an extinct magma chamber beneath Syrtis Major, Mars: a look at the magmatic plumbing system. Earth Planet Sci Lett 222:349
67. Schubert G, Russell CT, Moore WB (2000) Timing of the Martian dynamo. Nature 408:666
68. Lillis RJ, Manga M, Mitchell DL, Lin RP, Acuna MH (2005) Evidence for a second Martian dynamo from electron reflectometry. Lunar Planet Sci XXXVI:1578
69. Plescia JB (2004) Moprhometric properties of Martian volcanoes. J Geophys Res 109. https://doi.org/10.1029/2002JE002031
70. Grott M, Wieczorek M (2012) Density and lithospheric structure at Tyrrhena Patera, Mars, from gravity and topography data. Icarus 221(1):43–52
71. Konopliv A, Asmar S, Folkner W, Karatekin O, Nunes D et al (2011) Mars high resolution gravity fields from MRO, Mars seasonal gravity, and other dynamical parameters. Icarus 211(1):401–428
72. Gregg TKP, Crown DA, Sakimoto SEH (2002) Volcanic evolution and erosion at Tyrrhena and Hadriaca Paterae, Mars. Lunar Planet Sci XXXIII, abstract 1560
73. Werner S (2009) The global Martian volcanic evolutionary history. Icarus 201(1):44–68
74. Mest S, Crown D (2001) Geology of the Reull Vallis Region, Mars. Icarus 153(1):89–110
75. Gregg TKP, Crown DA, Greeley R (1998) Geologic map of part of the Tyrrhena Patera region of Mars. U.S. Geol. Surv. Misc. Invest. Ser., I-2556
76. Williams D, Greeley R, Werner S, Michael G et al (2008) Tyrrhena Patera: geologic history derived from Mars Express High Resolution Stereo Camera. JGR Planets 113(E11):E11005
77. Farley M, Gregg T, Crown D (2004) The great extent of Tyrrhena Patera, Mars: mapping MTM quadrangle—20257 in western Hesperia Planum using multiple data sets. Abstract #7–2, GSA annual meeting

78. Young A (1972) Slopes. Longman Group, New York, pp 72–74
79. Crown D, Porter T, Greeley R (1991) Physical properties of lava flows on the southwest flank of Tyrrhena Patera, Mars. NASA Tech Memo 4300–4301:167
80. Albin E (1986) Mars: volcanic features in the cratered uplands and possible tectonic associations. Masters thesis, Arizona State University, Tempe, AZ, 84 pp
81. Mandt KE, de Silva SL, Zimbelman JR, Crown DA (2008) Origin of the Medsae Fossae Formation, Mars: insights from a synoptic approach. J Geophys Res 113:E12011. https://doi.org/10.1029/2008JE003076
82. Gregg TKP, Krysak DJ (2011) Apollinaris Mons, Mars: a new name and a new past. In: 42nd lunar and planetary science conference, Abstract 1608.pdf
83. Kerber L, Head JW, Madeleine J-B, Forget F, Wilson L (2011) The dispersal of pyroclasts from Apollinaris Patera, Mars: implications for the origin of the Medsae Fossae formation. Icarus 216:212–220
84. Hood LL, Harrison KP, Langlais B, Lillis RJ, Poulet F, Williams DA (2010) Magnetic anomalies near Apollinaris Pater and the Medusae Fossae formation in Lucus Planum, Mars. Icarus 208:118–131
85. Thornhill G, Rothery D, Murray J, Cook A, Day T et al (1993) Topography of Apollinaris Patera and Ma'adim Vallis: automated extraction of digital elevation models. JGR Planets 98(E12):23581–23587
86. El-Maary MR, Heggy E, Dohm JM (2014) Radar investigations of Apollinaris Mons on Mars: exploring the origin of the fan deposits. Planet Space Sci 103:262–272
87. Lang NP, Kelley RJ, Farrell AK (2010) Can Apollinaris-related pyroclastic deposits be attributed to changing water: magma interaction ratios on Mars? Geol Soc Am Abstr Programs 42(5):225
88. Bandfield JL, Hamilton VE, Christensen PR (2000) A global view of Martian surface compositions from MGS-TES. Science 287:1626–1630
89. Wyatt MB, McSween HY (2002) Spectral evidence for weathered basalt as an alternative to andesite in the northern lowlands of Mars. Nature 417:263–266
90. Langlais B, Purucker M (2007) A polar magnetic paleopole associated with Apollinaris Patera, Mars. Planet Space Sci 55(3):270–279
91. Boynton WV, Feldman WC, Squyres SW, Prettyman TH, Bruckner J, Evans LG, Reedy RC, Starr R, Arnold JR, Drake DM, Englert PAJ, Metzger AE, Mitrofanov I, Trombka JI, d'Uston C, Wanke H, Gasnault O, Hamara DK, Janes DM, Marcialis RL, Maurice S, Mikheeva I, Taylor GJ, Tokar R, Shinohara C (2002) Distribution of hydrogen in the near surface of Mars: evidence for subsurface ice deposits. Science 297:81–85
92. Mellon MT, Feldman WC, Prettyman TH (2004) The presence and stability of ground ice in the southern hemisphere of Mars. Icarus 169(2):324–340
93. Byrne S, Dundas CM, Kennedy MR, Mellon MT, McEwen AS, Cull SC, Daubar IJ, Shean DE, Seelos KD, Murchie SL, Cantor BA, Arvidson RE, Edgett KS, Refer A, Thomas N, Harrison TN, Posiolova LV, Seelos FP (2009) Distribution of mid-latitude ground ice on Mars from new impact craters. Science 325:1674–1676
94. Dundas CM, Byrne S, McEwen AS, Mellon MT, Kennedy MR, Daubar IJ, Saper L (2014) HiRISE observations of new impact craters exposing Martian ground ice. J Geophys Res Planets 119:109–127
95. Hynek B, Beach M, Hoke M (2010) Updated global map of Martian valley networks and implications for climate and hydrologic processes. JGR Planets 115(E9):E09008
96. Chevrier V, Mathe PE (2007) Mineralogy and evolution of the surface of Mars: a review. Planet Space Sci 55(3):289–314
97. Di Traglia F, Cimarelli C, de Rita D, Geimeno Torrente D (2009) Changing eruptive styles in basaltic explosive volcanism: examples from Croscat complex scoria cone, Garrotxa Volcanic Field (NE Iberian Peninsula). J Volcanol Geoth Res 180:89–109
98. Houghton BF, Wilson CJN, Del Carlo P, Coltelli M, Sable JE, Carey R (2004) The influence of conduit processes on changes in style of basaltic Plinian eruptions: Tarawera 1886 and Etna 122 BC. J Volcanol Geoth Res 137:1–14
99. Christensen PR, McSween HY, Bandfield JL, Ruff SW, Rogers AD, Hamilton VE, Gorelick N, Wyatt MB, Jakosky BM, Kieffer HH, Malin MC, Moersch JE (2005) Evidence for magmatic evolution and diversity on Mars from infrared observations. Nature 436:504–509
100. Kennedy B, Stix J, Vallance JW, Lavallee Y, Longpre M-A (2004) Controls on caldera structure: results from analogue sandbox modeling. GSA Bulletin 114: 515–524. https://doi.org/10.1130/B25228:1
101. Wilson L, Head JW (1981) Morphology and rheology of pyroclastic flows and their deposits, and guidelines for future observations. In: Lipman PW, Mullineaux DR (eds) The 1980 eruptions of Mount St. Helens, Washington. U.S. Geol. Surv. Prof. Pap. 1250, pp 489–512
102. Calder ES, Sparks RSJ, Gardeweg MC (2000) Erosion, transport and segregation of pumice and lithic clasts in pyroclastic flows inferred from ignimbrite at Lascar Volcano, Chile. J Volcanol Geotherm Res 104:201–235
103. Limaye A, Aharonson O, Perron J (2012) Detailed stratigraphy and bed thickness of the Mars north and south polar layered deposits. JGR Planets 117 (E6)
104. Gullickson AL, Edgar LA, Rumpf ME, Skinner JA (2019) USGS Astrogeology terrestrial analog

program: development of multipurpose guides. Geol Soc Am Abstr Programs 51(5):2020. https://doi.org/10.1130/abs/2019AM-339116
105. Head JW (2007) The geology of Mars: new insights and outstanding questions. In: Chapman M (ed) The geology of Mars: evidence from earth-based analogs. Cambridge University Press, pp 1–46
106. Hauber E, Bleacher J, Gwinner K, Williams D, Greeley R (2009) The topography and morphology of low shields and associated landforms of plains volcanism in the Tharsis region of Mars. J Volc Geotherm Res 185(1–2):69–95
107. Byrnes JM, de Silva SL (2003) Formation of Martian Paterae: insights from terrestrial ignimbrite shields. In: 34th lunar and planetary science conference, Abstract 1175.pdf
108. Pabst S, Worner G, Civetta L, Tesoro R (2008) Magma chamber evolution prior to the Campanian Ignimbrite and Neapolitan Yellow Tuff eruptions (Campi Flegrei, Italy). Bull Volcanol 70:961–976
109. Francis P, Wood C (1982) Absence of silicic volcanism on Mars: implications for crustal composition and volatile abundance. JGR Solid Earth 87(B12):9881–9889
110. Baker MCW (1981) The nature and distribution of upper Cenozoic ignimbrite centres in the Central Andes. J Volcanol Geotherm Res 11:293–315
111. de Silva SL (1989) Geochronology and stratigraphy of the ignimbrites from the 21° 30′ S to 23° 30′ S portion of the Central Andes of northern Chile. J Volcanol Geoth Res 37(2):93–131
112. Rymer H, van Wyk de Vries B, Stix J, Williams-Jones G (1998) Pit crater structure and processes governing persistent activity at Masaya Volcano, Nicaragua. Bull Volcanol 59:345–355
113. Williams H, McBirney AR (1979) Volcanology. San Francisco Publ., Berkeley, CA, Freeman, Cooper and Co, p 397
114. Wilshire HG (1992) Geologic map of the Cow Cove Quadrangle, San Bernardino County, California: U. S. Geological Survey Open-File Report 92–179, 26 p
115. Andronico D, Cristaldi A, Del Carlo P, Taddeucci J (2009) Shifting styles of basaltic explosive activity during the 2002–03 eruption of Mt. Etna, Italy. J Volcanol Geotherm Res 180:110–122

Infrasound Propagation on Mars Atmosphere

G. Lacanna, E. Pace, and M. Ripepe

Abstract

The presented work relies on a two-dimensional finite-difference time domain (2D-FDTD) simulation to study infrasonic waves on Mars. The analysis was carried out considering the real infrasonic records on Earth for long-range propagation (>100 km) and comparing with those would have been recorded for Martian atmosphere. As well known, infrasound propagates on Earth at ranges of hundreds or thousands kilometres mainly due to the combination of low absorption coefficient of atmosphere and of stratospheric duct. Here, we show an example of this long-range propagation generated by major explosion at the Stromboli volcano, which was recorded in the near field at 500 m from the vent and, as stratospherical arrivals, at a distance of 120 km by infrasonic array located at ETNA volcano. The recorded waveform in near field was used to estimate the source time function applied both for Earth and Mars numerical simulations in order to analyse the different propagation behaviour in two atmospheres. Effects of atmospheric structure and absorption are included into the model. The Earth simulation successfully reproduced the stratospherical arrivals at ETNA volcano from the infrasound signal generated by the Stromboli explosion. The comparison between Earth's and Mars' numerical waveforms along the path well depicts the differences in infrasonic wave propagation caused by the strong differences of the two atmospheric properties and structures. Our numerical results highlight the strongly limited infrasound long-range propagation, even though at low frequency (1 Hz), in the Martian atmosphere with respect to Earth due to the lack of stratospherical dust and of larger atmospherical absorption.

1 Introduction

The technology for detecting acoustic pressure waves in the atmosphere has been successfully used for several decades in various research fields from physics of the atmosphere to volcanology. Any pressure source when coupled with the atmosphere generates acoustic waves whose low frequency components (infrasound) propagate very efficiently even for thousands of kilometres owing to the low absorption of the Earth's atmosphere [1].

Acoustic waves produced by natural events, such as volcanic eruptions, earthquakes, meteorites or atmospheric perturbations, are currently

G. Lacanna · M. Ripepe (✉)
Department of Earth Sciences,
University of Florence, Florence, Italy
e-mail: Maurizio.ripepe@unifi.it

E. Pace
Department of Physics and Astronomy,
University of Florence, Florence, Italy

G. Leone (ed.), *Mars: A Volcanic World*,
https://doi.org/10.1007/978-3-030-84103-4_10

used to study the dynamics of the Earth's atmosphere with the aim of improving climate models and consequently weather forecasting (ARISE Project,).[1]

More accurate wave propagation models in the atmosphere also can be exploited to monitor natural phenomena such as volcanic eruptions, meteorite impacts, and cyclones, on a regional and global scale [2].

Therefore, measuring acoustic waves on other planets would pave the way to the analysis of the dynamic behaviour of their atmospheres as well as to identify possible natural infrasonic sources. Planetary atmospheres are, in fact, complex dynamic systems whose structure and composition directly influence the speed and mode of propagation of the acoustic waves. However, the use of pressure sensors in planetary exploration has never been carried out.

The other planets of the Solar System have been explored for decades; however, we still do not have any records or scientific evidence of how sound propagates in extra-terrestrial atmospheres. Most of the extra-planetary journeys have focused on the analysis of radar data, visible and infrared images, but even today no space mission has brought instruments for measuring acoustic waves on other planets.

Although the Mars' atmosphere has only 1% of the Earth's pressure, laboratory experiments and theoretical models [3–5] have shown that it is possible to record sound on Mars with sensors similar to those used on Earth. The first attempt to measure sound on Mars was made in the 1990s by the Space Sciences Laboratory at Berkeley University[2] with the ill-fated NASA Mars Polar Lander (MPL) mission that lost contacts with the Earth immediately after touching the Martian ground [6].

In 2007, the Centre National d'Etudes Spatiales (CNES) created a microphone to be installed on board of landers in the NetLander mission [7]. Research has confirmed that measuring sound on Mars is also possible using inexpensive (< 70,000 Euro), small footprint (25 cm^3), low consumption (0.1 W) and very light (50 g) sensors, but unfortunately the mission was cancelled due to the high costs.

The Mars InSight mission, designed primarily for seismic prospecting, has infrasonic sensors on board and therefore is unable to distinguish seismic waves produced by tectonic or volcanic activity from those produced by a meteorite impact. Currently, also the Mars2020 mission has a microphone installed on a rover to measure the wind speed on Mars [8].

1.1 Long Range Infrasound Propagation on Earth

Infrasound is defined as the sound wave below 20 Hz or the acoustic energy too low in frequency to be detected by human hearing. Infrasound waves can be generated by different kind of sources such as mining explosions, industrial activities, as well as natural phenomena such as volcanic eruptions [9], earthquakes [10], large tsunami, and meteorite impact [2]. Because infrasound on Earth can propagate through the atmosphere at ranges of hundreds or thousands of kilometres, depending on strengths and frequency content of the sources [11], long-range infrasound propagation has been exploited to monitor nuclear weapons testing [12] and volcanic activity [13–15] Here, to analyse the propagation differences between Earth and Mars, we focus on an example of this long-range infrasound propagation generated by Stromboli volcano explosion. On July 19th 2020 at 03:00:45 UTC a major explosion occurred on Stromboli which was recorded at MIC5 station [16] with a differential sensor pressure at a sensitivity of 1 mV/Pa that are able to detect atmospheric fluctuations in the range of ±2500 Pa located at 500 m from the vent. The infrasound signal was formed by two impulsive transients separated by 6 s and with amplitude of 2489 Pa and 1555 Pa (Fig. 1a), respectively. Infrasound of this major explosion was detected at ETN infrasound array deployed at a source-to-receiver distance of 120.7 km (Fig. 1b,c). The ETN infrasound array is a small-aperture (250 m) four-element array, deployed at ETNA volcano at an

[1] http://arise-project.eu.

[2] http://sprg.ssl.berkeley.edu/marsmic.

elevation of 2100 m above sea level and a distance of 5500 from summit craters [9] (Fig. 1d). The array is equipped with differential pressure sensors, with a sensitivity of 25 mV/Pa and pressure range of ±100 Pa. The ETN array analysis indicates a continuous arrival of infrasound detected with a stable back-azimuth of ~330°N consistent with the direction of ETNA craters up to 03:07:43 UTC (Fig. 1b), when the back-azimuth change at ~10° N pointing to the Stromboli Volcano (Fig. 1). The timing of the first infrasound detection at ETN array (03:07:43 UTC) with respect to the major explosion onset time (03:00:45 UTC) appears consistent with a celerity of ~290 m/s (Fig. 1a, b), thus suggesting stratospheric arrival at ETN array of infrasound wave generated by the Stromboli explosion.

1.2 Planetary Acoustics

Several planets and/or moons of our solar system have an atmosphere and would lend themselves very well to the use of acoustic waves to distinguish endogenous activities (volcanism, tectonics) from exogenous ones (meteorite impacts, atmospheric storms). Mars, along with Venus and Titan, is certainly among those planets on which this technique could be applied with considerable success. Infrasound would provide useful information on the atmospheric dynamics, on the evaluation of the energy associated with sandstorms, on ground wind speed, and the incidence of meteorite impacts. This is a dataset of fundamental importance for the assessment of natural risk on the Martian ground.

At the surface level of these planets, in fact, the speed of sound varies between ~200 m/s on Titan; up to ~410 m/s on Venus, (on Earth it is 340 m/s at 293 K and 1 atm). The vertical velocity profile depends on the presence of thermal discontinuities and on the molecular composition of the atmosphere (Fig. 2).

The chemical composition of the Martian atmosphere (95% CO_2, 2.7% N_2, 1.6% Ar, 0.13% O_2) is dominated by CO_2 with a smaller amount of nitrogen (N_2), argon, and a very small amount (ppm) of H_2O. Using this chemical composition, we can evaluate the degree of sound absorption on Mars [4]. Considering that the atmospheric pressure at the ground level is about 6 millibars (600 Pa or 0.006 atm), although variations between 2 and 6 millibars have been measured within the Gale crater [17], and the mean temperature is 210 K (133 K the minimum and 293 K the maximum), from the thermal and viscous conduction of CO_2 it is possible to obtain the absorption of infrasound waves, which should be 0.03 Np/m at a temperature of ~220 K; this is about 100–500 times higher than on Earth.

The Martian atmosphere has the greatest absorption of acoustic waves among the analysed planets, while Titan has the most efficient sound propagation conditions (Fig. 3).

This may be a sensitivity constraint to the pressure sensors to be developed for Mars, as well as a limitation to the minimum observable pressure, but we will see that it will not have significant effects on infrasonic waves. However, both issues must still be carefully considered in the development of the sensors and assessment of their sensitivity.

1.3 Long Range Infrasound Propagation on Mars

The possibility to measure infrasound from sources at a given distance depends on various parameters including molecular effect which affects the acoustic attenuation coefficient, geometric attenuation (amplitude decreasing by the inverse of the distance from the source) and acoustic impedance $Z = \rho c$ of the atmosphere, where ρ is the fluid density and c the speed of sound in the medium. For plane waves, the Earth's acoustic impedance is 407 kg/m^{2}s while for Mars it decreases up to 4.4 kg/m^{2}s, which means that the Mars' atmosphere is almost two orders of magnitude less transmissive than that of Earth. Sound in the audio band (>2 kHz) produced by the same source will be 20 dB weaker on Mars than on Earth [18]. This means that acoustic monitoring systems on Mars will not be able to record the weakest acoustic sources and the dynamic range

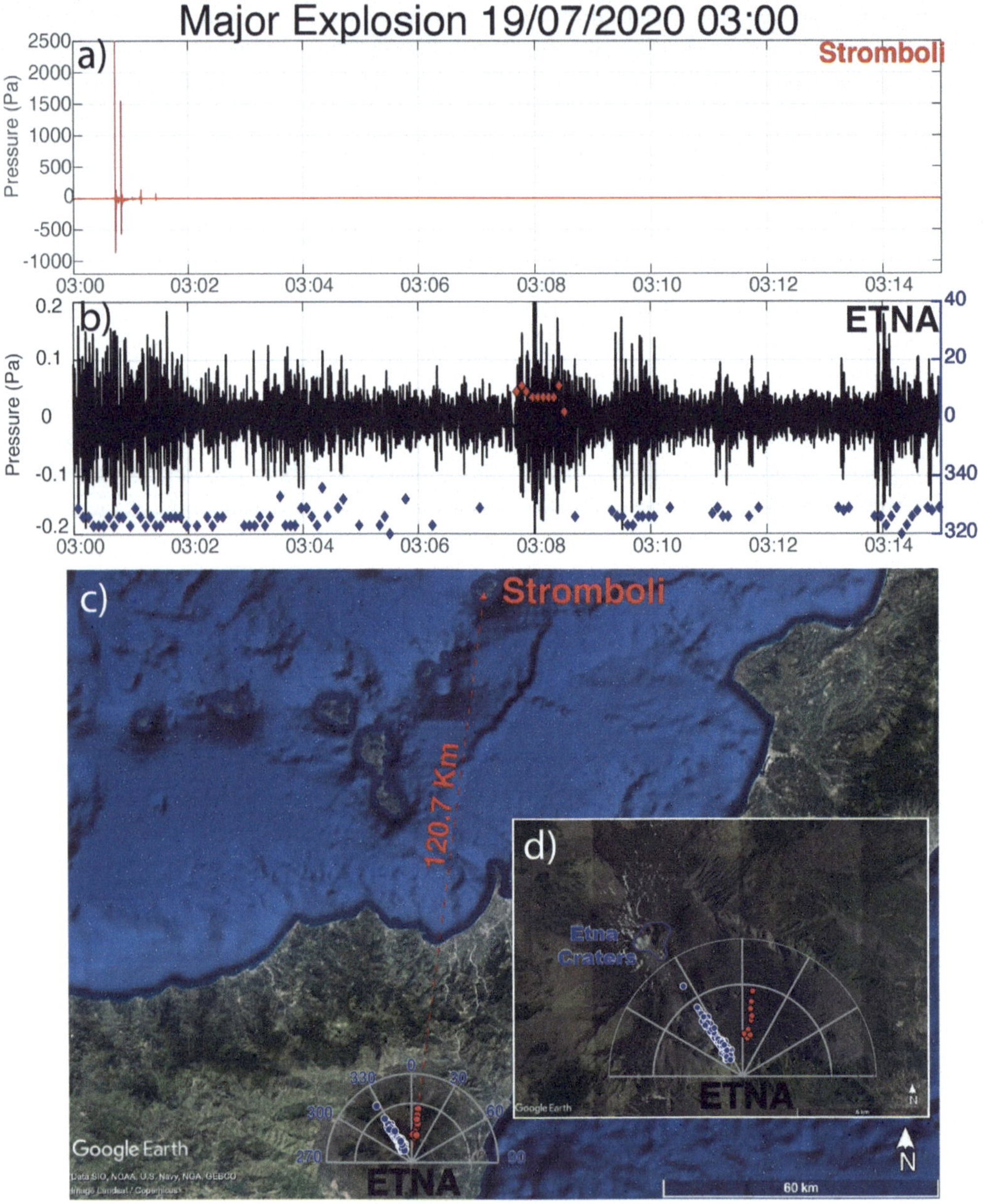

Fig. 1 **a** Pressure waveform recorded at MIC5 station located at 500 m from the Stromboli vent. **b** Pressure waveform recorded at ETN array (black line), diamonds are results of ETN array analysis (blue diamonds indicate a back-azimuth of ~330°N, red diamonds indicate a back-azimuth of ~10°N). **c** Map showing the locations of ETNA distant from Stromboli of 120.7 km with a direction of ~10°N and the rose diagram of infrasound detection. **d** zoom of the ETN array with infrasound detections

of natural phenomena observable in the audible frequencies will be reduced by at least 20 dB.

Considering a source with an intensity of 50 dB, the sound at frequencies around 20 kHz will be severely attenuated due to the effect of the molecular relaxation time.

In the infrasound field (<20 Hz), on the other hand, the molecular absorption effect will be

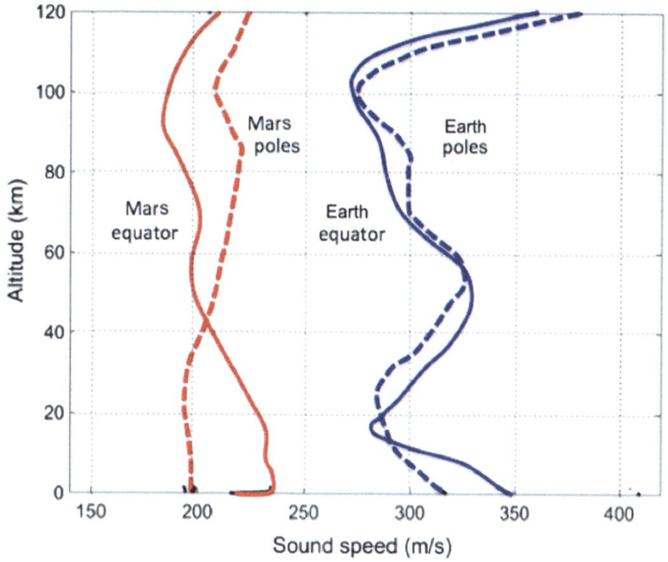

Fig. 2 Variation of the speed of sound on Earth and Mars as calculated using the thermal profile and the molecular model of the atmosphere [18]

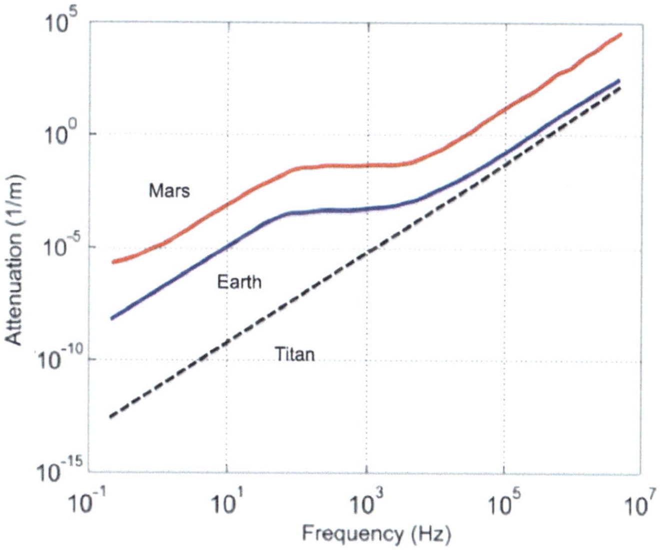

Fig. 3 Absorption of acoustic waves on Earth (blue), Mars (red) and Titan (dashed)

negligible and the attenuation of the acoustic waves will be dominated by the geometric attenuation (Fig. 4). Infrasonic waves (<20 Hz) also on Mars will be poorly attenuated and can be used as on Earth to monitor natural events even on a regional scale (>500 km).

Gravity waves (>200 s period) are an important mechanism for transporting energy and momentum within stable but stratified atmospheres. For years, Atmosphere Physics has considered the Earth's atmosphere as a stratified medium within which the interactions between the troposphere (where we live) and the outermost layers were practically absent. However, the dynamics of the outermost layers of our atmosphere have a strong influence on the troposphere and therefore on the climate. This interaction is only possible thanks to the formation of gravity waves which can be monitored through pressure sensors. Direct observations and theory indicate that the Mars' atmosphere is also affected by gravity waves which influence

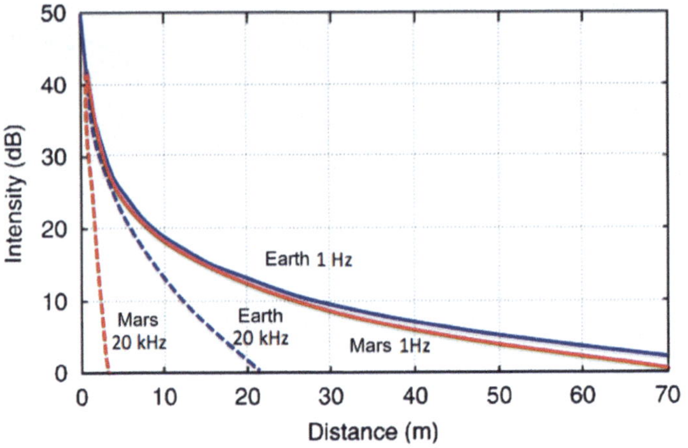

Fig. 4 Decrease in acoustic intensity with the distance calculated considering a standard atmosphere for the Earth (blue) and Mars (red). It can be seen how molecular effects of the atmosphere on the propagation of sound are larger on Mars than on the Earth for frequencies around the audible (20 kHz), while geometric attenuation at infrasound frequencies dominates (<20 Hz)

the global circulation facilitating the formation of CO_2 ice in the central layers of the atmosphere with immediate influences on the dynamics of the Martian thermosphere [19].

1.4 FDTD—Model for Absorption Atmosphere

The pressure waves and their propagation through the earth and mars atmosphere have been modelled using 2D-Finite Difference Time Domain (FDTD) numerical simulation to better understand the amplitude distribution in long range and its detectability by an infrasound sensor. Here we used the FDTD method for acoustic propagation in an absorbing and no moving atmosphere proposed and applied by de Groot-Hedlin (2008) [20]. The grid and time sampling rate in a FDTD simulation depend on the maximum frequency range of the wave used in the propagation models and on the sound speed. Here, we consider a highest frequency propagated through the grid of 1 Hz and a lower sound speed of 200 m/s recorded on the Mars atmosphere (Fig. 2a). We set a square lattice grid size Δ of 10 m corresponding to 20 nodes for wavelength and a time discretization Δt of 0.037 s, which thoroughly satisfy the stability condition for accurate FDTD simulation[20]. The vertical length of the calculation space was 60 km and the horizontal length was up to 130 km in order to include the path from Stromboli to Etna and the calculation domain is confined by the perfectly matched layer (PML) [21]. Calculations were performed until the infrasonic wave passed through the calculation space in the section (up to 650 s).

To explain the FDTD scheme we use the notation proposed by de Groot-Hedlin (2008) [20] for acoustic propagation in absorbing and non-moving atmosphere. The acoustic wave pressure and velocity perturbations are

$$p_{I,J}^{n} = p(I\Delta x, J\Delta z, n\Delta t) \tag{1}$$

$$v_{x,I,J}^{n+1/2} = v_x((I - 1/2)\Delta x, J\Delta z, (n + 1/2)\Delta t) \tag{2}$$

$$v_{z,I,J}^{n+1/2} = v_z(I\Delta x, (J - 1/2)\Delta z, (n + 1/2)\Delta t) \tag{3}$$

where z is the vertical axis, x is the horizontal axis in the calculation plane, t is the time, p is the acoustic pressure, and v_x and v_z are the horizontal and vertical components of the velocity perturbation due to the acoustic wave respectively. Using this notation, the update equation for the pressure fluctuations, horizontal and vertical components of the particle velocity are:

$$p_{I,J}^{n+1} = p_{I,J}^{n} e^{-\xi_{I,J}\Delta t} - \rho_{I,J} c_{I,J}^{2} \Delta t \left[\frac{v_{x_{I+1,J}}^{n+1/2} - v_{x_{I,J}}^{n+1/2}}{\Delta x} + \frac{v_{z_{I,J+1}}^{n+1/2} - v_{z_{I,J}}^{n+1/2}}{\Delta z} \right] e^{-\xi_{I,J}\Delta t/2} \tag{4}$$

$$\frac{v_{x_{I,J}}^{n+1/2} - v_{x_{I,J}}^{n-1/2}}{\Delta t} = -\frac{1}{\rho_{I-1/2,J}} \left[\left(1 + \frac{\mu_{I-1/2,J}}{\Delta t}\right) \left(\frac{p_{I,J}^{n} - p_{I-1,J}^{n}}{\Delta x}\right) - \frac{\mu_{I,J-1/2}}{\Delta t} \left(\frac{p_{I,J}^{n-1} - p_{I-1,J}^{n-1}}{\Delta x}\right) \right] \tag{5}$$

$$\frac{v_{z_{I,J}}^{n+1/2} - v_{z_{I,J}}^{n-1/2}}{\Delta t} = -\frac{1}{\rho_{I,J-1/2}} \left[\left(1 + \frac{\mu_{I,J-1/2}}{\Delta t}\right) \left(\frac{p_{I,J}^{n} - p_{I,J-1}^{n}}{\Delta z}\right) - \frac{\mu_{I,J-1/2}}{\Delta t} \left(\frac{p_{I,J}^{n-1} - p_{I,J-1}^{n-1}}{\Delta z}\right) \right] \tag{6}$$

We assume a model that is temporally invariant and vertically stratified, sound speed, ambient density, shear viscosity and damping values at the center of the *[I,J]*th cell represented as$\rho_{I,J}$,$c_{I,J}$, $\mu_{I,J}$ and $\xi_{I,J}$ respectively, the half-integer subscripts denote the average values between two adjacent cells. The atmospheric (i.e., pressure, temperature and viscosity coefficient (η)) and the molecular properties of the gas species of Mars were taken from the Mars Climate Database[3] using the data measured by the Mars Curiosity lander to estimate the vertical sound speed profile, density, and the total attenuation coefficient of sound (Fig. 5, red curve). For the Earth, we use the atmospheric characteristic (i.e., pressure, temperature and viscosity coefficient(η)) and molecular properties for a standard midlatitude summer atmospheric profile (Fig. 5, blue curve). The absorption mechanism is cumulative. Thus, the total attenuation coefficient of sound, in dB/km, is given by the linear sum of the attenuation components associated with each individual loss mechanism (i.e. diffusion, heat conduction, and viscosity). The attenuation coefficients for Earth and Mars were computed using the expressions in Sutherland and Bass (2004, 2006) [22, 23] and are shown in Fig. 5c as a function of altitude for a frequency of 1 Hz. De Groot-Hedlin (2008) [20] shows that for small viscosity coefficients $\xi \ll \omega$ and $\mu\omega \ll 1$ the attenuation coefficient of sound (α) is related to the shear viscosity and damping coefficient as follows:

$$\alpha(\omega) \approx \left(\xi + \mu\omega^2\right)/(2c) \tag{7}$$

From Eq. (7) we estimated the damping ξ (Fig. 5d), for an angular frequency of $\omega = 6.28 rad/s$, shear viscosity $\mu = 4\eta/(3\rho c^2)$ (Fig. 5d), attenuation coefficient α (Fig. 5c) and sound speed c (Fig. 5a) for Earth and Mars.

We used in the FDTD model the same source time function for Earth and Mars numerical simulation expressed in terms of volumetric flux (Fig. 6a). The volumetric flux in volcano infrasound is conventionally evaluated from the acoustic linear theory, where the overpressure generated by an acoustic source expanding into atmosphere is proportional to the first derivative of volumetric flux of the source [24]. From low-pass filtered pressure signal with a corner frequency of 1 Hz $\Delta P\left(t + \frac{r}{c}\right)$ recorded at MIC5 located at 500 m (r) from the Stromboli vent we estimated the volumetric flux at source (Fig. 6a, black line) as follows:

$$q(t) = \frac{2\pi r}{\rho} \int_{0}^{t} \Delta P\left(t + \frac{r}{c}\right) dt \tag{8}$$

1.5 Results and Conclusions

The calculations were performed in two dimensions. The major advantage of a two-dimensional code for simulating large spatial scales is that fewer computational resources are required than for a full three-dimensional simulation. The two-dimensional cartesian coordinates only gives an approximation for actual geometric spreading. Van Rentergem et al. (2005) [25] showed that the amplitude decay of the acoustic pressure signal due to geometric spreading in a two-dimensional cartesian coordinate system is proportional to $\sqrt{d_{rif}/d}$, where d and d_{rif} are the source-receiver

[3] (http://www-mars.lmd.jussieu.fr/mcd_python/).

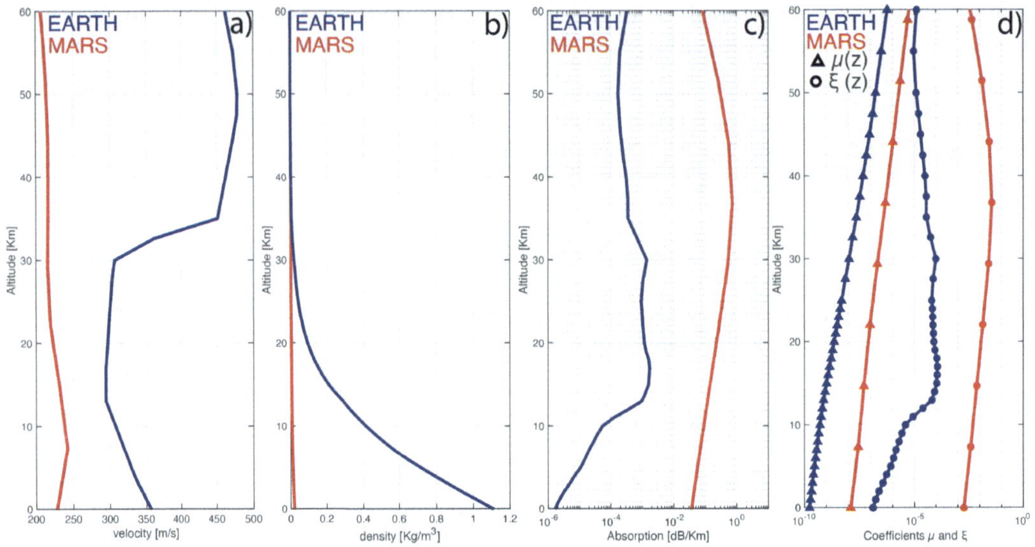

Fig. 5 **a** Sound speed profile, **b** density profile, **c** attenuation coefficient and **d** shear viscosity (μ triangles) and damping (ξ, circles) of Earth (blue lines) and Mars (red lines)

distances for a station and the reference station, respectively. Therefore, each numerical waveform is scaled by multiplying the ratio by the square root of the inverse ratio of the distance between the source and receiver so as to correctly approximate for geometric spreading in a three-dimensional model using a two-dimensional approximation [1, 26]. Here, we used as a reference the MIC5 station located a 500 m from the Stromboli vent.

The first main difference between Earth and Mars acoustic propagation is the pressure amplitude at the same distance close to the source (500 m) generated by the same source (Fig. 6a–c). As expected by almost two order of magnitude discrepancy between their acoustic impedance, we estimated a maximum pressure of 802 Pa on Earth and 9.2 Pa on Mars (Fig. 6b, c) induced by the same volumetric flux at the source (Fig. 6a).

The second main difference is the effect induced by the larger attenuation coefficient on Mars than on Earth. The waveforms in Fig. 6d, e are normalized to 10 Pa for Earth and 0.1 Pa for Mars propagation, respectively. The notable features observed in Fig. 6d, e is that the amplitude of direct waves on Mars, in the distance range between 30 and 70 km from the source, are strongly attenuated if compared with the amplitude calculated at same distance on the Earth's atmosphere. To visualize the amplitude distribution, we stored the time evolution of the sound pressure at all grid points to calculate the RMS amplitude (Fig. 7b, d). The RMS ratio, which correspond to the relative sound pressure level with respect to MIC5 station located at 500 m from the Stromboli vents, $RSPL = 20\log(RMS/RMS_{MIC5})$, is shown on a logarithmic scale. Figures 7b, d show the result for the atmospheric structures corresponding to Fig. 7a, c, respectively. The RSPL distribution for Mars's atmosphere shows waveguide into the lower atmosphere (Fig. 7d) due to increasing of sound speed profile up to 8 km (Fig. 7c). On the contrary, on Earth the amplitude distribution is bent upward (Fig. 7b) due to decreasing of sound speed profile up to 12 km (Fig. 7a). The other important feature observable on the Earth's RSPL map distribution is the increasing of amplitude at a distance larger than 70 km from the source (Fig. 7b). This latter phenomenon is generated by the refraction of the acoustic ray downward from the top of the stratosphere where the acoustic sound speed increases up to 450 m/s (Fig. 7a). These rays are considered forming the secondary wavefront (stratospheric arrival) observed at the numerical

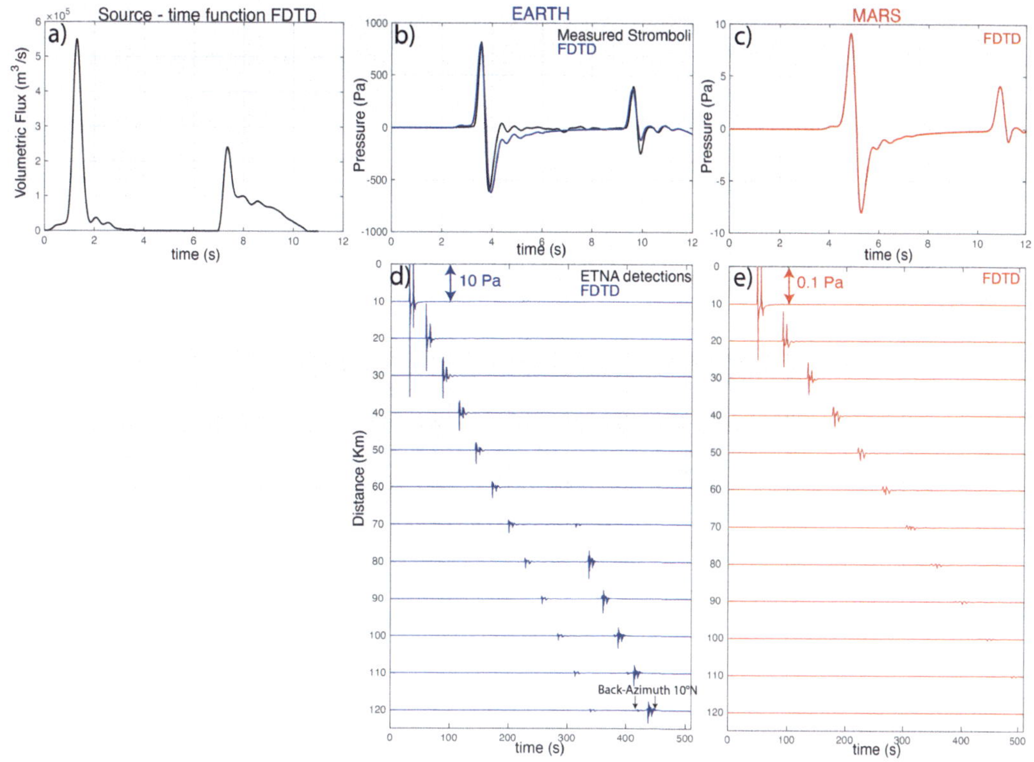

Fig. 6 **a** Volumetric flux used as source function in the FDTD numerical model estimated from low-pass filtered pressure signal with a corner frequency of 1 Hz recorded at 500 m from Stromboli vent (black line in **b**), blue line in (**b**) is the numerical waveform at 500 m from the source for Earth's atmosphere. **c** Numerical waveforms estimated at 500 m from the source for Mars's atmosphere. **d** Numerical waveforms calculated from 10 km to 120.7 km for Earth's atmosphere, the black arrows indicate the time of infrasound detections at ETN array with a direction of ~10°N consistent with a Stromboli direction. **e** Numerical waveforms calculated from 10 to 120 km for Mars's atmosphere

waveforms calculated for Earth's atmosphere for a distance larger that 70 km (Fig. 6d).

The good timing match observed between infrasound detections, with a back-azimuth at 10° N at ETN array consistent with Stromboli direction, and the numerical waveforms calculated at a distance of 120.7 km from the source (Fig. 6d) confirms that these detections are compatible with the stratospheric arrivals generated by the Stromboli explosion.

Summarizing, the characteristics of Earth's atmosphere make more favourable the acoustic propagation than Mars's atmosphere for three main reasons:

(1) larger acoustic impedance, with the same source, generates a larger acoustic amplitude almost of two order of magnitude;
(2) lower attenuation coefficient
(3) increase of sound speed profile between stratosphere and mesosphere at an altitude ~40 km, which produces a downward of acoustic ray, which make the propagation for long-range distances (hundreds or thousands of kilometers) possible.

2 A Glance to the Future

The development of instrumentation with suitable sensitivity for the detection of infrasound in planetary environments would represent a major step ahead to qualify them at operating temperatures (i.e., compatible with those detected firstly

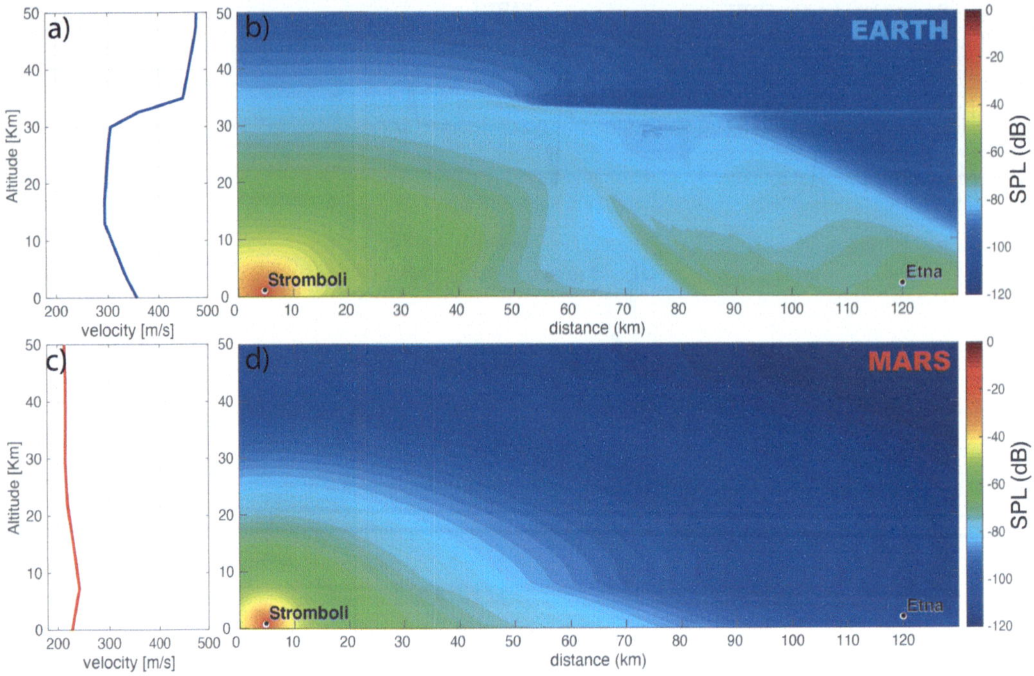

Fig. 7 **a** Sound speed profile for Earth's atmosphere. **b** Sound pressure level in decibel with respect to values at 500 m from the source for Earth acoustic propagation. **c** Sound speed profile for Mars's atmosphere. **d** Sound pressure level in decibel with respect to values at 500 m from the source for Mars acoustic propagation

on the Mars' surface and then on Titan's) at a more advanced technological stage. Even disentangling seismic activity from meteoritic impact on planetary surfaces would improve the estimation of impact rates allowing a better understanding not only of the inter-planetary environment but also of the evolution of the Solar System, and generally of planetary systems.

The technological challenges of these new tools are:

(1) operation in extreme conditions of heat and cold (volcanic and polar environments);
(2) reception sensitivity in atmospheres where the acoustic propagation is slower as occurs on Mars and Titan (~240 and ~210 m/s, respectively);
3) extension of the operating frequency band from 0.1 mHz to 100 Hz to cover both the infrasonic frequency range and those of gravity waves, thus increasing the ability to detect distant sources.

The propagation speed of acoustic waves depends on the pressure and temperature profile; therefore, it is an excellent indirect tool for analyzing the atmospheric structure. The great success of this technique lies in the ability to locate moving sources even at great distances; for this reason, it is widely used to track the atmospheric entry of meteorites [27], the movement of large and deep areas having low pressure, sandstorms and volcanic activity.

The scientific implications, in addition to the primary interest in expanding knowledge on the planetary atmospheres, are related to the possibility of distinguishing signals produced by volcanic or tectonic (seismic) activity from that produced by the impact of meteorites, increasing knowledge on the Martian environment, that a normal seismometer is unable to do. A system of at least four infrasonic sensors is able to recognize the pressure signals generated by meteorites and trace the fall trajectory with feedback on the safety standards for any planetary base camp.

On Mars, a network of wide-band infrasonic sensors could also provide valuable information on the dynamics of sandstorms and their movements.

References

1. de Groot-Hedlin C, Hedlin MAH, Walker K (2011) Finite difference of infrasound propagation through a windy, viscous atmosphere: application to a bolide explosion detected by seismic networks. Geophys J Int 185:305–320. https://doi.org/10.1111/j.1365-246X.2010.04925.x
2. Campus P, Christie DR (2010) The IMS infrasound network: worldwide observations of infrasonic waves. In: Le Pichon A, Blanc E, Hauchecorne A (eds) Infrasound monitoring for atmospheric studies. Springer Geosciences. ISBN: 978–1–4020–9507–8, 745
3. Bass HE, Sutherland LC, Piercy J, Evans L (1984) Absorption of sound by the atmosphere. In: Physical acoustics: principles and methods, vol XVII. Academic Press, Inc., Orlando, FL, pp 145–232
4. Bass HE, Chambers JP (2001) Absorption of sound in the Martian atmosphere. J Acoust Soc Am 109 (6):3069–3071. https://doi.org/10.1121/1.1365424
5. Williams J-P (2001) Acoustic environment of the Martian surface. J Geophys Res 106(E03):5033–5041. https://doi.org/10.1029/1999JE001174
6. Delory GT, Luhmann J, Curtis DW, Friedman L, Primbish JH, Mozer FS (1998) Development of the first audio microphone for the Mars Surveyor 98 Lander. In: First international conference on mars polar science and exploration, held at the episcopal conference center at Camp Allen, TX. LPI Contribution No. 953, Lunar and Planetary Institute, Houston
7. Marsal O, Venet M, Counil J-L, Ferri F, Harri A-M, Spohn T, Block J (2002) The NetLander geophysical network on the surface of Mars: general mission description and technical design status. Acta Astronaut 51(1–9):379–386. https://doi.org/10.1016/S0094-5765(02)00069-3
8. Maurice S, Wiens RC, Rapin W, Mimoun D, Jacob X, Betts B, Bell III JF, Delory G, Clegg SM, Cousin A, Forni O, Gasnault O, Lasue J, Meslin P-Y (2016) A microphone supporting LIBS investigations at Mars. #3044, 47th Lunar and Planetary Science Conference, Houston, TX
9. Ripepe M, Marchetti E, Delle Donne D, Genco R, Innocenti L, Lacanna G, Valade S (2018) Infrasonic early-warning for explosive eruption. J Geophys Res 123. https://doi.org/10.1029/2018JB015561
10. Marchetti E, Lacanna G, Le Pichon A, Piccinini D, Ripepe M (2016) Evidence of large infrasonic radiation induced by earthquake interaction with alluvial sediments. Seismol Res Lett 87:3. https://doi.org/10.1785/0220150223
11. Drob DP, Picone JM, Garcés M (2003) Global morphology of infrasound propagation. J Geophys Res 108(D21):4680. https://doi.org/10.1029/2002JD003307
12. Evers LG, Haak HW (2010)Worldwide observations of infrasonic waves, In: Pichon A, Blanc E, Hauchecorne A (eds) Infrasound monitoring for atmospheric studies. Springer Geosciences. ISBN: 978–1–4020–9507–8, 745
13. Dabrowa AL, Green DN, Rust AC, Phillips JC (2011) A global study of 559 volcanic infrasound characteristics and the potential for long-range monitoring. Earth Planet Sci Lett 310:369–379. https://doi.org/10.1016/j.epsl.2011.08.027
14. Matoza RS, Le Pichon A, Vergos J, Herry P, Lalande JM, Lee H, Il-Young C, Rybin A (2011) Infrasonic observations of the June 2009 Sarychev Peak eruption, Kuril Islands: implications for infrasonic monitoring of remote explosive volcanism. J Volcanol Geothermal Res 200:35–38
15. Marchetti E, Ripepe M, Campus P, Le Pichon A, Vergoz J, Lacanna G, Mialle P, Héreil P, Husson P (2019) Long range infrasound monitoring of Etna volcano. Sci Rep 9:1. https://doi.org/10.1038/s41598-019-54468-5
16. Ripepe M, Marchetti E, Ulivieri G (2007) Infrasonic monitoring at Stromboli Volcano during the 2003 effusive eruption: insights on the explosive and degassing process of an open conduit system. J Geophys Res 112:B09207. https://doi.org//10.1029/2006JB004613
17. Guzewich SD, Newman CE, de la Torre Juárez M, Wilson RJ, Lemmon M, Smith MD, Kahanpää H, Harri AM (2016) Atmospheric tides in Gale Crater, Mars, Icarus. 268:37–49, https://doi.org/10.1016/j.icarus.2015.12.028
18. Petculescu A, Lueptow RM (2007) Atmospheric acoustics of Titan, Mars, Venus, and Earth. Icarus 186(2):413–419. https://doi.org/10.1016/j.icarus.2006.09.014
19. Yiğit E, Medvedev AS (2012) Thermal effects of internal gravity waves in the Martian upper atmosphere, Geoph. Res Let 39(5):L05201. https://doi.org/10.1029/2012GL050852
20. de Groot-Hedlin C (2008) Finite-difference time-domain synthesis of infrasound propagation through an absorbing atmosphere. J Acoust Soc Am 124 (3):1430–1441. https://doi.org/10.1121/1.2959736
21. Berenger JP (1994) A perfectly matched layer for the absorption of electro-magnetic waves. J Comput Phys 114:185–200. https://doi.org/10.1006/jcph.1994.1159
22. Sutherland LC, Bass HE (2004) Atmospheric absorption in the atmosphere up to 160 km. J Acoust Soc Am 120:2985. https://doi.org/10.1121/1.1631937
23. Sutherland LC, Bass HE (2006) Erratum: atmospheric absorption in the atmosphere up to 160 km. J Acoust Soc Am 120:2985. https://doi.org/10.1121/1.2355481

24. Lightill MJ (1978) Waves in fluid. Cambridge University Press, New York
25. Van Renterghem T, Salomons EM, Botteldooren D (2005) Efficient FDTD-PE model for sound propagation in situations with complex obstacles and wind profiles. Acta Acust United Acust 91:671–679. 1854/4796
26. Lacanna G, Ichihara M, Iwakuni M, Takeo M, Iguchi M, Ripepe M (2014) Influence of atmospheric structure and topography on infrasonic wave propagation. J Geophys Res Solid Earth 119(4): 2988–3005. https://doi.org/10.1002/2013JB010827
27. Le Pichon A, Ceranna L, Pilger C, Mialle P, Brown D, Herry P, Brachet N (2013) The 2013 Russian fireball largest ever detected by CTBTO infrasound sensors. Geophys Res Lett 40:3732–3737. https://doi.org/10.1002/grl.50619

Potential Futures in Human Habitation of Martian Lava Tubes

Nicola Mari, Gernot Groemer, and Nina Sejkora

Abstract

Lava tubes exists as volcanic features not only on Earth, but even on other worlds. Lava tubes are the most practical and effective places where to install the first human bases on Mars. We review the human to Mars architectures currently being developed, subsequently we describe a context for the use of lava tubes for human habitation. The architecture for the use of in-situ resources as fuel for a return rocket, the use of a landing vehicle as the initial habitat are discussed. Also, the core landing site selection criteria are discussed, including how the access to a lava tube might change the landing approach. Additionally, the challenges to constructing a larger scale habitat for a growing population necessary to achieve a permanent base and the possible solutions are proposed. Whether lava tubes are present on the Martian surface, the shielding effect that they can provide is exceptionally advantageous in both logistical and economical terms for a planetary mission, since they do not require a huge mass load to be brought on a space mission and minimize considerably the work to be done on the dangerous Martian surface after landing—however, habitats on the surface still need to be considered where lava tubes are absent. Lava tubes can naturally protect from deadly radiations, extreme temperature variations, dust storms, and micrometeorite impacts. Thermal mining inside lava tubes is proposed as a realistic way to extract water ice from below the surface and use it for living purposes. Robots and humans should work together in order to detect and access the best lava tube close to the selected landing site. Lava tubes on Earth or the Moon can result in a great opportunity for conducting analog Martian missions.

1 Introduction to Human Missions from Mars Architecture

According to current NASA's Artemis plans [1], the next humans will land on the Moon within this decade. This mission will be followed by further surface missions of increasing sojourn duration. Those, prepared with intermediate stays on a lunar orbital station dubbed "the Deep Space Gateway (DSG)" [2, 3], are meant to build the knowledge base and facilitate the telerobotics operations [4] required for a successful human Mars mission in the 2030s. NASA's concepts of how the early Mars missions could look like are

N. Mari (✉)
Department of Earth and Environmental Sciences, University of Pavia, Via A. Ferrata 1, 27100 Pavia, Italy
e-mail: nicola.mari@unipv.it

G. Groemer · N. Sejkora
Austrian Space Forum (OeWF), Etrichgasse 18, Innsbruck, Austria

G. Leone (ed.), *Mars: A Volcanic World*,
https://doi.org/10.1007/978-3-030-84103-4_11

formulated in the Design Reference Architecture (DRA), currently in its fifth edition [5, 6]. These documents aim at providing a common framework describing the current vision for a future human exploration mission to Mars. It is not a formal plan, but describes a likely scenario based on the current state of knowledge. It covers the first three human missions to Mars. Those will each visit a different location on Mars and start in three consecutive launch windows (i.e. approx. 2 years apart).

The scenario for each of the missions would be the following: prior to the start of the crew, a few cargo transports would be sent to the red planet. Those carry, amongst others: (i) in-situ resource utilization equipment (ISRU) which is producing fuel for the crew's return trip from the Martian atmosphere; (ii) the habitat lander, a spacecraft containing the habitat, which is waiting for the crew's arrival in Mars orbit; and (iii) the Mars Ascent Vehicle (MAV) which is the spacecraft to bring the crew from the Martian surface back to orbit after the end of the mission. Once the equipment is deployed, the crew, consisting of six people of mixed gender, launches from Earth on their 180-day journey to Mars. When arriving in Mars orbit, they transfer to the habitat lander and descend to the surface, where they stay for approximately 500 days. During that time, they live in a central habitat from where they start scientific sorties in a pressurized rover [6]. When the time comes to return, the crew uses the MAV, filled with the locally produced fuel, to get back to orbit. There, the transfer vehicle, with which they already travelled to Mars, awaits them for orbital rendezvous. After another 180 days, they arrive back to Earth, finishing their about 900-day mission.

For the first exploration missions, the DRA envisions a mission architecture where the crew lives in a central habitat during their surface stay. For short excursions, unpressurized rovers are provided (similar to the Apollo Lunar Roving Vehicle LRV). In addition to those, the crew will also have two pressurized rovers, carrying two astronauts for longer ($\sim$1 week) scientific excursions. In this early phase, the habitat will be modest in size—about 250 m^3 of pressurized volume [6]. This includes storage space, outfitting, etc.…; the habitable volume can be estimated to be about 60% of the total pressurized volume [7], i.e., 150 m^3, which would imply an approximate usable floor area of 70 m^2. A visualization of such a habitat together with a pressurized (foreground) and an unpressurized rover (background) can be seen in Fig. 1.

The effort of sustaining a human presence on Mars shall be warranted by the missions' scientific return. The basic planetary science issues humans on Mars will address, include [6]:

- *Geology*: Characterization of the geologic system and its history.
- *Geophysics*: Characterization of the Martian interior including its dynamics and evolution.
- *Climatology*: Martian climate history, especially the modern climate and weather.
- *Astrobiology*: Study of early Mars habitability and possible pre-biotic processes. Search for ancient (extinct) life and determine whether life forms still persist (extant life).

To accomplish those objectives, the Mars mission will carry various scientific equipment. The following list shall give a general overview and is not meant to be either exhaustive or a specific suggestion. Some tasks are better suited for human operators than for robotic missions, like collecting optimal geological samples or setting up and operating deep drilling equipment. Also, the characterization of the subsurface using ground-penetrating radar is projected. For geophysical measurements, networks of sensor stations including, e.g., seismometers or heat flow probes. Some argue that setting up such devices would better be left to robots to ensure a precise setup and relieve astronauts of this task [8]. A device which is highly likely to be onboard a human mission is a meteorological station; this serves both for climatology purposes as well as operational aspects—weather forecasts will also be essential on Mars to ensure the safety of astronauts and equipment. Apart from these devices for use in outside environment, an important asset for scientific investigations on Mars will be a laboratory inside the habitat to

Fig. 1 Artist's impression of a potential Mars surface base for the scenario described in the text. Courtesy of NASA

process and analyze the collected samples. Particularly for biological studies, a suite of equipment is required: microscopes, spectrometers, gas chromatograph mass spectrometer, biomolecule screening capabilities, equipment for nucleic acid extraction and sequencing, etc., just to name a few desirable technologies [5].

Mission architectures, like the one described above, are based on years of experience, collected in multiple studies, modeling efforts and experiments. One important such pillar is analog planetary research. The term covers studies performed in locations which are in some respects analogous to conditions encountered in planetary exploration scenarios. The analogy can, for example, refer to terrain and/or geology which is similar to features on the respective planetary body; or a habitat simulator where research is focusing on human factors investigations. Analog research prepares for future planetary missions through, for example: the development and testing of equipment, like spacesuits, robotic elements, or scientific devices; the study of the impact of human factors on mission success and, derived from that, guidelines for training future astronauts to improve such effects; the fine-tuning of workflows among and between ground and space segments of mission operations.

Planetary analog studies include, e.g., the NASA DESERT-RATS field tests [9], the NASA HI-SEAS [10, 11] and BASALT missions [12], the MOONWALK project [13, 14], ESA's CAVES [15] and PANGAEA [16] or the Austrian Space Forum's field missions of the PolAres and AMADEE programs (e.g. [17, 18]); Fig. 2.

1.1 Getting There and Back Again

Once the crew landing is successful, it is expected that the landing vehicle may also serve as a habitation module (expanded by inflatable structures to augment the living and working space for the astronauts) during their first missions,. This was one of the design drivers, e.g. for the geometry of the original Australian Mars-Oz concept station [19] or the Mars Desert Research Station of the Mars Society—to allow for a representative model of a human habitation module, later expanded by a greenhouse, observatory and an additional laboratory module [20]. However, these simulations are not embedded in comprehensive mission architectural considerations and therefore lack a representativeness beyond the volumetric modelling from the habitat design perspective [21].

To minimize the landing masses required for a crewed mission, the concept of in-situ resource utilization (ISRU) is generally considered an

Fig. 2 Analog Astronauts of the Austrian Space Forum with the Husky rover (Graz University of Technology) during the AMADEE-18 Mars analog mission. *Image credits* OeWF/Voggeneder

efficient method. This significantly reduces the amount of material transported from Earth and may enable—in the long-term perspective—both independence from Earth and export opportunities for future human settlements. In 1978, a ground breaking research [22] suggested to manufacture ascent propellants from Martian atmospheric carbon dioxide to reduce launch mass. ISRU can be the key to implementing a sustained and affordable human and robotic program to explore the Solar System and beyond. Potential Mars resources include water, vast quantities of metals and minerals, atmospheric constituents, abundant solar energy. Suitable processing can transform these raw resources into useful materials and products. Studies have shown that for every kg of payload delivered to the Mars surface 3.5 to >5 kg are required in Low Earth Orbit, depending on propulsion technology used [23].

ISRU can be utilized for various purposes, including radiation shielding, local construction materials, crew consumables (such as water or oxygen), but also fuel manufacturing. Notably, transforming components of the Martian atmosphere into methane is of interest, as it can be used as a propellant [22]. More generally, regarding the type of resource used, the following advantages and risks are identified:

	Advantages and strengths	Weaknesses and risks
Atmospheric processing	• Atmospheric resources are globally obtainable (independence from landing site) • Production of Oxygen from CO_2 makes >75% of ascent propellant mass • High level of technological readiness, including an in-situ demonstration for the MSL Perseverance mission in 2021	Production of methane requires delivery of hydrogen from Earth which is volume inefficient or needs water harvesting from the subsurface
Soil water harvesting	• Surface material characteristics well studied • Water (in the form of hydrated minerals) identified	• Risk associated with the complexity of the required infrastructure is yet to be determined

(continued)

	Advantages and strengths	Weaknesses and risks
	globally near the surface • Regolith excavation and thermal processing techniques can be utilized for the Moon and Mars • Even low concentrations of water in surface hydrated mineral soil (3%) still provide mass benefits	• Autonomous operations required • Location dependency on water resource concentration and form • Concerns from planetary protection
Metal ore harvesting	• Allows for solid structures and raw materials for metal 3D printing, extending the hardware robustness and spare part strategy	• Low technology readiness in terms of prospecting, harvesting and processing

On Mars, atmospheric CO_2—which constitutes roughly 95% of the atmospheric components—can be harvested by several methods to produce oxygen [24]. The obtained oxygen can then be combined with hydrogen—extracted from the Martian regolith [25] or having been transported from Earth—to obtain water. The first step will be to obtain the CO_2 from the atmosphere—although available at high concentration, the gases still need to be compressed and transferred into a storage. This is possible, e.g. via a rapid cycle adsorption workflow, these temperature-swing adsorption pumps have been proposed as a method of acquiring and compressing Martian atmospheric CO_2 for downstream processing [26]. Such an adsorption process can concentrate a gas species far beyond its initial density, whilst avoiding high-speed pumps operating over extended periods having a high risk of mechanical failure. This technology that has been demonstrated on a laboratory scale, is based upon the basic reaction:

$$3\ CO_2 + 6\ H_2 \rightarrow CH_4 + 2\ CO + 4\ H_2O$$

Once the CO_2 is rarefied, it would be split into the carbon and oxygen, being the basis for the synthesis of methane (e.g., with hydrogen to be either acquired in-situ or brought along from Earth) for fuelling the return vehicle. The chemical engineering is based upon the Sabatier process, which is a well characterized process [27–29], yielding methane and water from a reaction of hydrogen with carbon dioxide. Usually, relatively high temperatures at around 300–400 °C and high pressure at 30 bar are required, however, the reaction can be significantly cooler with metal catalysts. A variation of the basic Sabatier methanation reaction may be used via a mixed catalyst bed and a reverse water gas shift in a single reactor to produce methane from the raw materials available on Mars, utilizing carbon dioxide in the Martian atmosphere. A 2011 prototype test operation that harvested CO_2 from a simulated Martian atmosphere and reacted it with H_2, produced methane rocket propellant at a rate of 1 kg/day, operating autonomously for 5 consecutive days, maintaining a nearly 100% conversion rate. An optimized system of this design weighing 50 kg is projected to produce 1 kg/day of O_2:CH_4 propellant while consuming ~17 kWh per day of electrical energy at a continuous power of 700 W [30].

1.2 Landing Site Selection Criteria for Focused Lava Tube Access

Landing on Mars is a complex engineering challenge, given re-entry speeds of around Mach 25 and dissipating the kinetic energy of an entry-vehicle within less than 10 min [31]. The technologies needed for this include the capability of maintaining structural integrity during the passage of atmospheric interfaces, deploying parachutes at supersonic speeds, and providing stability despite shear winds and gusts, and finally challenges posed by the terrain, such as steep slopes, boulders and sand. For landing masses above several hundred kilograms, propulsive landing—using thrusters—is generally considered the best approach [31]. For comparison, the rover mass of the Mars Science Laboratory Curiosity was roughly 1 ton, whereas the landing mass for a first crewed Mars mission, as described in the NASA DRA 5.0 [5], is projected to be in the order of 392 tons for all the key elements (which will probably be distributed over several landings [32]). Further, the required landing precision for minimizing the deviation from the intended landing location is a technology yet to be developed for large landing masses.

There are well established processes for the landing site selection for Mars surface missions, sometimes starting with hundreds of potential landing locations which then are narrowed down based upon scientific merit and engineering capabilities [33–35]. The candidate sites are usually determined based upon orbital reconnaissance data, like the Mars Reconnaissance Orbiter (MRO), providing a ground resolution of <60 cm for selected areas, offering multi-spectral observations, providing a first understanding of the surface mineralogy, or subsurface information provided by ground-penetrating radar. Long duration observations, such as acquired by the Mars Global Surveyor and Mars Odyssey [36] or the Mars Climate Sounder [37] also allow for studying local atmospheric effects. The engineering landing requirements may include parameters like projected slope inclinations, probability of gusts, maximum geographical latitude dictated by the propulsion system and the orbital geometry before landing, rock and boulder size distribution, diurnal variations in illumination conditions (which are relevant for vision-based landing algorithms) and many others [38, 39].

For comparison, the landing site selection process for the MSL2020 Perseverance rover started in 2013—8 years before the landing. Engineering constraints included [35]:

- Geography: between ±30° latitude, elevation below −0.5 km, <16 km landing ellipse.
- Terrain: on 2–10 km length scale, less than 20° terrain inclination. At 1 m-1000 m baseline length scale less than <100 m relief undulations and finally, at 2–5 m length scale less than 25° inclination; rock height must be smaller than 0.6 m.
- Atmosphere: 25 m/s horizontal and 20 m/s vertical winds; during operations: <15 m/s steady, <30 m/s gusts.
- Other: Ka radar band reflectivity, load bearing surface not dominated by dust to avoid landing structure sinking.

Once these engineering parameters are met, the discussion of the scientific merits of a particular candidate site follows. Such a discussion also reflects contemporary scientific priorities, such as the search for traces of water like phyllosilicates or olivine deposits of sediments that might preserve putative biomarkers, like clay deposits.

Thus, in summary, the engineering criteria for the Entry-Descent-Landing (EDL)-capabilities define which sites are accessible, whereas the scientific discussion indicated which of those are desirable.

When it comes to landing sites with access to lava tubes, this poses a challenge, as these are associated with volcanic activity, hence often at high altitudes, like in the Tharsis region. Based upon measurements of the THEMIS thermal camera onboard the Mars Global Surveyor spacecraft, several skylights have been identified potentially leading into lava tubes [40–42]. The issue with areas located at higher altitudes is the reduction of the efficiency of atmospheric drag for dissipating entry velocity with parachutes, which leads to a significant increase in landing mass, for enabling a high-power propulsive landing. Hence, one might argue that landing in a low-altitude location but instead offering long-distance (e.g. >1000 km) mobility for accessing lava tubes might be a more efficient approach.

Currently there are seven known skylight candidates, based upon thermal inertia measurements. The skylights that have been identified so far have typical openings sizes of 50–100 m in diameter on Pavonis Mons and Ascraeus Mons.

These dimensions would allow for deploying infrastructures, such as powered elevators or similar. Or, depending on the topography of the skylight, might even allow for an inclined access path, although on surfaces that might be weak in load-bearing.

Examples of potentially interesting locations may include the following sites:

Region	Lava tubes?	Scientific potential
Alba Patera	High density of lava tubes	Subsurface ice with astrobiological potential
Arsia Mons	Lava tubes with skylights identified	Potential subsurface ice in lobed features, layered lava flows
Elysium Fossae	Lava tubes identified	Extensive ancient fluvial processes
Hellas Eastern Rim	Lava tubes identified along the rim	Debris aprons indicate glacial features, potentially present-day glacial activity
Nili Fossae	Potential for lava tubes	Large olivine outcrops indicate previous contact with water
Olympus Mons, northern flank	Lava tubes identified	Igneous petrology and recent glacial deposits
Syrtis Major Planum	Lava tubes identified	Igneous petrology, lava flows, potential for recent hydrothermal activity
Tyrrhena Patera	Lava tubes identified	Igneous petrology, ancient river channels on flanks
Pavonis Mons	Lava tubes identified	Igneous petrology and recent glacial deposits

Lava tubes are a well-known feature of the Martian surface. Assuming that the lava flow physics is comparable to their terrestrial counterparts, pāhoehoe lava flows are generally very smooth and billowy and are also expected for Mars. For instance, on the boundary between the northern lowlands and southern highlands of Mars (constituting its dichotomy) is a relief-shield volcano Syrtis Mons in the Libya Montes region. This feature is considered of high scientific interest due to the abundant minerals next to the Isidis impact basin [43].

Although the first crewed missions are generally believed to be surface sojourns, later long-term habitats or even settlements might consider lava tubes as a suitable environment due their favourable properties, including:

- Geological stability and protection from local, regional, and global dust storms
- Radiation protection
- Shielding from meteoritic influx
- Relative thermal stability with dampened diurnal fluctuations.

Within the lava-tubes, inflatable structures can be used, as opposed to the heavy-structure surface buildings. For this reason, studies assessed the potential of caves as human habitats as an alternative for an initial settlement of Mars [32]. Details on Martian lava tubes and their possible habitation will be discussed in the next sections.

1.3 Radiation Hazard During the Surface Operations and Possible Shielding Strategy

Outside of the Earth's protective magnetosphere, ionizing radiation presents a constant danger to astronauts [44–46]. The radiation environment in free space is dominated by two contributions [47]: the low-level, but permanently present Galactic Cosmic Radiation (GCR) and sporadic but very intense Solar Particle Events (SPE).

The GCR is composed of highly energetic ions, mainly protons ($\sim$90%) and helium ions ($\sim$10%). They originate from outside the Solar System, probably accelerated by supernova shocks in the Milky Way galaxy. Their flux inside our solar system is influenced by the 11-year activity cycle of our Sun: the Sun generates a magnetic field which is stronger during solar maximum and weaker during minimum; the stronger the magnetic field, the better it can shield the Solar System from incoming radiation. Therefore, the GCR flux is higher at solar minimum and lower at solar maximum [48].

Solar Particle Events are particles accelerated at the Sun by shocks from Coronal Mass Ejections (CMEs). They are mainly composed of protons (90%). Although they can occur at any time, they are more frequent during solar maximum. Typically, there are about ten SPEs per year, each lasting between a few hours and a couple of days. Of special concern for astronaut safety is the sudden onset of SPEs and the fact that their occurrence cannot currently be predicted—therefore a SPE warning may reach the astronauts only minutes to hours before the particles arrive [48]. Thus, astronauts could be accidentally exposed to dangerous high levels of radiation when they are caught outside during an extra-vehicular activity [49]. Most scenarios assume that there would be a section of the habitat, which is more strongly shielded, to be used as a "storm shelter" where astronauts can stay during a SPE.

However, the radiation as measured in orbit around Mars is not what is experienced by the astronauts on the surface: when the particles collide with matter, they may fragment into smaller particles or knock out particles from the traversed material. Thus, the radiation at the surface is altered by traversing the Martian atmosphere; in addition, particles may enter the soil and produce secondary particles there, such as neutrons (so called "albedo neutrons"), which then contribute also to the radiation exposure above the ground. Providing adequate radiation shielding for a human habitat on Mars is of fundamental importance. If the crew lived in a habitat without any dedicated shielding, it would be subjected to radiation doses beyond permissible exposure limits. Following recommendations of the International Commission on Radiological Protection [50], the occupational limit for the annual effective dose is 20 mSv/year. Based on measurements by the MSL Curiosity rover, the dose equivalent from GCR on the Martian surface is about 230 mSv per year [47].

An alternative option is to go underground and use the Martian regolith as a shielding material. However, the necessary soil cover has to be a few meters thick, which shows the appeal of cave-like structures, such as lava tubes. In fact, the effective shielding of radiation in the Martian subsurface, based on the Curiosity measurements [47], are:

Depth below surface (m)	Effective shielding mass (g/cm^2)	GCR dose-equivalent rate (mSv/year)
0 (surface)	0	232
–0.1	28	295
–1	280	81
–2	560	15
–3	840	2.9

Therefore, considerable shielding is of fundamental importance to get down to the desired radiation levels. Manufacturing shielding for habitat (against radiation, micrometeoroid, and exhaust plume debris) and surface nuclear power (radiation) from in-situ materials significantly reduces delivered mass to surface. For radiation, as surface mission durations increase, the accumulated risk to crew also increases. For instance, a radiation shelter using a water tank surrounding the crew quarters requires an optimal water thickness of at least 6 cm. Assuming that the crew quarters fit in a cylinder 5 m in diameter and 3 m tall, the water volume for this isolation is roughly 5.1 m^3, equalling 5100 kg. If the hardware and infrastructure used to create habitat shielding or shielding around emplaced nuclear reactors or landing pads was the same as that used for in-situ propellant production, the 'additional' launched mass to enable much greater protection is negligible.

1.4 Mitigation Strategy from the Martian Dust

The severity of the issue of planetary dust is another issue and became apparent during the Apollo missions. Dust tore through the fabric layers of the spacesuits and jammed their joints, obscured the astronauts' vision and let them lose their footing. The dust that was brought back into the capsule after the EVA led to irritation of eyes and lungs [51]. Dust on Mars is expected to be slightly less abrasive, as particles are more rounded due to aeolian erosion. However, it will still be a major source of material fatigue especially considering that Apollo missions lasted a few days whilst a human Mars mission will last many months. Mars's surface is largely covered by regolith soil and dust, which is easily picked up and transported by the wind. The constant presence of dust in the atmosphere is what causes the (transient) reddish appearance of the Martian sky [52]. The level of dust strongly increases during dust storms which can occur on a local, regional or global scale. In several regions of the 2018 global dust storm, less than 5% of the incident sunlight reached the surface [53]. In addition to these extended dust storms, which can last for months, dust devils are a common occurrence.

An end-to-end dust mitigation strategy will be crucial for any human mission to the Red Planet. This starts with the type and size of surface power generators; solar panels would need to be over dimensioned to power the habitat and auxiliary equipment during a storm. Also, nuclear power sources do not free up from dust concerns entirely. Dust-coated radiators (not only for power plants but also for other applications) lose efficiency and can lead to hardware malfunctions or failures due to overheating. This problem also occurred during the Apollo missions, the rover had problems due to an overheating battery caused by the dust cover. Abrasion caused by dust can be an issue for spacesuits (fabrics, joints, etc....) or robotic equipment (especially moving parts), but can also obscure vision for critical systems if windows or camera lenses are not sufficiently scratch-resistant [54]. A major issue, which has not been well characterized yet, is the physiological impact that dust entering into the habitat can have. It is highly oxidative and chemically reactive and contains substances known to be harmful to humans. The various mineral dusts might cause various lung diseases when inhaled, like silicosis. The dust is also known to contain trace amounts of toxic metals: e.g. chromium, which in the form of hexavalent chromium is genotoxic, or perchlorate, which blocks thyroid function, as well as arsenic, cadmium, and beryllium [55]. In addition, the Martian soil is thought to be acidic, which could be damaging the lung tissue and also lead to corrosion of equipment [52].

The first missions to Mars will be carried out with a small crew staying in a very basic habitat. When attempting to start a more sustainable presence, a larger and more permanent habitat infrastructure will be required. However, the specifications of this habitat will depend on its use: will it be a research station occupied by alternating crews, like, for example, the Concordia Station in Antarctica? Will it be more commercially oriented, as the center for mining activities, also occupied by rotating crews of miners? Or will it be a settlement of pioneers, attempting to start a new life in a new world, like the European settlers emigrating to the Americas? Such factors will influence the setup and requirements for the base. Astronauts who stay there only for a short time (maximum a few years) before returning to Earth will require a different infrastructure compared to permanent settlers; they would for example need a broader medical care system, also involving care for an aging population and, on the other hand, infrastructures for the next generations like schools. Independent of such considerations, a larger base with more occupants will need to be more self-sufficient than the first explorative missions. Especially food will have to be produced locally for a large fraction of the total need. This requires the necessary infrastructure to grow plants and maybe someday even farms with animals. Additionally, large and permanent structures require a high amount of material to build, some of which has a high mass per volume, such as steel. Launching all of these materials from Earth is inefficient, as overcoming Earth's gravity requires a significant amount of fuel, and it is thus costly (a launch from Earth's surface to low Earth orbit needs a Δv of ~ 8 km/s, whereas the transfer from Mars needs only ~ 3.7 km/s [56]).

Alternatively, the resources could be mined locally on Mars—needing heavy equipment in turn. Resources might come from a different source altogether: asteroid mining operations; extracting volatiles and metals from near-Earth asteroids, or Mars-crossing asteroids, which could support the Martian development [57]. In the case of commercial ventures or permanent emigration, the mining of Mars and nearby asteroids could be a potential source of income for a Mars colony. Another resource that a large Martian habitat structure will require in abundance is electrical power—it will be even more crucial for the inhabitants' survival than it is on Earth: maintaining a breathable atmosphere, a human-friendly temperature, running hydroponic farms—all will depend on electricity. The Martian base will therefore need an electrical power supply that is highly reliable—it will be a decision to take, whether to rely on nuclear power or identify other solutions.

1.5 Lava Tubes: A Possible Option for Subsurface Mars Habitation

The presence of effusive volcanism on Mars, implied the generation of low-viscosity lava on the surface [58]. This type of lava can form peculiar features that resemble caves or tunnels in morphology: these are called lava tubes (or pyroducts).

A lava tube is, in fact, a lava cave; i.e., a cave made of mostly basaltic rock [59, 60]. Lava tubes form only during (or a few hours after) effusive eruptions. The subsequent cooling of a low-viscosity lava, which is in contact with an atmosphere once erupted from the vent of a volcano, enhances the development of a hard crust above the flow—while the lava is continuously flowing in the interior of it. This is possible because lava is injected underground during an effusive eruption. A lava tube keeps lava hot enough to continue flowing in its interior, even if the exterior cool down forming a basaltic crust and walls. This is an effective process to insulate the interior hot lava of a tube from the cold external environment [60]. In other words, lava contained into this hard crust continues to flow until the interior drains out and becomes completely empty. When the entire flow process ends, a basaltic roof is formed above the lava tube (Fig. 3). Lava tubes have been found both on the Moon [61] and Mars [62]. The final size of a lava tube depends on the gravity of the planetary body on which they forms, but for Earth they can be up to 15 m wide, and can exists at 1–15 m below the surface [63]. Their lengths strongly depend on their planetary body physical characteristics; in fact, they can be exceptionally long.

Pāhoehoe lava flows are in particular the best types of low-viscosity lava that can form series of lava tubes [59]. In fact, they allow hot lava to break out at a certain point in order to continue flowing towards new directions, building at end

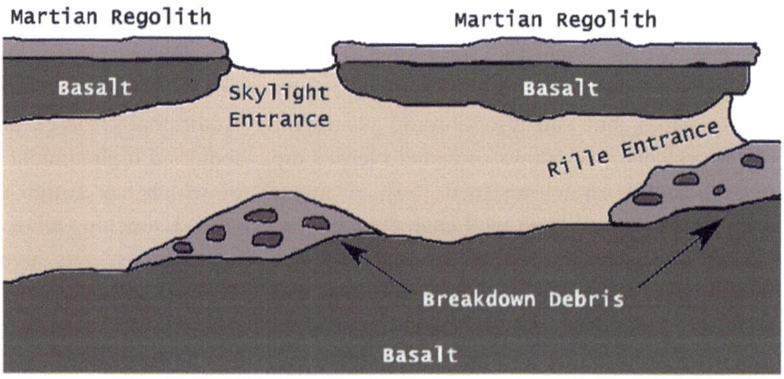

Fig. 3 Sketch of a Martian lava tube, with labeled all the classical features that can be found in its interior

interconnected lava tubes once the main flow cools down. This whole process forms long underground labyrinths. Once cooled, lava tubes exhibit several unique features among the volcanic structures. They have a basaltic ceiling (or roof) commonly filled with small speleothems of basaltic composition that resemble the drip of the basaltic droplets during the last instants of cooling of the roof (Fig. 4d) [64]. These morphologies include stalactites (or lavacicles), tubular forms, and 'shark tooth' forms. The roof of a lava tube can often be characterized by a window to the exterior called 'skylight' (Fig. 4e). Skylights generally form after the end of the process of lava tube formation, even many years later, because they are collapse features [59]. This leaves breakdown debris just below the

Fig. 4 Lava tube features on different Earth locations. **a** A lava tube entrance at Etna volcano (Southern Italy). **b** Flow lines on the wall of a lava tube, representing the passage of the lava flow; it is possible to notice two large 'fossils' of gas slug imprinted in the wall (Grotta dei Tre Livelli, Etna volcano, Italy). **c** Lava flow lines imprinted on the wall of the Corona Lava Tube (Lanzarote island, Spain). **d** Basaltic speleothems on the ceiling of the Corona Lava Tube (Lanzarote island, Spain). **e** Skylights with access to multiple tunnels in the Gruta do Carvao lava tube (Azores islands, Portugal). **f** Sample of breakdown debris of the roof of the Gruta do Carvao, where it is possible to notice speleothems (Azores islands, Portugal)

collapsed roof—composed of the basaltic material that was once part of the roof of the lava tube (Fig. 4f) [65]. In addition to skylights, even lateral entrances due to collapsed trenches can be found in lava tubes. They are also characterized by a breakdown debris (Fig. 4a). The walls of a lava tube are also mainly composed of basalts and they display other unique morphological features. During the lava drain process, step marks are traced on the walls of the lava tubes. These step marks are commonly called 'flow lines', and record the various depths at which the lava flowed (Fig. 4b, c) [66]. The floor of a lava tube generally looks like a pāhoehoe flow in its morphology. Mineral deposits can be present on the floor in the form of crust or even speleothems [67]. Stalagmites may also be present on the floor of a lava tube, and some of them may have a central conduit (hornitos).

The first lava tubes on Mars were recognized after photogeological analysis of Viking orbiter images, but they were subsequently investigated in details through images acquired from Mars Odyssey, Mars Global Surveyor, Mars Express, and Mars Reconnaissance Orbiter in particular [62, 68, 69]. They are close to different Martian volcanoes, mainly Olympus Mons, Pavonis Mons, and Arsia Mons (Fig. 5) [68] and were recognized using two different approaches: by looking for rilles (peculiar morphologies resembling long sinuous troughs), which could be debris derived by the collapse of ancient lava tubes, or by detecting skylights, which appear like dark circles in the orbiter images. Martian lava tubes are from 1 to 3 orders of magnitude larger than the terrestrial ones [63], gravity on Mars is only 38% of the Earth's gravity (3,711 and 9,807 m/s^2, respectively), and this has a positive influence on the spreading of low-viscosity lava flows [70]. A detected series of lava tubes near Pavonis Mons shows a skylight of 190 $\times$ 160 m wide and more than 115 m deep [68]; skylights with these characteristics are nearly impossible to find on Earth. However, it is expected that, even at different planetary conditions, Martian lava tubes retain many of the same geomorphological features of the terrestrial ones [58]. They have rill entrances, skylights (or pit craters), breakdown debris on the floor, and may also have basaltic speleothems and lava step marks on the walls.

More than 1000 lava tube entrance candidates have been identified on Mars, with the vast majority of them located in the volcanic region of Tharsis—they are more than a thousand and are mainly located close to the three Tharsis Montes (Ascraeus Mons, Pavonis Mons, and Arsia Mons) [71]. These lava tubes entrances include skylights or pit craters of various morphologies, deep fractures, and other kinds of voids (Fig. 6).

Lava tube exploration on Mars might have both a scientific importance and a strategic value for the future human colonization of Mars. Scientifically, lava tubes can reveal if possible forms of life may have existed in the past. Thus, surveys and samplings in Martian lava tubes might confirm whether potential ancient or present-day microbiology exists on Mars. Extremophile hypogeum Martian life, if exists, is expected to be chemolithotroph or lithoautotroph [72], i.e. not requiring to live on the surface in order to survive and with nutrients deriving from the volcanic minerals present in lava tubes [67].

The importance of lava tubes for the human colonization of Mars is relatively to the vicinity of useful resources. They can help in different ways: first of all, lava tubes are natural shelters from the deadly radiations that continuously hit the Martian surface [47]. They can also serve as a protection from frequent dust storms in the open environments [73] or micrometeorite impacts [74]. In addition, they do not require a large quantity of building material to be brought on Mars, in order to build a cover for a base; and they can be a rich source of geo-resources (as ice present in the subsurface; [75, 76]). In other words, they can be seriously used as the first human habitats on Mars. The pros and cons will be discussed in detail in the next section.

1.6 Benefits of a Lava Tube Habitat

It could be common to think that the first human settlements on Mars will be located on its surface. Thinking in terms of first outpost, the answer is positive because a temporary shelter is

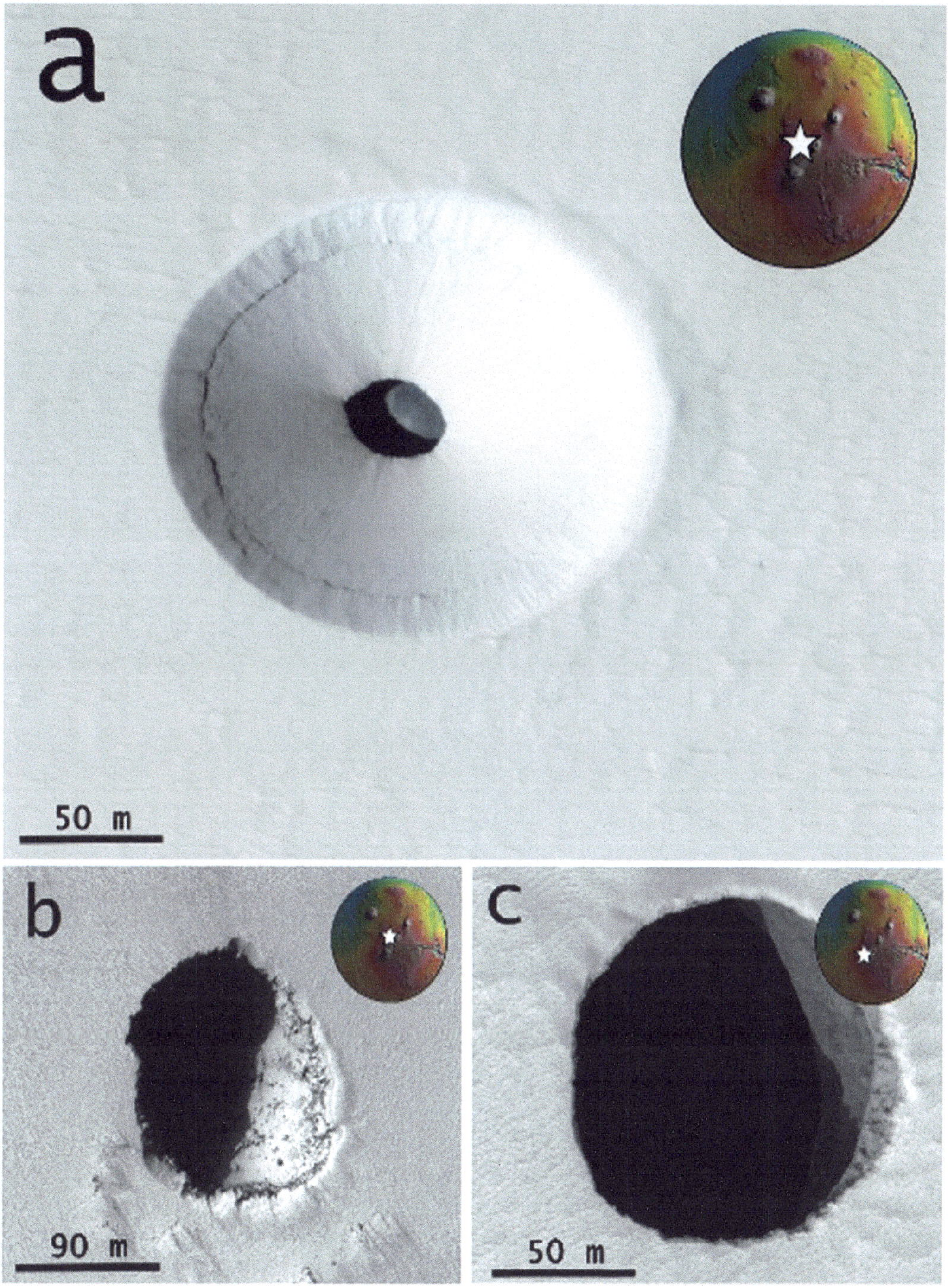

Fig. 5 Lava tube skylight entrances detected with HiRISE on Mars. **a** A 35 m wide skylight on the flank of Pavonis Mons; the hole is surrounded by a showy collapsed pit crater with a relatively steep slope. **b** A skylight on the southern flank of Pavonis Mons, 180 m wide. **c** An around 150 m wide hole (called "Jeanne") on the flank of Arsia Mons; the depth of this cavity is at least 178 m. Courtesy of HiRISE/NASA-JPL/University of Arizona

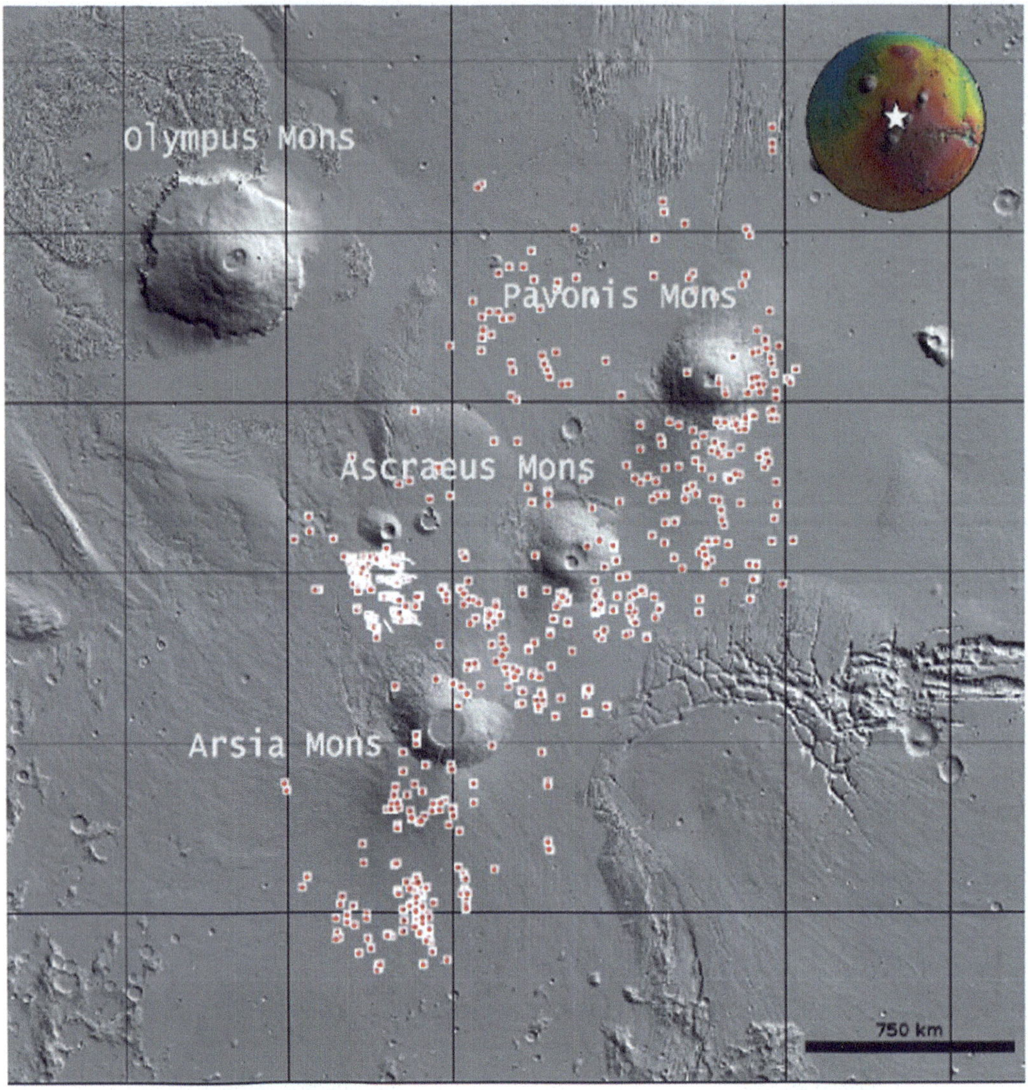

Fig. 6 All lava tube entrance candidates in the Tharsis region on Mars (red squares). It is possible to notice that there is an high density of lava tube entrances on the southern and northern areas of Arsia Mons and on the southern areas of Pavionis and Ascraeus Mons. Courtesy of USGS—Astrogeology Science Center

necessary before constructing an underground base. The surface of Mars is a dangerous place due to many reasons: ionizing radiations coming from space [47], extreme temperature fluctuations [77], dust storms [73], micrometeorite impacts [74]. Lava tubes may be the immediate solution to the traditionally envisioned surface habitats provided they are close to or offer easy access to vital resources like water and minerals. They can in fact help saving time and effort for the digging process that would be required for a large and safe underground Martian base. Additionally, the habitation of a lava tube could serve as an entrance and processing facility for subsurface mining of ice and possibly other raw materials to support a self-sustaining base.

Extreme temperature fluctuations generate a temperature range of about 100 °C on a single Martian sol (since heat energy can be radiated away more easily without a thick atmosphere; [78]), but this does not apply to the interior of lava tubes. In fact, as we know from Earth, the lava tube environments are able to retain a sort of stable microclimate [79], especially due to their effective insulation from the outer environment. And it is reasonable to think that lava tubes on Mars may have almost the same climate phenomena [80, 81]. The thick roof of a lava tube would act as a very effective shelter for the deadly fluxes of UV radiations, alpha particles, and cosmic rays coming from space and hitting the Martian surface [82]. The thickness of basaltic roofs in lava tubes can be tens of meters, resulting problematic or impossible for the ionizing radiation to pass through them. The effective shielding from UV radiation has been directly measured in Martian basaltic material of a meteoritic sample believed to be part of the roof/wall of a Martian lava tube [83]. Of course, fluxes of high energetic radiation can still be present if habitats are built close to skylights, since them are openings directly to the surface, and thus to the atmosphere.

Dust storms, winds, and micrometeorite impacts are also negligible when inside a lava tube environment [65, 82]. The instruments needed in an habitat would be protected from regolith microparticles that, with time, could generate system malfunctions or, even more dramatic, damage to the astronauts' suits during surface operations. However, even in the case of a lava tube, it is always valid the rule to build habitats as much far as possible from the skylights or lateral entrances. A small amount of particles coming from dust storms can still enter the lava tube from skylights or lateral entrances.

Another one among the benefits in building the first Martian human habitats inside lava tubes is pretty logistical and economical [84]. The significant reduction in payload mass needed from Earth in order to construct our first shelters would save a considerable amount of money. In addition, astronauts are not required to expose themselves to the dangers of the Martian surface during operations conducted from the landing vehicle. At the same time, construction material can be found and directly used inside lava tubes. In fact, one of the most practical solutions in manned (or robotic) missions is the use of a new generation 3D printers, which will use regolith material as the 'ink', to artificially construct various objects from planetary surfaces [85]. In this sense, Martian basaltic materials found in lava tubes (the breakdown debris from collapsed roofs forming skylights, for example) can be used for such a purpose. Additionally, useful minerals and other important resources can be present inside lava tubes [67].

One of the best types of resources that might be found in Martian lava tubes is water or dry ice [75, 86, 87]. Basaltic caves are able to trap volatile elements (such as H and O, that compose water) and ice can remain stable under the special microclimate of a lava tube provided that the local pressure will be enough to avoid sublimation. Temperature stability is always assured by the interior cycling of cold air [80]. In some cases, it would be possible to extract ice present in the subsurface of Martian lava tubes [88] and it can be used and recycled for a lot of purposes related to food, drink and, oxygen.

1.7 Criteria of Selection for Lava Tubes

Access to Martian lava tubes would be possible by entering from one of their terminations (normally at the end of a rille) or by using ropes in order to descend through skylights [89]. Another possibility would be drilling directly their roof—but this may be problematic, because it needs a precise drill to be brought on the mission, and also dangerous, because sections of the roof may collapse during the drilling operations. Thus, entering through lateral entrances or skylights is the best option.

It is possible to construct a level of criteria for selecting the best lava tubes on Mars, in which the first habitat might be installed. These criteria, ordered by the one that need to come first in the selection, are: difficulty of access, internal roughness of the tube, and presence of subsurface ice or mineral resources.

Regarding the difficulty of access, both end sections of the lava tubes (lateral entrances) do not have to present too many obstacles that may be mostly due to roof collapse debris [65]. In fact, this may prevent an easy access to the interior of the lava tube, especially when carrying all the instruments needed in the first habitats. In addition, this may be a threat for the astronauts themselves, they will have to go scrambling the rough terrain while wearing a heavy spacesuit and this is absolutely hard and dangerous because there might be a lot of rubble piles and debris below them (due to the previous roof material collapse; [90]). Thus, when selecting lava tubes we need to take into account the presence of an easy entrance, i.e. not blocked by too many debris. Another major problem may arise when accessing lava tubes from skylights, the depth of the cave. Many Martian lava tubes are very deep (for example, the "Jeanne" hole on the flank of Arsia Mons is 178 m deep; Fig. 5c) and it is crucial to know exactly the depth at which the ground is located before trying to access it using ropes. Too deep lava tubes may not be the best option in terms of system communications and frequent travels from the subsurface to the surface. The roofs of lava tubes also need to be thick enough. Thin roofs may be very dangerous, because they can break up and collapse at any time. For example, a known 35 m wide skylight on the flank of Pavonis Mons (in Fig. 5a) is surrounded by a large collapsed pit crater. This would not be the best entrance for a lava tube since it is an indication that other pit craters may form [91]. At the same time, too many skylights in a lava tube means less protection from radiations or microparticles coming from outside. In addition, all entrances need to be large enough (at least 5×5 m) in order to ease the movements of instruments and materials in all safety.

The roughness of the tube is another factor that needs to be taken into account since it may be problematic for the installation of the first inflatable habitat. Roughness in lava tubes is mainly due to the excessive presence of collapse debris on the floor [65]. Such a roughness, due to the differing cooling properties of the volcanic material, may cut or damage the astronauts' spacesuits. For the same reason, the excessive presence of speleothems on the ground (basaltic stalagmites) is not welcome.

The final point is related to the possible presence of subsurface ice or mineral resources. First of all, it is convenient to look for areas of accumulation of mineral deposits that, to the observer, may look different in colour and physical characteristics from the surrounding areas. Minerals commonly found inside lava tubes on Earth are sulphate and calcite [62, 92]. Calcite is very rare on Mars, thus only sulphates are expected to be found inside Martian lava tubes. Sulphates may be hydrated in case of presence of subsurface water ice [75]. This would be a perfect marker of ice beneath the subsurface, another aspect to take into account when selecting lava tubes. In this sense, sulphates may be present as gypsum ($CaSO_4 \cdot 2H_2O$). Thus, where to look to have the best chances to find ice beneath the ground in lava tubes on Mars? Surely, the best areas would be located close to the Martian poles where ice caps are present (even if they are especially composed by CO_2 ice; [93]). However, lava tubes close to the Martian poles were not detected yet and the possibility to find some with the right characteristics described above could be scarce. Another possibility is to look for lava tubes close to volcanoes that were affected by extensive glacial periods during the Martian geological history. The most recent of the Martian glacial periods occurred from 0.5 to 2.1 Ma, while today Mars is considered to be in an interglacial period [94]. Thus, almost every Martian volcanoes could have the characteristics to have retained ice into lava tubes, if these were kept in the right environmental conditions. However, even if the majority of the lava tubes were found in the Tharsis region, the Tharsis Bulge may not allow the retention of ice into lava tubes (if not at great depths inside them). The reason stands in the fact that the Tharsis Bulge is more elevated than the other areas located along the Martian equator and thus ice would sublimate easier being affected by less atmospheric pressure [95].

In conclusion, the best lava tubes to be selected for a manned mission to Mars need to have the following characteristics: entrance with relatively scarce breakdown debris, large enough and easy to access while carrying all the necessary instruments and materials, with a thick roof, and not too deep if a rope is used to access the lava tube from a skylight; relatively plain floor without a large density of speleothems; presence of mineral deposits and/or subsurface ice useful for mining purposes.

1.8 Accessing and Preparing a Lava Tube for Human Activities

Once the best lava tube for installation of the first human settlement has been selected, it is time to access it and prepare it for human activities. 'Prepare' in which sense? Lava tubes are not "ready to use" houses. Raw lava tubes, even the ones with relatively few breakdown debris and speleothems on the floor, are almost never ready to the hosting of an habitat [65]. In fact, the floor where the habitat is expected to be installed still needs to be cleaned up of all the debris properly until a completely plain surface is reached. Also, the entrance(s) need(s) to be cleaned up as much as possible.

The best way to clean a lava tube from debris is to use a robotic approach, before human go down by themselves. Also, robots are able to map in 3D the entire structure before humans can enter the tube. Robotic technologies that are being tested and proposed in terrestrial lava tubes, for the purpose of helping astronauts during a reconnaissance phase before entering a planetary cave, are mainly quadruped robotic "pack mules" [96]—as the Multi-Utility Legged Explorer (MULE™) or the Multilegged Autonomous eXplorer (MAX). These looks like "robotic dogs", ultra-light, and can help astronauts not only in themapping of the caves but even in carrying equipment necessary for habitat installation. In particular, the six-legged robot MAX is extremely adaptive to challenging terrains, as could be the debris on the floor of a lava tube [97]. The best quadruped robot used in planetary exploration are, however, the DARPA-funded BigDogs. Developed also for military use, the DARPA BigDog quadruped robot has an exceptional management system for locomotion in all types of terrains (even by climbing slopes of 35°): a gyroscope, a stereo vision system, and an incorporated LIDAR (useful for 3D mapping of lava tubes); it can also carry up to 150 kg of equipment [98]. Even better, the DARPA Legged Squad Support System (LS3) can function autonomously and is adapted to extreme environmental conditions (both very hot or very cold), thus making it perfect for the exploration of Martian lava tubes (Fig. 7); it can carry even

Fig. 7 The Legged Squad Support System (LS3). This special quadruped robot is fully autonomous and can work in extreme environments. It would be perfect for the exploration of planetary lava tubes. Courtesy of DARPA

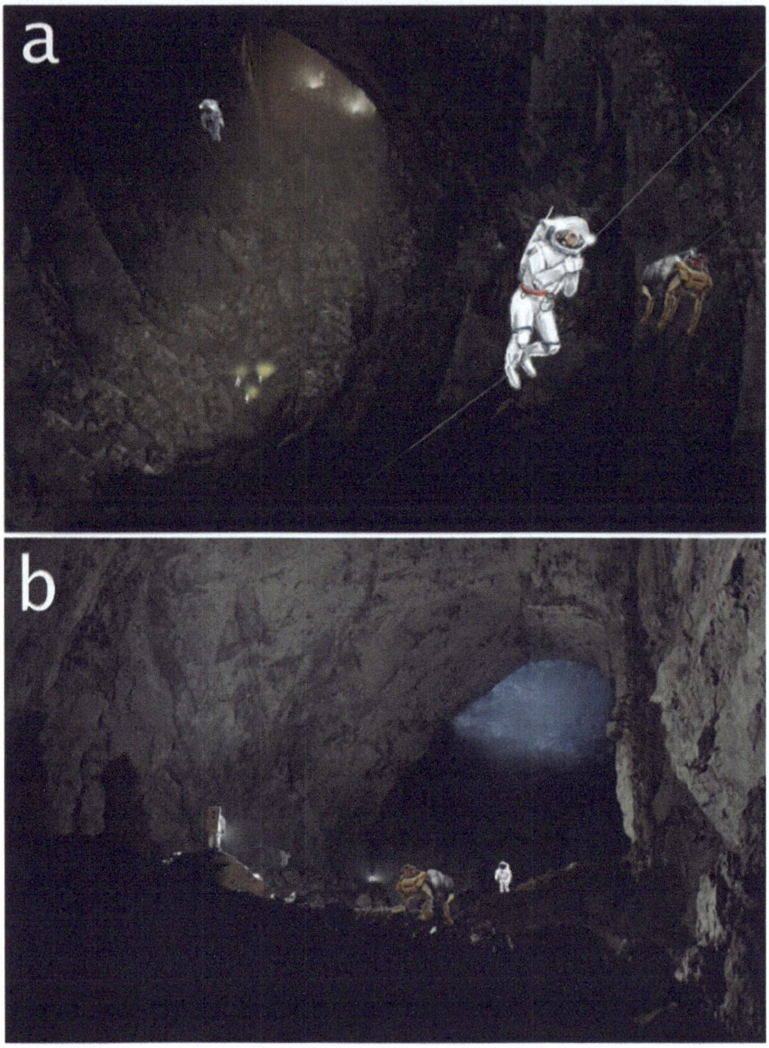

Fig. 8 Artistic view of a descent and initial reconnaissance in a Martian lava tube. **a** During descent from a skylight entrance, ropes should be used by astronauts. At the same time, quadruped robots should be dropped. **b** Once on the floor of the lava tube, quadruped robots can still help in evidencing dangers on the ground and in the mapping of the area. Courtesy of XArc

more equipment for the astronauts, up to 180 kg [99]. Drones could also be used to explore planetary caves, but they might be very difficult to control in a lava tube [89]. Robots can descend into the lava tube through skylights or through one of the rille entrances (Fig. 8).

Once the robots ends their reconnaissance of the lava tube, reporting 3D mapping and terrain evaluation, it is time for the humans to access it. Rappelling could be the best option, but, in future, we might be able to use also jetpack systems. The first things to do once astronauts reach the lava tube floor are: (a) establishing an easy communication link with the surface/outdoor and (b) install a temporary pressurized tent.

A lot of hand work is expected for the first astronauts that will need to clean the floor of the lava tube from all debris. One by one, every piece of basaltic material, deriving from collapsed pit craters or from the cooling of the lava tube, needs to be removed from the proposed area of installation of the first habitat. This work may require some days, during which the astronauts may have to exit the lava tube every time they need to return to the landing vehicle. The

Fig. 9 Artistic view of the interior of a planetary lava tube. On the left, it is possible to see all the breakdown debris on the floor directly below a skylight, due to a previous collapse. This debris need to be cleaned up in order to obtain a plain surface where to install an initial habitat, as can be seen on the right of the image. *Image credits* Emilia Rosselli Del Turco and Irene Zaccara

best plain surface inside the lava tube must be selected. Not only all the debris, but even eventual speleothems present on the floor need to be cut and removed . Debris can be left on the side of the interested area, but not put in the middle of the way of access to other parts or to the entrance/exit of the cave (Fig. 9).

Analysis of the resistance of the walls or part of the roof must be taken in consideration and should be seriously conducted using sclerometers (as the Schmidt Hammer, for example, easy to carry in a manned mission) for an in situ test of the mechanical strength of the most fractured parts of the lava tube [100]. In fact, a lava tube is always formed after lava cooling and thus an high level of contraction may have characterized its structure developing many fractures in walls and ceilings [90]. It is much better to destroy walls/parts of the roof where instabilities are found in order to prevent any possible collapse or rockfalls that will be dangerous once the habitat will be installed. Basaltic roofs close to skylights may collapse very easily and such sclerometric test are crucial in order to anticipate these unwanted events.

The smooth surface created after the debris removal process will be the first place where the habitat will be installed in the subsequent days.

1.9 Constructing a Habitat and a Mining Base

The idea when designing a human habitat inside a lava tube is to seal and pressurize the whole cave, in order to isolate as much as possible the internal from the external environment [65, 101]. The sealing process could be achieved by using the loose basaltic material available on the floor of the lava tube as an "ink" for 3D printers [85]. Once the material is printed, every hole, skylight, or fracture linked with the exterior, should be covered, in order to obtain an artificially pressurized cave. In this way, facilities as greenhouses could be built into the lava tube once it is completely sealed. However, this idea has a disadvantage because it is not easy to pinpoint all the existing holes or cracks inside a lava tube and thus the environment would not be totally sealed.

A much more feasible design idea is to isolate only the main habitat present in the lava tube by creation of an inflatable habitat (Fig. 10). In this way, the skylight can be left open allowing the sunlight to enter the cave with a benefit on the psychologic condition of the astronauts. However, proper shielding should be installed directly below the skylight because radiations could be dangerous in the long term [82]. The inflation of the habitat is quick and could be ready even within 24 h. Once the habitat is inflated, it needs to be protected from possible collapse of parts of the roof or debris coming from the ceiling. These possible events might break the habitat and harm its pressurization effect. A cover needs to be created with the correct size of the habitat. How to do this? Using the basaltic material in 3D printers it is possible to print exactly the size of the cover that the astronauts needs for their habitat. The inflatable habitat will then be covered with insulated material—exactly like a shell [101].

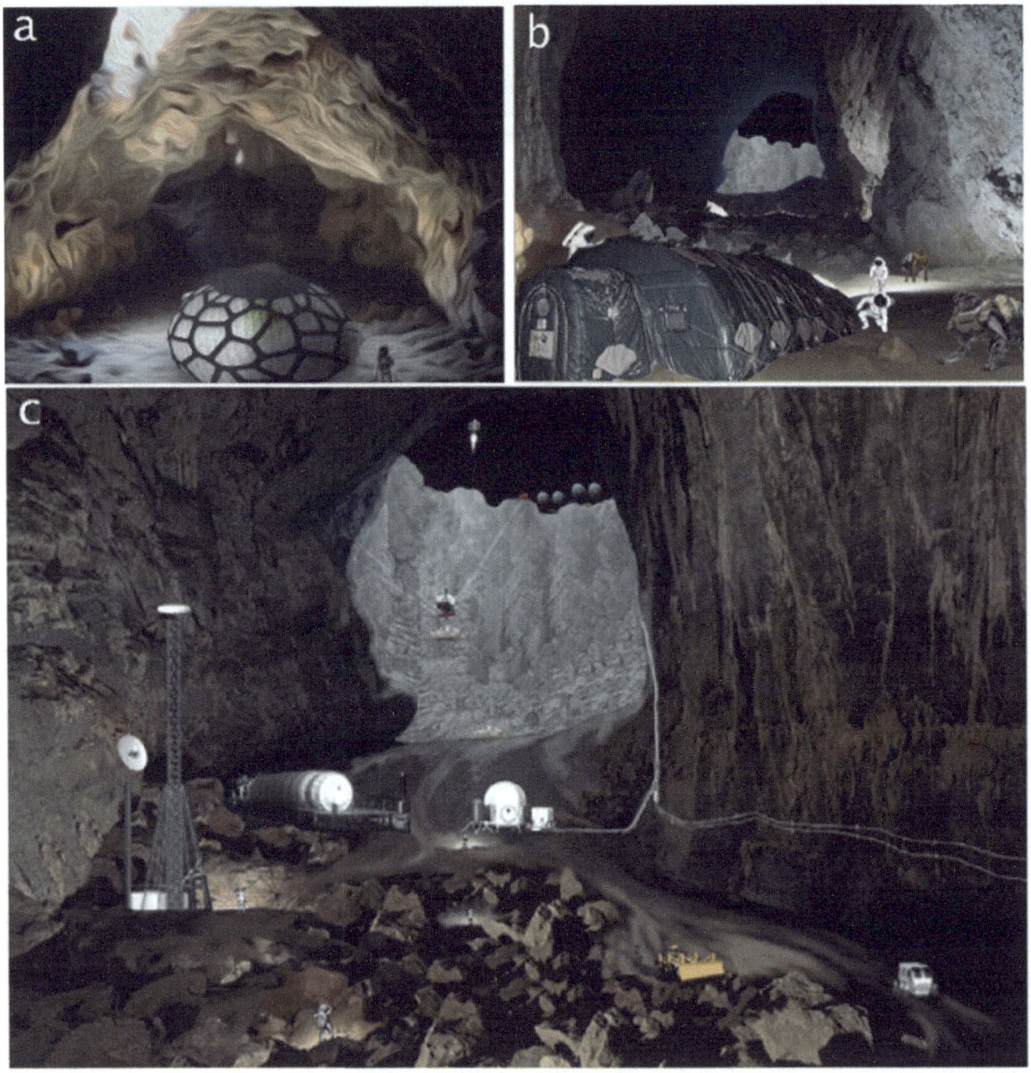

Fig. 10 Design of an habitat into a planetary lava tube. **a** Art sketch of an inflatable habitat covered with basaltic regolith derived from 3D printing. Image credits: Emilia Rosselli Del Turco and Irene Zaccara. **b** Temporary inflatable tent to be used during the installation of the main habitat. **c** Installation of communication systems, elevators, and mining facilities in the interior of the lava tube. Courtesy of XArc

The interior design of the habitat will require an airlock to be established early. Then, inner walls and partitions into internal chambers can be decided (bedrooms, living areas, laboratories). A central greenhouse in the interior of the habitat is highly suggested because this can reduce stress and other psychological diseases in the astronauts [102]. Also, the installation of a pump that may sometime drive Martian atmosphere inside the greenhouse can be useful for plants because they can naturally convert CO_2 into O_2. Obviously, solar energy is needed to support these processes and to keep lit and warm the environment. Solar power units need to be installed outside of the cave, or below the skylight where possible [85]. At the same time, communication systems (as antennae) must be installed outside of the lava tube. These systems (solar power units and communication devices) are then connected with the lava tube habitat through cables. Energy will also serve to install an illumination system along the whole lava tube, which is essential in dark areas.

The presence of deposits of frozen water may be desirable into a Martian lava tube [75]. In fact, human and mining operations require liquid water to be recycled and re-used for separating minerals, drinking, and for producing food in greenhouses—the molecule of water can also be split to obtain oxygen and hydrogen for fuel [103]. Other useful volatiles for life-support systems that might be recovered from below the surface in a lava tube are N_2 and He [67]. It is economically unsustainable to bring all the massive instruments used for mining on Earth to another planet, thus a novel method need to be used to extract ice and minerals on Mars. Thermal mining can be the ideal solution for the extraction of ice and other volatiles in Martian lava tubes [88]. Heat can be used to increase the volatility of the compounds below the surface in order for them to be automatically extracted. How the process of thermal mining works? Two ways can be chosen: (1) heat can be directly applied on the floor of the cave (through heaters dropped into boreholes) or (2) using minimum energy, heat can be created by deviating the sunlight from the skylights and concentrating this light on the ground. In either way, once heat is actively present on the ground ice the production of water vapour will start. This process automatically extracts water molecules from below the surface. At this point, water vapour can be captured into a specific tent. This tent has two cold traps attached to it, the volatiles will be refrozen and collected. The extracted ice can subsequently be processed in a refinement plant to be then used for production of water and/or oxygen. Other volatiles can be trapped in the tent, CO_2 for example, which can be used as propellant although less effective than water.

1.10 Earth and Moon Analogues for Martian Lava Tubes

Analogies of lava tubes on Earth with their Martian counterparts are important in better defining, in all safety under terrestrial conditions, what can realistically be done in a manned mission in a lava tube on Mars and how this can be achieved. Hundreds of kilometres of terrestrial lava tubes have been extensively investigated during speleological explorations around several shield volcanoes [104, 105]. The best examples of lava tubes on Earth have been investigated in Hawai'i [59, 106], in Iceland [107], in the Canary Islands (Spain) [108], and in Sicily (Southern Italy) [109]. Many of these sites can potentially be considered as analogues for Martian lava tubes [63].

The NIAC (the original NASA Institute of Advanced Concepts) program is, to date, the most accurate mission design and study for lava tube habitation in planetary environments [110], which is useful for application to Martian caves. However, a complete mission architecture for lava tube exploration and habitation, to achieve in a simulated Mars mission on Earth, is still lacking and should represent the next step.

In the meantime, the ongoing ESA program PANGAEA is designed at preparing the next generation of astronauts with expertise on Earth and planetary geology in order to evaluate them as possible partners for future exploration missions on planetary environments [111]. This program is especially willing at making humans and robots to work together during the mission. Into this program, the CAVES (Cooperative Adventure for Valuing and Exercising human behaviour and performance Skills) course serves, in particular, to prepare astronauts in working in extreme environments, as can be planetary caves and lava tubes. In this course, astronauts must explore and map caves working as a group, while also conduct scientific experiments. They have to experience also communication, safety procedures and emergencies, speleology techniques, and sampling. Monitoring the results during and at the end of the simulated expeditions is important in determining what worked and what still need to be better implemented for real planetary cave exploration missions.

The Mars Lava Tube Pressurization Project (MLTPP) explores an analog mission scenario in which a base facility is installed in a lava tube on Mars [112]. This project is still under development and is expected to be conducted in the Stefánshellir lava tube, which is located in the Hallmundarhraun lava field (Iceland). This project not only aims at creating an internal habitat but also at pressurizing a lava tube by creating barriers with the scope to reduce or neutralize the external atmosphere of the planetary environment. This process is crucial for astronauts who on Mars need to maintain facilities as habitats or greenhouses completely enclosed into basaltic caves under a stable pressure. In the MLTPP project, astronauts will use 3D printers and basaltic material found in the cave as an "ink" to construct the shell parts of the structure in order to isolate the lava tube. This process is meant to be used in Martian lava tubes for future manned missions.

In the near future, lunar lava tubes could also be used for further development and refinement of the technology for eventual use on Mars. These lava tubes are typically found at the boundaries between the lunar maria and the lunar highlands [61]. Using lava tubes on the Moon as a training exercise would be much better than terrestrial analogues due to the lower gravity and the real isolation. Lunar lava tubes are much larger than those on Earth. Their widths can reach 500 m size before a gravitational collapse can happen [90]. However, the Moon is even more affected by micrometeoroid impacts, due to the almost total absence of an atmosphere, and even to seismic events [113]. These phenomena can create instabilities in lava tubes. The depth of lunar lava tubes can reach 40 m below the basalt and the interior environment could have a stable −20° C temperature [65, 91]. Thus, these deeper lava tubes might be chosen for lunar exploration because they offer a better protection from the external environment and a stable microclimate. The vast majority of the lunar lava tubes were found using the NASA's Lunar Reconnaissance Orbiter (LRO) imagery, which has allowed the discovery of more than 200 entrance candidates (especially skylights) from about 5 to 900 m in diameter [101] (Fig. 11). The Marius Hills region shows a collapsed pit 65 m wide with the floor at a depth of 36 m [61], which might be regarded as a good candidate (Fig. 11b). However, the best lava tubes for a manned mission to the Moon should be the ones located near the lunar equator where the KREEP terrane could be explored for rare earths. The ISRO Chandrayaan-1 orbiter mission reported a rille entrance from which it can be possible to enter a 2 km long and 360 m wide uncollapsed lava tube at the lunar equator [114]. This might be considered as a main target for future lunar missions.

1.11 Volcano-Speleology on Mars

After the installation of an habitat within a Martian lava tube, the astronauts may still want to explore it further. Astronauts can set up additional habitats at various points in the lava tube in order to extend the exploration and the use of it. Lava tubes are kilometres long below the surface [63] and the full exploration of them may reserve interesting new findings, as for example: better areas where to install another habitat, additional mineral or ice deposits, uncommon stratifications or samples useful to better understand volcanological processes on Mars [115].

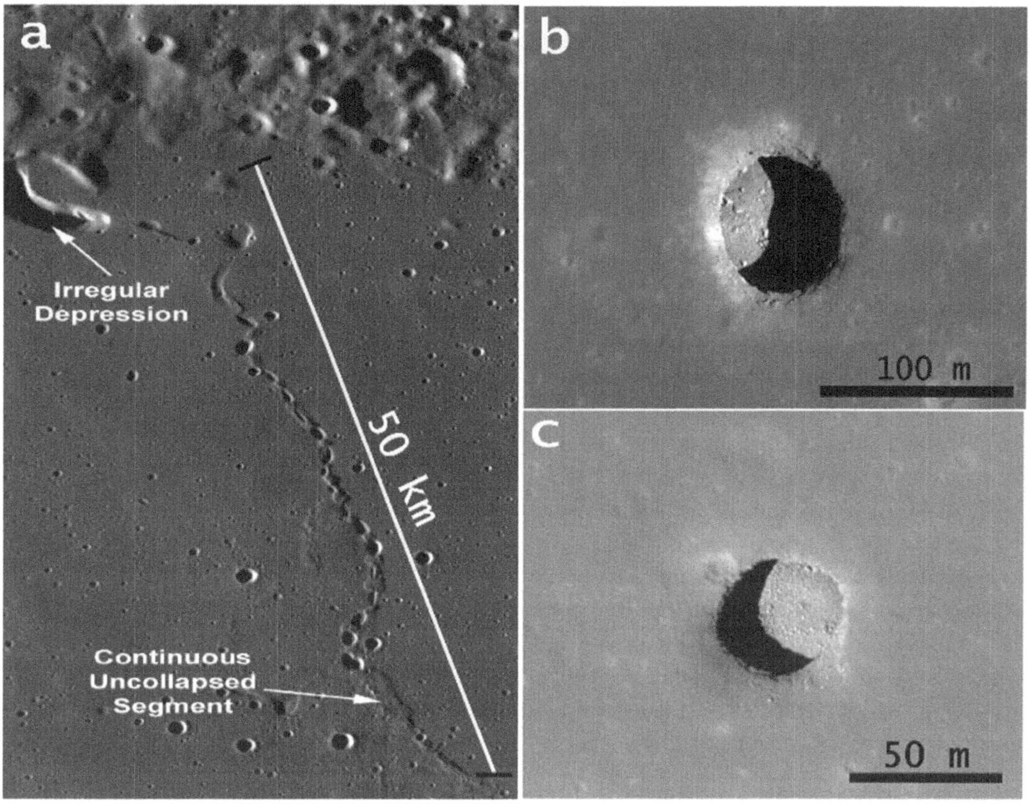

Fig. 11 Lunar lava tubes detected by the Lunar Reconnaissance Orbiter. **a** A sinuous 50 km long chain of collapsed pit craters that continues in an uncollapsed segment of a lava tube in the Marius Hills region. The irregular depression on the north of the area could be the main source of the lava that flowed in order to generate the lava tube morphology. **b** A 100 m deep skylight located in Mare Tranquillitatis. **c** A 50 m wide skylight in the Marius Hills region. Courtesy of NASA/GSFC/Arizona State University

Speleological techniques have been used on Earth to fully explore the lava tube in all safety [85]. However, specific speleological approaches on Mars might be needed with respect to the methods used on Earth's lava tubes, especially because of the different gravity. Human exploration will take advantage of robotic exploration [96]. Quadruped robots, for example, can help in carrying all the equipment on the rough terrain of the lava tube. They may also serve as 3D mappers of the tube, step after step. In addition, ESA developed technologies for 3D mapping of a tube at a resolution of just a few centimetres using a LIDAR system that can be installed in a backpack and thus easy to carry on a planetary mission [89]. Robots are much more effective than humans in reaching inaccessible or dangerous places across the complex rough floors of lava tubes. One of the best robotic concepts is the snake robot. This type of machine is developed by SINTEF and uses hydraulics mechanisms to move the different parts of the 'body' of the snake—simulating a real snake [116]. In fact, snake robots will be able to climb up and down, squeeze around corners, and adapt to every impossible terrain scenarios and hidden obstacles found in lava tubes. Remotely controlled drones are another exciting possibility to explore lava tubes, especially because they can fly in the artificial atmosphere created in the tube and thus can easily scan the areas near their ceilings [89].

Common cases to encounter in a deep section of a lava tube is CO_2-rich layers since CO_2 is the main constituent of the Martian atmosphere.

1.12 Summary

During the course of this chapter we saw that lava tubes exists as volcanic features not only on Earth, but even on other worlds. Where available, lava tubes are the most practical and effective places where to install the first human bases on Mars. The shielding effect that they can provide is exceptionally advantageous in both logistical and economical terms for a planetary mission, since they do not require a huge mass load to be brought on a space mission and minimize considerably the work to be done on the dangerous Martian surface after landing. The thick basaltic cover in a lava tube (on a planet with thin atmosphere thickness, as is Mars) can naturally protect from deadly radiations, extreme temperature variations, dust storms, and micrometeorite impacts. In addition, lava tubes need to be considered a fortune if water ice is detected below their floor—as water can be extracted by thermal mining and is needed for all essential life-supporting systems inside a human habitat. Robots and humans should work together in order to detect and access the best lava tube close to the scrupulously selected landing site. A selected lava tube must undergo initial preparation and cleaning of the basaltic floor before an inflatable habitat can be constructed within it. Lava tubes on Earth (or, in the future, on the Moon) can result in a great opportunity for conducting analogue Martian missions, for the development and refining of the necessary technology for eventual use on Mars. Further direct exploration of Martian lava tubes after habitat installation is crucial to investigate, as never before, the possibility of past or present hypogeum Martian life, development of mining facilities, laboratories of research, and new hidden aspects about the volcanology of Mars.

References

1. NASA (2020) Artemis Plan—NASA's Lunar Exploration Program Overview [WWW Document] [Internet]. https://www.nasa.gov/sites/default/files/atoms/files/artemis_plan-20200921.pdf
2. Foust J (2019) Gateway or bust: NASA's plan for a 2024 lunar landing depends on a much-criticized orbital outpost. IEEE Spectr 56(7):32–37
3. Murtazin R, Sevastiyanov N, Chudinov N (2020) Fast rendezvous profile evolution: from ISS to lunar station. Acta Astronaut 173:139–144
4. Burns JO, Mellinkoff B, Spydell M, Fong T, Kring DA, Pratt WD et al (2019) Science on the lunar surface facilitated by low latency telerobotics from a Lunar Orbital Platform—gateway. Acta Astronaut 154:195–203
5. Drake BG, Hoffman SJ, Beaty DW (2010) Human exploration of Mars, Design Reference Architecture 5.0. In: 2010 IEEE aerospace conference [Internet]. IEEE, pp 1–24. http://ieeexplore.ieee.org/document/5446736/
6. Drake BG (2009) Human exploration of Mars Design Reference Architecture 5.0 (No. NASA/SP–2009–566). Human exploration of Mars Design Reference Architecture. NASA, Houston, TX
7. Rudisill M, Howard R, Griffin B, Green J, Toups L, Kennedy K (2008) Lunar architecture team: phase 2 habitat volume estimation: "'caution when using analogs.'" In: Earth & space [Internet]. American Society of Civil Engineers, Reston, VA, pp 1–11. http://ascelibrary.org/doi/10.1061/40988%28323%29101
8. Leidner D, Music S, Wedler A (2015) Robotic deployment of extraterrestrial seismic networks. In: 13th symposium on Advanced Space Technologies in Robotics and Automation (ASTRA)
9. Abercromby AFJ, Chappell SP, Gernhardt ML (2013) Desert RATS 2011: human and robotic exploration of near-Earth asteroids. Acta Astronaut [Internet] 91:34–48. https://linkinghub.elsevier.com/retrieve/pii/S0094576513001525
10. Binsted K, Hunter J, Halpern B, Caldwell B (2013) HI-SEAS: a long-duration human spaceflight analog in Hawaii. In: 64th International Astronautical Congress
11. Häuplik-Meusburger S, Binsted K, Bassingthwaighte T, Petrov G (2017) Habitability studies and full scale simulation research: preliminary themes following HISEAS mission IV. In: 47th international conference on environmental systems
12. Lim DSS, Abercromby AFJ, Kobs Nawotniak SE, Lees DS, Miller MJ, Brady AL et al (2019) The BASALT research program: designing and developing mission elements in support of human scientific exploration of Mars. Astrobiology [Internet] 19(3):245–259. https://www.liebertpub.com/doi/10.1089/ast.2018.1869

13. Imhof B, Hogle M, Davenport B, Weiss P, Urbina D, Røyrvik J, Vögele T, Parro V, Nottle A (2017) Project moonwalk: lessons learnt from testing human robot collaboration scenarios in a lunar and Martian simulation. In: 69th International Astronautical Congress (IAC)
14. Imhof B, Hoheneder W, Ransom S, Waclavicek R, Davenport B, Weiss P et al (2015) Moonwalk—human robot collaboration mission scenarios and simulations. In: AIAA SPACE 2015 conference and exposition [Internet]. American Institute of Aeronautics and Astronautics, Reston, Virginia. http://arc.aiaa.org/doi/10.2514/6.2015-4531
15. Bessone L, Sauro F, Stevenin H (2015) Training safe and effective spaceflight operations using terrestrial analogues. In: Space safety is no accident [Internet]. Springer International Publishing, Cham, pp 313–318. http://link.springer.com/10.1007/978-3-319-15982-9_37
16. Bessone L, Sauro F, Maurer M, Piens M (2018) Testing technologies and operational concepts for field geology exploration of the Moon and beyond: the ESA PANGAEA-X campaign. EGU Gen Assem 2018
17. Groemer G, Gruber S, Uebermasser S, Soucek A, Lalla EA, Lousada J et al (2020) The AMADEE-18 Mars analog expedition in the Dhofar Region of Oman. Astrobiology [Internet] 20(11):1276–1286. https://www.liebertpub.com/doi/10.1089/ast.2019.2031
18. Groemer G, Soucek A, Frischauf N, Stumptner W, Ragonig C, Sams S et al (2014) The MARS2013 Mars analog mission. Astrobiology [Internet] 14 (5):360–376. http://www.liebertpub.com/doi/10.1089/ast.2013.1062
19. Willson D et al (2005) MARS-OZ: a design for a simulated Mars base in the Australian Outback. JBIS—J Br Interplanet Soc 58(9–10):282–293
20. Hargitai HI, Gregory HS, Osburg J, Hands D (2007) Development of a local toponym system at the Mars Desert Research Station. Cartogr Int J Geogr Inf Geovisualizat 42(2):179–187
21. Doule O (2015) Systems architecture complexity in definition of human spaceflight simulators, analogs and human spaceflight design process dependent on mission goals and strategy. In: AIAA SPACE 2015 conference and exposition [Internet]. American Institute of Aeronautics and Astronautics, Reston, Virginia. http://arc.aiaa.org/doi/10.2514/6.2015-4400
22. Ash R, Dowler W, Varsi G (1978) Feasibility of rocket propellant production on Mars. Acta Astronaut [Internet] 5(9):705–724. https://linkinghub.elsevier.com/retrieve/pii/0094576578900498
23. Sanders GB, Paz A, Oryshchyn L, Araghi K, Muscatello A, Linne DL et al (2015) Mars ISRU for production of mission critical consumables—options, recent studies, and current state of the art. In: AIAA SPACE 2015 conference and exposition [Internet]. American Institute of Aeronautics and Astronautics, Reston, Virginia. http://arc.aiaa.org/doi/10.2514/6.2015-4458
24. Zubrin RM, Muscatello AC, Berggren M (2013) Integrated Mars in situ propellant production system. J Aerosp Eng 26(1):43–56
25. Gayen P, Sankarasubramanian S, Ramani VK (2020) Fuel and oxygen harvesting from Martian regolithic brine. Proc Natl Acad Sci 117(50):31685-31689
26. Berg JIAHH (2019) Experimental configuration and preliminary results of testing a rapid cycle adsorption pump for Martian CO_2 acquisition. NASA Tech Rep 1–13
27. Brooks KP, Hu J, Zhu H, Kee RJ (2007) Methanation of carbon dioxide by hydrogen reduction using the Sabatier process in microchannel reactors. Chem Eng Sci 62(4):1161–1170
28. Jürgensen L, Ehimen EA, Born J, Holm-Nielsen JB (2015) Dynamic biogas upgrading based on the Sabatier process: thermodynamic and dynamic process simulation. Bioresour Technol 178:323–329
29. El Sibai A, Rihko-Struckmann L, Sundmacher K (2015) Synthetic methane from CO_2: dynamic optimization of the Sabatier process for power-to-gas applications. In: Computer aided chemical engineering. Elsevier B.V., pp 1157–1162
30. Muscatello A et al (2005) Integrated Mars in-situ propellant production system. In: Space resources roundtable VII (2005)
31. Li S, Jiang X (2014) Review and prospect of guidance and control for Mars atmospheric entry, vol 69, Progress in aerospace sciences. Elsevier Ltd, pp 40–57
32. Fratini E, Amendola R (2014) Caves and other subsurface environments in the future exploration of Mars: the absence of natural background radiation as biology concern. Rend Lincei [Internet] 25 (S1):91–96. http://link.springer.com/10.1007/s12210-013-0270-0
33. Golombek MP, Cook RA, Moore HJ, Parker TJ (1997) Selection of the Mars pathfinder landing site. J Geophys Res Planets 102(E2):3967–3988
34. Grant JA, Golombek MP, Grotzinger JP, Wilson SA, Watkins MM, Vasavada AR et al (2011) The science process for selecting the landing site for the 2011 mars science laboratory. In: Planetary and space science. Pergamon, pp 1114–1127
35. Grant JA, Golombek MP, Wilson SA, Farley KA, Williford KH, Chen A (2018) The science process for selecting the landing site for the 2020 Mars rover. Planet Space Sci [Internet] 164:106–126. https://linkinghub.elsevier.com/retrieve/pii/S0032063318301077
36. Fritts DC, Wang L, Tolson RH (2006) Mean and gravity wave structures and variability in the Mars upper atmosphere inferred from Mars Global Surveyor and Mars Odyssey aerobraking densities. J Geophys Res 111(A12):A12304

37. McCleese DJ, Heavens NG, Schofield JT, Abdou WA, Bandfield JL, Calcutt SB et al (2010) Structure and dynamics of the Martian lower and middle atmosphere as observed by the Mars Climate Sounder: seasonal variations in zonal mean temperature, dust, and water ice aerosols. J Geophys Res E Planets 115(12):E12016
38. Djachkova MV, Litvak ML, Mitrofanov IG, Sanin AB (2017) Selection of Luna-25 landing sites in the South Polar Region of the Moon. Sol Syst Res 51(3):185–195
39. De Rosa D, Bussey B, Cahill JT, Lutz T, Crawford IA, Hackwill T et al (2012) Characterisation of potential landing sites for the European Space Agency's Lunar Lander project. In: Planetary and space science
40. Jung J, Yi Y, Kim E (2014) Identification of Martian cave skylights using the temperature change during day and night. J Astron Sp Sci [Internet] 31(2):141–144. http://koreascience.or.kr/journal/view.jsp?kj=OJOOBS&py=2014&vnc=v31n2&sp=141
41. Leone G (2014) A network of lava tubes as the origin of Labyrinthus Noctis and Valles Marineris on Mars. J Volcanol Geotherm Res 277:1-8
42. Leone G (2016) Alignments of volcanic features in the southern hemisphere of Mars produced by migrating mantle plumes. J Volcanol Geotherm Res 309:78–95
43. Mustard JF, Murchie SL, Pelkey SM, Ehlmann BL, Milliken RE, Grant JA et al (2008) Hydrated silicate minerals on Mars observed by the Mars Reconnaissance Orbiter CRISM instrument. Nature [Internet] 454(7202):305–309. http://www.nature.com/articles/nature07097
44. Kim MHY, De Angelis G, Cucinotta FA (2011) Probabilistic assessment of radiation risk for astronauts in space missions. Acta Astronaut 68(7–8):747–759
45. Elgart SR, Little MP, Chappell LJ, Milder CM, Shavers MR, Huff JL et al (2018) Radiation exposure and mortality from cardiovascular disease and cancer in early NASA astronauts. Sci Rep 8 (1):8480
46. Chancellor J, Scott G, Sutton J (2014) Space radiation: the number one risk to astronaut health beyond low earth orbit. Life 4(3): 491-510. https://doi.org/10.3390/life4030491
47. Hassler DM, Zeitlin C, Wimmer-Schweingruber RF, Ehresmann B, Rafkin S, Eigenbrode JL et al (2014) Mars' surface radiation environment measured with the Mars Science Laboratory's Curiosity Rover. Science (80-) [Internet] 343(6169):1244797. https://www.sciencemag.org/lookup/doi/10.1126/science.1244797
48. National Research Council (2008) Managing space radiation risk in the new era of space exploration [Internet]. National Academies Press, Washington, DC. http://www.nap.edu/catalog/12045
49. Wilson JW, Anderson BM, Cucinotta FA, Ware J, Zeitlin CJ (2006) Spacesuit radiation shield design methods. In: 36th international conference on environmental systems, pp 277–293
50. Valentin J (2007) The 2007 recommendations of the International Commission on Radiological Protection. ICRP. Elsevier, Oxford
51. Gaier JR (2005) The effects of lunar dust on EVA systems during the Apollo missions (No. NASA/TM—2005-213610). NASA, Cleveland, Ohio
52. Levine JS, Winterhalter D, Kerschmann RL (2018) Dust in the atmosphere of Mars and its impact on human exploration. Cambridge Scholars Publishing, 293 pp
53. NASA/JPL-Caltech (2018) PIA22737: Mars climate sounder studies 2018 dust storm [WWW Document] [Internet]. https://photojournal.jpl.nasa.gov/catalog/PIA22737
54. Rucker MA (2017) Dust storm impacts on human Mars mission equipment and operations. In: Dust in the atmosphere of Mars and its impact on human exploration workshop
55. Sim PA (2017) Martian dust and its interaction with human physiology: an emergency physician's perspective. In: Dust in the atmosphere of Mars and its impact on human exploration workshop
56. Sonter MJ (1997) The technical and economic feasibility of mining the near-earth asteroids. Acta Astronaut [Internet] 41(4–10):637–647. https://linkinghub.elsevier.com/retrieve/pii/S0094576598000873
57. Leone G (2020) The absence of an ocean and the fate of water all over the Martian history. Earth Sp Sci 7(4):e2019EA001031
58. Wilson L, Head JW (1994) Mars: review and analysis of volcanic eruption theory and relationships to observed landforms. Rev Geophys [Internet] 32(3):221. http://doi.wiley.com/10.1029/94RG01113
59. Peterson DW, Holcomb RT, Tilling RI, Christiansen RL (1994) Development of lava tubes in the light of observations at Mauna Ulu, Kilauea Volcano, Hawaii. Bull Volcanol [Internet] 56(5):343–360. http://link.springer.com/10.1007/BF00326461
60. Valerio A, Tallarico A, Dragoni M (2008) Mechanisms of formation of lava tubes. J Geophys Res Solid Earth [Internet] 113(B8). http://doi.wiley.com/10.1029/2007JB005435
61. Greeley R (1971) Lava tubes and channels in the lunar Marius Hills. Moon 3(3):289–314
62. Léveillé RJ, Datta S (2010) Lava tubes and basaltic caves as astrobiological targets on Earth and Mars: a review. Planet Space Sci 58(4):592–598
63. Sauro F, Pozzobon R, Massironi M, De Berardinis P, Santagata T, De Waele J (2020) Lava tubes on Earth, Moon and Mars: a review on their size and morphology revealed by comparative planetology. Earth-Science Rev [Internet] 209

(July):103288. https://doi.org/10.1016/j.earscirev.2020.103288
64. López-Martínez R, Barragán R, Beraldi-Campesi H, Lánczos T, Vidal-Romaní J, Aubrecht R et al (2016) Morphological and mineralogical characterization of speleothems from the Chimalacatepec lava tube system, Central Mexico. Int J Speleol [Internet] 45(2):111–122. http://scholarcommons.usf.edu/ijs/vol45/iss2/2/
65. Theinat AK, Modiriasari A, Bobet A, Melosh HJ, Dyke SJ, Ramirez J et al (2020) Lunar lava tubes: morphology to structural stability. Icarus [Internet] 338:113442. https://doi.org/10.1016/j.icarus.2019.113442
66. Atkinson A, Griffin TJ, Stephenson PJ (1975) A major lava tube system from Undara Volcano, North Queensland. Bull Volcanol [Internet] 39 (2):266–293. http://link.springer.com/10.1007/BF02597832
67. Boston PJ, Spilde MN, Northup DE, Melim LA, Soroka DS, Kleina LG et al (2001) Cave biosignature suites: microbes, minerals, and mars. Astrobiology [Internet] 1(1):25–55. http://www.liebertpub.com/doi/10.1089/153110701750137413
68. McEwen AS, Eliason EM, Bergstrom JW, Bridges NT, Hansen CJ, Delamere WA et al (2007) Mars Reconnaissance Orbiter's High Resolution Imaging Science Experiment (HiRISE). J Geophys Res [Internet] 112(E5):E05S02. http://doi.wiley.com/10.1029/2005JE002605
69. Malin MC, Bell JF, Cantor BA, Caplinger MA, Calvin WM, Clancy RT et al (2007) Context camera investigation on board the Mars Reconnaissance Orbiter. J Geophys Res [Internet] 112(E5): E05S04. http://doi.wiley.com/10.1029/2006JE002808
70. Leone G (2017) Mangala Valles, Mars: a reassessment of formation processes based on a new geomorphological and stratigraphic analysis of the geological units. J Volcanol Geotherm Res 337:62–80
71. Cushing GE (2017) U. SGSu. Mars Global Cave Candidate Catalog (MGC3). Astrobiol Sci Cont 2017 2017(1965):86001
72. Fairén AG, Dohm JM, Uceda ER, Rodríguez AP, Baker VR, Fernández-Remolar D et al (2005) Prime candidate sites for astrobiological exploration through the hydrogeological history of Mars. Planet Space Sci 53(13):1355–1375
73. Leovy CE, Zurek RW, Pollack JB (1973) Mechanisms for Mars dust storms. J Atmos Sci [Internet] 30(5):749–762. http://journals.ametsoc.org/doi/abs/10.1175/1520-0469%281973%29030%3C0749%3AMFMDS%3E2.0.CO%3B2
74. Wilson AP, Genge MJ, Krzesińska AM, Tomkins AG (2019) Atmospheric entry heating of micrometeorites at Earth and Mars: implications for the survival of organics. Meteorit Planet Sci [Internet] 54(9):1–19. https://onlinelibrary.wiley.com/doi/abs/10.1111/maps.13360
75. Williams KE, McKay CP, Toon OB, Head JW (2010) Do ice caves exist on Mars? Icarus [Internet] 209(2):358–368. https://linkinghub.elsevier.com/retrieve/pii/S0019103510001491
76. Lee P (2020) Ice-rich caves on the Moon and Mars: prospects and pragmatic recommendations for exploration. 3rd Int Planet Caves Conf 2197 (2197):1066
77. Spanovich N, Smith MD, Smith PH, Wolff MJ, Christensen PR, Squyres SW (2006) Surface and near-surface atmospheric temperatures for the Mars Exploration Rover landing sites. Icarus [Internet] 180(2):314–320. https://linkinghub.elsevier.com/retrieve/pii/S0019103505003313
78. Banfield D, Spiga A, Newman C, Forget F, Lemmon M, Lorenz R et al (2020) The atmosphere of Mars as observed by InSight. Nat Geosci [Internet] 13(3):190–198. http://www.nature.com/articles/s41561-020-0534-0
79. Schörghofer N, Businger S, Leopold M (2018) The coldest places in Hawaii: the ice-preserving microclimates of high-altitude craters and caves on tropical island volcanoes. Bull Am Meteorol Soc [Internet] 99(11):2313–2324. https://journals.ametsoc.org/bams/article/99/11/2313/70246/The-Coldest-Places-in-Hawaii-The-IcePreserving
80. Richardson MI, Wilson RJ, Rodin AV (2002) Water ice clouds in the Martian atmosphere: General circulation model experiments with a simple cloud scheme. J Geophys Res E Planets 107(9):1–29
81. Benson JL, James PB, Cantor BA, Remigio R (2006) Interannual variability of water ice clouds over major martian volcanoes observed by MOC. Icarus 184(2):365–371
82. De Angelis G, Wilson JW, Clowdsley MS, Nealy JE, Humes DH, Clem JM (2002) Lunar lava tube radiation safety analysis. J Radiat Res 43 (Suppl):41–45
83. Mari N, Riches AJV, Hallis LJ, Marrocchi Y, Villeneuve J, Gleissner P et al (2019) Syneruptive incorporation of martian surface sulphur in the nakhlite lava flows revealed by S and Os isotopes and highly siderophile elements: implication for mantle sources in Mars. Geochim Cosmochim Acta 266:416-434
84. Perkins S (2020) Core concept: lava tubes may be havens for ancient alien life and future human explorers. Proc Natl Acad Sci [Internet] 117 (30):17461–17464. http://www.pnas.org/lookup/doi/10.1073/pnas.2012176117
85. von Ehrenfried M (2019) "Dutch." Terrestrial cave and lava tube research. In: From cave man to cave Martian [Internet]. Springer International Publishing, Cham, pp 49–81. http://link.springer.com/10.1007/978-3-030-05408-3_4
86. Maltagliati L, Titov DV, Encrenaz T, Melchiorri R, Forget F, Garcia-Comas M et al (2008) Observations of atmospheric water vapor above the Tharsis volcanoes on Mars with the OMEGA/MEx imaging spectrometer. Icarus 194(1):53–64

87. Moudden Y, Forbes JM (2009) Mars W cloud: evidence of nighttime ice depositions. Geophys Res Lett 36(14):1–5
88. Sowers GF, Dreyer CB (2019) Ice mining in lunar permanently shadowed regions. New Sp 7(4):235–244
89. Whittaker W (2012) Technologies enabling exploration of skylights, lava tubes and caves. NASA, US, Report, no NNX11AR42G
90. Blair DM, Chappaz L, Sood R, Milbury C, Bobet A, Melosh HJ et al (2017) The structural stability of lunar lava tubes. Icarus [Internet] 282:47–55. https://doi.org/10.1016/j.icarus.2016.10.008
91. Wagner RV, Robinson MS (2014) Distribution, formation mechanisms, and significance of lunar pits. Icarus [Internet] 237:52–60. https://doi.org/10.1016/j.icarus.2014.04.002
92. Huerta P, Martín-Pérez A, Martín-García R, Rodríguez-Berriguete Á, La Iglesia Fernández Á, Alonso-Zarza AM (2019) Gypsum speleothems in lava tubes from Lanzarote (Canary Islands). Ion sources and pathways. Sediment Geol [Internet] 383:136–147. https://linkinghub.elsevier.com/retrieve/pii/S0037073819300375
93. Forget F, Pollack JB (1996) Thermal infrared observations of the condensing Martian polar caps: CO_2 ice temperatures and radiative budget. J Geophys Res Planets [Internet] 101(E7):16865–16879. http://doi.wiley.com/10.1029/96JE01077
94. Head JW, Mustard JF, Kreslavsky MA, Milliken RE, Marchant DR (2003) Recent ice ages on Mars. Nature [Internet] 426(6968):797–802. http://www.nature.com/articles/nature02114
95. Wise DU, Golombek MP, McGill GE (1979) Tharsis province of Mars: geologic sequence, geometry, and a deformation mechanism. Icarus [Internet] 38(3):456–472. https://linkinghub.elsevier.com/retrieve/pii/0019103579902008
96. Wait KW, Goldfarb M (2014) A pneumatically actuated quadrupedal walking robot. IEEE/ASME Trans Mechatronics [Internet] 19(1):339–347. http://ieeexplore.ieee.org/document/6410036/
97. Elfes A, Steindl R, Talbot F, Kendoul F, Sikka P, Lowe T et al (2017) The Multilegged Autonomous eXplorer (MAX). Proc—IEEE Int Conf Robot Autom 1050–1057
98. Raibert M, Blankespoor K, Nelson G, Playter R (2008) BigDog, the Rough-Terrain Quadruped Robot [Internet]. IFAC Proc Vol. IFAC 41:10822–10825. https://doi.org/10.3182/20080706-5-KR-1001.01833
99. Michael K (2012) Meet Boston dynamics' LS3—the latest robotic war machine. Faculty of Engineering and Information Sciences - Papers: Part A. 2773. https://ro.uow.edu.au/eispapers/2773
100. Brozovsky J (2013) Comparsion of compressive strengths of concrete testing by different of sclerometers. Procedia Eng [Internet] 65:254–259. https://linkinghub.elsevier.com/retrieve/pii/S1877705813015464
101. Ximenes SW, Elliott JO, Bannova O (2012) Defining a mission architecture and technologies for lunar lava tube reconnaissance. Earth Sp 2012—Proc 13th ASCE Aerosp Div Conf 5th NASA/ASCE Work Granul Mater Sp Explor 344–354
102. Manzey D (2004) Human missions to Mars: new psychological challenges and research issues. Acta Astronaut [Internet] 55(3–9):781–790. https://linkinghub.elsevier.com/retrieve/pii/S0094576504001705
103. Bard AJ, Fox MA (1995) Artificial photosynthesis: solar splitting of water to hydrogen and oxygen. Acc Chem Res [Internet] 28(3):141–145. https://pubs.acs.org/doi/abs/10.1021/ar00051a007
104. Kempe S (2012) Volcanic rock caves. In: Encyclopedia of caves [Internet]. Elsevier, pp 865–873. https://linkinghub.elsevier.com/retrieve/pii/B9780123838322001250
105. Kempe S (2019) Volcanic rock caves. In: Encyclopedia of caves [Internet]. Elsevier, pp 1118–1127. https://linkinghub.elsevier.com/retrieve/pii/B978012814124300131X
106. Kauahikaua J, Cashman V, Mattox N, Christina C, Hon A, Mangan T et al (1998) Observations on basaltic lava streams in tubes from Kilauea Volcano, island of Hawai'i. J Geophys Res 103:303–323
107. Hróarsson B, Jónsson SS (1991) Lava caves in the Hallmundarhraun lava flow, Western Iceland. Proc 6th Int Symp Vulcanospeleol 85–88
108. Wilkens H, Iliffe TM, Oromí P, Martínez A, Tysall TN, Koenemann S (2009) The Corona lava tube, Lanzarote: geology, habitat diversity and biogeography. Mar Biodivers [Internet] 39(3):155–167. http://link.springer.com/10.1007/s12526-009-0019-2
109. Calvari S, Pinkerton H (1999) Lava tube morphology on Etna and evidence for lava flow emplacement mechanisms. J Volcanol Geotherm Res [Internet] 90(3–4):263–280. https://linkinghub.elsevier.com/retrieve/pii/S0377027399000244
110. Boston PJ, Frederick RD, Welch SM, Werker J, Meyer TR, Sprungman B, Hildreth-Werker V, Thompson SL Murphy, DL (2003) Human utilization of subsurface extraterrestrial environments. Gravitat Sp Biol Bullett 16(2):121–131
111. Bessone L, Sauro F, Maurer M, Piens M (2018) Testing technologies and operational concepts for field geology exploration of the Moon and beyond: the ESA PANGAEA-X campaign. Geophys Res Abstr 20

112. Gasser M, Dunn MC (2018) Moon and Mars habitation in lava tubes: the first explorers will be cave men again. Stefánshellir test site in Iceland. In: European planetary science congress 12.
113. Nakamura Y (2005) Farside deep moonquakes and deep interior of the Moon. J Geophys Res [Internet] 110(E1):E01001. http://doi.wiley.com/10.1029/2004JE002332
114. Arya AS, Rajasekhar RP, Thangjam G, Ajai, Kiran Kumar AS (2011) Detection of potential site for future human habitability on the Moon using Chandrayaan-1 data. Curr Sci 100(4):524–529
115. Kerber L et al (2018) Moon diver: a discovery mission concept for understanding the history of the mare basalts through the exploration of a lunar mare pit. In: 49th lunar and planetary science conference
116. Transeth AA, Leine RI, Glocker C, Pettersen KY, Liljeback P (2008) Snake robot obstacle-aided locomotion: modeling, simulations, and experiments. IEEE Trans Robot [Internet] 24(1):88–104. http://ieeexplore.ieee.org/document/4456759/

Future Perspectives of Mars After the Latest Studies

Giovanni Leone, Caitlin Ahrens, Giuseppe Calabrese, Hiroyuki K. M. Tanaka, and Piero D'Incecco

Abstract

Although we have seen that Mars has little (if not) water for a human colonization on a large scale, it still offers many opportunities for future explorations. These are related to the potential richness of its minerals, both on the surface and its subsurface, and a challenging environment where to test technological solutions for the exploitation of inhospitable worlds. We explore new methodologies for a drier extraction of minerals, with respect to current methods used in the mining industry on Earth, and existing concepts of human colony based on in-situ resource utilization (ISRU). 3D printing techniques will allow construction of buildings from the volcanic materials available on the surface of Mars.

1 Introduction

After more than two decades of fluvial Mars, we go back to the original NASA views of a dry and volcanic Mars returned by the Mariner missions and supported by the various studies and data mentioned in this book. This return to the original (and realistic) view does not necessarily mean that we must give up the exploration of Mars as it happened for twenty years after the Viking missions. The inevitable dismay following the discovery of an inhospitable world that likely never hosted life and never had running water on its surface should not prevent us from further exploring it. We only need to adjust the way we want to do it because current NASA plans for the exploration of Mars are still based on "follow the water" [1–4], a >20 years misconception related to the believed presence of past surface water [5], which has already been widely discussed in chapter "History of Scientific Studies and Current Views of Mars". At last, despite several papers still claiming past oceans and rivers

G. Leone (✉)
Instituto de Investigacion en Astronomía y Ciencias Planetarias, Universidad de Atacama, Copiapó, Chile
e-mail: giovanni.leone@uda.cl

C. Ahrens
NASA Goddard Space Flight Centre, Greenbelt, MD, USA

G. Calabrese
Space and Moon Habitat Architecture, Sydney, Australia

G. Leone · H. K. M. Tanaka
Virtual Muography Institute, Global, Tokyo, Japan

H. K. M. Tanaka
International Muography Research Organization (MUOGRAPHIX), The University of Tokyo, Tokyo, Japan

H. K. M. Tanaka
Earthquake Research Institute, The University of Tokyo, 1-1-1 Yayoi, Bunkyo, Tokyo 113-0032, Japan

P. D'Incecco
Dipartimento di Ingegneria e Geologia (INGEO), Universitá D'Annunzio, Chieti, Italy

G. Leone (ed.), *Mars: A Volcanic World*,
https://doi.org/10.1007/978-3-030-84103-4_12

on Mars [6–8], we have now realized how the Earth is (and was) the only planetary body with liquid water running on its surface in the inner solar system and how valuable is this water [5]. Clearly, if someday life and water will be discovered in the underground of Mars, we will all be happy, but the search for life should not be the priority of the space agencies or the main reason that justifies the public funding for planetary exploration. There is still a wealth of minerals on Mars that will be particularly useful once those on Earth will be inexorably depleted in a few hundreds of years or even less, according to available estimates [9–12]. Understanding the Martian regolith would also help expand other branches of planetary science, such as muography to investigate the Martian atmospheric environment of the past (see Sect. 3.2). Recycling of rare earths elements (REE) might play a key role in lowering the geopolitical consequences of a shortage in the supply chain [13–15] but will not solve the problem in the long term [10–12]. New sources must be found and mined responsibly to protect the environment [15].

Looking for new sources in space is an attractive and intriguing idea but going to Mars opens an assortment of problems. The big challenge is related to the scarcity of water, the lack of protection from radiations, the operations and the health of the astronauts in a lower (than Earth) gravity environment, the logistics of going back and forth from our terrestrial or lunar or even asteroidal bases, just to mention a few. Overcoming these problems will turn out into an opportunity for the development of new technologies that will be very useful on Earth as well as on Mars. Saving water for all the populations on Earth who suffer drought and for all our daily uses has now become a must. All the economic sectors will have an advantage in saving water but for the space industry is of fundamental importance because carrying water from Earth in the various human missions planned is terribly expensive and certainly not sustainable in the long term. The mining industry will also benefit from new technologies that will extract minerals saving valuable water, as well agriculture will be modernized with new systems of irrigation, and so on with many other economic sectors. We must think that all the human activities carried out on Earth will be needed on Mars and on other planetary bodies as well. This requires an accurate planning of the operations based on the real situation of the resources present on Mars. For this reason, there is still a lot of exploration that we must do to understand where the valuable resources are located on Mars.

2 Human Exploration of Martian Volcanism

Given complications of cost and risk, explorations on Mars have previously and currently been achieved by robots with the human input made from Earth. Eventually, humans will re-enter into the focus of exploration, especially with the goal of investigating the questions regarding a vast and diverse Mars. To answer such questions, new technologies would be invaluable (as discussed in this chapter). The Human Exploration of Mars Science Analysis Group (HEM-SAG), now a part of the Mars Exploration Program Analysis Group (MEPAG) has discussed several fundamental (geological) questions, specifically the volcanic history of the planet and the present volcanic activity on Mars today.

As duly noted throughout this book, the complexity of Martian volcanism has wide-ranging implications regarding not only the planet Mars, but also the nature and feasibility of future human exploration and technologies (e.g., 3D printing, mining exploration, biodomes). Thus, none of these exploratory questions could be answered by a single landing site. The diversity of surface morphology and diverse volcanic formations (including lava plains and post-volcanic morphologies) demands an array of landing sites that potentially span the various geologic histories (and volcanic processes) that are representative of Martian geologic history.

To evaluate the pyroclastic processes, their distribution, igneous evolution, and characterization of mineralogical composition of the crust (as observed along the Martian dichotomy), geologic approaches would involve drilling, in-situ composition analyses, and traverse sampling. Human technologies on the Martian surface, especially at volcanic regions, can prove useful not only from a geologic standpoint, but also from an economic point of view (e.g., mining resources). As noted in several chapters, the variety of mineralogy found in many volcanic regions would certainly be a priority for mining endeavors. If not for excavation purposes, the basaltic rocks can also be useful in potential 3D printing as a strong cementation resource.

2.1 Potential Landing Sites

As a preliminary analysis, we specify 11 distinct volcanic regions that would potentially benefit the presence of human explorers. These are also potential priority landing sites from reports by MEPAG [16].

- *Jezero Crater* (18.4°N, 77.7°E): this 45-km crater on the NW margin of the Isidis impact basin in the Nili Fossae region of Mars. This Noachian region is important for understanding the formation of the basin and subsequent volcanism (and modification thereof). To the southwest are Hesperian lava flows from Syrtis Major. Sampling from Noachian versus Hesperian- type lava sediments would prove useful in gaining insight to the evolution of magma composition.
- *Mangala Valles* (11.32°S, 208.61°E): a Noachian-aged valley that emanates from the Arsia Mons volcanic field.
- *Arsia Mons* (4.8°S, 126°W): the broadest lava fields are found on Arsia Mons formed during the Noachian period. This area also could provide insight into late-stage volcanism and dike processes, especially at the northwest flank.
- *Gusev Crater—Columbia Hills* (14.6°S, 175.4°E): crater infilling of volcanic deposits lying along the crustal dichotomy provides a wealth of information on the crustal evolution and potential clues as to the Hesperian ridged plains volcanism.
- *Eastern Olympus Mons* (17.7°N, 128.2°W): old volcanic and tectonic activity, perhaps within the first few hundreds of millions of years. Certainly, the substantial amount of lava flows observed around Olympus Mons would prove useful for lava processes and mineralogical evolution. The caldera of Olympus Mons would also be a priority in discovering caldera processes and subsurface volcanic interactions.
- *Elysium Planitia* (5.0°N, 150°E): Late Noachian volcanic lava flows with platy units would be an interesting geological investigation into lava morphology and subsequent processes.
- *Alba Patera* (40.7°N, 109.6°W): such a diverse volcanic region with young (Late Hesperian) mantle deposits and ancient (Noachian) volcanic valley networks on the northern flanks.
- *Medusae Fossae* (1.6°N, 173.2°W): especially in areas where the Medusae Fossae Formation and Tharsis lava flows interconnect. This could establish some interesting views on varying volcanic processes and subsequent interactions.
- *Valles Marineris* (6.2°S, 70°W): the sulfate-rich deposits would potentially only be accessible by human exploration, which could give insight to the volcanic mineralogic evolution of this volcanic valley.
- *Syrtis Major Planum* (7.0°N, 69.0°E): Not only would this area help confirm possible SNC meteorite ejection localities, but also provide clues to the silicate-rich Hesperian flows and deposits within the calderas.
- *Hecates Tholus* (32.0°N, 150.3°E): offers a unique view of volcanic structures on Mars, including the concentration of valley networks from extensive Noachian volcanic activity, though there are potential interactions from Hesperian activity.

We admit that this list is not exhaustive and that indeed a large array of landing sites around Mars would be ideal. From this list alone, plus the chapters presented in this book, one can see the diversity (and therefore challenge) of exploring these volcanic regions for a variety of geologic (and technological) research.

3 Technologies for the Earth and the Moon That Could Be Applied to Mars

3.1 Mineralogical Technologies

The future colonization of Mars will be very challenging, especially given the distance of the red planet from the Earth and its adverse surface environment conditions. For this reason, the Moon will be used as a "launch pad", to test the necessary facilities and technologies for a sustainable human habitat outside of Earth.

Several space missions are currently focused on the exploration of the Moon, such as the NASA ARTEMIS program [17, 18] to look for the most suitable sites to build a human base. The short-term main objective of these missions is that of detecting regions relatively richer in water ice and other exploitable mineral resources (i.e., iron oxides, methane, deuterium) which are crucial for the self-sustainment of a future human colony on the Moon. In a long-term perspective, the technologies developed to colonize the Moon will be then used for the colonization of Mars.

The Chandrayaan-1 Moon Mineralogy Mapper (M^3) instrument has recently identified a number of clusters of so-called ice exposure points in the permanently shadowed regions (PSRs) of the south pole area of the Moon [19]. The PSRs are areas which are never reached by the sunlight, so that the interstitial ice eventually contained in the lunar regolith wouldn't have the chance to sublimate [e.g., 20]. Even though the M^3 instrument could only detect the spectral signatures of the first few millimeters of lunar surface, chances are hopefully high to find certain quantities of water ice in depth. However, some studies also investigated the possibility to extract water by reducing the ilmenite contained in the lunar regolith [e.g., 21]. Recently, it has been further demonstrated how it is possible to contemporaneously extract oxygen and produce metal alloys from the lunar regolith through the metalysis "Fray, Farthing, Chen" (FFC) approach (Fig. 1) [22]. This methodology of oxygen extraction from the surface regolith, as well as other metalysis processes, can be also used for the future colonization of Mars.

One of the greatest challenges that future colonizers will have to face is that of minimizing the negative effect of the cosmic radiation and the lower gravity on the human body. As mentioned in this chapter, lava tubes may constitute a natural shelter, as they can at least partially act as a shield against the cosmic rays, but they cannot solve the problem of the lower gravity.

However, when thinking of a future human colony on a planetary body, it is very important to select the right site on the basis of the availability of the resources (e.g., water, minerals) which are essential for the self-sustainment of the colony, rather than on the presence of natural shelters. For this reason, building an underground facility using borehole excavating machines directly where those resources are available would constitute a more advantageous solution. Preliminary tests of mineral extraction,

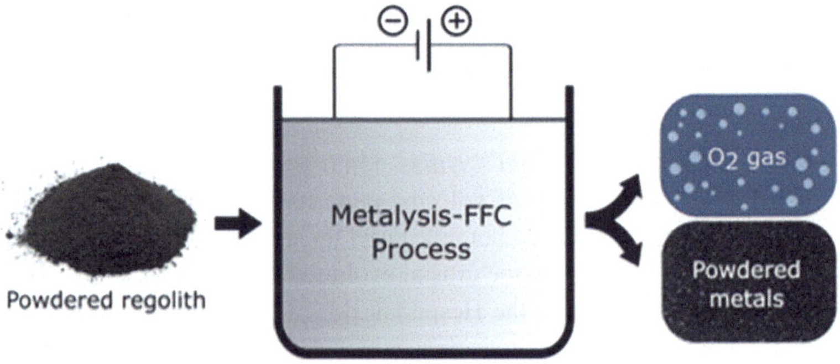

Fig. 1 Graphical illustration of the metalysis FFC process to extract oxygen and metal alloys from the powdered regolith. Figure from Lomax et al. [22]

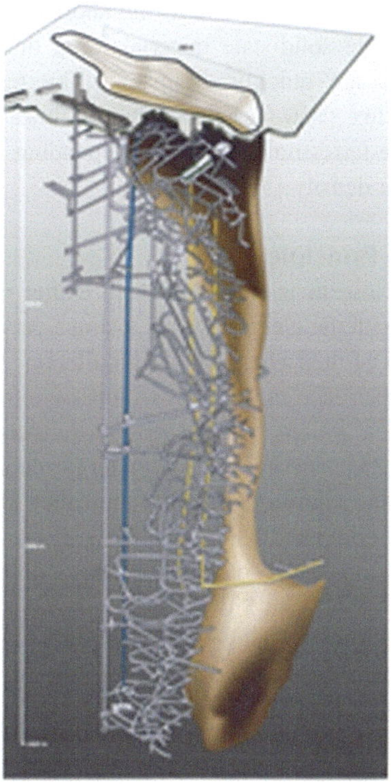

Fig. 2 Illustrated cross-section of the CallioLab underground facility with available spaces/caverns at several depths all the way to the main level at 1400 m (*Source* www.calliolab.com)

lunar and martian simulants, and personnel habitability in Moon-like simulated environment will be performed at CallioLab, an underground facility located in Finland, in the Pyhäsalmi Mine, a former copper and zinc mine (Fig. 2). Underground facilities like CallioLab will allow simulations to further colony-related field activities on the surface of extraterrestrial environments such as those of the Moon and Mars, but also constitute a possible example of how the structure of a future human colony on the Moon and Mars should be arranged.

3.2 Technologies with Muography

Muography, the technique of using high-energy (relativistic) muons as a radiographic probe, can be used to image the internal structure of hectometric to kilometric-scale objects on the Earth. The term muography literally means "muon rendering" in ancient Greek. While photography utilizes photos (the word for "light" in ancient Greek), muography utilizes the characteristics of relativistic muons. The way that muography works can be explained by using the analogy of medical radiographic imaging that also utilizes relativistic particles as probes. Muons have a stronger penetration power than X-rays. Muons can therefore be used to generate similar images of the inside structure of objects with much larger proportions than the human body, showing various densities as shadows.

An application of cosmic ray muons to the investigation of environmental sciences on Mars was first proposed in 2007 [23]. The muon, also indicated by the Greek letter μ, is a charged lepton, which is free from strong force interaction but sensitive to electromagnetic force. This particle has the same properties as an electron, but its mass is approximately 200 times larger than the electron. Therefore, the possibilities to produce photons are suppressed in comparison to lighter particles, and as a consequence, relativistic muons have a stronger penetration power.

On the Earth, high-energy (relativistic) muons are produced continuously via the interactions of galactic cosmic-ray (GCR) particles with the upper atmosphere of the Earth. The GCRs collide with atmospheric nuclei such as nitrogen or oxygen to produce mesons such as pions and kaons that eventually decay into muons. These muons can be easily detected with suitable instruments and they have a stronger penetration power than X-rays. These muons are called cosmic-ray muons. Expressions for the differential cosmic-ray muon spectrum on the Earth at various zenith angles have been proposed by several researchers [24–27], and they have been validated and improved by comparison with the spectrum of muons taken from experiments performed at sea level [25, 28–31].

GCRs also arrive at other planets, comets and asteroids. On these stellar bodies, muons are generated either in the atmosphere or within the planet itself, depending on the existence and thickness of the atmosphere. Tanaka [23] took advantage of the penetration strength of the cosmic ray muons to investigate the Martian atmospheric environment of the past. The muon arrival rate had been reflected by the modulation of atmospheric pressures on the Martian surface in the past, and is possibly recorded underneath the Martian regolith as the muogenic radionuclide, which has sufficient concentration for current mass spectroscopy.

On the Earth, the muon flux reaches its maximum ($\sim$200 m^{-2} $s^{-1}sr^{-1}$) at an atmospheric depth of 300 g cm^{-2}, and then slowly decreases as the muons pass through the additional atmospheric depth. As a consequence, the muon flux is $\sim$90 m^{-2} $s^{-1}sr^{-1}$ at sea level. However, on Mars, the surface atmospheric pressure is only $\sim$1/1000 that of the Earth, and thus, many of the primary particles reach the surface before undergoing a primary interaction. This results in both a lower meson production rate and a lower meson decay rate than in the Earth's atmosphere. Tanaka [23] indicated that based on the steep power law dependence in the early stages of the shower cascade, the meson flux would be strongly dependent on the atmospheric pressure, zenith angle and muon energy. As a consequence, the horizontal muon rate is higher on Mars than the vertical muon rate due to the additional atmospheric mass through which the muons transverse. Considering an horizontal atmospheric depth of Mars of 100 g cm^{-2}, a sufficient number of mesons are generated in the horizontal direction and their energies are sufficiently high because the probability of their decay before further interactions is higher than it would be on Earth. This is in contrast to the muon generation on Earth. (On the Earth, the horizontal muon rate is suppressed compared to the vertical muons because there is an appreciable probability that pions will collide with atmospheric nuclei before decaying into muons, and at larger angles (from the zenith) low-energy muons decay before reaching the Earth's surface).

Even if there is no atmosphere on a planetary body, muons are generated. GCRs directly collide with the surface and generate muons inside it. The average energy of these muons is much lower than the energy of the atmospheric muons because the secondary mesons are more copiously produced, due to the short mean free path inside solid-state materials, and thus their energies are quickly dissipated. The resultant muon flux (>50 GeV) is 10^3–10^4 times lower than the terrestrial muon flux, depending on the medium density [32].

3.2.1 Principle

Muons lose their energy through matter by colliding with the electrons, thus losing a very small fraction of their energy in each collision. Muons also experience the radiative processes. In these processes, muons lose a large but random fraction of their energy. For high-energy muons (>1 TeV), the contribution of the ionization process is small compared with the other three processes. For example, the radiative process dominates above 708 GeV in SiO_2 [33]. For the calculation of the analytical range for high-energy muons, the energy loss relationship presented by Adair and Kasha [34] is convenient:

$$\frac{dE}{dx} = \left[1.88 + 0.77\ln\left(\frac{E}{M}\right) + bE\right] \times 10^{-6}\,[\mathrm{TeVg^{-1}cm^2}] \qquad (1)$$

where M is the muon mass, E is the muon energy, x is the densimetric thickness (path length times average density along the muon path) [35] along the path traversed by the muon, and the energy-loss coefficient $b = 3.9$ (for $E = 1$ TeV) is obtained by adding the probabilities per unit thickness for the three radiative processes and integrating it over the energy fraction that a muon loses. The first two terms represent the energy loss by ionization process and the third term represents the loss by radiative processes.

Since the range is a function of the incident muon energy, it can be incorporated in the open-sky muon spectrum. Once both muon path length and average density along the path are known, the densimetric thickness (x) can be calculated by multiplying them, and thus the minimum energy (E_c) of muons that can penetrate through a material with this thickness can be determined using Eq. (1). By integrating the open-sky spectrum $I(E, \theta)$ from E_c to infinity, we obtain the integrated muon intensity $N(E_c)$, which represents the number of muons that have enough energy to escape from the target of interest:

$$N(E_c, \theta) = \int_{E_c}^{\infty} I(E, \theta) dE \qquad (2)$$

where $I(E, \theta)$ is the zenith-angle-dependent open-sky muon energy spectrum. While the muon spectrum on the Earth has been well measured at various zenith angles, we need to speculate the spectrum on Mars by utilizing the Monte-Carlo-based extended air shower simulations [23].

3.2.2 Muography on Mars

Although the concept of muography has existed for decades [36], recent improvements in detector technology and target matching have made it now usable reliably for subsurface imaging of the Earth. The first volcano image was captured in 2007 [23] by showing the empty space underneath the crater floor and subsequently bubbly magma was imaged in 2009 [37].

Muography offers a new possibility for planetary exploration, providing a new type of measurement to interpret composition and evolution of the Martian surface and to detect potentially habitable subsurface structure. The technique has the potential to reveal the internal geological structures located on the surface of Mars to address geologic history, climate, and biologic possibilities. Kedar et al. [38] proposed the following possible Martian targets that can be studied with muograpy: (1) rootless cones on platy lava flows in Elysium Planitia (HiRISE); (2) enigmatic edifices in Hydraotes region that may be rootless, or connected to a vent system (HiRISE); (3) chains of collapse pit on Ascraeus Mons, areas between pit chains may contain lava tubes or caves in the subsurface; (4) complex terrain of collapse structures on the lower flanks of Arsia Mons, this structure likely contains lava tubes and caves in the subsurface [39, 40]. Muography could be used to map the internal structure of these objects by combining high-resolution digital topography data.

However, the following themes have to be investigated and overcome to realize Martian muography: (A) landing site, a landing site for muography cannot be easily chosen, some of the most interesting geological targets on the Martian surface are located at high altitudes; (B) temperature, Mars average temperature is below −50 °C, the detector and detector electronics must operate at low temperatures; (C) radiation, an extensive shielding is required to avoid degradation of detector performance, the background events that degrade the quality of muographic images can be originated from non-muon charged particles. Primary GCRs produce not only muons but also cascades of protons, neutrons, mesons, and other electromagnetic particles. While the number of these particles rapidly decreases as they travel in the atmosphere of the Earth, due to the thin atmosphere on Mars most of them arrive the ground surface of Mars. These backgrounds substantially degrade the image contrast. In particular, the protons are heavier than muons; hence longer MFP is present within the material ($\sim$10 cm in Pb), and thus could be a serious source of the background. The electromagnetic components such as electrons and gammas can be rejected relatively simply by requesting a linear trajectory through multiple layers of detectors and radiation shields. The radiation length of electrons is less than 6 mm in Pb. However, (D) on the Earth there is no mass constraints for detectors, but the detector weight is a critical condition in any Mars lander mission. (E) Power consumption. Muography does not require an active power source. However, at least a few watts of power are required for operating the detectors in the current design, the further power reduction is an essential for a Mars mission. (F) Mechanical strength. The detector must be designed to physically tolerate Entry, Descent, and Landing (EDL) processes. Recent improvements in navigation and EDL technology as demonstrated by the Mars Science Laboratory (MSL) [41] will provide a more practical solution for the placements of muographic apparatus in the future Mars mission.

4 Water-Saving Methods of Mineral Extraction

In the search for new water saving methods of mineral extraction, the discovery of an iron-oxidizing bacterium in bituminous coal mines since the 50s [42] showed how microorganisms can extract many important elements (copper, cobalt, nickel, and zinc) occurring in sulfide minerals, a real breakthrough towards the practice called "biomining" [43]. Despite this method is more respectful for the environment compared to traditional methods, it is not widely used yet because of a general mistrust in "bio" processes and because of the maximization of existing investments in smelters; furthermore, biomining also needs CO_2-producing processes such as tank leaching of mineral concentrates [43]. Application to space mining is also affected by the knowledge that different gravity conditions from those present on Earth influence the metabolism and thus the growth of microbes [44–46]. However, a recent experiment on the International Space Station (ISS) simulating both microgravity and Mars gravity showed no significant difference from Earth in the bioleaching and subsequent yield of Rare Earth Elements (REEs) from basaltic rock made by microorganisms [47]. Microorganisms can effectively extract REEs through bioleaching without the use of toxic compounds such as cyanides used in traditional leaching methods [48]. Given that these experiments were successfully carried out on the ISS, it might be feasible to carry them out on Mars as well.

5 ISRU Locations, Prospections, and Possibilities

In chapters "General Concepts on Porous-Media Flow on Mars—Thermal H_2O–CO_2—fers" and "Buried Volcano-Hydrothermal Systems and Minerals on Mars" we have seen how the hydrothermal systems transport valuable minerals like the rare earths elements (REE) to form ore. Understanding these processes has given to us an idea of where these minerals might be located. Obviously, with hydrothermal systems related to volcanic processes, volcanic areas are the best candidates where to find valuable minerals. So far, we have approached the study of the surface of Mars with spectroscopic methods mainly from orbit and by rovers in a few selected locations where it was possible to land in security according to NASA protocols. Also, radar methods were used to explore the shallow underground searching for water. The results in the search for water were quite scarce because water is absent in the shallow underground of the volcanic areas so that we need new or improved traditional methods to explore the deeper underground in the hope to find some trapped fossil water remaining after the volcanic activities. Depending on the quantities that will be found, we might keep into consideration the option of additional sources of supply coming from nearby asteroids or even plan well in advance tug ships of water coming from the outer solar system. Relying on Martian fossil water is not recommendable unless it will be used (if and when it will be found) for the initial operations. A plan must be accounted for the sustainability of long-term operations.

5.1 3D Printing with In-Situ Resource Utilization (ISRU)

3D printing of cementitious material is a relatively new additive manufacturing process whose growing interest and fast development is mainly due to the digitalized manufacturing, allowing the disposition of material where it pleases. Several possibilities of 3D printed structures from the analysis of historic strategies have been developed to build cantilevers, vaults and domes previously built-in masonry, together with some guidelines for the design of printable geometries and the generation of robotic toolpaths [49].

3D printing is indeed a very interesting sector that automates the entire process of construction but still needs to be more studied. On Mars, automation is welcome especially at the beginning when the inhabitants will have several issues to deal with the first outpost. As much In-Situ Resource Utilization (ISRU) as possible will be necessary to avoid debris and waste, thus not falling into the same problems of disposal we have currently on planet Earth. Limiting waste, cost and speed of fabrication are the priorities in analyzing the feasibility in using such a method. A robot would be able to 3D print autonomously and possibly without human assistance until the structure is finished. Studies highlight the difficulties uncovered during the 3D printing of cantilevered structures on Earth due to the fluidity of the fresh material forced down by gravity before it has a chance to set [50] Due to the Martian reduced gravity there is an advantage in this aspect. Printing continuously without stopping and restarting each layer is a necessity, this to improve bonding and avoid inaccuracy due to settlements of fresh materials during extrusion.

3D printing is a largely automated process done by a 3D printer connected to a gantry bridge, with windows in a structure also robotically placed. The gantry bridge would have to be a very stiff structure to ensure high precision of the nozzle position at high speeds if necessary. One could raise the issue of the transport of such a facility, but we now consider the advantages and disadvantages of 3D printing with ISRU:

- Advantages
  - *Speed in design and production*: Allows for the design and print of complex designs than traditional manufacturing; 3D printing can manufacture minute parts within hours, which speeds up the prototyping process. This allows each stage to complete faster.
  - *Rapid Prototyping*: 3D printing can manufacture very small parts within hours, which speeds up the prototyping process and particularly suits fabrication of parts that wear out due to wear and tear.
  - *Cost effective*: When compared to machining prototypes, 3D printing is inexpensive and quicker at creating parts allowing each design modification to be completed at a much more efficient rate.
  - *Print on demand*: This does not need a lot of space to stock inventory, unlike traditional manufacturing of materials such as bricks, cement, gravel etc.; this saves space & costs, as there is no need to print in bulk unless required. The 3D design files are all stored in a virtual library and they are printed using a 3D model, this means they can be located and printed when needed and repeated indefinitely. Edits to designs can be made at very low costs by editing individual files without wastage of out-of-date inventory and investing in tools. This also reduces the likeliness of design mistakes.
  - *Tailored materials*: Parts can be created from tailored materials to provide specific properties such as heat resistance, higher strength or water repellency.
  - *Environment friendly*: The use of basalt fibers is a successful candidate for 3D printing. A biopolymer basalt fiber composite, possibly biodegradable, recyclable and derived from natural materials that is found on the surface of Mars has been used by NASA for the successful 3D printing of New York-based AI SpaceFactory's MARSHA prototype [51]. As this technology reduces the amount of material wastage used in the production process, it is inherently environmentally friendly. While using local basalt fibers and other materials indigenous to Mars, we are not creating any waste. The trip to Mars would be the only part of the process technically non-environmentally friendly.
  - *Flexible Design*: 3D printing allows for more complex designs than traditional manufacturing processes.
- Disadvantages
  - *Limited Materials*: Not all materials can be temperature controlled enough to allow 3D printing. The palette of materials though will likely grow and be adapted to indigenous Mars materials.

- *Restricted Build Size*: 3D printers currently have small print chambers, which restrict the size of parts that can be printed. Anything bigger will need to be printed in separate parts and joined together after production. Imagine 3D printing of a whole building, say 4–5 m high if not more, it would require substantial structures erected to allow stability and accuracy of the printing itself. Printing separate parts and joining them could increase costs and time for larger parts due to the printer needing to print more parts before manual labor could join the parts together. There would also be a risk of separation in the joints.
- *Post Processing*: The parts need some form of cleaning up to remove support material from the build and to smooth the surface to achieve the required finish. The post processing would be of great importance as on Mars we would have at least two machines constantly printing and the downtime in case of a machinery failure would cost significantly. Post processing methods on Earth use water jets, sanding, and chemical soak and rinse, air or heat drying, assembly and others. Some of these methods may not be readily available on Mars and it would also mean having to transport additional payload for maintenance. The amount of post processing required depends on factors including the size of the part being produced, the intended application and the type of 3D printing technology used for production. Therefore, while 3D printing allows for the fast production of parts, the speed of manufacture can be slowed by post processing.
- *Delamination*: 3D printing parts are produced layer-by-layer. Although these layers adhere together, it also means that they can delaminate under certain stresses or orientations. There are potential unknowns of how such a structure would react with the fierce and constant temperature fluctuations as well as micrometeorite impact.
- *Inaccuracies*: Some printers have lower tolerances; final parts may differ from the original design. This is directly related to the type of machine or process used and can be fixed in post processing, but this will further increase the time and cost of production.
- *Effort*: The effort is excessive. 3D printing is fast for small parts that require to be substituted in a machine due to wear and tear as mentioned earlier but relatively slow for a habitat scale, it would take almost two years to print something that could be occupied in a matter of days with an inflatable structure for the same square meters. The payload also to transfer all the processing and energy equipment, maintenance to the Moon would require several trips being in the order of tons that would definitely not fit in one trip.

6 Possible Shelters for Astronauts in Ground Operations

We have seen how the debate between the use of existing lava tubes or artificial shelters built on purpose on Mars shows pros and cons in both capacities. However, it is also clear that lava tubes cannot be present in every place where a base is needed and thus the construction of an artificial shelter is in order. The point is understanding which type of construction fits better all the safety and operational requirements. Mars is a planet with a thin atmosphere and lacks a magnetic field, this situation exposes the surface at risk of impacts and radiation, so that a surface construction must be hardened enough to resist such risks. Obviously, this kind of protection enormously increases the construction costs that can instead be saved with the construction of a subterranean base in which the rocks offer a cheap and natural protection from these risks. However, at the beginning, humans supposed to build the underground base will need to carry out surface operations. Thus, a surface shelter is required anyway.

The solution starts with materials that are harvested directly from the Martian surface via ISRU to create a powerful shield against the Martian surface. Artificial Intelligence (AI) extractors will initially head towards the mapped sites of assumed smectite, other toward extracting the basalt fibers from Martian rock, especially those found at Martian volcanic regions. In terms of a sequence of events that would lead to the construction of such a structure of habitats and greenhouses (like those in Fig. 3), we would heavily rely on AI due to the sheer amount of work. Due to this the methods proposed reduce the number of variables such that would be for example in fabricating a dome with compressed bricks (CB), in that case all CB would be slightly different to accommodate for the angle and positioning (Fig. 4). This is something we want to avoid. Repetition and simple structures are faster and less problematic.

Fig. 3 A proposed "Martian city" using ISRU techniques for habitation and greenhouse complexes. *Image credit* Giuseppe Calabrese & Mars City Design

Fig. 4 Compressed Bricks with interlocking block designs. *Image credit* Giuseppe Calabrese & Mars City Design

Fig. 5 Illustration of the 3D printing setup used for the construction of the Nubian vaults. *Image credit* Giuseppe Calabrese & Mars City Design

Fig. 6 Interlocking bricks by AI and a Nubian vault capped with glass termination. *Image credit* Giuseppe Calabrese & Mars City Design

AI would dig site trenches for foundations and commence the 3D printing in basalt fiber reinforced plastic of a so called central Green Power House (GPH) geothermal foundation and pools to include algal raceways for an algae aquaculture technology and house bio-processors with photosynthetic collector cells. CB would be microwaved and locked together without mortar to form walls for a central core green powerhouse non-pressurized building.

After completion of the central core walls and walled structure arms, the walls would be covered with 3D printed basalt Nubian vaults at an angle of 40° (Fig. 5). Nubian vaults do not necessitate any scaffold and are perfect for AI guided construction remotely. Extracted silica will be shaped into glazing to close each end of the Nubian vaults (Fig. 6). The 3D printing will proceed to the lower out wall arms that will house International Space Station (ISS)-type greenhouse cylinders and habitations. The lower walls will support a series of Nubian vaults each capped with a glazing lid, each to correspond to one ISS greenhouse cylinder it houses. Each ISS greenhouse cylinder (Fig. 7) has skylights that will receive light from the Nubian vault end glazing above it. All plants will also be receiving ample light from the LEDs from the solar collectors.

The Martian inhabitants will be able to slowly convert the regolith into a soil able to sustain microbe life and at an accelerated rate without having to wait decades for terraforming efforts. All waste will need to be immediately dealt with

Fig. 7 ISS cylinders with pre-planted seeds inserted inside the shielding structure of ISRU. *Image credit* Giuseppe Calabrese & Mars City Design

and it will be longed-for as it will be the power that will sustain and produce more life on planet Mars with the green powerhouse. Waste currently ending in landfill & contributing to global warming on Earth could be a precious resource for Mars to fuel the pyrolysis processing machine. Imagine if we could save our planet and at the same time fuel another with payloads of excess waste from Earth to Mars.

7 Future Perspectives

In this chapter we summarized the need for further exploration at the volcanic regions of Mars, particularly with the further implication of utilizing the volcanic surface and material based on two main objectives: (i) habitation via caving, either through pre-existing lava tubes or construction of a deep, subsurface laboratory, or the prospective of using muographic techniques; and (ii) use of the basaltic and silica minerals found in abundance at the volcanic plains for 3D printing of materials. We could also utilize the lunar surface for such technologies and essentially use the Moon as a base for resources onward to Mars and back to Earth. Such 3D printing from basaltic rocks can be further advanced for compact brick construction, Nubian vaults, and green houses for eventual human habitation. Although the technologies are rapidly evolving for the vision of Martian prospecting and in-situ resource utilization, a lot of work has to be done before a human foot can be solidly placed on Mars.

References

1. Williford KH, Farley KA, Stack KM et al (2018) The NASA Mars 2020 Rover mission and the search for extraterrestrial life. In: From habitability to life Mars. Elsevier, pp 275–308
2. Hubbard GS, Naderi FM, Garvin JB (2002) Following the water, the new program for Mars exploration. In: Acta astronaut. Pergamon, pp 337–350
3. Smith JA, Onstott TC (2011) "Follow the water": Steve Squyres and the Mars exploration rovers. J Franklin Inst 348:446–452
4. Pla-García J, Rafkin SCR, Martinez GM et al (2020) Meteorological predictions for Mars 2020 Perseverance Rover landing site at Jezero crater. Space Sci Rev 216:203–231
5. Leone G (2020) The absence of an ocean and the fate of water all over the Martian history. Earth Space Sci 7(4):e2019EA001031
6. Carr M, Head J (2019) Mars: formation and fate of a frozen Hesperian ocean. Icarus 319:433–443
7. Rodriguez JAP, Fairén AG, Tanaka KL et al (2016) Tsunami waves extensively resurfaced the shorelines of an early Martian ocean. Sci Rep 6:25106
8. Carr MH (2012) The fluvial history of Mars. Philos Trans R Soc A Math Phys Eng Sci 370:2193–2215
9. Nassar NT, Du X, Graedel TE (2015) Criticality of the rare earth elements. J Ind Ecol 19:1044–1054
10. Habib K, Wenzel H (2014) Exploring rare earths supply constraints for the emerging clean energy technologies and the role of recycling. J Clean Prod 84:348–359
11. Zhou B, Li Z, Chen C (2017) Global potential of rare earth resources and rare earth demand from clean technologies. Minerals 7:203
12. Wang J, Guo M, Liu M Wei X (2020) Long-term outlook for global rare earth production. Res Policy 65:101569
13. Binnemans K, Jones PT, Blanpain B, Van Gerven T, Yang Y, Walton A, Buchert M (2013) Recycling of rare earths: a critical review. J Clean Prod 51:1–22
14. Eggert R, Wadia C, Anderson C, Bauer D, Fields F, Meinert L, Taylor P (2016) Rare earths: market

disruption, innovation, and global supply chains. Annu Rev Environ Resour 41:199–222
15. Wang X, Lei Y, Ge J, Wu S (2015) Production forecast of China's rare earths based on the generalized Weng model and policy recommendations. Resour Policy 43:11–18
16. Johnson JR (ed) (2010) Mars scientific goals, objectives, investigations, and priorities: Mars Exploration Analysis Group (MEPAG). White Paper accessed http://mepag.jpl.nasa.gov/reports/index.html
17. Sibeck DG, Angelopoulos V, Brain DA, Delory GT, Eastwood JP et al (2011) ARTEMIS science objectives. In: The ARTEMIS mission, pp 27–59
18. von Ehrenfried MD, von Ehrenfried MD (2020) Artemis influence on Mars planning. In: The Artemis lunar program: returning people to the Moon, pp 184–199
19. Li S, Lucey PG, Milliken RE, Hayne PO, Fisher E et al (2018) Direct evidence of surface exposed water ice in the lunar polar regions. Proc Natl Acad Sci 115 (36):8907–8912
20. Mazarico E, Neumann GA, Smith DE, Zuber MT, Torrence MH (2011) Illumination conditions of the lunar polar regions using LOLA topography. Icarus 211(2):1066–1081
21. Li Y, Li X, Wang S, Tang H, Gan H et al (2012) In-situ water production by reducing ilmenite. In: Moon. Springer, Berlin, Heidelberg, pp 189–200
22. Lomax BA, Conti M, Khan N, Bennett NS, Ganin AY, Symes MD (2020) Proving the viability of an electrochemical process for the simultaneous extraction of oxygen and production of metal alloys from lunar regolith. Planet Space Sci 180:104748
23. Tanaka HKM (2007) Monte-Carlo simulations of atmospheric muon production: implication of the past Martian environment. Icarus 191:603–615
24. Bull R, Nash WF, Rustin BC (1965) The momentum spectrum and charge ratio of mesons at Sea-Level. Nuovo Cimento. XLA 2:365–384
25. Matsuno S, Kajino F, Kawashima Y, Kitamura T, Mitsui K, Muraki Y, Ohashi Y, Okada A, Suda T, Minorikawa Y, Kobayakawa K, Kamiya Y, Nakamura I, Takahashi T (1984) Phys Rev D 29:1–23
26. Bugaev EV, Misaki A, Naumov VA, Sinegovskaya TS, Sinegovsky SI, Takahashi N (1998) Phys Rev D 58:054001. https://doi.org/10.1103/PhysRevD.58.054001
27. Gaisser T, Stanev T (2008) Particle astrophysics and high-energy cosmic rays. Phys Lett B 667:254–260
28. Jokisch H, Carstensen K, Dau W, Meyer H, Allkofer O (1979) Cosmic-ray muon spectrum up to 1 TeV at 75° zenith angle. Phys Rev D 19 (5):1368–1372
29. Allkofer OC, Bella G, Dau WD, Jokisch H, Klemke G, Oren Y, Uhr R (1985) Cosmic ray muon spectra at sea-level up to 10 TeV. Nucl Phys B 259 (1):1–18
30. Achard P, Adriani O, Aguilar-Benitez M, Van den Akker M, Alcaraz J (2004) Studies of hadronic event structure in e+ e- annihilation from 30 to 209GeV with the L3 detector. Phys Rep 399(2):71–174
31. Haino S, Sanuki T, Abe K, Anraku K, Asaoka Y, Fuke H, Imori M et al (2004) Measurements of primary and atmospheric cosmic-ray spectra with the BESS-TeV spectrometer. Phys Lett B 594(1–2):35–46. https://doi.org/10.1016/j.physletb.2004.05.019
32. Prettyman TH, Koontz SL, Pinsky LS, Empl AM, Ittlefehldt DW, Reddell BD, Sykes MV, NIAC Phase I Final Report, The National Aeronautics and Space Administration (2013) Deep mapping of small solar system bodies with galactic cosimic ray secondary showers (Grant Number NNX13AQ94G). https://www.nasa.gov/sites/default/files/files/Prettyman_2013_PhI_MuonDeepMapping.pdf
33. Groom DE et al (2001) Muon stopping-power and range tables: 10 MeV–100 TeV. At Data Nucl Data Tables 78:183–356
34. Adair RK, Kash H (1977) Cosmic-ray muons. In: Hughes VW, Wu CS (eds) Muon physics, volume 1: electromagnetic interactions. Elsevier. https://doi.org/10.1016/B978-0-12-360601-3.X5001-4
35. Tanaka HKM (2020) Development of the muographic tephra deposit monitoring system. Sci Rep 10, 14820. https://doi.org/10.1038/s41598-020-71902-1
36. Alvarez LW, Anderson JA, Bedwei FE, Burkhard J, Fakhry A, Girgis A, Goneid A, Hassan F, Iverson D, Lynch G, Miligy Z, Mousaa AH, Sharkawi M, Yazolinio L (1970) Search for hidden chambers in the pyramids. Science 167:832–839
37. Tanaka HKM et al (2009) Cosmic-ray muon imaging of magma in a conduit: degassing process of Satsuma-Iwojima Volcano, Japan. Geophys Res Lett 36:L01304
38. Kedar S, Tanaka HKM, Naudet CJ, Jones CE, Plaut JP, Webb FH (2013) Muon radiography for exploration of Mars geology. Geosci Instrum Method Data Syst 2:157–164. www.geosci-instrum-method-data-syst.net/2/157/2013/. https://doi.org/10.5194/gi-2-157-2013
39. Cushing GE, Titus TN, Wynne JJ, Christensen PR (2007) THEMIS observes possible cave skylights on Mars. Geophys Res Lett 34:L17201. https://doi.org/10.1029/2007GL030709
40. Williams KE, McKay CP, Toon OB, Head JW (2010) Do ice caves exist on Mars? Icarus 209:358–368. https://doi.org/10.1016/j.icarus.2010.03.039
41. Grotzinger JP, Crisp J, Vasavada AR, Anderson RC, Baker CJ, Barry R, Blake DF, Conrad P, Edgett KS, Ferdowski B, Gellert R, Gilbert JB, Golombek M, Gomez-Elvira J, Hassler DM, Jandura L, Litvak M, Mahaffy P, Maki J, Meyer M, Malin MC, Mitrofanov I, Simmonds JJ, Vaniman D, Welch RV, Wiens RC (2012) Mars Science Laboratory mission and science investigation. Space Sci Rev 170:5–56

42. Colmer AR, Temple KL, Hinkle ME (1950) An iron-oxidizing bacterium from the acid drainage of some bituminous coal mines. J Bacteriol 59:317–328
43. Johnson D (2018) The evolution, current status, and future prospects of using biotechnologies in the mineral extraction and metal recovery sectors. Minerals 8:343
44. Moissl-Eichinger C, Cockell C, Rettberg P (2016) Venturing into new realms? Microorganisms in space. FEMS Microbiol Rev 40:722–737
45. Horneck G, Klaus DM, Mancinelli RL (2010) Space microbiology. Microbiol Mol Biol Rev 74:121–156
46. Chen B, Wang Z, Huang L, Wu F, Chen J, Xu W (2000) An experimental study on the effects of microbes on the migration and accumulation of REE in the weathering crust of granite. Chinese J Geochem 19:280–288
47. Cockell CS, Santomartino R, Finster K et al (2020) Space station biomining experiment demonstrates rare earth element extraction in microgravity and Mars gravity. Nat Commun 11:1–11
48. Gu T, Rastegar SO, Mousavi SM, Li M, Zhou M (2018) Advances in bioleaching for recovery of metals and bioremediation of fuel ash and sewage sludge. Bioresour Technol 261:428–440
49. Carneau P, Mesnil R, Roussel N, Baverel O (2020) Additive manufacturing of cantilever—from masonry to concrete 3D printing. Automat Construct 116:103184
50. Khoshnevis B (2004) Automated construction by contour crafting—related robotics and information technologies. Autom Constr 13(1):5–19
51. Roman M, Yashar M, Fiske M, Nazarian S, Adams A, Boyd P, … Ballard J (2020, July) 3D-Printing lunar and Martian habitats and the potential applications for additive construction. In: 2020 international conference on environmental systems

Index

G. Leone (ed.), *Mars: A Volcanic World*,
https://doi.org/10.1007/978-3-030-84103-4